Crystallization Processes in Fats and Lipid Systems

Crystallization Processes in Fats and Lipid Systems

edited by

Nissim Garti

The Hebrew University of Jerusalem
Jerusalem, Israel

Kiyotaka Sato

Hiroshima University
Higashi-Hiroshima, Japan

MARCEL DEKKER, INC. NEW YORK • BASEL

ISBN: 0-8247-0551-3

This book is printed on acid-free paper.

Headquarters
Marcel Dekker, Inc.
270 Madison Avenue, New York, NY 10016
tel: 212-696-9000; fax: 212-685-4540

Eastern Hemisphere Distribution
Marcel Dekker AG
Hutgasse 4, Postfach 812, CH-4001 Basel, Switzerland
tel: 41-61-261-8482; fax: 41-61-261-8896

World Wide Web
http://www.dekker.com

The publisher offers discounts on this book when ordered in bulk quantities. For more information, write to Special Sales/Professional Marketing at the headquarters address above.

Current printing (last digit):
10 9 8 7 6 5 4 3 2 1

PRINTED IN THE UNITED STATES OF AMERICA

Preface

More than 10 years ago we published our first book, *Crystallization and Polymorphism of Fats and Fatty Acids*. The book presented the basic concepts and techniques available at that time associated with the physical behavior of fats and oils with relation to their crystallization and polymorphic transformations. The book was well accepted by the scientific community as well as by technologists and engineers.

Significant progress has been made in recent years in this research area, both in fundamental aspects of crystallization and in applications. We felt that it was an appropriate time to update our book with relevant new developments in the area. Of special interest are the advanced analytical tools such as the microprobe techniques, high-power Synchrotron radiation X-ray diffraction, vibrational spectroscopy and infrared adsorption, Raman scattering and nuclear magnetic resonance, computer modeling, and sophisticated calculations. These advanced techniques have provided significant new information resulting in cutting-edge research and a better understanding of molecular structures, morphology and texture of fats and lipids, kinetic-phase properties of phase transformations, mixing behavior, and interactions with seeding materials, especially with surfactants.

The past 10 years have brought new aspects and new concepts of fat crystallization and polymorphism of new fats, complex blends of specialty fats, fatty acids, non-saponifiable lipids and mono- and diglycerides. This book is bringing to light new and emerging techniques, processes, and mechanisms, stressing what is new and different, promising and revolutionary. We are strongly impressed by the new dimensions that some scientists have brought to this field. Interfacial crystallization in emulsions and microemulsions, lyotropic liquid crystals, and other self-assembled structures are only some of these new areas. Crystallization in confined volumes is an innovative approach that leads to interesting options for preparation of nano-sized crystals with unique properties, as well as new polymorphic forms that are interfacially mediated. Other promising studies are related to pharmaceuticals, cosmetics, and food applications.

Lipids and fats are drawing strong attention from nutritionalists and life science researchers. The needs for specialty fats and oils with strong emphases on health, balanced nutrition, and functionality are a strong motivation for better selection of fat blends in the solid form. These dictate narrow particle sizes, flowing properties, and new physical properties derived from a better control of polymorphism and crystallization. Industrial crystallization of fats and lipids is an important issue mainly for food applications. Some progress has also been made in this area.

We believe that this new book brings a well-balanced structure of fundamental and applied developments to the field of fats and lipids crystallization. We hope that the reader will enjoy this book, as well as find it useful and rewarding.

Nissim Garti
Kiyotaka Sato

Contents

Contributors

R. Adleman Pillsbury Co., Minneapolis, Minnesota

Dino Aquilano Department of Mineralogy and Petrology, University of Torino, Torino, Italy

Hidetoshi Arima Faculty of Pharmaceutical Sciences, Kumamoto University, Kumamoto, Japan

P. Bennema RIM Department of Solid State Chemistry, University of Nijmegen, Nijmegen, The Netherlands

S. X. M. Boerrigter RIM Department of Solid State Chemistry, University of Nijmegen, Nijmegen, The Netherlands

Heike Bunjes Department of Pharmaceutical Technology, Institute of Pharmacy, Friedrich Schiller University, Jena, Germany

Nissim Garti Casali Institute of Applied Chemistry, The Graduate School of Applied Science, The Hebrew University of Jerusalem, Jerusalem, Israel

R. F. P. Grimbergen* RIM Department of Solid State Chemistry, University of Nijmegen, Nijmegen, The Netherlands

* Current affiliation: DSM Research, Geleen, The Netherlands

R. W. Hartel Department of Food Science, University of Wisconsin, Madison, Wisconsin

Masamichi Hikosaka Faculty of Integrated Arts and Sciences, Hiroshima University, Higashi-Hiroshima, Japan

F. F. A. Hollander RIM Department of Solid State Chemistry, University of Nijmegen, Nijmegen, The Netherlands

Fumitoshi Kaneko Graduate School of Science, Osaka University, Toyonaka, Japan

K. E. Kaylegian Wisconsin Center for Dairy Research, Madison, Wisconsin

William Kloek† Department of Food Science and Technology, Wageningen University, and Wageningen Centre of Food Sciences, Wageningen, The Netherlands

Tetsuo Koyano Confectionery Research and Development Laboratory, Meiji Seika Kaisha, Ltd., Saitama, Japan

Niels Krog Danisco Cultor, Brabrand, Denmark

Hajime Matsuda Shiseido Laboratories, Yokohama, Japan

H. Meekes RIM Department of Solid State Chemistry, University of Nijmegen, Nijmegen, The Netherlands

Koji Nozaki Department of Physics, Faculty of Science, Yamaguchi University, Yamaguchi, Japan

Malcolm J. W. Povey Proctor Department of Food Science, University of Leeds, Leeds, England

Kiyotaka Sato Faculty of Applied Biological Science, Hiroshima University, Higashi-Hiroshima, Japan

Giulio Sgualdino Department of Chemistry, University of Ferrara, Ferrara, Italy

† Current affiliation: DMV International, Veghel, The Netherlands

Kevin W. Smith Unilever Research Colworth, Sharnbrook, Bedfordshire, England

Satoru Ueno Faculty of Applied Biological Science, Hiroshima University, Higashi-Hiroshima, Japan

J. van de Streek RIM Department of Solid State Chemistry, University of Nijmegen, Nijmegen, The Netherlands

Ton van Vliet Department of Food Science and Technology, Wageningen University, and Wageningen Centre of Food Sciences, Wageningen, The Netherlands

Pieter Walstra Department of Food Science and Technology, Wageningen University, Wageningen, The Netherlands

Kirsten Westesen Department of Pharmaceutical Technology, Institute of Pharmacy, Friedrich Schiller University, Jena, Germany

Michihiro Yamaguchi Shiseido Laboratories, Yokohama, Japan

Junko Yano Faculty of Applied Biological Science, Hiroshima University, Higashi-Hiroshima, Japan

Crystallization Processes in Fats and Lipid Systems

1

Fundamental Aspects of Equilibrium and Crystallization Kinetics

Dino Aquilano
University of Torino, Torino, Italy

Giulio Sgualdino
University of Ferrara, Ferrara, Italy

I. INTRODUCTION

The early stage of crystallization is the formation of three-dimensional (3D) nuclei in the mother phase (vapor, solution, melt); the second step is represented by the advancement of the faces bounding these nuclei. Nucleation and growth are not equilibrium processes: crystal nucleation can occur within a supersaturated bulk phase without the intervention of foreign particles or substrates (homogeneous nucleation), but this event is not easily found in natural and in laboratory environments where preexisting solid surfaces can promote the formation of a new crystal phase (heterogeneous nucleation). According to Boistelle [1], we define, as a first step, the equilibrium and supersaturation conditions of a mother phase in contact with its own crystals; later we confine our attention to the factors determining the advancement rate R of the crystal faces under fixed values of the crystallization temperature (T_c) and the supersaturation ($\Delta\mu$). These factors can be roughly divided into two categories.

1. *Factors depending on the bulk and surface crystal structure*
   a. The surface structure of a given crystallographic form depends on the bulk crystal structure and on the character [2] associated with the thickness of the slice allowed by the space group extinction rules.
   b. The perfection, or not, of the surface: if one (or more) screw disloca-

tions crosses the crystal surface, the growth is favored, especially at low supersaturations.

2. *Factors depending on the mother phase*
   c. The structure of the medium surrounding the crystal affects the equilibrium properties of the interface. For a given interface, the value of its specific surface energy γ depends on the interaction between the two bulk phases and from the impurity adsorption as well.
   d. Transport properties influence the growth rate of a face, particularly when growth takes place from solution or melt, some parameters have to be taken into account: density (ρ), viscosity (η), and the relative velocity (u) of the flowing matter with respect to the growing face. Furthermore the effect of the size of the face and the thickness (δ) of the diffusion boundary layer can be practically neglected only in vapor growth. Surface diffusion plays a relevant role (with respect to diffusion in the volume of the growth medium) in growth from pure and impure vapor and from solution. These main factors determine the growth of a given crystal face, as summarized in Fig. 1. The competition between the growth rates of the different forms determines the growth morphology and hence the final habit of a crystal. We later show that the growth mechanism(s) of crystals can be established with good reliability; at different T_c the growth rate R is experimentally determined as a function of varying supersaturations $\Delta\mu$; then, the growth mechanism is that corresponding to the theoretical law which better fits the experimental data.

A. Atomistic Approach to the Structure of Crystal Faces

Figure 2 shows an idealized picture of a perfect simple cubic crystal (Kossel model). Each small cube represents a growth unit (GU): for the sake of simplicity the interaction energy (φ) between GUs is confined to the nearest neighbors. According to that, three kinds of faces may be distinguished:

{100} Within the uppermost layer of these faces two uninterrupted and periodic chains of bonds are easily seen, along two perpendicular rows ([010] and [001] for the face (100), as an example). The slices of thickness d_{100} are compact and very stable; according to the Hartman-Perdok definition [2], these faces are called *flat (F) or equilibrium faces*.

{110} In a slice of thickness d_{110}, the GUs form only one uninterrupted chain of bonds ([001] for the face (110)). These faces are less compact compared to the previous ones, since there is no correlation (in the first-neighbors approximation) between two adjacent

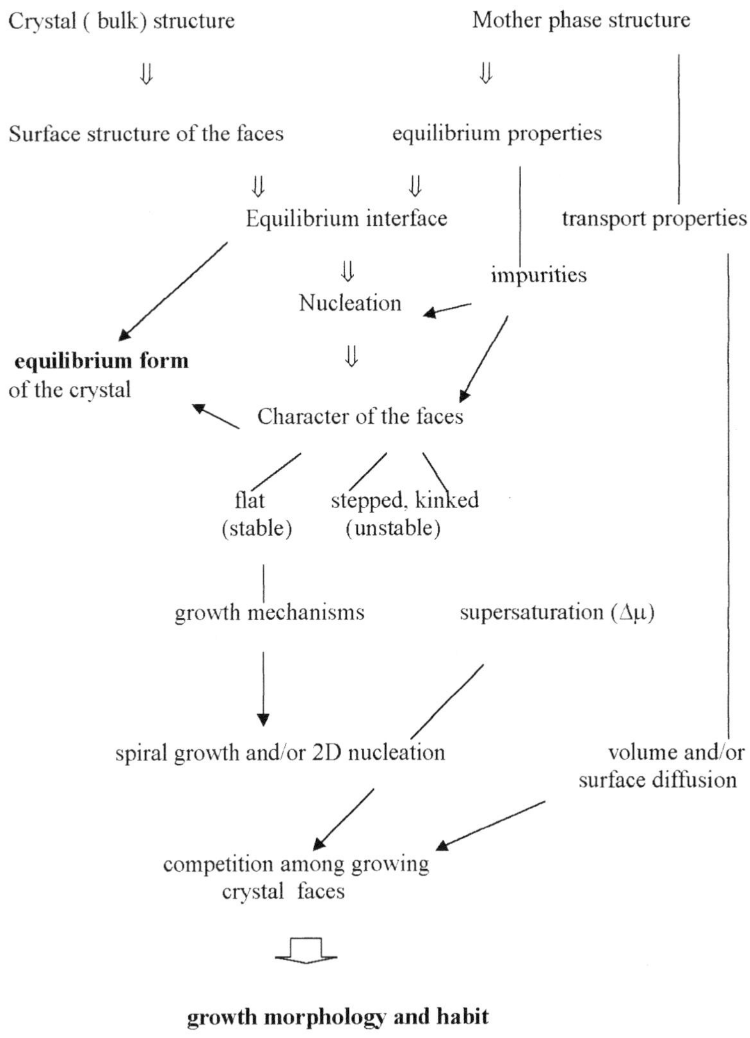

Fig. 1 Flowchart of the main parameters intervening in determining the equilibrium form of crystals along with their growth morphology and habit. Impurity (solvent and additives) adsorption can play a significant role for the equilibrium interface and for kinetics of nucleation and growth.

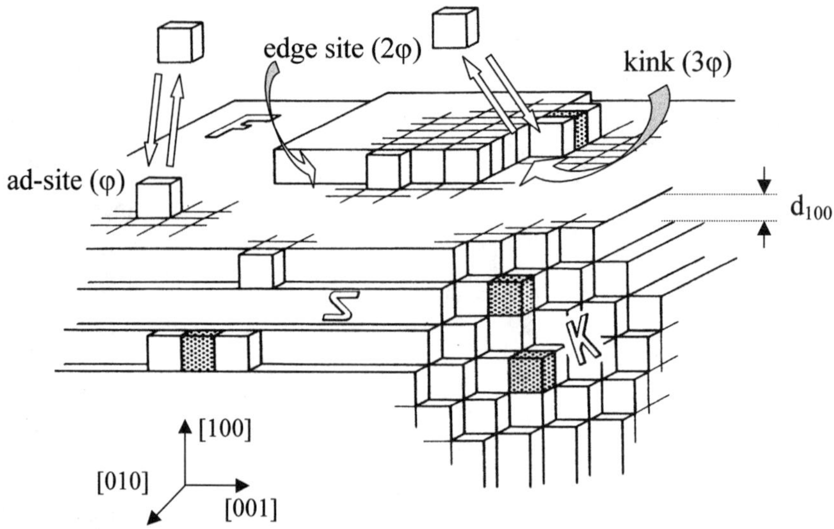

Fig. 2 Idealized picture of a perfect simple cubic crystal (Kossel model). Each small cube represents a growth unit (GU). Within the first-neighbors approximation, φ represents the energy to be spent in order to separate two adjacent GUs. Three kinds of faces are described: (a) Flat (F) faces, as (100): in their uppermost layer two periodic φ-bond chains (PBC) develop, running along [010] and [001] directions. (b) Stepped (S) faces, as (110): within the thickness d_{110} only one PBC runs along the [001] direction. (c) Kinked (K) faces, as (111): in the uppermost d_{111} layer no bond exists between two neighboring GUs. The separation work is also shown, according to whether the GU is detached from an adsorption surface site (φ), an adsorption edge site (2φ), or a kink site (3φ). Dotted cubes represent impurities.

chains. They are called *stepped (S) faces*, because it results from their profile.

{111} No bond exists between two adjacent GUs within a d_{111} slice. These are *kinked (K) faces.*

This classification is rather simple and rough; nevertheless it is useful if one aims at understanding the behavior of a crystal face at equilibrium and during growth.

B. The Equilibrium between a Crystal and Its Mother Phase

Let us consider the elementary evaporation-condensation processes on the crystal surfaces (Fig. 2). Each cube has mass *m*; a mean vibration frequency ν is assumed for every surface site. We will show that finite and infinite crystals are not in equilibrium at the same vapor pressure (concentration or temperature, according to whether the fluid phase is a vapor, solution, or melt). A crystal is considered infinite when its surface atoms are negligible with respect to those of the bulk (its evaporation energy is then independent on size).*

1. Equilibrium between a Vapor and a Crystal of Infinite Size

The equilibrium pressure p_{eq}^{∞} between a monoatomic vapor and its infinite crystal phase depends on its evaporation energy $\varepsilon_v - \varepsilon_{c\infty}$ [3]:

$$p_{eq}^{\infty} = (2\pi m)^{3/2}(kT)^{-1/2}\nu^3 \exp\left[-\frac{\varepsilon_v - \varepsilon_{c\infty}}{kT}\right] \tag{1}$$

where $\varepsilon_v - \varepsilon_{c\infty} = w_\infty$ is the work to keep an atom fixed in a ''mean site'' of an infinite crystal and put it in the vapor; ε_v and $\varepsilon_{c\infty}$ represent the potential energy of an atom in the vapor and in the infinite crystal, h and k are Planck and Boltzmann constants, respectively. When considering the mean evaporation enthalpy $\langle \Delta H_{evap} \rangle$, in which the term $p\,dV$ can be assumed negligible with respect to the energetic term dU, and when assuming that $\varepsilon_v = 0$, the result of the footnote is recovered, since $\langle \Delta H_{evap} \rangle = -\varepsilon_{c\infty} = 3\varphi$.

2. Equilibrium between a Vapor and a Crystal of Finite Size

In an infinite crystal the number of atoms belonging to the crystal surface is negligible with respect to those lying in the crystal bulk; thus, the surface atoms do not contribute to the $\langle \Delta H_{evap} \rangle$ of the infinite crystal. This is no longer true when the condensed phase has finite dimensions, since the percentage of surface atoms dramatically increases with decreasing crystal size. It is easy to see that the mean potential energy of an atom in a finite n-sized crystal (ε_n) should be higher than $\varepsilon_{c\infty}$, because the surface atoms are less bounded to the crystal with respect to those of the bulk. Consequently, the mean evaporation work w_n will be lower than w_∞, as shown in Fig. 3. For finite crystals a new equilibrium pressure, $p_{eq}(n)$, has to be considered in Eq. (1), in which the potential energy ε_n replaces $\varepsilon_{c\infty}$. Then the relation between the two pressures reads

$$p_{eq}(n) = p_{eq}^{\infty} \exp\left[-\frac{w_n - w_{c\infty}}{kT}\right] \quad \text{or} \quad w_{c\infty} - w_n = kT \ln\left[\frac{p_{eq}(n)}{p_{eq}^{\infty}}\right] \tag{2}$$

---

* Within the first neighbors, the work to be spent to evaporate a cube of finite size (n^3 atoms) is $3\varphi(n^3 - n^2)$. The mean evaporation work for the n-sized cube is $\langle \Delta H \rangle_n = 3\varphi(1 - 1/n)$. Then for a cube of infinite size, $\langle \Delta H \rangle_\infty = 3\varphi$.

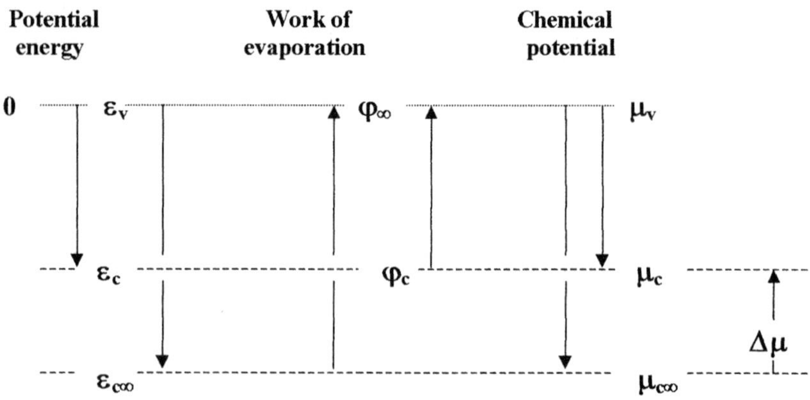

Fig. 3 The potential energy of a growth unit in a mean site of an infinite crystal ($\varepsilon_{c\infty}$) is lower than that of the same GU in a mean site of a finite crystal (ε_c), both being lower than its potential energy (arbitrarily put equal to zero) in the vapor phase. The same applies to chemical potentials (μ_c, $\mu_{c\infty}$). The work of separating a GU from a mean crystal site to the vapor (evaporation work) behaves oppositely to potential energies and chemical potentials.

Recalling that $w_{c\infty} - w_n > 0$, it follows that $p_{eq}(n) > p_{eq}^{\infty}$, which means that *finite crystals, contrary to the infinite ones, are in equilibrium only with a supersaturated vapor.*

3. The Thermodynamic Definition of Supersaturation

The equilibrium can also be viewed in terms of chemical potential. Let μ_∞ and μ_n be the chemical potential (per molecule) of the infinite and finite crystals, respectively. From the Helmholtz free energy it follows that $\mu_\infty = -w_\infty - Ts_\infty$ and $\mu_n = -w_n - Ts_n$, where s_∞ and s_n are the vibrational entropy per molecule. The entropy s_∞ being close to s_n (the vibration frequency being the same for finite and infinite crystals), the difference $\mu_n - \mu_\infty$ reduces to $w_{c\infty} - w_n$. Then expression (2) allows us to find a relation between the pressure ratio $p_{eq}(n)/p_{eq}^{\infty}$ and $\Delta\mu = \mu_n - \mu_\infty$. Thus we can define the *master equation for the equilibrium*

$$\Delta\mu = \mu_n - \mu_\infty = w_\infty - w_n = kT \ln\left[\frac{p_{eq}(n)}{p_{eq}^{\infty}}\right] = kT \ln \beta_v \tag{3}$$

where $\Delta\mu$ is defined as the "thermodynamic supersaturation" and

$$\beta_v = \frac{p_{eq}(n)}{p_{eq}^{\infty}} = \frac{p_{eq} + \Delta p}{p_{eq}^{\infty}} = 1 + \sigma_v \tag{4}$$

is the supersaturation ratio of the bulk phase and $\Delta p = [p_{eq}(n) - p_{eq}^{\infty}]$.

4. More Considerations on the Equilibrium between a Small and Large Phase and on the Thermodynamic Supersaturation in Different Growth Media

It was just stressed that the equilibrium between two infinite phases is reached when the chemical potential of a particle is equal in both of them. Let us consider a single component in two infinite phases (1 and 2), in equilibrium at given p and T. One can write

$$\mu_1^\infty(p_{eq}, T_{eq}) = \mu_2^\infty(p_{eq}, T_{eq}) \tag{5}$$

Let us imagine now that a new equilibrium can be reached when the spatial distribution of the two phases has been changed: phase 1 is still infinite, but the other is small and, for the sake of simplicity, spherical, as a liquid droplet in equilibrium with its vapor. The new equilibrium can then be written

$$\mu_1(p_1, T) = \mu_2(p_2, T) \tag{6}$$

The *capillarity pressure* p_γ, due to the capillarity forces acting on the curved vapor-liquid interface, represents the difference between the internal and external pressures of the droplet: $p_\gamma = p_2 - p_1$. From the master equation $d\mu = \Omega\, dp - s\, dT$ for reversible transitions, we express explicitly the finite difference

$$\begin{aligned}\Delta\mu &= \int_{p_{eq}}^{p_2} \Omega_2\, \mathbf{dp} - \int_{p_{eq}}^{p_1} \Omega_1\, \mathbf{dp} = \int_{T_{eq}}^{T} (\mathbf{s}_2 - \mathbf{s}_1)\, \mathbf{dT} \\ &= \Delta\mathbf{s}(\mathbf{T} - \mathbf{T}_{eq}) = \Delta\mathbf{H}\, \frac{\mathbf{T} - \mathbf{T}_{eq}}{\mathbf{T}_{eq}}\end{aligned} \tag{7}$$

which holds for every kind of phase transformation. In Eq. (7), Ω and s are molecular volume and entropy, respectively, and ΔH is the enthalpy variation in the system during the transformation. We may choose two extreme examples:

a. The transition of a saturated to a supersaturated vapor in equilibrium with a liquid droplet having the same chemical composition (vapor-liquid transformation), or the transition from a saturated vapor (or solution) to its supersaturation value in equilibrium with its small, spherical crystal phase
b. The transition of a pure melt to an undercooled one in equilibrium with its small (spherical) crystals (nucleation from melts)

In both cases p_γ is related to the interfacial specific energy (γ) and to the radius

r of the condensed phase by the Laplace equation:

$$p_\gamma = \frac{2\gamma}{r} \tag{8}$$

Case a. $\Delta\mu$ for vapor $\Rightarrow$ liquid and vapor (solution) $\Rightarrow$ solid transitions.

In practice, nonideal gases and solutions are encountered; then fugacity and activity should be used. Here we confine our treatment to pressures and concentrations. Both these transformations occur under isothermal conditions; hence, $\Delta T = T_{\text{eq}} - T = 0$, while the vapor pressure changes from $p_{\text{eq}} = p_\infty$ to p_r, Ω_2 being constant, due to the incompressibility of the condensed phase. Equation (7) transforms to

$$\Delta\mu = kT \ln\left(\frac{p_r}{p_{\text{eq}}}\right) = 2\Omega_2\left(\frac{\gamma}{r}\right) \tag{9}$$

where γ means $\gamma_{\text{vapor/liquid}}$ or $\gamma_{\text{vapor/solid}}$. When solutions are concerned,

$$\Delta\mu = kT \ln\left(\frac{c_r}{c_{\text{eq}}}\right) = 2\Omega_2\left(\frac{\gamma_{\text{l,s}}}{r}\right) \tag{10}$$

where $\gamma_{\text{l,s}}$ represents the specific energy of the crystal-solution interface. Relations (9) and (10) are known as Gibbs-Thomson equations.

Case b. $\Delta\mu$ *for melt* $\Rightarrow$ *crystal transformation.* We can reasonably assume that the pressure does not change in the melt; i.e., $p_1 \cong p_{\text{eq}} = p_\infty$, whereas the temperature should change from the equilibrium (melting) temperature ($T_m = T_{\text{eq}}$) to T. This means that Eq. (7) becomes

$$\frac{1}{T_{\text{eq}}}\int_{T_{\text{eq}}}^{T} T_{\text{eq}}\,(s_2 - s_1)\,dT = \frac{\Delta H_{\text{cr}}}{T_f}\,(T_{\text{eq}} - T) \tag{11}$$

Recalling that Ω_2 does not vary during the phase transition and that the enthalpy of crystallization is opposite to the melting enthalpy, the Gibbs-Thomson equation reads

$$\Delta\mu = \Omega_2(\mathbf{p}_2 - \mathbf{p}_1) = \Omega_2\mathbf{p}_\gamma = \Omega_2\,\frac{2\gamma_{\text{ls}}}{r} = \frac{\Delta\mathbf{H}_{\text{m}}}{\mathbf{T}_{\text{m}}}(\mathbf{T}_{\text{m}} - \mathbf{T}) \tag{12}$$

From this expression we know the driving force that must be imposed, under constant pressure, to obtain two phases in equilibrium: one of them is an infinite melt at the actual temperature; the other consists of crystalline droplets (spherical or nearly spherical) the size of which depends on the value of Ω in the crystalline phase and on the excess energy created between the liquid and solid phases.

Moreover, $\Delta\mu$ is connected to the supercooling $\mathbf{\Delta T} = \mathbf{T}_m - \mathbf{T}$ through a factor depending on ΔH_{cr} and T_m; this means that *supercooling* (as for the supersaturation in solution or for the overpressure in vapor growth) *cannot be confused with the true driving force, which is* $\Delta\mu$, *which must be taken into account in all growth processes.*

C. Crystal Size and Solubility

The Gibbs-Thomson equation has an important practical meaning. If we consider, as an example, the solution growth, we may say that solubility values quoted in the literature refer to infinite crystals; whereas if we look at expression (10), it appears that the crystal size determines, for a given interfacial tension, the equilibrium concentration (c_r) of a solution. Hence, there is a correspondence between solubility and crystal size. Thus, for given crystal and solvent, two classes of crystal particles (having radii r_1 and r_2) are in equilibrium with solutions saturated at different concentrations c_{r_1} and c_{r_2}, respectively. From Eq. (10) it follows that

$$kT\ln\left(\frac{c_2}{c_1}\right) = 2\Omega_m\gamma_{l,s}\left(\frac{1}{r_2} - \frac{1}{r_1}\right) \tag{13}$$

This means that if $r_1 > r_2$ then $c_2 > c_1$: *small grains are more soluble than large ones* (supersolubility). Equations (9)–(13) have been written under the tacit assumption that γ does not vary with crystal size. On the other hand, crystal growth processes are generally modeled within the frame of nearest-neighbors interaction only. Nevertheless this can be an adequate approximation for nonpolar and molecular crystals, whereas for ionic ones this model is rather poor.

D. The Equilibrium Form of a Crystal

In Fig. 4a a convex crystal is represented; its volume V is bounded by faces having areas S_i and specific surface free energy γ_i. According to the Gibbs phenomenological treatment, the total surface free energy ($\Sigma\gamma_i S_i$) reaches, at equilibrium, a minimum value, for any given and constant crystal volume (minimum of the free Helmholtz energy):

$$\delta(\Sigma\gamma_i S_i) = 0, \qquad \delta V = 0 \tag{14}$$

Expression (14) may be written in terms of the distances (h_i) between a central arbitrary point of the polyhedron and the corresponding i faces. Then, from a

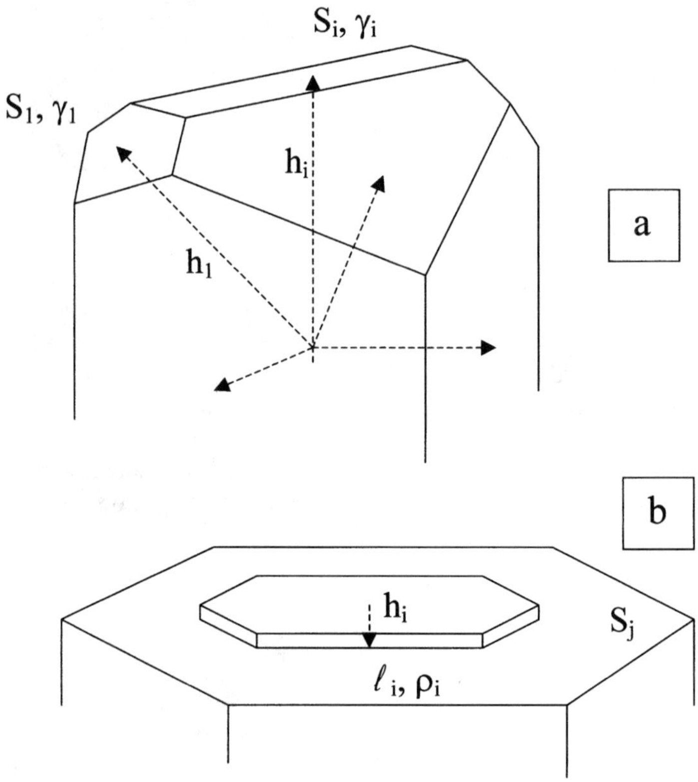

Fig. 4 The equilibrium form of crystals: (a) In the 3D case the form is a convex polyhedron bounded by the faces having the lowest values of the surface free energy γ. The distances h_i of the faces S_i from the center of the form are proportional to their γ_i values. (b) In the 2D case the equilibrium shape of the nucleus lying on a given crystal surface S_j will depend, as for the 3D crystal, on the ratios between the edge free-energy values ρ_i associated to the edges of length l_i.

calculation shown in detail by Toschev [4] and Kern [5], the Gibbs-Wulff theorem is obtained in the simple form

$$\frac{\gamma_1}{h_1} = \frac{\gamma_2}{h_2} = \cdots = \frac{\gamma_i}{h_i} = \text{constant} \tag{15}$$

which implies

$$\gamma_1 : \gamma_2 : \cdots : \gamma_i = h_1 : h_2 : \cdots h_i \tag{16}$$

Two simple and important consequences arise:

1. The equilibrium shape of a crystal is a convex polyhedron, which is, in turn, bounded by faces having the lowest γ values.
2. The shape of the equilibrium form depends on the ratio among the γ_i values (and not on their absolute values).

Nevertheless, absolute values of the γ_i, and hence of $\langle\gamma\rangle$, play a fundamental role not only at equilibrium but also when dealing with the nucleation of a crystal phase, as we will see in Sec. II. The Gibbs-Wulff theorem applies also to the equilibrium shape of 2D nuclei forming on the faces of preexisting 3D crystals. In this case the surface energy has to be replaced by the specific free energy (ρ_i) of the i ledges bounding the 2D nuclei (Fig. 4b). The equilibrium condition of these nuclei is

$$\delta(\Sigma\rho_i l_i) = 0 \qquad \text{and} \qquad \delta S = 0 \tag{17}$$

where l_i corresponds to the length of the ith edge and S is the area of the 2D nucleus. Then the equilibrium condition is expressed by a relation analogous to (16):

$$\rho_1 : \rho_2 : \cdots : \rho_i = h_1 : h_2 : \cdots h_i \tag{18}$$

The equilibrium shape of a crystal (3D or 2D) is affected by the physical properties of the surrounding medium since each γ_i and ρ_i value can vary, according to whether the fluid phase is a vapor, a solution, or a melt. Foreign adsorption can modify these values as well. A phenomenological quantity which takes into account the features of the medium surrounding a crystal is the *adhesion energy* (β_{adh}), which modifies the value of the interfacial energy between two condensed phases. Then the modifications in the equilibrium form due to the β_{adh} values will be treated, along with the β_{adh} role in the heterogeneous nucleation. Twinning, epitaxy, and all other ways of making contact between two (or more) growing crystals are very interesting cases as well in which heterogeneous nucleation is concerned.

E. Adhesion Energy (the Dupré Formula) and Wetting

The energy balance resulting from the generation of an interface between two phases (i) and (s) is $W_{is} = W_i + W_s - 2E_{is}$, where W_i and W_s are the amount of work necessary to separate each phase along a common area S, and $-2E_{is}$ is the energy recovered when the two phases are put in contact through the new interfaces generated (Fig. 5). By dividing each member of the preceding relation by the total area generated ($2S$), one obtains the Dupré formula:

$$\gamma_{is} = \gamma_i + \gamma_s - \beta_{\text{adh}} \tag{19}$$

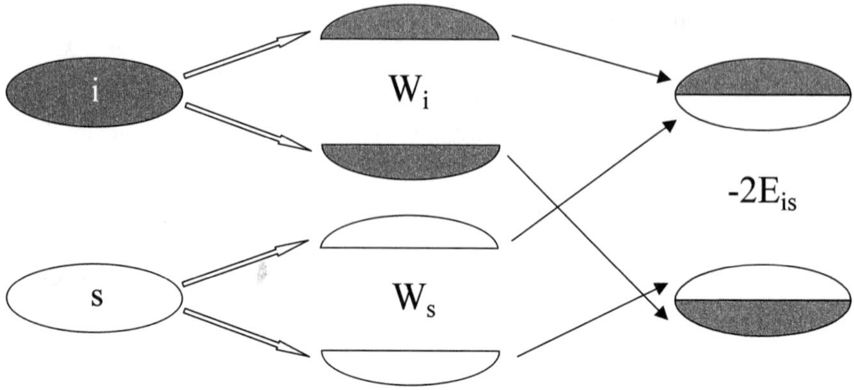

Fig. 5 Scheme to obtain Dupré's formula. Work W_i and W_s are spent to separate the two phases along an area S. The energy $-2E_{is}$ is gained when two new interfaces are generated. The energy balance yields the excess work $W_{is} = W_i + W_s - 2E_{is}$. From this equation the specific interfacial energy (γ_{is}) is obtained in terms of the specific surface energies (γ_i, γ_s) and the specific adhesion energy (β_{adh}) (see text).

where the excess term γ_{is} represents the specific interfacial energy, i.e., the energy expended to increase (a unit area) the surface common to the two phases, γ_i and γ_s are the specific surface energies of the phases i and s, and $\beta_{adh} = -2\ E_{is}/2S$ is the adhesion energy.

Wetting originates at a liquid-solid interface, where one out of the two phases is capable of being deformed; it can be measured through the angle α that forms at the contact between the two condensed phases and the vapor (Fig. 6a), where α depends on the mutual relations between the specific surface energies γ_{sl}, γ_{sv}, γ_{lv}, of the solid-liquid, solid-vapor, and liquid-vapor, interfaces, respectively. At equilibrium the liquid drop assumes the shape of a spherical cup and α fulfills the Young relation: $\gamma_{sl} = \gamma_{sv} + \gamma_{lv} - \gamma_{lv}(1 + \cos\alpha)$; thus, from (19), α can be defined through the adhesion energy between the liquid and solid phases: $\cos\alpha = \beta_{adh}/\gamma_{lv} - 1$, which shows that the contact angle depends only on the ratio between the adhesion energy (drop/substrate) and the specific surface energy of the drop and its vapor. Due to the limits of $\cos\alpha$, three cases may be distinguished, two of them usually interesting natural and laboratory systems.

1. *No wetting*: $\beta_{adh} = O \Rightarrow \alpha = 2\pi$; the contact is limited to a geometrical point and mechanical equilibrium is indifferent.
2. *Imperfect wetting*: $\beta_{adh} < 2\gamma_{lv} \Rightarrow 2\pi > \alpha > 0$; the shape of the drop is a spherical cap, the height of which decreases with wetting.

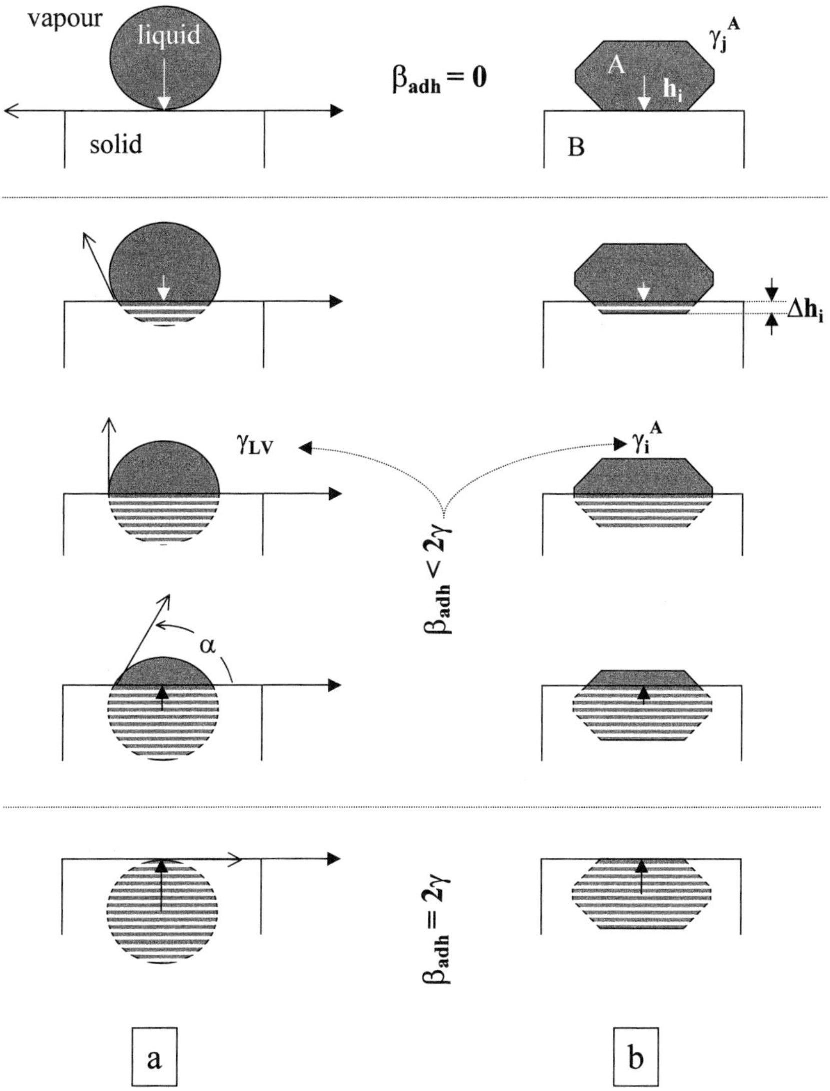

Fig. 6 The equilibrium form of a liquid drop (a) and of a crystal (b) lying on a solid substrate, according to the wetting (i.e., to the adhesion energy). The contact angle α is a measure of the adhesion energy through the Dupré and Young equations (see text).

3. *Perfect wetting*: $\beta_{adh} = 2\gamma_{lv} \Rightarrow \alpha = 0$: an extremely thin liquid layer forms on the solid substrate, and the liquid layer equilibrium on the substrate is indifferent.

The case $\beta_{adh} > 2\gamma_{lv} \Rightarrow \alpha < 0$ can no longer be treated as the preceding ones, because of the cos α limits. Nevertheless this condition corresponds to a physically meaningful situation (as the epitaxial nucleation of 2D crystal phases) which is beyond the scope of this chapter.

Young's equation represents the ''isotropic case'' of the Wulff theorem. Its consequences can be easily transferred to the equilibrium of the anisotropic phases (Fig. 6b). In an interesting review paper [6] it is shown how, on the grounds of these simple concepts, a general treatment allows one to describe homogeneous 3D and 2D nucleation, heterogeneous 3D and 2D nucleation, and epitaxy (and its special cases like twinning and polytypism). Here we confine our attention to the variation introduced in the equilibrium form of crystals when a given face A, for instance, is in adhesion with a given substrate. With reference to Fig. 6b the Wulff theorem is

$$\frac{\gamma_j(\mathrm{A})}{h_j} = \gamma_j(\mathrm{A}) - \frac{\beta_{adh}}{h_s} = \cdots = \text{constant} \tag{20}$$

the parallelism between the isotropic and anisotropic cases being evident. In the latter case the wetting reflects in the truncation Δh_i of crystal A, amounting to

$$\frac{\Delta h_i}{h_i} = \frac{\beta_{adh}}{\gamma_i} \tag{21}$$

II. NUCLEATION

When a crystal forms from another unstable phase, the GUs building it meet to generate 3D and/or 2D ordered aggregates which can set up in the bulk of the mother phase or on a foreign substrate. These two kinds of events are named 3D and 2D nucleations, respectively.*

The transition from equilibrium to the instability state in which the crystalline phase originates is a complex process involving the setting up, in a finite time, of a supersaturation σ_v. As σ_v increases until it reaches the wanted value, the crystalline phase forms under the effect of statistical chemicophysical fluctuations and it turns out to be localized into volumes (or areas) of nanosizes. In each volume, n GUs coalesce to form clusters of size n; increase the cluster sizes

* At times one may meet the term 1D nucleation used to describe the advancement mechanism of a step on a growing crystal face; however, close examination of this aspect is beyond the aim of this chapter.

lowers the σ_v value since they become richer in new GUs, transferred from the mother phase to the clusters through a diffusion halo whose thickness depends on environmental characteristics. The n^*-sized clusters, which are unstable, are named ''nuclei of critical size''; nuclei larger than n^*-sized clusters grow spontaneously and then attain macroscopic sizes, while those smaller than n^* tend to be reabsorbed. The central problem of 3D nucleation theory is predicting the nucleation rate J_{3D} (the number of critical nuclei forming in the unit of time per unit of volume) as a function of the phenomenological parameters of crystallization. In this section we discuss the basic thermodynamic and kinetic aspects indispensable to understanding the mechanisms of crystallization.

A. Nucleation Thermodynamics

1. Heterogeneous and Homogeneous 3D Nucleation

Let us consider the case shown in Fig. 7a, where n_c GUs are transferred from a fluid phase (A) to a crystal cluster, under given T and $\Delta\mu$values. Clusters A form upon a solid substrate B, the adhesion energy being β_{adh} and the common interface having area S_{AB}. The specific surface energies between the substrate and the vapor and between each face of a cluster and the vapor are labeled γ_B and $\gamma_j^{(A)}$, respectively; the cluster/substrate specific energy is γ_{AB}. The GUs belonging to the supersaturated phase undergo a spontaneous transition; for each of them the

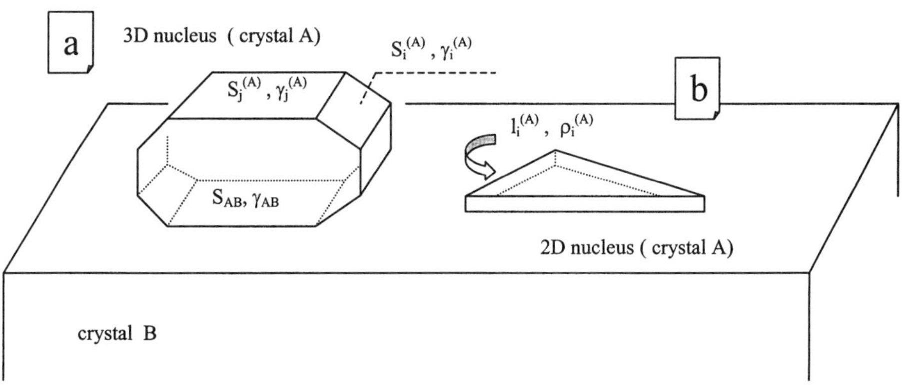

Fig. 7 Thermodynamic parameters intervening in (a) heterogeneous 3D nucleation and (b) 2D nucleation. Growth units of phase A, coming from the mother phase, can group, forming 2D and/or 3D nuclei. The common area S_{AB} forms at the interface between the new crystal phase and the substrate (B). The free surface $S_j^{(A)}$ and edges $l_i^{(A)}$ are associated with their specific free energies $\gamma_i^{(A)}$ and $\rho_i^{(A)}$, respectively.

chemical potential decreases; then the variation of the bulk Gibbs free energy G_V, corresponding to their condensation, is negative ($-n\,\Delta\mu$). New free surfaces (S_j) of the solid phase A are created (for each cluster), while the interface A/B generates at the expense of the free surface (S_{AB}) of B. The total G variation can be expressed in terms of volume and surface contribution:

$$\Delta G = \Delta G_V + \Delta G_S = -n_c\Delta\mu + \sum_{j\neq i} \gamma_j^{(A)} S_j^{(A)} + (\gamma_{AB} - \gamma_B)S_{AB} \tag{22}$$

Crystal areas $S_j^{(A)}$ and S_{AB} can be expressed as functions of n_c through the geometrical constants c_j and c_i, depending on the embryo shape: $S_j^{(A)} = n_c^{2/3}c_j$, $S_{AB} = n_c^{2/3}c_i$. Hence, from (19),

$$\Delta G = -n_c\Delta\mu + n_c^{2/3}\left[\sum_{j\neq i} \gamma_j^{(A)}c_j + (\gamma_i^{(A)} - \beta_{adh})c_i\right] \tag{23}$$

This final equation relates the function ΔG to $\Delta\mu$ and to the equilibrium parameters $\gamma_j^{(A)}$, $\gamma_i^{(A)}$, and β_{adh}. Under the natural condition of imperfect wetting ($2\gamma_i^{(A)} < \beta_{adh}$), Eq. (23) shows that the function ΔG reaches a maximum for the critical heterogeneous nucleus size:

$$\boldsymbol{n}_{3D}^{het*} = \frac{8}{27}\frac{\left[\sum_{j\neq i} \gamma_j^{(A)}c_j + (\gamma_i^{(A)} - \beta_{adh})c_i\right]^3}{(\Delta\mu)^3} \tag{24}$$

This maximum exists only because the contribution of the surface term ΔG_S is everywhere positive; i.e., when $\Delta\mu = 0$ or $\Delta\mu < 0$ any maximum can exist, and hence nuclei formation is forbidden. When even one GU is added to this critically sized nucleus, it transforms into a continuously growing aggregate, the corresponding activation energy for nucleation being

$$\mathbf{\Delta G}_{3D}^{*het} = \frac{4}{27}\frac{\left[\sum_{j\neq i} \gamma_j^{(A)}c_j + (\gamma_i^{(A)} - \beta_{adh})c_i\right]^3}{(\Delta\mu)^2} \tag{25}$$

The higher the values of γ of the free faces of the nucleus, the higher its activation energy and the larger its size; thus good wetting (high adhesion energy) and low specific surface energies promote the occurrence of large critical nuclei able to grow on the guest substrate.

From Eqs. (24) and (25) and *when the adhesion energy vanishes* (no wetting), *the peculiar case of homogeneous nucleation is obtained.* The embryo can reach in this case an unstable equilibrium without the "help" of a substrate. The corresponding activation energy changes and becomes

$$\Delta G_{3D}^{*\text{homo}} = \frac{4}{27} \frac{\left[\sum_j \gamma_j^{(A)} c_j\right]^3}{(\Delta\mu)^2} \tag{26}$$

Relations (24)–(26) are generalized expressions of the Gibbs-Thomson equation and can be applied to any closed and convex polyhedron; in the most elementary and hypothetical case of spherical crystals, the constant c_j, relating the spherical surface to the number n_c of GUs (each of volume Ω) inside the sphere, takes the value $c_j = (3\Omega)^{2/3}(4\pi)^{1/3}$. Since γ is constant on the sphere, $\Delta G = -n_c\Delta\mu + 4\pi r^2\gamma$, from which it is possible to derive both the size of the critical embryo, $r^* = 2\gamma\Omega/\Delta\mu$, and the related activation energy for the homogeneous nucleation, $\Delta G^*_{\text{sphere}} = 16\pi\Omega^2\gamma^3/3(\Delta\mu)^2$. It is easy to see that $\Delta G^*_{\text{sphere}} = \Phi/3$ (where $\Phi = 4\pi\gamma(r^*)^2$ gives the total surface energy of the critical nucleus), this result being absolutely independent of the shape of the nucleus. A comparison of (24) and (26) suggests that at given T, β_v, and γ of the crystallizing material the *3D heterogeneous nucleation is favored with respect to the homogeneous one*.

2. *Heterogeneous and Homogeneous 2D Nucleation*

Let us consider now a 2D nucleus, built by n_c molecules of the phase A (each having area a^2), forming onto a substrate B. The lateral surface of the embryo reduces to a rim of sides of length l_i and of specific energy ρ_i, as shown in Fig. 7b. The formation of this nucleus implies (a) the creation of a free surface of A having area $S_{AB} = n_c a^2$ and the creation of the same area at the A/B interface and (b) the loss of area S_{AB} from the free surface of B. Thus, by applying Eq. (19), we get

$$\Delta G_{2D} = -n_c[\Delta\mu - a^2(2\gamma_A - \beta_{adh})] + n_c^{1/2}\sum_i c_i\rho_i^A \tag{27}$$

The term $a^2(2\gamma_A - \beta_{adh})$, which represents an energy, depends on the properties of the A/B interface. Let us label $a^2(2\gamma_A - \beta_{adh}) = \Delta\mu_0$ and follow the procedure that allowed us to deduce the expressions of type (26). When considering different wetting conditions, one obtains:

1. *Imperfect wetting*: $2\gamma_A > \beta_{adh} \Rightarrow \Delta\mu_0 > 0$: (a) No possibility of 2D heterogeneous nucleation if $\Delta\mu < \Delta\mu_0$; (b) possibility of heterogeneous nucleation of A on B when $\Delta\mu > \Delta\mu_0$.
2. *Perfect wetting*: $2\gamma_A = \beta_{adh} \Rightarrow \Delta\mu_0 = 0$. Homogeneous nucleation (A/A) occurs.

3. *More than perfect wetting*: $2\gamma_A < \beta_{adh} \Rightarrow \Delta\mu_0 < 0$: This implies heterogeneous nucleation of A on B can occur even under conditions of undersaturation.

From all the above results it may be concluded that heteronucleation is always favored in the 3D case and that homonucleation is favored in the 2D case, if the wetting is imperfect, while under the exceptional conditions of absolutely perfect wetting heteronucleation becomes favored. The 2D nucleation is much less costly than 3D since, in the 3D case, a surface between two bulk phases must be created, whereas in the other case new ledges are to be generated which only increase the area of the separation surface already in existence.

B. Nucleation Kinetics

Thermodynamics allows us to deduce the work necessary to nucleus formation; from thermodynamics it is always possible to know that the shape of the unstable critical nuclei is the same as if they were in a stable equilibrium with their genetic environment [5]. However, this information alone does not allow prediction of how many nuclei do form per unit of time and per unit of volume of the growth medium. That is the nucleation frequency J_{3D}. Hereafter we delineate the scheme of the evaluation of J_{3D}. For each $\Delta\mu$ value there is a distribution of clusters with varying size n; the concentration in critical embryos complies with the Boltzmann distribution:

$$C^*_{3D} = N_0 \exp - \left(\frac{\Delta G^*_{3D}}{kT} \right) \tag{28}$$

where, in the case of 3D (2D) homogeneous nucleation, N_0 may be approximated to the reciprocal of the molecular volume Ω (molecular area). In the case of heterogeneous nucleation, N_0 must be replaced by the concentration of the substrate embryos which are present and active in the mother phase. On considering that the only nuclei capable of growing spontaneously are those which attain critical size, the nucleation frequency J_{3D} turns out to be proportional to C^*_{3D}. Then

$$J_{3D} = A \exp - \left(\frac{\Delta G^*_{3D}}{kT} \right) \tag{29}$$

where, for practical uses, $A = n_1\nu$ represents the global kinetic coefficient, n_1 being the GU concentration and ν the frequency at which the GUs add to the critical embryos. Generally A is assumed to be a constant (Arrhenius approximation). Equation (29) shows that the nucleation frequency is a function of two factors: thermodynamic, related to the ΔG^*_{3D}, and kinetic, related to the term A, out of the exponent. If the 3D nucleus is cubic, it follows that, substituting into (29) the value of ΔG^*_{3D} inferred from (26) and remembering that, for a cube, $c_j = a^2 = \Omega^{2/3}$,

$$J_{3D} = A \exp\left(-\frac{32\Omega^2\gamma^3}{(kT)^3(\ln \beta)^2} \right) \tag{30}$$

This last equation very clearly shows how the nucleation frequency depends on the experimental variables: the exponential factor increases as the third power of T and as the square of $\ln \beta = \ln(1 + \sigma_v)$, whereas it decreases as the third power of γ. Figure 8 shows the trend of J_{3D} evaluated for a spherical embryo, together with that of its derivative $\partial J_{3D}/\partial\beta$. The plot shows how J_{3D} increases from a negligible to a catastrophic value as soon as the β_{cr} values are crossed and the J_{3D} versus β curves all have the same slope, irrespective of the γ values. The plot of the derivative, however, shows that, for given increments in β beyond the critical value, its variation gradient greatly differs according to different γ values. This is a further evidence of the sensitivity of the nucleating system from a metastable state to spontaneous nucleation. However, the determination of the thermodynamic factor alone does not allow evaluation of the absolute value of J_{3D}; the only practical way out is to employ kinetic models of the nucleation processes, both in 3D and 2D instances. The examination of this aspect is beyond our purpose, and therefore we refer the reader to some fundamental reviews [7,8].

A final consideration has to be made on the main factors, apart from the impurities, affecting the nucleation frequency. Equation (30) was obtained under the assumption that the GUs can reach the nucleus without crossing activation barriers (as for a crystal-vapor system). But when the growth medium is condensed (solution, melt, solid), the GUs have to cross a more-or-less thin interface between the nucleating crystal and the mother phase. Eq. (29) is then modified, because the activation energy for volume diffusion must also be taken into account:

$$J_{3D} = A \exp - \left(\frac{\Delta G^*_{3D} + \Delta G^*_{vd}}{kT} \right) \tag{31}$$

C. Nucleation of Twins and Polytypes

1. Twins

For the geometric definition of twins and polytypes (Fig. 9), see Refs. 9 and 10. From the point of view of growth, a twin can be defined as a faulted nucleation, since a nucleus starting a twin is something different from the other normal nuclei.

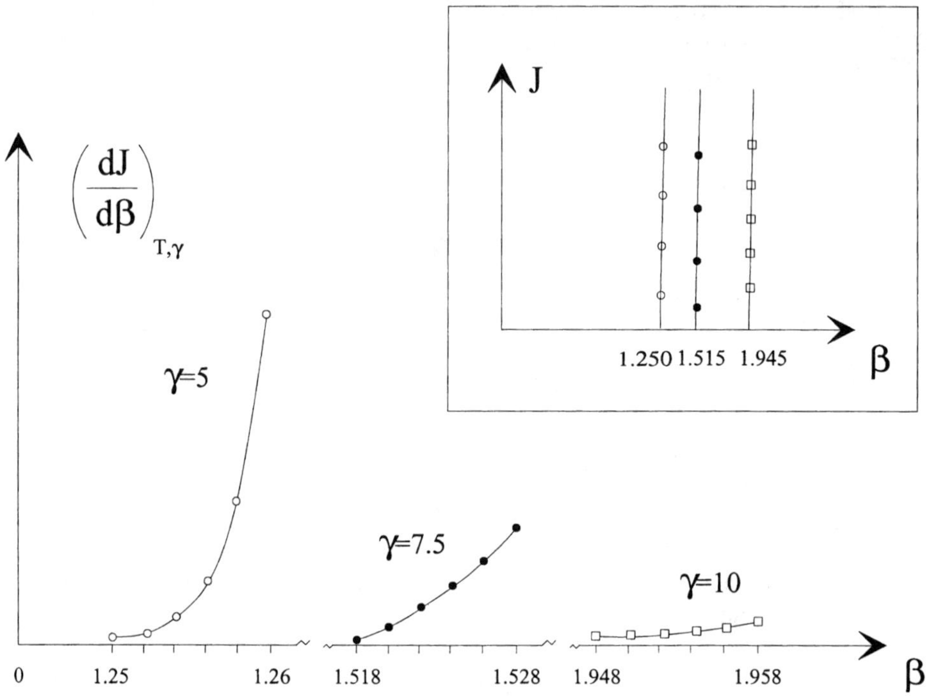

Fig. 8 The 3D nucleation rate J as a function of the supersaturation β. All parameters of Eq. (30) are fixed except the specific surface energy γ which assumes three values (5, 7.5, and 10 erg cm^{-2}). The small figure shows how a small increase in γ does not change the shape of the curve (J vs. β), but dramatically increases the width of the metastable zone. In the large figure the slope of the nucleation rate is represented as a function of β to outline how the γ value influences the transition between the metastable and precipitation zones.

Generally the faulted nucleus is not fully coherent with the normal crystal (at the interface level), since, after geometrical twin operation, the bond energy between first neighbors remains practically constant, whereas farther-neighbors energy changes slightly [11,12]. Owing to the lack of geometrical coherence, a strain energy is created at the interface between a faulted nucleus and a regularly oriented substrate: thus a strain energy E_s (per molecule) is needed to bring the faulted nucleus into full coherence. Once the coherence has been achieved, the faulted nucleus is in a higher potential energy state compared to a regular one. Labeling E_0 the bonding energy per molecule within the regular nucleus and E_T that for the twinned one, one may define the fault energy of the twin (per mole-

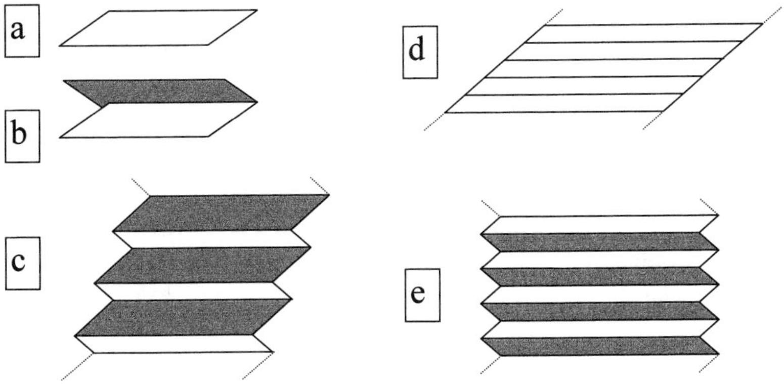

Fig. 9 Twins and polytypes. The black and white domains represent two individuals of the same type symmetry related by twinning operation: (a) normal crystal, (b) twin, (c) periodic polysynthetic twin. Two polytypes are represented in (d) and (e): the same basic structure layer repeats either periodically (d) without changing orientation (stacking . . . AAAAA . . .) or by alternating twinning (stacking . . . ABABAB . . .), giving rise (e) to another period perpendicular to the basic layer.

cule): $\Delta E = E_0 - E_T > 0$. After defining this quantity, it is easy to refer the nucleation of a twinned nucleus to the general case, namely that of heterogeneous nucleation; this amounts to considering the sum of strain and fault energies equivalent to the term which modifies the nominal supersaturation in Eq. (27); that is,

$$E_s + \Delta E = a^2(2\gamma_A - \beta_{adh}) \tag{32}$$

Here we referred to the equation of 2D nucleation since the fault energy is small compared to the bond energies which come into play at the interface between the twinned nucleus and its support (about 1/20 of the interaction energy φ between first neighbors [11,12]), and this implies $2\gamma_A \cong \beta_{adh}$. The introduction of this condition in (27) amounts to assuming that the twinned crystal reduces to a 2D island, as if a 2D embryo, regularly oriented, was nucleating. Experience confirms [13] that the nucleation of a twin on the original composition plane has mostly a 2D character. But, contrary to the normal 2D nucleation, the nonnull term $a^2(2\gamma_A - \beta_{adh})$ has to be added to the critical supersaturation needed to nucleate a regular 2D embryo:

$$\Delta\mu^*_{twin} = \Delta\mu^*_{regular} + (E_s + \Delta E) \tag{33}$$

Even if there is no strain energy, the supersaturation needed to nucleate a 2D twin is not negligible; in fact, assuming $\Delta E \cong (1/20)\varphi \cong 0.1$ kcal mol^{-1} and recalling that $\Delta\mu = kT \ln \beta$, it turns out that to nucleate 2D twin nuclei at room temperature we must have $\beta \geq 1.25$, having assumed, as an example, $\varphi = 2kT$. Once the threshold supersaturation has been exceeded, twins appear suddenly. From Sec. II.A.2 it can be deduced that the nucleation rate for regular 2D embryos is

$$J^{2D}_{\text{regular}} \approx \exp\left(-\frac{\Omega^{2/3}\rho^2}{\Delta\mu(kT)}\right) \tag{34}$$

where ρ is the specific free energy of the step. As mentioned, the effective supersaturation for twinned embryos reduces to $\Delta\mu_{\text{twin}} = \Delta\mu - (E_s + \Delta E)$. Hence,

$$J^{2D}_{\text{twin}} \approx \exp\left[\frac{-\Omega^{2/3}\rho^2}{[\Delta\mu - (\Delta E + E_s)]kT}\right] \tag{35}$$

and then

$$\frac{J_{\text{twin}}}{J_{\text{regular}}} = \exp\left[-\frac{\Omega^{2/3}\rho^2}{kT}\left(\frac{kT\ln\beta}{\Delta E + E_s} - 1\right)^{-1}\right] \tag{36}$$

Figure 10 shows the ratio $J_{\text{twin}}/J_{\text{regular}}$ as a function of $\Delta\mu/kT = \ln\beta$, having assumed the following values for the parameters: $\Omega = 6 \times 10^{-23}$ cm^3, $\rho = 5 \times 10^{-5}$ erg cm^{-1}, $E_s + \Delta E = (1/20)\varphi$, where $\varphi = 2kT$ and $T = 300$ K. It is evident that the ratio tends to the limiting value 1 and has a physical significance only when $\ln\beta \geq 0.1$, the increase in $\Delta\mu$ being dramatic once the threshold is passed. Figure 11 shows a twinning often occurring during the crystallization of normal alkanes [14,15]. The two interlaced spirals are mutually rotated nearly 90° and reveal that a screw dislocation crossing the (001) face of a monoclinic crystal over which a twinned layer has been crystallized promoted the lateral growth of both individuals (regular and twinned) so as to generate a periodic polysynthetic twin as illustrated in Fig. 9.

2. *Polytypes*

Though the concept of polytypism was historically applied first to minerals and synthetic inorganic crystals, here we deal only with some examples relevant to the subject of this book. Polytypism can be considered as a higher-order structural variation with respect to polymorphism (see the next section). Stearic acid, one of the most common biolipid constituents, reveals both polymorphism and polytypism. The former originates from different molecular conformations and packings, resulting in different 3D unit cell structures, whereas polytypism arises from alternative stacking modes of the same long-chain (001) lamellae. Figure 12 shows the first-order structural difference between C and B polymorphs of the stearic acid (different chain conformations) and the higher-order difference be-

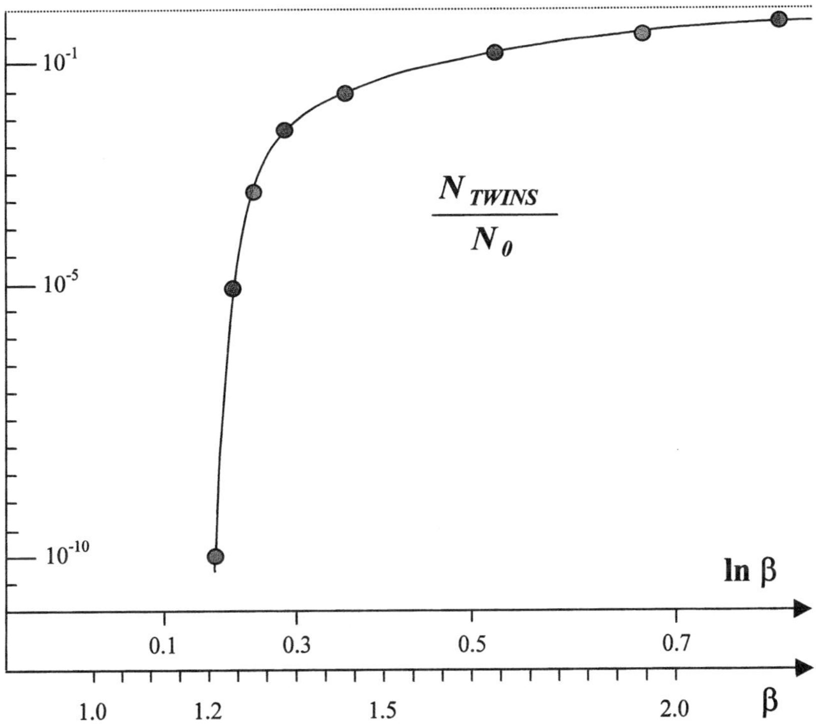

Fig. 10 The ratio between the number of twins and the number of normal crystals (N_0) as a function of supersaturation for a given set of energetic parameters (see text). The curve shows that below the critical supersaturation value ($\beta = 1.2$) twinning does not occur, while under a small increase of β their percentage becomes appreciable.

tween B_{mon} and B_{ortho} polytypes, characterized by the . . . AAA . . . and . . . ABABAB . . . sequences of the same lamella, respectively. Experimental evidences of polytypism, for long-chain compounds, were found from X-ray structure determination [16] and from the comparison between surface structures generated by growth (or dissolution) spirals on (001) forms of normal alkanes [14,17] and fatty acids [18,19].

A polytype can also be considered as a periodically repeated twin, from both the structural and energetic points of view. According to that, the difference between the Gibbs free energy (G) values of different polytypes must be very close each other. This has been proved for stearic acid: from low-frequency Raman spectroscopy and Brillouin spectra, the identification of stearic acid polytypes was obtained along with the evaluation of their thermodynamic and me-

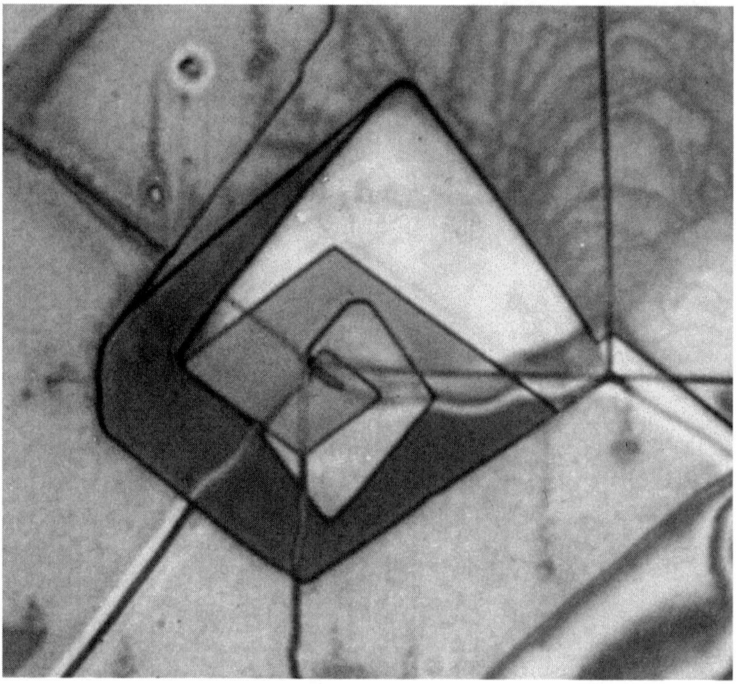

Fig. 11 Optical micrography (reflected light) of two interlaced growth spirals mutually rotated by nearly 90°. The pattern reveals the mirror symmetry of each individual and the structure of a periodic polysynthetic twin.

chanical properties[20]. Then the difference $G(B_{mon}) - G(B_{ortho}) = 0.7$ kJ mol^{-1} at 25°C is one order of magnitude lower than the difference between the G values of the B and C polymorphs (6.4 kJ mol^{-1}). The thermodynamic prediction is qualitatively substantiated by comparing the solubility and the crystal growth behavior (the nucleation frequency from solution) between the two polytypes: when single crystals of B_{mon} and B_{ortho} polytypes were put in a nearly saturated decane solution (at $T = 30.0 \pm 0.05$°C), the growth of B_{ortho} took place at the expense of the B_{mon} crystals.

Stability and occurrence of polymorphs and polytypes have been investigated on stearic acid through the step morphology of the as-grown crystal surfaces. Monolamella step morphology revealed composite polymorphic/polytypic transformations by means of observed overgrowth of B_{ortho} on B_{mon} or of B_{mon} on B_{mon} [21] in the same way as composite growth spirals showed syntactic coalescence between monoclinic and orthorhombic polymorphs of n-$C_{34}H_{70}$ paraffin crystals [22]. Summing up, in order to get a full understanding of the phase

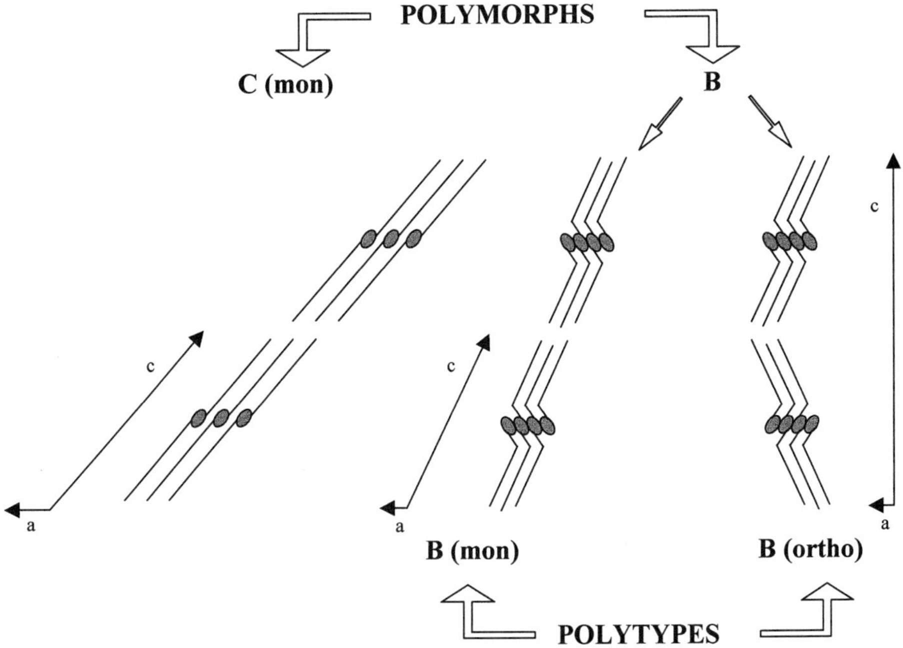

Fig. 12 The idealized stearic acid structure projected along the [010] direction. Chain conformations and cell vectors are represented for the polymorph C (monoclinic) along with the two polytypes B (monoclinic) and B (orthorhombic) both belonging to the polymorph B.

relation of polymorphs, we have to build up the phase diagram by taking both polymorphism and polytypism into account.

III. THE NUCLEATION OF THE POLYMORPHS AND THE OSTWALD RIPENING

A. The Ostwald Step Rule

Many solids can show two or more phases with the same composition but different crystal structures (*polymorphs*): this term can be stricto sensu employed only in this case. However, such a term more generally indicates other cases, as the existence of different solvates (hydrates), which should be better named *pseudopolymorphs*, and amorphous phases. For a fixed set of conditions (T, p, composition x_i), enantiotropic transition points excepted, only one solid phase will be consistent with a minimum free energy of the system: this will be the sole equilibrium phase stable under the given conditions, while all other phases will be metastable with respect to it. The appearance of metastable phases is associated with environmental conditions at the time of their formation, and, as a result, it is of considerable importance in several fields such as geochemistry, biomineralization, and industrial chemistry. For any polymorphic system phase transitions toward the stable phase are unavoidable, due to thermodynamic drive to energy minimization. However, several metastable phases appear quite stable despite the thermodynamics: for them the kinetic hindrances to phase transition are quite big. A metastable crystal can change into the stable one via rearrangements of its structural unities, until a complete transformation occurs (*solid-state phase transformation*), or via melting and recrystallization (*melt-mediated phase transformation*); otherwise, if a suitable solvent is involved, the metastable phase dissolves and a new stable phase is allowed to form from its supersaturated solution (*solvent-mediated phase transformation*). Phase transition kinetics can be quite different, ranging from almost instantaneous to extremely slow for some solid-state transition, where solid diffusion with high activation barriers occurs.

When considering industrial processes, both polymorphism and formation of metastable phases involve the production of materials for commodity products and fine chemicals with high added value. In general, only one phase is required for its technological properties, and hence the crystallization conditions must be controlled to obtain the expected structure only. The isothermal continuous steady-state crystallizations are the easiest controllable processes, thanks to the time-independent constant set of conditions fixed in the crystallizer (T, p, x_i). On the contrary, the control of the discontinuous processes, even isothermal, is much more difficult: here the system is closed, and its composition then undergoes time-dependent changes. Owing to the β_v decrease during crystallization, the system, which at the beginning is kinetically controlled, tends to thermodynamic control: precipitations of more phases can occur because the wide condition range can favor the formation of metastable phases. Otherwise a change of solvent can raise problems: the stability of polymorphs does not depend on the solvent, but a change of solvent can force phase transitions. To optimize the precipitation conditions, knowing the thermodynamic driving force $\Delta\mu$ would be helpful. Unfortunately the phase diagrams are a poor guide to foresee which phase will form, since the equilibrium lines concern stable phases; calculations of the actual driving forces become complicated, especially for multicomponent systems far from equilibrium. As a rule, information about stability of polymorphs can be drawn both from melting points and solubilities: the *metastable phases show lower melting points and higher solubilities*. Hence, we can understand why polymorphism is a serious problem in several chemical industrial processes.

The Ostwald step rule [23] is strictly related to the polymorphism. It states: ‘‘When a number of phase transformations from a less stable state to more and more stable states are possible, usually the closest more stable modification is formed and not the most stable one corresponding to the least amount of free energy.'' The meaning of the rule can be better illustrated by the phase diagram for a monotropic system (Fig. 13 left). A solution of concentration x_B at the temperature T_x is supersaturated only with respect to the stable phase. On the contrary, from a solution of composition x_A both phases can precipitate. According to the rule, phase I crystallizes first and the solution composition drops to x_I (the solubility concentration of this phase at T_x). From this time on, the solution becomes undersaturated only with respect to phase I; crystals of this phase dissolve, thereby increasing the supersaturation for phase II. Hence, nuclei of the stable phase can form and grow at the expense of the metastable one if the system remains for a fairly long time in these conditions (Fig. 13 right).

Lipids represent a heterogeneous class of polymorphic substances for which insolubility in water is a common characteristic, due to the hydrophobic moieties of their molecules mainly constituted by fatty acids. The multiple polymorphic modifications in the crystalline and liquid-crystal states of the fatty acids arise from the different molecular conformation of the fatty acid moieties of lipid molecules connected to their polar groups.

Stearic acid proved a good model in the study of saturated fatty acids. It crystallizes in four polymorphic modifications: triclinic A, monoclinic B, monoclinic C, and monoclinic E. In addition, the B and E phases show polytypism as

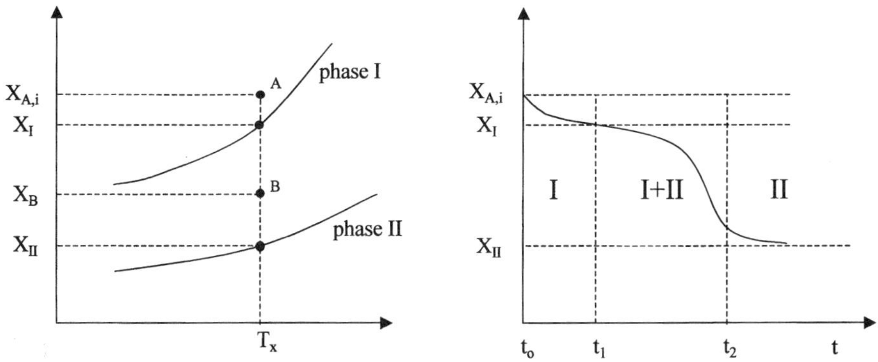

Fig. 13 (left) Concentration (X) vs. temperature (T) diagram for a monotropic substance. Solutions of composition $X_{A,i}$ and X_B are both supersaturated with respect to phase II, while the solution of composition X_B is supersaturated with respect to phase II and undersaturated with respect to phase I. (right) The evolution of the concentration and Ostwald's step rule: from the initial supersaturation ($X_{A,i}/X_I$) phase I begins to crystallize until, at time t_1, the solution becomes saturated with respect to this phase and supersaturated only with respect to phase II. In the time interval ($t_2 - t_1$) both phases coexist (phase II growing at the expense of phase I, which progressively dissolves). Phase II stops growing at time.

well (see above). The E and B phases, obtained by precipitation in various organic solvents, are the stable ones at low *T*, while the C phase, obtained from melt, becomes stable at T > 32°C. B phase has lower Gibbs energy than C below 32°C, and vice versa [24]. As the solid-state C → B transformation is kinetically hindered, the solution-mediated (*n*-hexane) transformation is the only method for achieving the actual C → B transition. The transitions E or B → C occur by heating and are irreversible. Hence, the Ostwald step rule can be observed below 32°C only.

Oleic acid, the most typical unsaturated fatty acid, shows three polymorphs (α, β, γ) : β is the most stable (m.p. at 16.2°C); α and γ are metastable and undergo a reversible solid-state transformation at −2.2°C on heating. Moreover, α melts at 13.3°C and crystallizes more rapidly than β from the melt phase. The stability of the β phase is confirmed by solubility, β being less soluble than the other two forms in decane and acetonitrile. The transition α → β (at 6°C in acetonitrile) and γ → β (at −5°C in decane) are solution mediated [25].

Petroselinic acid (*cis*-6-octadecenoic acid, $C_{18}H_{34}O_2$) is a naturally occurring *cis*-monounsaturated fatty acid with a rather unusual crystal structure. It shows two polymorphs: a low-melting-point (28.5°C) phase (LM) and a high-melting (30.5°C) triclinic structure (HM). The LM polymorph can crystallize in two polytypic structures: the first one, monoclinic (LM_{mon}), less stable, represents a transient state to the orthorhombic one ($LM_{ortho\ II}$), more stable and generated onto the grown (LM_{mon}) phase. The polymorphic-polytypic LM → HM transitions observed in petroselinic acid are quite peculiar in fatty acids. The solid-state LM_{mon} → HM instantaneous transformation can be induced by mechanical stresses or by rapid heating (10°C/min), whereas the LM_{mon} → HM and $LM_{ortho\ II}$ → HM transformations occur by slow heating (1°C/min). Rapid LM_{mon} → HM transformation has been interpreted as a martensitic transition (cooperative and rapid displacements of numerous molecules at the growing crystal interface) and represents the first case in fatty acids and other lipid-related compounds. The slow LM_{mon} → HM and $LM_{ortho\ II}$ → HM transformations, on the contrary, take place through the rearrangement of molecules at the growing interface through thermally activated molecular jumps; moreover, the double-layered $LM_{ortho\ II}$ structure transforms into the more stable HM triclinic single-layered structure [26] (further details given in Chapter 2).

A typical melt-mediated transformation is that occurring in a polymorphic system revealing multiple melting behavior; in this case the transformation occurs via crystallization of the more stable form after melting of the less stable one. An excellent example is tripalmitoylglycerol (PPP), which reveals monotropic polymorphism among α (disordered aliphatic chain conformation), β′ (intermediate chain packing), and β (most dense chain packing). Solid-state transition occurs either via α → β′ → β or α → β at low and high incubation temperatures, respectively. The crystallization rate is extremely different between simple cooling and melt mediation [27]: in fact, the rates of the solid-state $\alpha \rightarrow \beta'$ and $\beta' \rightarrow \beta$ transitions increase with T, being lower (2 hr) for $\beta' \rightarrow \beta$ than for $\alpha \rightarrow \beta$ (1 hr) at 44°C. At 52°C, β form crystallizes after an induction time of 6 min (by simple cooling from melt), whereas the $\alpha \rightarrow \beta$ melt-mediated transformation occurs in 30 sec.

As regards the influence of impurities on the nucleation probability of a given polymorph, we mention two interesting cases.

1. The addition of span 60 and span 65 (at 5%) had a remarkable influence on the crystallization rate of PPP from melt. The induction time for nucleation was increased for all three polymorphs, yet β′ was affected more than the other two forms, thereby resulting in a decrease in the range of temperature at which only β′ was solidified, in comparison to the pure sample (Fig. 14).
2. It is well known that nanospheres of stearic acid can carry rather large amounts of drugs (phenothiazine and nifedipine), and this can be used for their systemic delivery. Solid lipid nanospheres were prepared [28] by spraying in cold water a warm oil/water (O/W) microemulsion whose internal phase was composed of solid lipophilic substances with low melting points (50–70°C), such as fatty acids and triglycerides. Aquilano et al. [29] and Cavalli et al. [30] showed

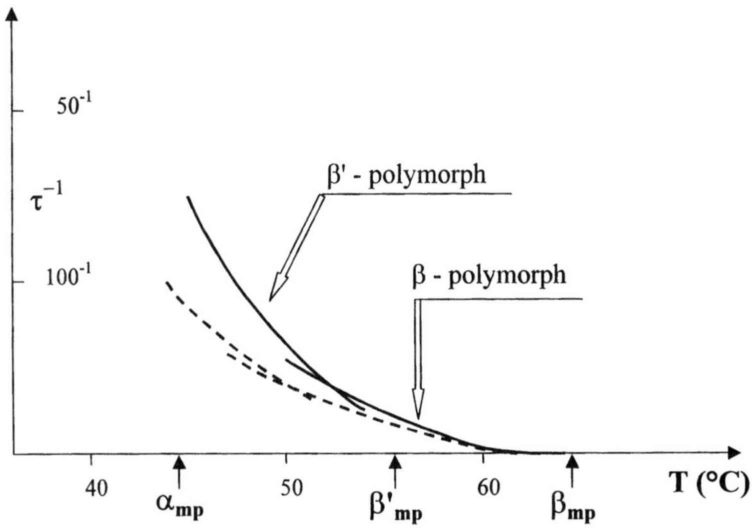

Fig. 14 The inverse of induction time (τ, sec) for melt crystallization β′ and β polymorphs of tripalmitin (PPP), as a function of the temperature of crystallization. Precipitation from pure melt (full line). Precipitation in the presence of an additive [5 wt% sorbitan monostearate (span 60)] (dotted line). Induction times are increased for both polymorphs.

that the droplet O/W microemulsion structure allows the formation of the stearic acid B-polymorph, within the concentration limit 1.65< butanol <2.02 mM. Butanol surely affects the interfacial properties of stearic acid, for it is experimentally evident that it does not enter the crystal lattice of either polymorph B or C of the crystalline phase of the lipospheres. Hence it promotes the lowering of the specific interfacial energy (B-polymorph/growth medium) so enhancing the nucleation frequency of the B phase. Then small amounts of butanol favor the formation of this polymorph, as found for the crystallization of B-stearic acid in the presence of polar solvents [24]. The same occurs with the crystallization of n-alkanes in the presence of linear solvents [31]. In both these cases the solvent mainly acts as a single molecule placed sideways to the crystallizing chains, forming bonds (temporary adsorption) perpendicular to the chain development.

B. Transition Kinetics

The mechanism of polymorphic crystallization is complicated since the nucleation of one polymorph is followed by other competing processes, such as growth, dissolution, and transformation in an another polymorph. A complete understanding of such aspects needs a knowledge of the kinetics of the corresponding partial processes. The transformation of a metastable phase into a stable one may occur starting from a pure solid state or from crystals immersed in a solution. In the first case the transition may happen either via melting of the initial polymorph and the successive recrystallization of the second one (melt-mediated) or via solid-solid transformation through diffusive mechanisms [32]. In the second case the process of dissolution-recrystallization obtains since the metastable phase undergoes the solution-mediated phase transition [33–35]. Cardew and Davey discussed the theoretical aspects of Ostwald step rule in solution in the light of classic nucleation and growth theories. Later they developed a detailed model for solvent-mediated transformations occuring in a slurry of monosized metastable crystals in contact with a saturated solution containing nuclei of the equilibrium phase; the overall transformation is assumed depending both on dissolution of the metastable phase and growth rate of the stable one. Then it was suggested that during some period of the transformation growth and dissolution of the two phases must be balanced, so the solution concentration maintains constant (Fig. 13).

Concerning the solid-solid transformation the most typical example is the in situ measurement of the aragonite $\rightarrow$ calcite nucleation kinetics [36]. In the very early stage the transformation rate is dominated by nucleation at surface sites, near slip lines, twin and deformation bands, while the later stages of the transition are determined by surface nucleation along with a lower rate of bulk volume nucleations.

From the aforementioned examples and perusal of the literature, we see a somewhat confusing picture. The phase precipitations of a large number of polymorphic systems follow the Ostwald step rule fairly well, although many exceptions may be quoted. Therefore the validity of the rule was largely questioned and its elucidation debated: it cannot be considered universally valid because it has no theoretical foundation. Van Santen [37] has used the theory of linear irreversible thermodynamics to show that, in some cases, the formation of intermediate states (the Ostwald step rule) results in a lower rate of entropy production than direct formation of the stable product. This conclusion, however, was criticized by Casey [38], who showed that it was based on an incorrect assumption—that is, by examining entropy production, one may eliminate possible reaction pathways. The correct approach, after Casey, is to express entropy production for an overall reaction in terms of all possible paths: from these, one obtains that the rates of entropy production are independent of the number of steps in a reaction sequence and that indirect pathways are not thermodynamically favored relative to more direct formation of stable product.

C. Polymorphism and Nucleation

As discussed, in the study of the formation of polymorphic modifications it is quite complicated to establish if the nucleation process is limited by thermodynamic rather than kinetic factors; nevertheless the prevailing opinion favors the latter. Evidences exists that the prevailing modification has the highest growth rate when nuclei of more than one phase can form under a given set of conditions [1]. On the other hand, we have seen that a similar interplay between the thermodynamic limitation to the stable nuclei formation of a new phase and a kinetic limitation to the same process is a characteristic difficulty of classic nucleation theory. The ambiguousness lies in the nature of nucleation phenomena, since, when nuclei occur, metastability changes to instability. The interplay between metastability or instability of a mother phase and its transformation rate in the daughter is the interplay on which nucleation theory is based.

It is spontaneous, indeed, to assume that the first precipitating phase is that with the highest nucleation rate [i.e., with the lowest activation barrier $\Rightarrow$ Eq. (21)]. Classical nucleation theory correlates the highest formation rate of the critical nuclei with the lowest γ value of the precipitating modification [see Eq. (30) and Fig. 8]; moreover the preexponential factor is either constant or increases with the solubility of all polymorphic modifications of the same material [39]. Finally, since, after Thomson-Gibbs, γ increases with decreasing solubility (c_{eq}), which, in turn, is a measure of the molecular free energy in the crystal lattice (higher the solubility, lower the stability), it follows that the first separating phase under given experimental conditions has the highest equilibrium solubility; i.e., it is metastable according to the Ostwald step rule. On the other hand, classical nucleation theory assumes that the nuclei of the new phase has an equilibrium-like structure rather than one corresponding to the metastable phase. This is a consequence of the ''capillarity'' approximation introduced to describe the equilibrium between critical nuclei and the supersaturated parent phase. For such an equilibrium, macroscopic thermodynamic properties continue to pertain also to the thinly dispersed cluster phase [40]; therefore, this implies nucleation of the stable phase only, which is inconsistent with the Ostwald step rule. Nonclassical treatments of nucleation [7,40] have revealed that other structures, having higher energy with respect to the equilibrium ones, can preferably nucleate in a supersaturated parent phase, as predicted by the Ostwald step rule [7].

On considering Eq. (30) as a good first approximation for a dimorphic system, one can evaluate the ratio between the nucleation frequencies J_A and J_B of two polymorphs A and B:

$$\frac{J_A}{J_B} = \frac{A_A e^{-f_A \Omega_A^2 \gamma_A^3/(kT)^2(\ln\beta_A)^2}}{A_B e^{-f_B \Omega_B^2 \gamma_B^3/(kT)^2(\ln\beta_B)^2}} = \frac{A_A}{A_B} e^{-1/(kT)^2[f_A \Omega_A^2 \gamma_A^3 (\ln\beta_A)^{-2} - f_B \Omega_B^2 \gamma_B^3 (\ln\beta_B)^{-2}]} \tag{37}$$

As mentioned, it can be reasonably assumed that $A_A \approx A_B$; thus the two polymorphs will compete only when $f_A \Omega_A^2 \gamma_A^3 (\ln \beta_A)^{-2} = f_B \Omega_B^2 \gamma_B^3 (\ln \beta_B)^{-2}$. Factor forms and molecular volumes of A and B polymorphs are surely different, but of the same order of magnitude; then we may conclude that the competition between polymorphs occurs under the condition

$$\left(\frac{\gamma_A}{\gamma_B}\right)^{3/2} = \frac{\ln \beta_A}{\ln \beta_B} \tag{38}$$

Let us assume now that phase A is less soluble than B; this implies $\gamma_A > \gamma_B$. The two nucleation frequencies will be comparable only when condition (38) is fullfilled, and phase A will prevail on B only if the supersaturation β of the parent phase, with respect to A, satisfies the relation

$$\ln \beta_A \geq \ln \beta_B \left(\frac{\gamma_A}{\gamma_B}\right)^{3/2} \tag{39}$$

That is what is expected from Eq. (30), having assumed that the two crystal phases are isotropic; this argument has been extensively treated for the nucleation of dimorphic (monotropic and enantiotropic) systems.

It may be of interest now to outline an aspect of the nucleation process which, usually neglected, affords a clue to the interpretation of nucleation frequencies of polymorphs. Real crystals cannot be treated as isotropic bodies, particularly when their low symmetry implies a marked variation, in the surface features, from one crystal form to another. If two polymorphs, A and B, nucleate homogeneously, that polymorph shall prevail to which the lowest value of activation energy ΔG^* may be associated. On assuming $\Delta G^*_B > \Delta G^*_A$, polymorph A shall be the prevailing one [41]. Otherwise, if A and B nucleate in a medium where a substrate C exists, the conditions may differ. If, for instance, the GUs adhere to the substrate C in such a way that the difference $(\gamma - \beta_{adh})$ for one or more faces of B reduces to a value that makes the numerator of Eq. (25), evaluated for B, smaller than that evaluated for A, then an inversion takes place and polymorphs B may have greater chances than A to nucleate. In addition, if substrate C has a definite orientation in space, then crystals of B may nucleate on C in a preferentially ordered way. The probability of this effect is greater as the difference between the lattice energies (ΔE_{latt}) of the involved polymorphs decreases; values of (ΔE_{latt}) range between 1.5–5 kcal $\cdot$ mol^{-1} [42], and hence they are small in comparison with the lattice energies of crystalline compounds, especially of ionic ones. The report of Falini et al. [43,44] on the quasi-ordered crystallization of aragonite on a collagen matrix oriented by stretching may be interpreted by such a mechanism of heterogeneous nucleation.

D. Ostwald Ripening

Prolonged contacts of a thinly crystal suspension with its parent phase can give rise to various transformations concerning crystals of the same phase. In these cases, the most relevant effect is the crystal size distribution (CSD) change and the crystal morphology modification with respect to the fresh precipitated phase. The CSD change is mainly caused by recrystallization and agglomeration [45]. The consideration of these effects is, for instance, fundamental for the design of commercial crystallizers, whereas their neglect may lead to gross performance errors.

Recrystallization implies both size and morphology changes for crystals precipitated through surface- or volume-diffusion-controlled mechanisms. Agglomeration is the coalescence of clusters of primarily formed microcrystals into more or less stable secondary particles held together mainly by physical forces. We can distinguish between isothermal and nonisothermal recrystallization. *Isothermal recrystallization*, known also as Ostwald ripening or coarsening, takes place in saturated systems at constant temperature. *Nonisothermal recrystallization* (non-Ostwald ripening or kinetic ripening) occurs only in the presence of finite supersaturation conditions, created either by temperature fluctuations or by impurities. The consequence of both recrystallization types is to shift CSD toward larger sizes, because such a process allows the growth of larger crystals at the expense of the smaller ones (which disappear). Then recrystallization implies both an increase in the average size and a drastic reduction in the number of freshly precipitated crystals.

Ostwald ripening is a spontaneous process reducing the total surface energy of the dispersed phase, this reduction being the effective driving force of the process. In order that the transformation be observed within reasonable time periods, the crystals have to be very small–below a few tens of microns [5]. In kinetic ripening the sizes of the involved crystals are, on the contrary, much larger, because the driving forces are stronger, resulting from supersaturation conditions. However, we will deal with Ostwald ripening only. Rather than closely discuss theoretical aspects and historical experiments, which have been reviewed elsewhere [46], we prefer to recall some fundamental results that may help to critically examine the literature.

Let us consider a solution formed by a crystallizing phase dispersed inside its parent phase. All parameters determining the growth rate of the crystals are to be considered as averaged parameters. So, even under steady conditions, the crystallization temperature T_c fluctuates around a fixed mean value, the same occurring for β_v, the bulk supersaturation. As an example, once the nucleation has proceeded, the growing crystals differ from each other, owing to these system fluctuations and to the density of dislocations, which randomly varies from one crystal to another. Consequently, it can be said that a closed system, when β_v tends to vanish and crystals no longer grow, can be described by a crystal population exhibiting different classes of size and shape. Such a system is unstable since the thermodynamic equilibrium is not fulfilled: in fact, for the given volume of crystallized matter, the total interfacial energy has not reached its minimum value, as required by the Gibbs conditions. Therefore, *the system must evolve spontaneously toward the equilibrium, through the progressive reduction of the total interfacial area between the crystals and the surrounding medium* (Ostwald ripening). The growing crystal population undergoes, in the closed system, a depletion of β_v, at time t_0, the supersaturation being $\beta_{v0} = c_{r0}/c_{eq}$. Only crystals that obey the Gibbs-Thomson equation are stable; i.e., those crystals (with size r_0) fulfilling the condition

$$r_0 = 2\Omega\gamma_{cl}\left[kT \ln\left(\frac{c_{r0}}{c_{eq}}\right)\right]^{-1} \tag{40}$$

With respect to all the others the mother phase is undersaturated (hence crystals can dissolve) or supersaturated (hence crystals can grow further). As an extreme consequence, Ostwald ripening would lead to the survival of only one crystal of finite size if the system should achieve true thermodynamic equilibrium! It is worth noting that Eq. (40) is strictly valid only in the case of spherical droplets suspended in a vapor or in another fluid. To extend its validity to microcrystals, a further hypothesis is needed, namely that their shape be in equilibrium, i.e., that they have already reached their equilibrium morphology according to Gibbs-Wulf [46]. Generally, Ostwald ripening is very fast for crystals of size $L = 2r_0 < 1\ \mu m$, fast for crystals of $L \geq 1\ \mu m$, but negligible as soon as L reaches about 100 μm [48]; the process occurs easily for sparingly soluble materials, owing to the large value of their interfacial energy (several tens of erg/cm^2) and resultant wide metastable zone. In precipitating systems the reduction in the number of the initial nuclei ranges from 10^{15} to 10^6 cm^{-3} or less [47,48].

IV. ELEMENTARY GROWTH MECHANISMS OF CRYSTAL FACES

A. Growth Kinetics of K and S Faces

A K face looks like an infinite population of kink sites (Fig. 2). In a real case $\alpha_K \approx 1$ (the sticking coefficient) represents the fraction of available kinks on the face; $n\downarrow$ and $n\uparrow$ are the number, per unit time, of GUs entering a kink and leaving it, respectively. The net balance is thus $N = n\downarrow - n\uparrow = n\downarrow(1 - n\uparrow/n\downarrow)$. The normal growth rate of the K face results in $R_K = \alpha_K d_K N$, where d_K is the equidistance of the lattice planes parallel to the face.

Since $n\uparrow = n\downarrow[\exp(-\Delta\mu/kT)]$, the growth rate can be expressed as

$$R_K = \alpha_K d_K n\downarrow\left[1 - \exp\left(\frac{-\Delta\mu}{kT}\right)\right] \tag{41}$$

where $\Delta\mu$ is a function of practical parameters, such as $\Delta p/p_{eq}$, $\Delta c/c_{eq}$, or $\Delta T/T_{eq}$. Then, for vaporgrowth one obtains

$$\frac{\Delta\mu}{kT} = \ln\frac{p_{actual}}{p_{eq}} = \ln\left(\frac{p_{eq} + \Delta p}{p_{eq}}\right) = \ln(1 + \sigma_v)$$

Moreover, the number $n\downarrow$ of GUs of mass m, reaching a surface site having area a^2, depends on the Knudsen formula: $n\downarrow = a^2 p_{actual}(2\pi mkT)^{-1/2}$. Thus, relation (41) finally gives

$$R_K \cong [\alpha_K d_K a^2 p_{eq}(2\pi mkT)^{-1/2}]\sigma_v \rightarrow R_K = b_K\sigma_v \tag{42}$$

showing *that the growth rate of a kinked face is proportional to the relative supersaturation* σ_v according to a linear law in which b_K represents the kinetic coefficient.

A stepped (S) face, from a kinetic viewpoint, is similar to a kinked one; its sticking coefficient α_S is smaller than that of a K face, since the number of kinks, per unit area, is certainly lower compared to that of a kinked face (Fig. 2).

However, the linear law of the growth mechanism holds, when going from a K face to an S face, and hence we can say that, for the same crystal, $R_S < R_K$

B. Mechanisms of Growth Kinetics of Flat (F) Equilibrium Faces

Crystal faces, even if structurally flat, are not flat at the ''atomic scale''; moreover, the transition between the crystal and surrounding phase is not sharp, since there is a population of adsorbed growth units and holes on the outermost crystal layers. In other words, the interface has a finite thickness, which depends mainly on the nature of the two media in contact, on T and σ_v. Among the interface models, in this chapter we adopted a two-level interface proposed by Mutaftschiev [49]; it will not be presented in detail, but its simple assumptions and consequences have to be illustrated, because it works well, especially for vapor and melt growth.

A face like that shown in Figs. 2 and 15 is flat and populated, at a given T, by species A_S and A_E (adsorbed on the surface and edges, respectively), while H_S and H_E represent the corresponding holes. The coverage degree (number of occupied sites over the total number of available sites for each species) is $0 \leq \vartheta \leq 1$. Thus, let ϑ_{s+} and ϑ_{g+} be the coverage for adsorbed sites on surface and ledges, respectively, and similarly ϑ_{s-} and ϑ_{g-} for holes. Mutaftschiev calculated the adsorption isotherms for all these species as a function of the supersaturation in the crystal-vapor system (Fig. 16). The coverage degree, for adsorbed species, is a continuously increasing function of $\Delta\mu$ (while it is decreasing, obviously, for holes). In other words, both ledges and surface become more and more rough, until a critical supersaturation value is reached, over which metastable states appear (dotted lines) and the face can grow without any supersaturation increase

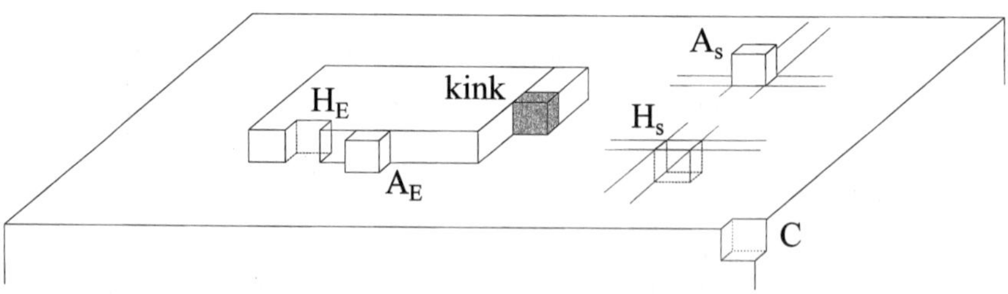

Fig. 15 Site population of a perfect flat face. Apart from the kink site (K), which is an equilibrium site (Sec. I), there are the adsorption sites A_S (on the surface) and A_E (on the edges) and the holes in the terrace (H_S) and in the edges (H_E). Corner sites (C) are numerically not relevant for face growth.

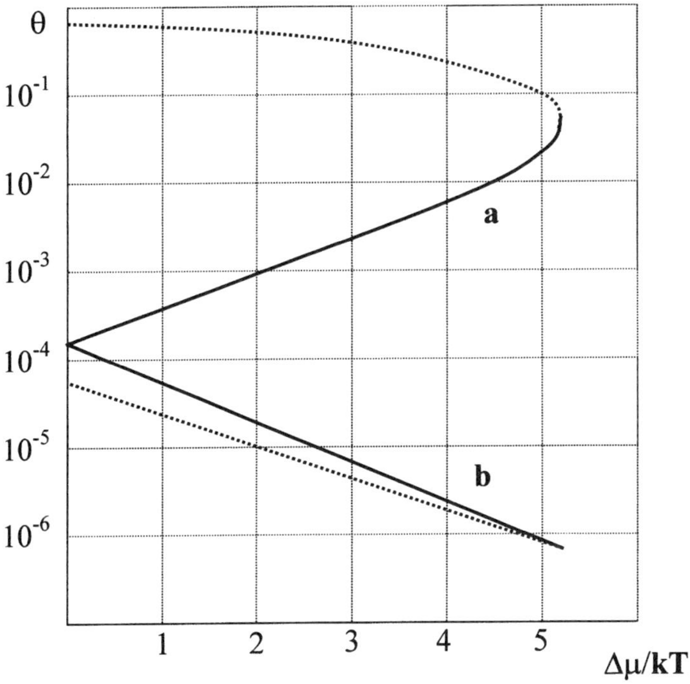

Fig. 16 The coverage degree (θ) for the adsorption sites A_S (on the surface) and for holes in the terrace (H_S) as a function of supersaturation, represented by the full curves *a* and *b*, respectively. (See text and Ref. 49 for a more complete explanation.)

(⇒ diffuse growth or roughening transition). This explains very well why crystal faces, initially flat at low T and σ_v values, no longer maintain their character when growing at high T and σ_v. However, such critical supersaturation values correspond, in vapor growth, to $\sigma_v \approx 10^5$ and 10^{10} for ledges and surface, respectively, both values being meaningless under normal growth conditions.

We must conclude that F faces cannot grow through this mechanism. Moreover, it is well known that flat crystal faces can grow even at very low supersaturations ($\sigma_v \approx 10^{-3}$–10^{-4}). Then it is worth finding some realistic and "less expensive" mechanism acting at low and very low values of the driving force.

1. Growth Mechanism of a Perfect Flat Face: 2D Nucleation

Let us now consider a flat surface, with a very low coverage degree for holes ($\vartheta_{s-} \to 0$) and a very dilute adsorption layer ($\vartheta_{s+} \ll 1$). An equilibrium exists, on the surface, between the adsorbed GUs and their "polymeric" 2D aggregations (2D nuclei). In Sec. II it was shown that the critical size of these nuclei (n^*_{2D}) is strictly related to the activation energy ΔG^*_{2D} for their formation. To obtain the advancement rate of a perfect F face a model is necessary which takes into account the 2D nucleation frequency (J_{2D}); moreover, assumptions are to be made on the competition, if any, among nuclei spreading and covering all the available area in steady state. In short, we have to decide whether to assume that a 2D nucleus, once is formed on a perfect face, may laterally grow until it fills the entire face before a new 2D embryo can form (mononuclear model) or that on the face several nuclei may form and grow at the same time (polynuclear model). For a better understanding of this complex problem see the exhaustive contribution by Söhnel and Garside [7]; here we simply:

1. Recall that the rate of growth of a perfect flat face through 2D nucleation (R^{2D}_F) should be proportional to J_{2D} and, hence, its expression contains the exponential term dependent on the Gibbs free energy of formation of critical 2D clusters.
2. Point out that the behavior of R^{2D}_F, as a function of the supersaturation $\beta_v = 1 + \sigma_v$, is, *mutatis mutandis*, that shown in Fig. 17. The figure shows that only for σ_v values exceeding 0.20 is the growth rate not null and that just beyond this value it becomes more or less catastrophic [as for 3D nucleation (Fig. 8)].

This means that the 2D mechanism is less expensive than ''diffuse growth'' and that a critical supersaturation has to be exceeded in order to observe an appreciable growth rate for a perfect F face.

2. Growth Mechanisms of Defective Flat Faces: Spiral Growth

Real crystals are not perfect: they contain numerous dislocation lines with different character (edge, screw, mixed). According to Frank [50], a crystal face can

---

Fig. 17 (left) Growth rate of {110} form of n-$C_{36}H_{74}$ paraffin crystals from petroleum ether solution as a function of supersaturation. The pure solution growth shows that the rate is not significant until supersaturation exceeds a critical value (the curve does not pass through the origin). The impurity (dioctadecylamine) effect is also shown: as the impurity concentration increases, the growth curves are continuously displaced toward higher values of the supersaturation. (right) Growth rate of {110} form of n-$C_{36}H_{74}$ paraffin crystals from heptane solution. The pure solution growth curve is compared with those obtained in the presence of different concentrations of a long-chain compound (the copolymer ethylene vinyl acetate). The pure solution growth curve does pass through the origin, while critical supersaturation values must be surmounted in order to observe regular growth [69].

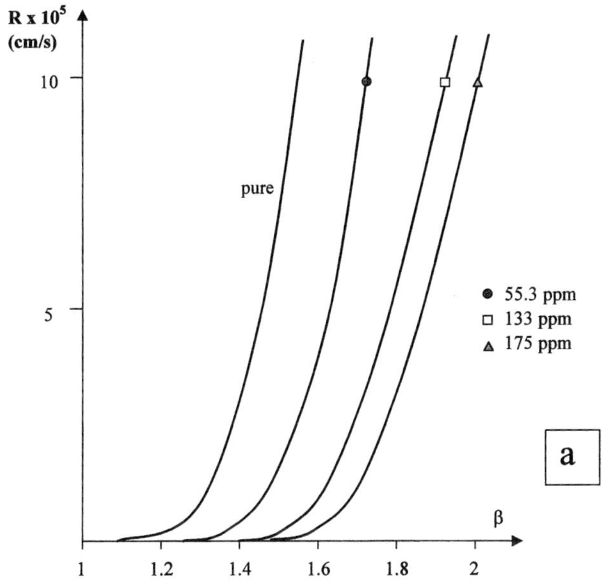
R x 10^5 (cm/s)
10
5
pure
55.3 ppm
133 ppm
175 ppm
a
β
1
1.2
1.4
1.6
1.8
2

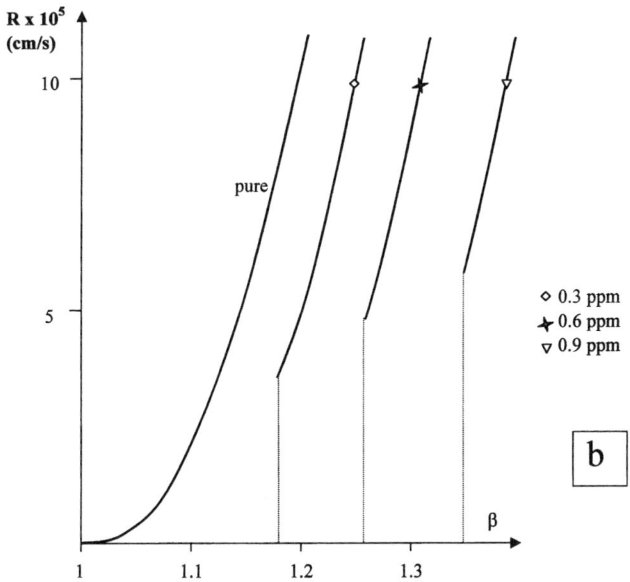
R x 10^5 (cm/s)
10
5
pure
0.3 ppm
0.6 ppm
0.9 ppm
b
β
1
1.1
1.2
1.3

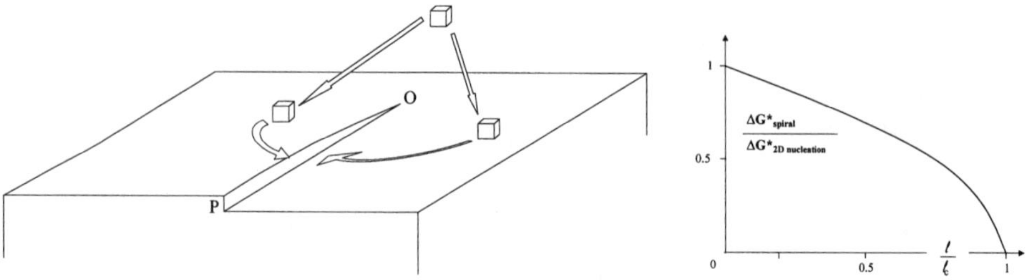

Fig. 18 (left) The exposed ledge OP, on a perfect surface, as generated by a screw dislocation having a component of its Burgers vector perpendicular to the surface. Growth units coming from the mother phase diffuse onto the surface and successively adsorb on the exposed ledge. (right) The activation energy for spiral growth is a decreasing function of the length $l = OP$ of the exposed ledge. When l reaches size l_c of the 2D nucleus that would be compatible with the supersaturation of the mother phase, the activation energy for spiral growth reduces to zero.

grow if just one screw dislocation crosses its surface, with its Burgers vector having a nonnull component perpendicular to it. In this case an exposed ledge, of length $OP = l$ and anchored in O will be observed on the surface (Fig. 18 left side). Burton, Cabrera, and Frank [51] calculated the activation energy ΔG^* required for the growth of the exposed ledge as a function of the ratio $l/2r^*$. Here r^* is the radius of the circular critical 2D nucleus in equilibrium with the supersaturated phase: $r^* = \rho a^2/\Delta\mu^*$ (Sec. II). When $l > l_c = 2r^*$, no activation energy is needed (Fig. 18 right side) and the step can advance with its own velocity (depending on T, $\Delta\mu$, . . .) even for very low values of $\Delta\mu$.

When the ledge advances in a steady state, new exposed ledges are generated by a self-perpetuating mechanism: the resulting pattern is a growth spiral (Fig. 19). Its shape reflects the symmetry of the face and the shape of the 2D nucleus in equilibrium on it. The equidistance y_0 between two successive steps depends on the shape of the critical nucleus and its size: $y_0 = A\rho a^2/\Delta\mu$, where A is the shape factor of the 2D nucleus: the higher the supersaturation the shorter the equidistance between the steps. If the step train advances with velocity v_T, the normal growth rate of the face is $R_F = dv_T/y_0$, where d represents the height of the step. Hence, the determination of R_F is restricted to the evaluation of v_T. The step train velocity v_T depends on the frequency with which molecules enter the steps, and hence the face kinetics is strictly related to the mechanisms experienced by GUs coming from the mother phase.

a. Surface Diffusion as the Rate-Determining Step (BCF Theory).

When a crystal grows from vapor the resistance due to the diffusion in the bulk is generally considered negligible compared with that encountered by the

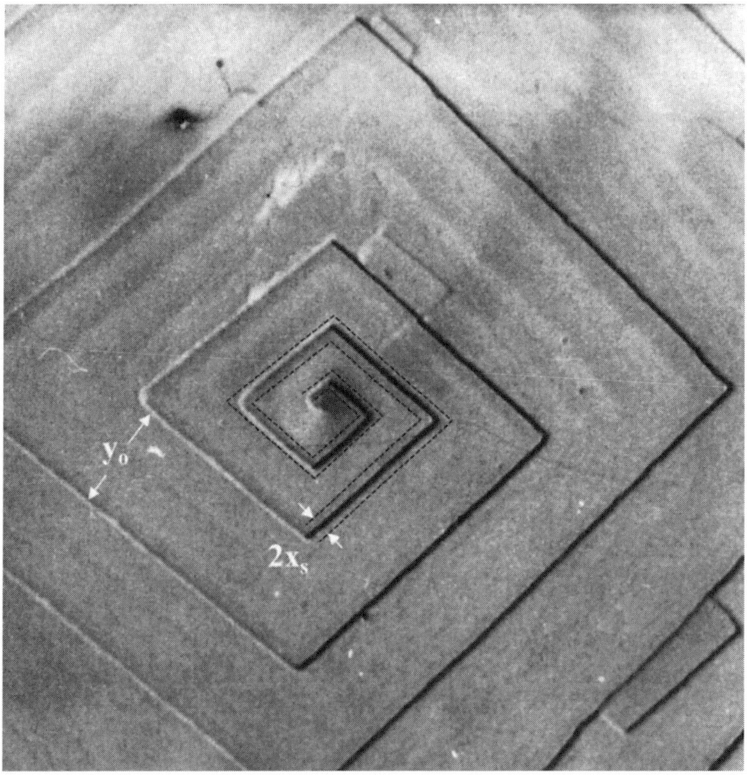

Fig. 19 Single spiral growing on the (001) face of an n-$C_{28}H_{58}$ paraffin crystal. The step equidistance y_0 and the width $2x_s$ of the capture area around each step are outlined.

GUs when they diffuse onto the surface (actually it is about 10^5 times lower). The diffusion paths being consecutive, the kinetics is determined by the slowest process (surface diffusion). Since a GU diffusing on the surface can achieve a mean path (x_s) before being reabsorbed in the bulk phase, two limiting cases occur, according to the relation $y_0 = A\rho a^2/\Delta\mu$, which can be approximated by $y_0 \approx 1/\sigma_v$.

1. *Very low and low supersaturations*: The equidistance between the steps (y_0) is very large and the capture areas of the steps do not overlap ($y_0 \gg 2\, x_s$), as schematized in Fig. 19: thus, those GUs impinging on the surface in between two consecutive capture areas cannot enter the crystal: the growth rate will not be proportional to the relative supersat-

uration, since the efficiency of the reaction is smaller than unity. The growth rate of the face can be approximated by the parabolic law

$$R_p \approx b_p(\sigma_v)^2 \tag{43}$$

where b_p is the kinetic coefficient for low supersaturation values.

2. *High supersaturations*: y_0 decreases with $\Delta\mu$, and when $y_0 \leq 2x_s$ capture areas overlap: then any GU arriving onto the surface enters one of the several steps. The growth rate of the face can be approximated, in this case, by the linear law

$$R_l \simeq b_l\sigma_v \tag{44}$$

where b_l is the kinetic coefficient for high supersaturation values.

Figure 20 describes these two extreme situations by two curves intersecting at a σ_v''' value, which is a measure of the interaction of diffusion fields around the steps. Its position on the abscissa axis, at a given T and for a given face, depends on the specific edge energy ρ of the spiral ledges: higher $\rho \Rightarrow$ larger parabolic regime. On the contrary, when either the temporary adsorption of impurities or the strong similarity between the crystal and the melt lowers ρ values, the linear law is favored even at low $\Delta\mu$ values.

b. Volume Diffusion as the Rate-Determining Step

Chernov [52] proposed a model of crystal growth from solution, assuming neither diffusion on the surface nor along the ledges. In his model the current of the growth units flows to a system of parallel crystal steps (the spiral arms), traveling through an immobile solution layer of thickness δ_c in which they diffuse (diffusion boundary layer). The value of δ_c depends on the relative velocity (u) between the crystal and the fluid medium surrounding it, δ_c decreasing with u. The calculation gives, for the growth rate, a result similar to that obtained when the BCF treatment is extended to the volume diffusion case [53]. This means that for low σ_v values a quadratic law is predicted, whereas a linear law controls the growth at high σ_v values. But, at variance with respect to the BCF treatment, the extrapolation of the curve corresponding to the linear law does not pass through the origin, intersecting the growth rate axis (R) in its ''negative'' part, as may be seen in Fig. 21. Within this model the slope depends on the structure of the face in the nonlinear region only. In the linear region the straight lines $R(\delta)$, for different faces and identical boundary layer thicknesses, have identical slopes. This happens since particles adsorbed on crystal faces play a secondary role in the growth process.

When the fluid phase has strong interactions with the growing crystals (solution or melt growth), all the interface energies (γ, ρ) are lowered; this implies that new interesting phenomena can occur. Two-dimensional embryos can nucleate between consecutive steps of a spiral even at low supersaturations, thanks to

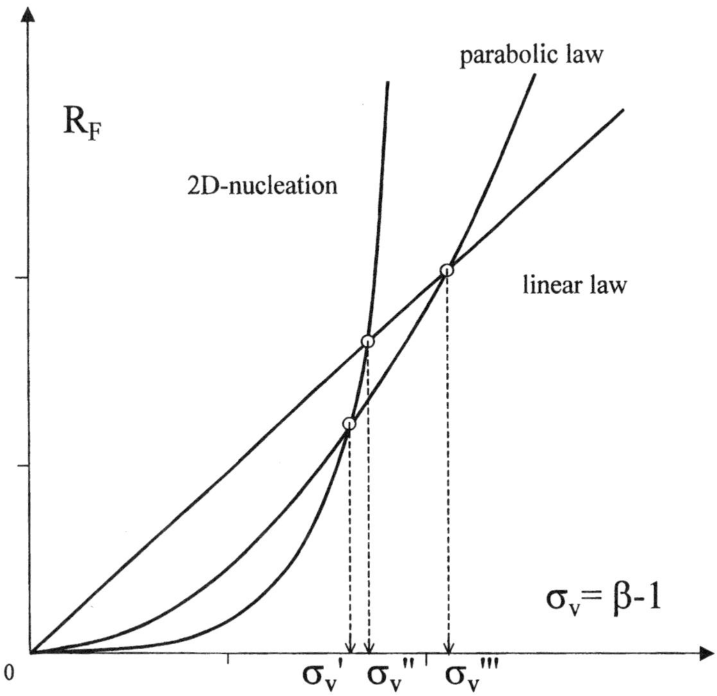

Fig. 20 The three fundamental laws of a growing flat crystal face when surface diffusion is the rate-determining step (see text). In the supersaturation range $0 < \sigma < \sigma_v'$ the parabolic law prevails because the two competing processes (2D nucleation and spiral growth) are in parallel. When $\sigma_v' < \sigma < \sigma_v''$ 2D nucleation dominates, anticipate the linear law. Finally σ_v''' represents the supersaturation value at which the parabolic and linear laws intersect, under the condition that 2D nucleation does not occur.

the lowered value of ρ; if this occurs, a competition is set up between spiral and nuclei ledges. The two growth unit currents (going both to spiral ledges and to nuclei) are not in series but in parallel: the more rapid mechanism dominates, and growth kinetics can be suddenly enhanced, thus anticipating the beginning of the linear regime (Fig. 20) [54].

V. CRYSTAL MORPHOLOGY AND CRYSTAL HABIT: THE EFFECT OF THE IMPURITIES

An {*hkl*} crystallographic form is the set of all equivalent faces mutually related by the symmetry elements of the crystal. The *crystal morphology* is the combina-

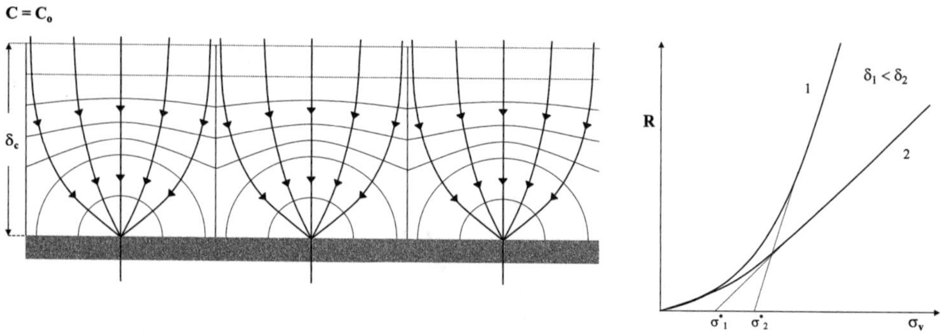

Fig. 21 (left) The scheme of growth of a spiral when the diffusion in the bulk of the mother phase is the rate-determining step. Within a boundary layer of thickness δ_c, growth units diffuse to the moving steps following the paths indicated by the arrows, the diffusion on the surface being negligible. (right) The corresponding growth rate of the face R vs. supersaturation for two different values of δ_c. The slopes of the linear part of the curves do not intersect the abscissa axis in the origin (see text).

tion of all $\{hkl\}$ forms bounding the crystal. The *crystal habit* denotes the shape (flattened, acicular, etc.) assumed by the crystal due to the dominant $\{hkl\}$ forms. Faces of a given form develop or disappear on crystal morphology according to their relative growth rates and geometric position. The slower the growth rate, the more *morphologically important* (more developed) is the form.

Two main ways exist for predicting the growth morphology of crystals.

1. The Hartman and Perdok original theory and its subsequent developments [55] relate the morphology to the bulk crystal structure through the energy released by a growth unit when adsorbed on a crystal surface (E_{att} = attachment energy). According to this theory, the crystal morphology is governed by chains of strong bonds (*periodic bond chains*, PBC) which run through the whole structure and determine the face characters F, S, and K (see Sec. I). The F faces grow layer by layer, so their growth rate is the slowest; as the attachment energy of a face is assumed to be proportional to its growth rate, E_{att} has to be considered as a morphology (habit) controlling factor. The growth habits of several crystals have been predicted on the basis of HP theory: naphthalene [56], sucrose [57], etc. Some drawbacks follow from the crude application of this theory: (a) the theory coherently applies to crystals growing from pure vapor only; (b) neither supersaturation nor temperature effects are considered; and (c) factors concerning the growth medium (interactions crystal surface/solvent and impurities) are not taken into account.

2. According to Jackson [58], microscopically rough or smooth surface morphologies can be classified by means of the *surface entropy factor* α, defined

by $\alpha = 4\varepsilon/kT$, where ε is the energy gained when a crystal-fluid bond generates at the interface. Monte Carlo simulations of growing surfaces indicate three approximate domains of α:

1. $\alpha < 3.0$: the surface is inherently rough ⇒ diffuse growth.
2. $3.0 < \alpha < 4.0$: the surface is smoother ⇒ 2D nucleation.
3. $\alpha > 4.0$: the surface is definitely smooth ⇒ growth by dislocations.

The α factor is also related to the solubility: $\alpha = (1 - x_s)^2 (\Delta G_{sf} - \ln x_s)$, where x_s is the solute fraction at the saturation and ΔG_{sf} is the free energy difference between solid and liquid particles.

Although these two approaches to the problem of crystal growth morphology yielded interesting results as regards the structure-morphology relationship, neither the Hartman-Perdok nor Jackson approaches consider the impurity action. The crystallization of natural or industrial products cannot neglect the presence of what we used to call ''impurities.'' Strictly speaking, any foreign substance other than crystallizing compound has to be considered as an impurity; its presence in the growth medium can play a decisive role on the crystal growth, and hence, on the final morphology and habit.

The first impurity is the solvent. Changing the solvent means changing the bulk supersaturation (σ_v) values; thus, since the growth rates of different $\{hkl\}$ forms do not necessarily behave linearly with σ_v (they can be parabolic, exponential, etc.), the ratios among different R_{hkl} vary with σ_v as well, and then solvents become a fundamental parameter for varying both growth morphologies and habits. The occurrence, during the growth, of faces that cannot exist on the equilibrium crystal form involves some critical σ_v value. Kern [59,60] experimentally showed that ionic crystals of NaCl type change their habit when σ_v exceeds a certain critical value. Here the K faces, like (111), are not stable due to their too high surface energy. However, surface reconstruction and very strong water adsorption allow these faces, not only to appear, but even to dominate growth habits. Water adsorption plays a fundamental role in determining the growth habit of sucrose crystals, on which it stabilizes the $\{101\}$ stepped form through a S → F character transition. Two water molecules indeed are strongly adsorbed between two consecutive PBCs [010] onto the face, thereby providing the missing PBCs to flatten its surface profile. Aquilano et al. [57,61] studied also the effects of σ_v and T on the growth habit of sucrose crystal. Figure 22 shows the growth habit of this crystal obtained from growth isotherms at 30° and 40°C: the $\{101\}$ form prevails, at 30°C, irrespective of the σ_v values, whereas its importance decreases at 40°C for all tested σ_v values except the largest one ($\sigma_v = 0.04$). In this last case, water adsorption becomes the growth-rate-determining step, as it does for ionic crystals, even if the σ_v effect does not completely compensate the T effect.

Finally, another interesting case is represented by the widening of the meta-

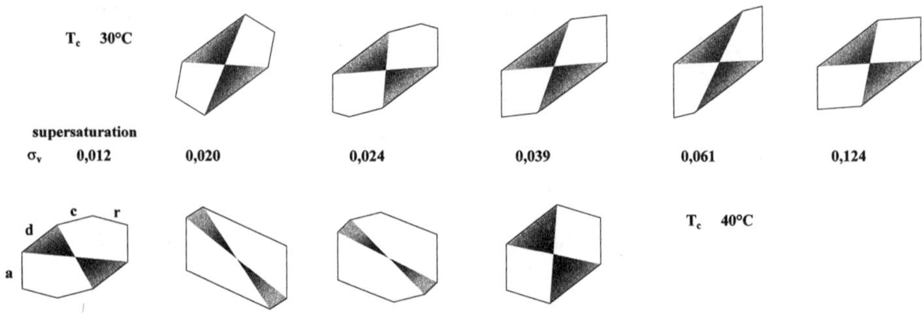

Fig. 22 The steady-state growth morphology of a sucrose crystal view along the [010] direction: according to two different crystallization temperatures the growth morphology strongly changes with supersaturation, as shown by the varying sizes of the growth sectors of the {101} form. This form has, structurally, S character, and its surface profile is very sensitive to water adsorption.

stable nucleation zone when *m*-xylene replaces *n*-dodecane as a solvent in the crystallization of *n*-alkanes [62].

The second type of impurity (that we will call additives) *consists of all compounds, other than solute and solvent, which are naturally present or intentionally added to a vapor, solution, or melt.* The manifold effects of additives on the crystal habit can be conveniently distinguished in two ways.

1. *Thermodynamics effects*. The thermodynamic parameter of interest in 2D nucleation and spiral growth is the edge free energy ρ of the edges running on the adsorbing face. By assuming (a) a Langmuir-type isotherm $\Gamma_A = bx_A$, where Γ_A and x_A are the edge concentration of adsorbed additive and its mole fraction in solution, respectively, and (b) an equilibrium between the impurity A adsorbed on the step and the bulk phase, when $\mu_A = \mu_A^\circ + kT \ln x_A$, we can write the variation of the step free energy as $\rho_A = \rho - kT \ln x_A$. Hence, ρ_A decreases with impurity concentration in the growth solution, and then the growth rate of the face should increase by either mechanism.
2. *Kinetic effects*. The kinetic term both in 2D nucleation and spiral growth model relates directly to the velocity of steps spreading across the crystal surface. Impurities adsorbed on the surface will impede this advance by different mechanisms, depending on the adsorption site; several references give detailed accounts of various theoretical relationships and their confrontation with experiments [63].

From the above, it follows that the theoretical growth models predict conflicting effects of impurity adsorption on the growth rate. Thermodynamic effects tend to increase it, whereas kinetic effects tend to decrease it. However, as in the case of 3D nucleation, kinetic effects are generally dominant; it is sufficient to poison a few kinks in order to slow the growth rate by several orders of magnitude. Depending on their nature, impurities may be active even at very low concentration, and the general trend is that their efficiency increases with increasing concentration; on the other hand, their efficiency decreases with increasing σ_v. Moreover, from the relation $\alpha = (1 - x_s)^2 (\Delta G_{sf} - \ln x_s)$, it follows that if an impurity can affect the crystal solubility, it can change the α factor of the faces as well, and hence their growth mechanism. An *increase in solubility* increases the availability of growth sites, and hence *accelerates growth*, while a *decrease in solubility* decreases the number of growth sites and *retards growth.*

According to a critical review on paraffin growth kinetics [64], the additives can act either as growth promoters or inhibitors. When added to the solution (prior to paraffin crystallization), some additives crystallize themselves, seeding the paraffin. This is the case of the polyalkylacrylate (having degree of polymerization $m > 30$), which acts as a preferential nucleation substrate either when it crystallizes as 2D hexagonal lamellae or when the lamellae form closed vesicles (so representing a classic example of heterogeneous 3D nucleation).

Other very recent and sound examples of heteronucleation assisted by impurities are (a) the formation of ''guest'' *n*-alcohol crystals from solution assisted by ''host'' fatty acids, previously structured in thin films obtained from vapor deposition [65], and (b) the acceleration of the nucleation of *n*-hexadecane crystals, in water-in-oil emulsion, due to the presence of a specific impurity such as the sucrose polyester of palmitic acid moiety (P-170) [66].

Apart from this specific effect additives can work as inhibitors, when they can mix, as soluble molecules, with solute. Two paths can be predicted:

1. When a competition is set up between impurity and solute molecules at kinks, with a relative high rate of dynamic exchange, the growth rate of the face is continuous, decreasing with impurity concentration (Fig. 17a) and depending only on the residence time of the impurity in the kink [67]. In this case Bliznakov's theory may be valid [68].
2. When the residence time of the impurity in the kink is so long that entrapment occurs, the flow of further layers can be hindered. This is the case of a copolymer (vinyl-ethylene-acetate) that dramatically increases the nucleation time and the growth kinetics (Fig. 17b) of the n-$C_{36}H_{74}$ paraffin [69] and of the raffinose molecule which stops the {110} form of sucrose crystal growing from aqueous solution [70]. These inhibition effects can be explained by the Cabrera and Vermilyea adsorption model [71]. This theory may also be invoked to explain the kinetic effect of L-phenylalanine (L-phe) on α and β polymorphs of L-glutamic acid (L-glu) [72]. Because the transformation $\alpha \rightarrow \beta$ being

solution-mediated, L-phe influences both polymorphs in a complex way. The growing (101) face of the β form is affected by morphological instability, whereas the growth rate of α crystals in the [111] direction is suppressed and the new (110) face appears. Fluctuations of the growth rate of both polymorphs is observed at higher L-phe concentrations; the critical L-phe concentration for stop growing steps (C_p^*), for the (110) face of α, is about twice that for the (101) face of β. The value of the critical relative supersaturation σ_v^* for β crystals is a few times that of α crystals, corresponding to C_p^*. All these sound single-crystal kinetic measurements indicate the preferential inhibition of the growth of the β polymorph.

Finally, we just note that great attention has been paid since the 1980s to the tailor-made additives aimed at designing crystal morphologies for industrial interest [73,74]. A *tailor-made additive* is a molecule very similar to the crystallizing species in its structure, but different in some specific way. The differences in the structural characteristic of the additive are such that, once incorporated in the structure of the host crystal, they will disrupt the correct bonding sequences in the lattice and interfere with the growth process at the crystal surface. In this manner these types of additives could dramatically affect crystal growth and habit.

To conclude, we point out once again that the final growth habits are conditioned by the interplay of factors internal and external to the crystals. Since knowledge of the structure of the growth solutions, particularly in the presence of impurities, is still unsatisfactory, it is understandable that the complex mechanisms of the impurity action during crystal growth may be difficult to define.

ACKNOWLEDGMENTS

The authors wish to thank professor Marco Rubbo (DSMP, University of Torino) for many valuable discussions, Professor Mariano Calleri who revised the language, and Dr. Linda Pastero for kind technical assistance.

REFERENCES

1. R. Boistelle, in *Crystallization and Polymorphism of Fats and Fatty Acids* (N. Garti and K. Sato, eds.), Marcel Dekker, New York, 1988, pp. 190–226.
2. P. Hartman, in *Morphology of Crystals*, Part A (I. Sunagawa, ed.), Terra Science Publishers, Tokyo, 1987, pp. 271–319.

3. B. Mutaftschiev, in *Interfacial Aspects of Phase Transformations*, NATO Advanced Study Institutes Series, Reidel, 1982, pp. 63–102.
4. S. Toschev, in *Crystal Growth: An Introduction* (P. Hartman, ed.), North-Holland, Amsterdam, 1974, pp. 328–341.
5. R. Kern, in *Morphology of Crystals* (I. Sunagawa, ed.), Terra Science Publishers, Tokyo, 1987, pp. 77–206.
6. R. Kern, G. Le Lay, and J. J. Métois, in *Current Topics in Materials Science*, Vol. 3 (E. Kaldis, ed.), North-Holland, Amsterdam, 1979, pp. 135–419.
7. O. Söhnel and J. Garside, in *Precipitation. Basic Principles and Industrial Applications* (O. Söhnel and J. Garside, eds.), Butterworth-Heinemann, 1992, pp. 1–391.
8. M. Jakubczyk and K. Sangwal, in *Elementary Crystal Growth* (K. Sangwal, ed.), Saan, Lublin, 1994, Chap. 3.
9. H. Curien and R. Kern, *Bull. Soc. Miner. Cristall. Franç.*, *80*: 111–132 (1957).
10. A. R. Verma and P. Krishna, *Polymorphism and Polytypism in Crystals*, Wiley, New York, 1966.
11. M. J. Buerger, *Am. Min.*, *30*: 469–482 (1945).
12. M. J. Buerger, *Am. Min.*, *32*: 593–606 (1947).
13. R. Kern, *Bull. Soc. Miner. Cristall. Franç.*, *84*: 292–311 (1961).
14. D. Aquilano, *J. Crystal Growth*, *37*: 215–228 (1977).
15. R. Boistelle and D. Aquilano, *Acta Cryst. A*, *33*: 642–648 (1977).
16. R. Boistelle, B. Simon, and G. Pepe, *Acta Cryst. B*, *32*: 1240–1243 (1976).
17. S. Amelinckx, *Acta Cryst.*, *9*: 16–23, 217–224 (1956).
18. A. R. Verma, *Proc. Roy. Soc. (London) A*, *228*: 34–42 (1955).
19. K. Inaoka, M. Kobayashi, M. Okada, and K. Sato, *J. Crystal Growth*, *87*: 243–250 (1988).
20. M. Kobayashi, T. Kobayashi, Y. Itoh, and K. Sato, *J. Chem. Phys.*, *80*: 2897–2903 (1984).
21. K. Sato, M. Kobayashi, and H. Morishita, *J. Crystal Growth*, *87*: 236–242 (1988).
22. C. Rinaudo, D. Aquilano, and R. Boistelle, *Acta Cryst. A*, *35*: 992–996 (1979).
23. W. Ostwald, *Z. Phys. Chem.*, *22*: 289–330 (1897).
24. K. Sato, in *Morphology of Crystals* (I. Sunagawa, ed.), Terra Science Publishers, Tokyo, 1987, pp. 513–530.
25. K. Sato and M. Suzuki, *J. Am. Oil Chem. Soc.*, *63*(10): 1356–1359 (1986).
26. F. Kaneko, M. Kobayashi, K. Sato, and M. Suzuki, *J. Phys. Chem.*, *101*(*2*): 285–292 (1997).
27. K. Sato and T. Kuroda, *J. Am. Oil Chem. Soc.*, *64*(*1*):124–127 (1987).
28. M. R. Gasco, S. Morel, and R. Carpignano, *Eur. J. Pharm. Biopharm.*, *38*: 7–12 (1992).
29. D. Aquilano, R. Cavalli, and M. R. Gasco, *Thermochem. Acta*, *230*: 29–37 (1995).
30. R. Cavalli, D. Aquilano, M. E. Carlotti, and M. R. Gasco, *Eur. J. Pharm. Biopharm.*, *41*: 329–333 (1995).
31. R. Boistelle, in *Current Topics in Materials Science*, Vol. 4 (E. Kaldis, ed.), North-Holland, Amsterdam, 1979, pp. 413–479.
32. K. Sato, *J. Phys. D: Appl. Phys.*, *26*: B77–B84 (1993).
33. P. T. Cardew and R. J. Davey, in *Proc. Symp. on Tailoring of Crystal Growth* (J. Garside, ed.), Manchester, 1982, pp. 1–7.

34. P. T. Cardew and R. J. Davey, *Proc. R. Soc. London A*, *398*: 415–428 (1985).
35. R. J. Davey, P. T. Cardew, D. McEvan, and D. E. Sadler, *J. Crystal Growth*, *79*: 648–653 (1986).
36. N. S. Brar and H. H. Schlössin, *Phase Transitions*, *1*: 299–324 (1980).
37. R. A. van Santen, *J. Phys. Chem.*, *88*: 5768–5769 (1984).
38. W. H. Casey, *J. Phys. Chem.*, *92*: 226–227 (1988).
39. M. J. J. M. Van Kamenade and P. L. de Bruyn, *J. Colloid Interface Sci.*, *118*: 564–572.
40. D. H. Rasmussen, *J. Crystal Growth*, *104*: 793–800 (1990).
41. M. Kitamura, S. Ueno, and K. Sato, in *Crystallization Processes* (H. Ohtaki, ed.), Wiley, New York, 1998, pp. 99–129.
42. C. B. Aakeröy and K. R. Seddon, *Chem. Soc. Rev.*, *22*: 397–407 (1993).
43. G. Falini, S. Albeck, S. Weiner, and L. Addadi, *Science*, *271*: 67–69 (1996).
44. G. Falini, S. Fermani, M. Gazzano, and A. Ripamonti, *Chem. Eur. J.*, *3*: 1087–1814 (1997).
45. J. Nývlt, O. Söhnel, M. Matuchova, and M. Broul, *The Kinetics of Industrial Crystallization*, Elsevier, Amsterdam, 1985, pp. 149–213.
46. R. Boistelle and J. P. Astier, *J. Crystal Growth*, *90*: 14–30 (1988).
47. J. A. Marquesee and J. Ross, *J. Chem. Phys.*, *79*: 373–378 (1983).
48. J. A. Marquesee and J. Ross, *J. Chem. Phys.*, *80*: 636–543 (1984).
49. B. Mutaftschiev, in *Adsorption et Croissance Cristalline*, Colloques Internationaux du CNRS (CNRS, ed.), No. 152, 1965, pp. 231–258.
50. F. C. Frank, *Disc. Faraday Soc.*, *5*: 33–40 (1949).
51. W. K. Burton, N. Cabrera, and F. C. Frank, *Phil. Trans. Roy. Soc. A*, *243*: 299–358 (1951).
52. A. A. Chernov, *Soviet Phys. Uspekhi*, *4*: 116–148 (1961).
53. P. Bennema and G. H. Gilmer, in *Crystal Growth: An Introduction* (P. Hartman, ed.), North-Holland, Amsterdam, 1974, pp. 263–288.
54. A. E. Nielsen and J. Christoffersen, in *Biological Mineralization and Demineralization* (G. H. Nancollas, ed.), Dahlem Konferenzen, Springer-Verlag, 1982, pp. 37–54.
55. P. Hartman and P. Bennema, *J. Crystal Growth*, *49*: 145–156 (1980).
56. P. Hartman, in *Physics and Chemistry in the Organic Solid State* (J. Fox, M. M. Labes, and A. Weissenberger, eds.), Interscience, New York, 1963, Vol. 1, 369; 1965, vol. 2, p. 873.
57. D. Aquilano, M. Rubbo, G. Vaccari, G. Mantovani, and G. Sgualdino, in *Industrial Crystallization 84* (J. Jancic and E. J. de Jong, eds.), Elsevier, Amsterdam, 1984, pp. 91–96.
58. K. A. Jackson, *J. Crystal Growth*, *5*: 13–23 (1969).
59. R. Kern, *Bull. Soc. Franç. Minér. Crist.*, *76*: 325–365 (1953).
60. R. Kern, *Bull. Soc. Franç. Minér. Crist.*, *78*: 461–497 (1955).
61. D. Aquilano, M. Rubbo, G. Mantovani, G. Sgualdino, and G. Vaccari, *J. Crystal Growth*, *83*: 77–83 (1987).
62. A. R. Gerson, K. J. Roberts, J. N. Sherwood, A. M. Taggart, and G. Jackson, *J. Crystal Growth*, *128*: 1176–1181 (1993).
63. K. Sangwal, *J. Crystal Growth*, *128*: 1236–1244 (1993).

64. R. Kern and R. Dassonville, *J. Crystal Growth*, *116*: 191–203 (1992).
65. H. Takiguchi, K. Iida, S. Ueno, J. Yano, and K. Sato, *J. Crystal Growth*, *193*: 641–647 (1998).
66. N. Kaneko, T. Horie, S. Ueno, J. Yano, T. Katsuragi, and K. Sato, *J. Crystal Growth*, *197*: 263–270 (1999).
67. B. Simon, A. Grassi, and R. Boistelle, *J. Crystal Growth*, *26*: 77–90 (1974).
68. G. M. Bliznakov, in *Adsorption et Croissance Cristalline*, Colloques Internationaux du CNRS (CNRS, ed.), No. 152, 1965, pp. 291–312.
69. J. C. Petinelli, *Rev. Inst. Franç. Pétrole*, *34*: 791–806 (1979).
70. F. Bedarida, L. Zefiro, P. Boccacci, D. Aquilano, M. Rubbo, G. Vaccari, G. Mantovani, and G. Sgualdino, *J. Crystal Growth*, *89*: 395–404 (1988).
71. N. Cabrera and D. A. Vermilyea, in *Growth and Perfection of Crystals* (R. H. Doremus, B. W. Roberts, and D. Turnbull, eds.), Wiley, New York, 1958, pp. 395–412.
72. M. Kitamura and T. Ishizu, *J. Crystal Growth*, *192*: 225–235 (1998).
73. L. Addadi, Z. Berkovitch-Yellin, N. Domb, E. Gati, M. Lahav, and L. Leiserowitz, *Nature*, *296*: 21–27 (1982).
74. Z. Berkovitch-Yellin, S. Ariel, and L. Leiserowitz, *J. Am. Chem. Soc.*, *107*: 3111–3122 (1985).

2

Polymorphism and Phase Transitions of Fatty Acids and Acylglycerols

Fumitoshi Kaneko
Osaka University, Toyonaka, Japan

I. INTRODUCTION

Fatty acids are widely distributed in biological tissues as building blocks of phospholipids and triacylglycerols, which are main constituents of biomembranes and fat tissues, respectively [1]. Furthermore, fatty acids and related lipid compounds have various applications in many industrial fields, such as foods, cosmetics, medicines, detergents, and so on.

One of the important properties of acyl chains is flexibility as to molecular conformation and lateral packing, which results in diverse solid and liquid crystalline states. The aggregation states of acyl chains in various conditions are highly important from a biological point of view. For instance, lateral packing and the conformation of acyl chains influence the fluidity of biomembranes, which closely relates to the functional activity of biomembranes [2].

The physical properties of solid states of long-chain compounds, such as melting point, heat capacity, elasticity, and so on, are also considerably dependent on crystal structure. Therefore, the technology to make a desirable solid state is of great importance in industry. To take an instance, triacylglycerols exhibit various crystalline phases [3], and their solid states closely relate to the quality of industrial products. Heat treatment is often utilized for reforming in production processing of oil and fats products.

Although the information about crystal structures of fats and complex lipids is fundamental for various fields, the structural analysis of these compounds often meets difficulty in preparing single crystals suitable for X-ray diffraction analysis. Furthermore, most metastable solid states of these compounds essentially contain several types of defects. Only the structures of a limited number of crystalline states have been determined. Therefore, the crystal structures of fatty acids are valuable references for understanding the packing, conformation, and dynamical properties of acyl chains in more complex systems.

Furthermore, fatty acids have interesting characteristics [4–6]. In spite of their simple molecular structures, fatty acids form a variety of crystalline states depending on occurrence conditions, and these crystalline states show many solid-state phase transitions between them. It has been found that various types of molecular motions are activated on these phase transitions: conformational disordering, large precessional motions, subcell rearrangement, and orientational change of functional groups. Such polymorphic behaviors of fatty acids receive remarkable influence of chain length, parity of carbon atoms, unsaturation and so on.

In this chapter, the crystal structures and solid-state phase transition mechanism of fatty acids and some related compounds will be described.

II. POLYMORPHISM AND CRYSTAL STRUCTURES OF FATTY ACIDS

A. Saturated Fatty Acids

The even-odd effect of carbon atoms has a significant influence on polymorphism and physical properties of fatty acids. The interfacial interactions between bimolecular layers result in a marked difference in crystal structures between even-numbered and odd-numbered fatty acids.

1. Even-Numbered Fatty Acids

The solid phases of the even-numbered fatty acids can be divided into four classes, named A, B, C, and E forms. The A form is characterized by the $T_{//}$ subcell (Fig. 1), where the zigzag planes are arranged parallel to each other, while the B, C, and E forms make the $O_{\perp}$ subcell where the acyl chains are set perpendicular to the nearest neighbors. Although these polymorphs can be obtained by solution crystallization, only the C form grows on melt crystallization. In addition to polymorphism, the B and E forms exhibit polytypism (Fig. 2)—a higher-order structural difference caused by a different stacking sequence of molecular layers [7].

a. A Form. So far several types of A forms have been reported, A-super [8,9], A_1 [10], A_2 and A_3 [11], each belonging to a triclinic system of space group $P\bar{1}$. Concerning lamellar structure, these crystal structures can be divided into two groups: ''segregated layer structure'' of A_2 and A_3, and ''nonsegregated layer structure'' of A-super and A_1. In the former the methyl terminals are separated

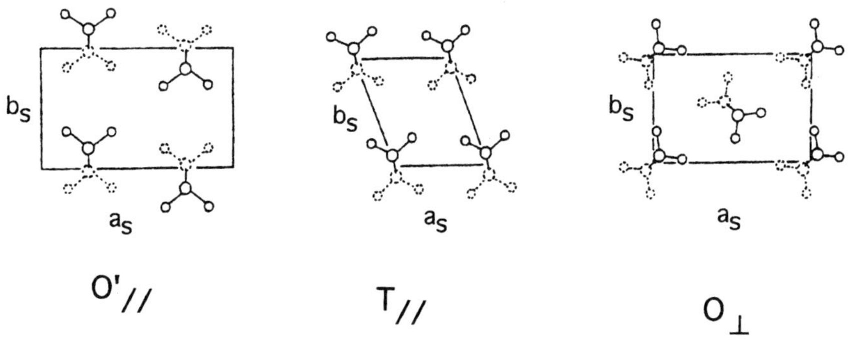

Fig. 1 Typical subcell structures of fatty acids.

from the carboxyl terminals, while in the latter the terminal plane (the surface of the monomolecular layer) consists of methyl and carboxyl terminals.

The A-super and A_1 forms of lauric acid (C_{12}) have been fully determined (Fig. 3). The unit cell of A-super contains six molecules that stand side by side in one monomolecular layer. Three molecules having methyl group on the upper side stand in a line, and then three molecules having carboxyl group on the upper side follow. The plane of the carboxyl group is set perpendicular with respect to the zigzag plane of the hydrocarbon chain in the central molecules of the trio, and that is slightly twisted in the rest of the molecules. On the other hand, the unit cell of A_1 contains two molecules, which stand side by side in one monomolecular layer. The carboxyl group of one molecule is at the top end, and that of another one is at the bottom end. The plane of the carboxyl group is slightly

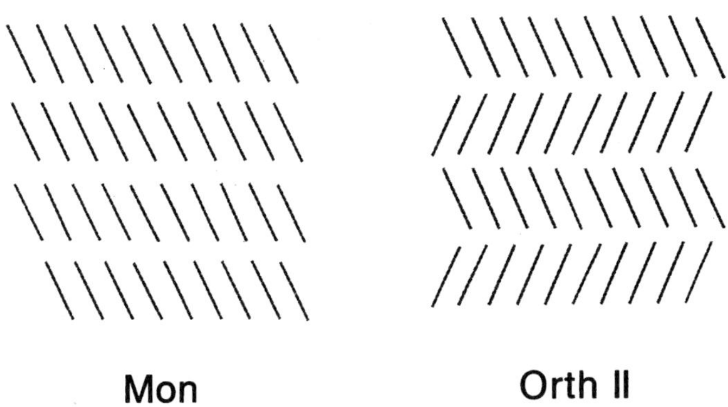

Fig. 2 Polytypic structures found in fatty acids.

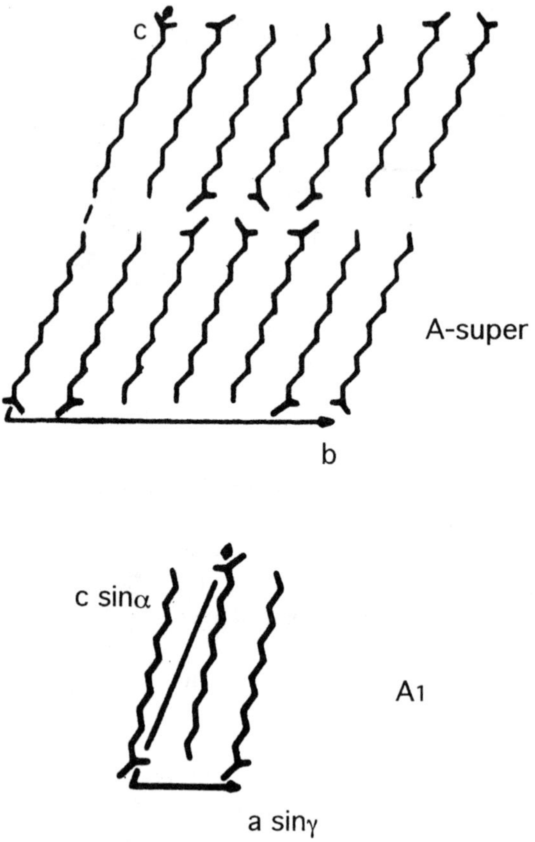

Fig. 3 Crystal structures of A-super and A_1.

deviated from the coplanar position with respect to the skeletal plane of the hydrocarbon chain.

The existence of the A_2 form has been confirmed in myristic acid (C_{14}), palmitic acid (C_{16}), and stearic acid (C_{18}) by X-ray diffraction, IR, and Raman spectroscopy, but its crystal structure has not been fully analyzed yet. Contrary to the A-super and A_1 forms, the A_2 form constructs a ''segregated lamellar structure,'' where the methyl and carboxyl groups form a terminal plane separately. The unit cell contains two types of dimers that are different from each other in the conformation of the carboxyl group. In a cooling process the A_2 form transforms reversibly to the A_3 form at about 140 K, accompanying a conformational change at the carboxyl groups.

The crystal structures belonging to the A form have a common feature.

The unit cell is larger than the $T_{//}$ subcell in the lateral directions; the cross-sectional area of the unit cell is equal to or more than two times that of the $T_{//}$ subcell. In other words, these crystal structures can be recognized as superstructures. The bulky carboxyl group is the cause for the formation of superstructure. To reduce their steric hindrance, carboxyl groups coexist with methyl groups in one terminal plane of the A_1 and A-super forms of lauric acid. Longer fatty acids such as palmitic acid take a different way for this purpose; in the A_2 and A_3 forms there are two kinds of carboxyl groups taking different conformation mutually in a terminal plane. The contribution of the van der Waals interaction between hydrocarbon segments becomes superior to the interaction between carboxyl groups as the chain length increases, which would relate to the conversion of lamellar structure from the nonsegregated type to the normal segregated one.

b. B Form. There are two polytypic structures in the B form [12–15]: single-layer structure, Mon [16] and double-layer structure, Orth II [17]. The crystallographic parameters of B(Mon) and B(Orth II) of stearic acid are summarized in Table 1, together with those of C, E(Mon), and E(Orth II).

The Mon type belongs to a monoclinic system of space group $P2_1/a$. There are four monomers in a unit cell. The acyl chain takes a *gauche* conformation at the C_2—C_3 bond near the carboxyl group and tilts toward the b_s axis of the $O_\perp$ subcell by 27°, as shown in Fig. 4 (the setting of the subcell axes is made according to orthorhombic polyethylene [18]). In this case, the (011) plane of the $O_\perp$ subcell is set parallel to the lamellar surface consisting of methyl terminals.

On the other hand, the Orth II type belongs to an orthorhombic system of space group *Pbca*, and the unit cell contains eight molecules. There are two bimolecular layers within a repeating period along the *c* axis (Fig. 5). The bimolecular layer of Orth II has the same structure as that of Mon. Contrary to the constant inclination in the Mon type, the acyl chain inverts its inclination direction at every stacking of bimolecular layers. It has been clarified that a systematic difference in thermodynamic and physical properties exists between Mon and Orth II.

Table 1 Crystallographic Parameters of the B, C, and E Forms of Stearic Acid

	B(Mon)	B(Orth II)	C	E(Mon)	E(Orth II)
Space group	$P2_1/a$	*Pbca*	$P2_1/a$	$P2_1/a$	*Pbca*
a (Å)	5.587	7.404	9.36	5.603	7.359
b	7.386	5.591	4.95	7.360	5.609
c	49.33	87.66	50.7	50.789	88.41
β (deg)	117.24		128.25	119.40	
V (Å^3)	1810	3629	1845	1825	3649
Z	4	8	4	4	8

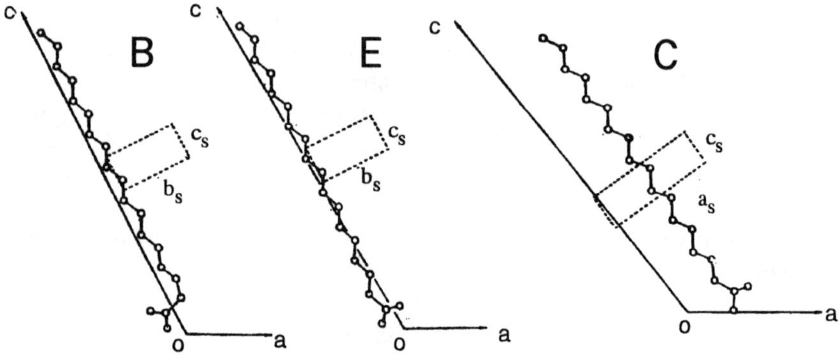

Fig. 4 Crystal structures of the B, C, and E forms.

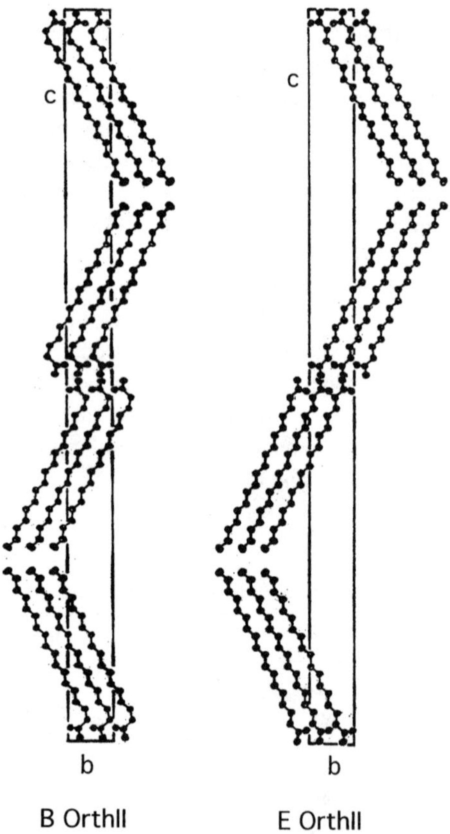

Fig. 5 Orth II type structures of the B and E forms.

As to stearic acid, the B form is the stable phase below 32°C. As shown in Table 1, the molecules are most densely packed in the B form. It seems that cohesive energy makes the B form stabler than the other phase.

c. C Form. The unit cell belongs to a monoclinic system of space group $P2_1/a$ and contains four molecules [19,20]. The acyl chain taking all-*trans* conformation inclines toward the a_s axis of the $O_\perp$ subcell by 35°, and the (201) plane of the $O_\perp$ subcell is set parallel to the lamellar interface.

From spectroscopic studies, it has been suggested that there are two types of configurations of carboxyl groups in the C form; the C=O bond takes *cis* or *trans* position against the C_2—C_3 bond in the C_3—C_2—C_1=O group [21,22]. These two kinds of configurations are in thermodynamic equilibrium through the simultaneous proton transfer mechanism [23]. In addition it was confirmed that the ratio of the *cis* conformation increases with temperature. The C form is the stable phase at the highest temperature region, which is ascribed to this molecular motion at carboxyl groups [5].

d. E Form. The E form exhibits two polytypic structures, single-layer structure Mon [24–27] and double-layer structure Orth II [28], as well as the B form. As to the molecular packing in Mon and Orth II, there is no essential difference between B and E forms. For the Mon type there are four molecules in a unit cell of space group $P2_1/a$, and for the Orth II type eight molecules in a unit cell of *Pbca*.

As to the structure of a bimolecular layer, the E form is analogous to the B form. The acyl chain inclines toward the b_s axis of the $O_\perp$ subcell by 27°, and its methyl terminal makes a plane parallel to the (011) plane of the subcell. However, the acyl chain takes all-*trans* conformation contrary to the bent conformation around the carboxyl terminal.

The E form is a metastable phase; it appears as the precursor of the B form on solution crystallization [29]. Single crystals of the E form that occur in the initial stage transform gradually to the B form through a solid-state phase transition and solution-mediated phase transition, as described in a later section.

2. *Odd-Numbered Saturated Fatty Acids*

The crystalline states of odd-numbered saturated fatty acids are roughly classified into three polymorphs, A′, B′, and C′ (Fig. 6). Polymethylene chains are packed in the $T_{/\!/}$ subcell for the A′ form, and in the $O_\perp$ subcell for the B′ and C′ forms. The C′ form appears in the narrow temperature region near melting points.

a. A′ Form. The crystal structure of the A′ form has been studied for tridecanoic acid [30] and pentadecanoic acid [31]. The unit cell belongs to a triclinic system of space group $P\bar{1}$ and contains one dimer. The acyl chain taking

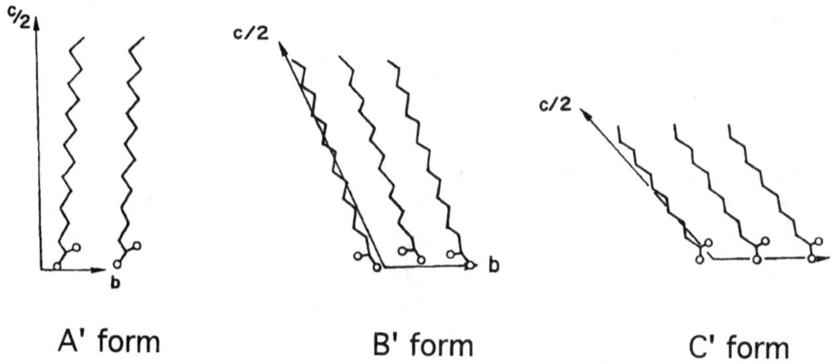

Fig. 6 Crystal structures of the A′, B′, and C′ forms.

all-*trans* conformation tilts toward the basal plane by about 17°. The A′ form performs rapid solid-state phase transitions, as described later.

b. B′ Form. The B′ form also belongs to a triclinic system of space group $P\bar{1}$ and its unit cell contains two dimers [32,33]. Two crystallographically independent molecules locate at an asymmetric unit. Although the two molecules take all-*trans* conformation and incline toward the basal plane by 33°, they are different in the way of chain inclination; one molecule tilts within a plane nearly parallel to the zigzag plane of hydrocarbon chain, and the other molecule tilts in the direction nearly perpendicular to that. The (111) plane of the $O_{\perp}$ subcell is set parallel to the methyl terminal plane.

c. C′ Form. The structure of the C′ form is similar to that of the C form. The unit cell is monoclinic, space group $P2_1/a$, and consists of two dimers [34]. The acyl chain inclines to the a_s axis of the $O_{\perp}$ subcell whose (201) plane is set parallel to the basal plane. Only in the arrangement of methyl terminals, is the C′ form different from the C form of even-numbered saturated fatty acids. Since the C′ form appears in the vicinity of melting point, the thermal motion of the methyl group is highly activated.

B. Unsaturated Fatty Acids

1. cis-Monounsaturated Fatty Acids

The polymorphic behavior of *cis*-monounsaturated fatty acids is quite sensitive to total chain length and position of double bond [35,36]. A variety of crystalline phases appear, depending on the crystallization condition as summarized in Table 2. Many crystalline phases are peculiar to each unsaturated fatty acid. For example, in spite of carbon atoms of the same number, oleic acid exhibits polymorphism that is completely different from that of petroselinic acid.

Table 2 Structural Data of Principal Monounsaturated Fatty Acids

					Subcell		Phase transition[b]		
Fatty acid	Form	Space group	Z	Olefinic conformation	Δ	ω	Form	T_{tr} (°C)	ΔH_{tr} (kJ/mol)
Palmitoleic acid	γ	$P2_1/a$	4	S—C—S′	$O'_{//}$	$O'_{//}$	→α	−18.4	7.5
[16:1(9)]	α	$P2_1/a$	4	S—C—T	$O'_{//}$	$O_{\perp}$ like	Melting	2.0	32.1
Asclepic acid	γ	$P2_1/a$	4	S—C—S′	$O'_{//}$	$O'_{//}$	→α	−15.4	7.8
[18:1(11)]	α	$P2_1/a$	4	S—C—T	$O'_{//}$	$O_{\perp}$ like	Melting	13.8	39.8
Oleic acid	γ	$P2_1/a$	4	S—C—S′(133°, *cis*,-133°)	$O'_{//}$	$O'_{//}$	→α	−2.2	8.8
[18:1(9)]	α	$P2_1/a$	4	S—C—T	$O'_{//}$	$O_{\perp}$ like	Melting	13.3	39.6
	β_2	Unclear	Unclear	Unclear	// type		Melting	16.0	48.9
	β_1	$P\bar{1}$	4	T—C—T(A:174°, *cis*, 173°)[a]	$T_{//}$		Melting	16.3	57.9
			4	T—C—T(B:175°, *cis*, 175°)[a]					
Petroselinic acid	LM	$Pbca$	8	157°, *cis*, −160°	$O_{\perp}$	$O_{\perp}$	Melting	—	—[c]
[18:1(6)]	HM	$P\bar{1}$	4	S—C—S(A: 91°, *cis*, 137°)[a]	$O_{\perp}$ like	$M_{//}$	Melting	30.5	47.5
				S—C—S(B:130°, *cis*, 119°)[a]					
Erucic acid	γ	$P2_1/a$	4	S—C—S′(129°, *cis*,-128°)	$O'_{//}$	$O'_{//}$	→α	−1.0	8.8
[22:1(13)]	α	$P2_1/a$	4	S—C—T (105°, *cis*,-172°)	$O'_{//}$	$O_{\perp}$ like	→α_1	31.2	5.4
	γ_1	$P\bar{1}$	2	S—C—S (98°, *cis*, 95°)	$T_{//}$	$T_{//}$	Melting	9.0	8.9
	α_1	$P\bar{1}$	2	S—C—S (96°, *cis*, 102°)	$T_{//}$	$T_{//}$	Melting	34.0	54.0
Elaidic acid	LM	$A2/a$	8	S—T—S′(118°, *trans*, −118°)	$O_{\perp}$	$O_{\perp}$	Melting	42.3	
[18:1(9)]	HM	Unclear	Unclear	Unclear	Unclear		Melting	42.8	

[a] A and B are two molecules in an asymmetric unit.
[b] T_{tr} = transition temperature; ΔH_{tr} = transition enthalpy.
[c] Melt-mediated transformation to the HM phase takes place.

The diversity of solid-state phase transition is another important characteristic of *cis*-monounsaturated fatty acids. In particular, two reversible phase transitions accompanied with conformational disordering, the $\alpha_1 \leftrightarrow \gamma_1$ and $\alpha \leftrightarrow \gamma$ transitions, are highly suggestive of the influence of *cis*-double bond on the dynamic properties of acyl chains. The crystal structures of the higher temperature phases, α and α_1, are described in the section dealing with phase transition mechanisms.

The marked diversity as to the conformation of *cis*-olefin group is the most important factor for the polymorphism of *cis*-monounsaturated fatty acids. In general, there are two stable rotation angles as to C—C bond adjacent to *cis*-C=C bond: *skew* (S = 120°) and *skew′* (S′ = −120°) (Fig. 7). However, its bond rotation energy increases slowly as the internal rotation angles change, which makes it possible for the C—C bond to adopt various rotation angles different from the standard S or S′. Indeed, the observed torsion angles of the solid states are distributed in the range from 91° to 175°. An acyl chain is divided by *cis*-C=C bond into two portions: methyl-sided chain (ω chain) and carboxyl-sided chain (Δ chain). The conformation of the C—C=C—C group determines the whole configuration of the molecule. In the case of S—C—S′ conformation, the Δ and ω chains are located on the same sides of the C—C=C—C plane and the molecule as a whole adopts a planar configuration. In the case of the S—C—S conformation, the Δ and ω chains are located on the opposite sides of C—C=C—C plane and the molecule has a twisted configuration. Accordingly, it is hard for S—C—S type molecules to make high-symmetry subcells.

a. γ Phase. The α and γ phases are the most popular polymorphs for *cis*-monounsaturated fatty acids. So far they have been found in gondoic, erucic, oleic, asclepic, and palmitoleic acids [37–40]. The γ phase that was previously reported as the low-melting form for oleic acid is monoclinic of space group $P2_1/a$ [41,42]. Figure 8 shows the crystal structure of erucic acid γ. The unit cell consists of two dimers. The *cis* double bond takes the S—C—S′ type conformation. With this conformation, the whole molecular shape becomes coplanar and the zigzag plane of two hydrocarbon chains is set perpendicular to the glide plane, which results in the $O'_{//}$ subcell characteristic of γ phase both in the methyl-sided chain and carboxyl-sided chain. The b_s axis of the $O'_{//}$ subcell coincides with the unique axis. The skeletal plane of polymethylene segments is set perpendicular to the glide plane. The acyl chain inclines toward the a_s axis by 34°; the (201) plane of the $O'_{//}$ subcell is parallel to the lamellar interface.

b. γ1 Phase. Figure 9 shows the structure of the γ1 phase of erucic acid, which belongs to a triclinic system of space group $P\bar{1}$ [43]. The unit cell consists of one dimer whose *cis*-olefin group takes the S—C—S conformation. Both the carboxyl-sided chain and the methyl-sided chain form the $T_{//}$ subcell, in which the former inclines by 43° and the latter by 42°.

a

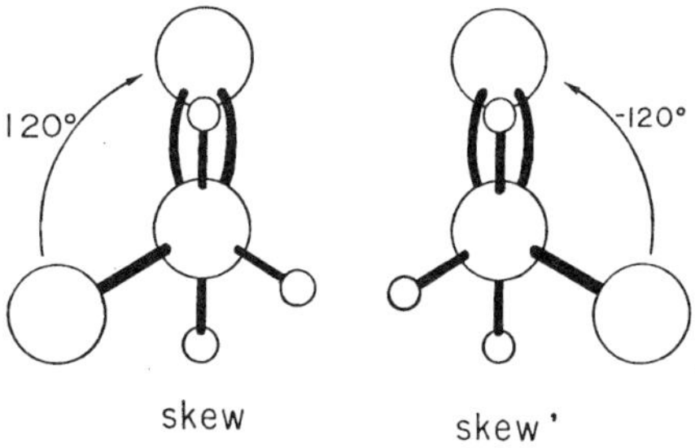

b

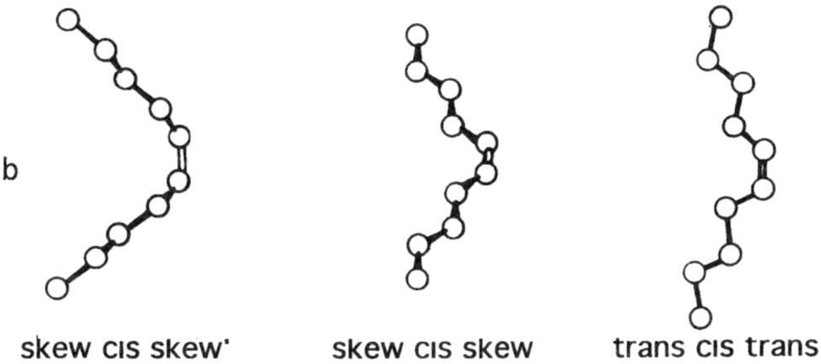

c

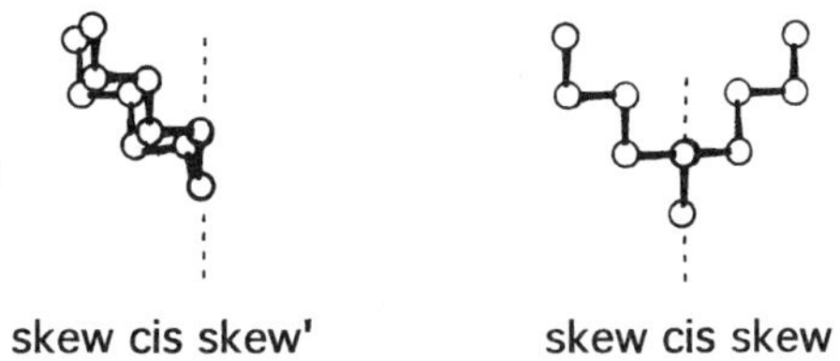

Fig. 7 Conformation of *cis*-olefin groups: (a) definitions of *skew* and *skew′* conformations; (b) side views of olefin groups taking *skew-cis-skew′*, *skew-cis-skew*, and *trans-cis-trans* conformation; (c) projections along C═C bond of *skew-cis-skew′* and *skew-cis-skew* conformations.

c. Low-Melting (LM) and High-Melting (HM) Phases of Petroselinic Acid. The polymorphism of petroselinic acid is completely different from that of other *cis*-monounsaturated fatty acids [44], which comes from its specific chemical structure. Although most *cis*-monounsaturated fatty acids have an odd number of carbon atoms in the methyl-sided chain, petroselinic acid contains an even number of carbon atoms (12) in the methyl-sided chain.

It has been clarified that the LM phase exhibits polytypism. In addition to the B and E forms of saturated fatty acids, there are two polytypes, Mon and

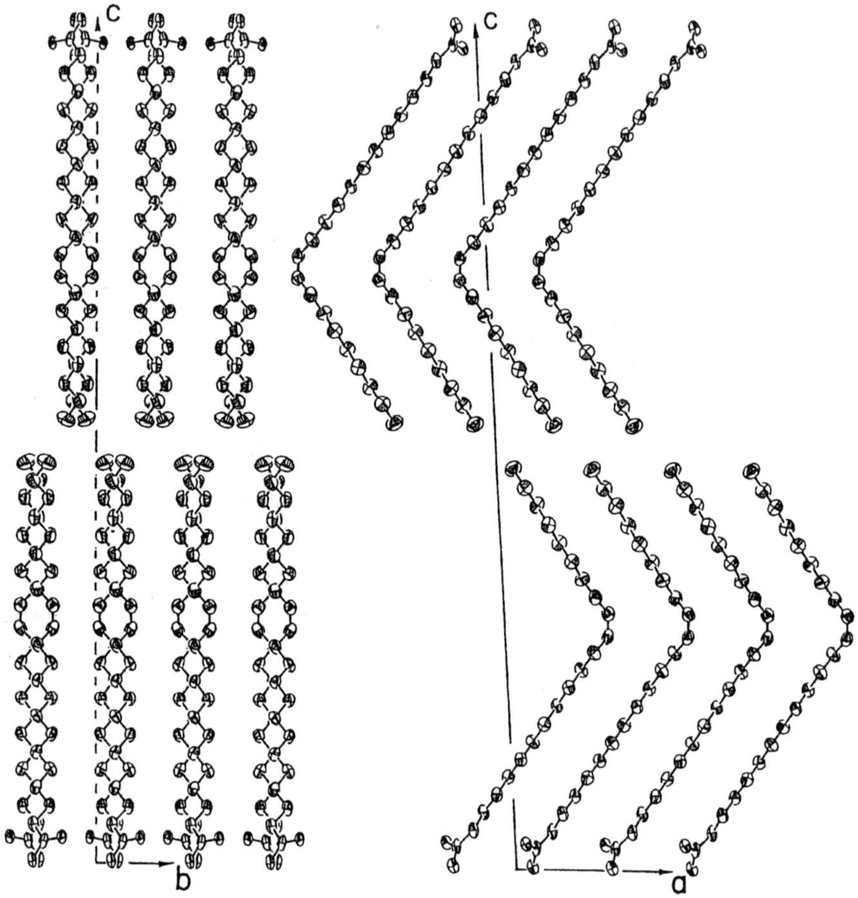

Fig. 8 Crystal structure of erucic acid γ.

Orth II [45]. The structure of the Orth II type was fully analyzed, as shown in Fig. 10. It belongs to an orthorhombic system of space group *Pbca* [46]. Two bimolecular layers are related to each other by twofold axis along the *c* axis. The methyl-sided and carboxyl-sided chains take all-*trans* conformation to form the $O_{\perp}$ subcell, and they incline toward the b_s axis by 26° and 22°, respectively. The (0 1 1) subcell plane of the methyl-sided chain is nearly parallel to the basal plane as in B and E forms of *n*-fatty acids. With respect to one bimolecular layer, the LM phase has the same symmetry as γ phase. By altering the dihedral angles of C—C=C—C portion significantly from the S—C—S′ conformation, the zig-

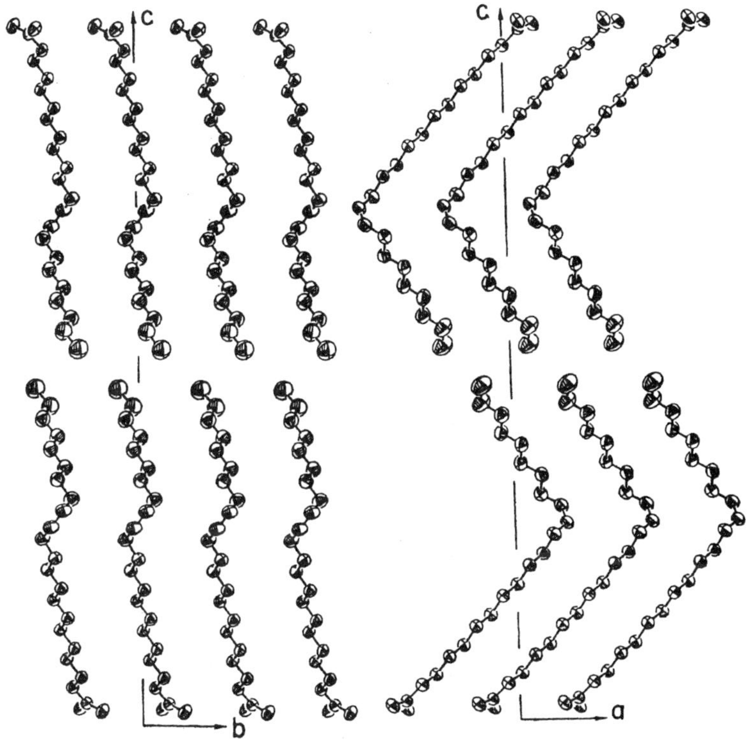

Fig. 9 Crystal structure of erucic acid γ_1.

zag planes of polymethylene chains make an angle of about 44° with the glide plane, which results in the formation of $O_\perp$ subcell.

The asymmetric unit of the HM phase contains two crystallographically independent molecules, as shown in Fig. 11 [47]. Although they can be recognized as S—C—S type, there is a large difference in the conformation of olefin group, enabling two types of subcell structures; the methyl-sided chains form the $M_{//}$ subcell, while the carboxyl-sided chains form a structure analogous to the $O_\perp$ subcell (Fig. 12).

d. β Phase. The β phase can be divided into two phases, β_1 and β_2, between which there is a slight difference in melting points; the former melts at 16.3°C and the latter at 16.0°C. The crystal structure of the β_1 phase was determined as shown in Fig. 13. It was clarified that the crystal structure of β_1 occupies a specific position for the polymorphism of *cis*-monounsaturated fatty acids [48]. The β_1 phase belongs to a triclinic system ($P\bar{1}$) containing two crystallographi-

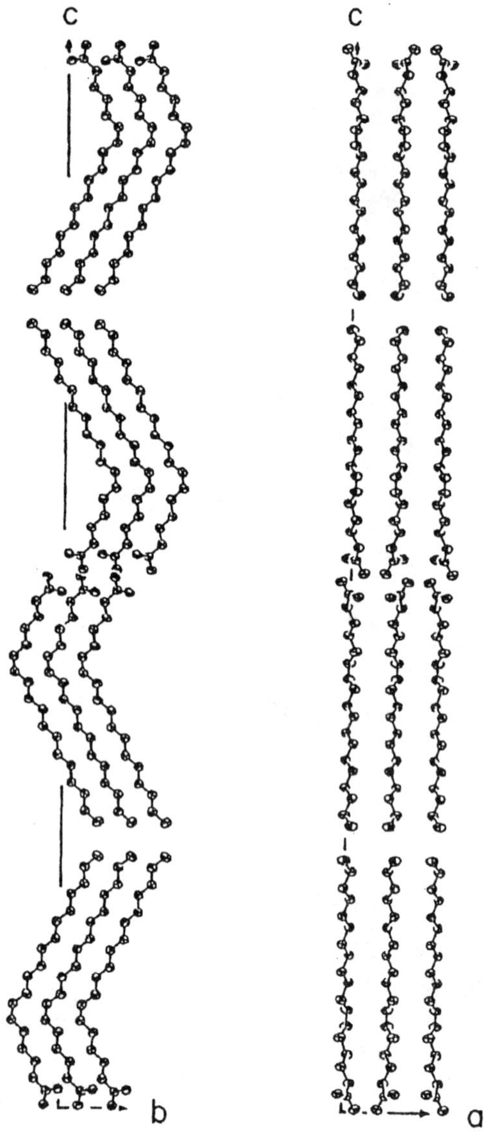

Fig. 10 Crystal structure of the low-melting phase of petroselinic acid (Orth II type).

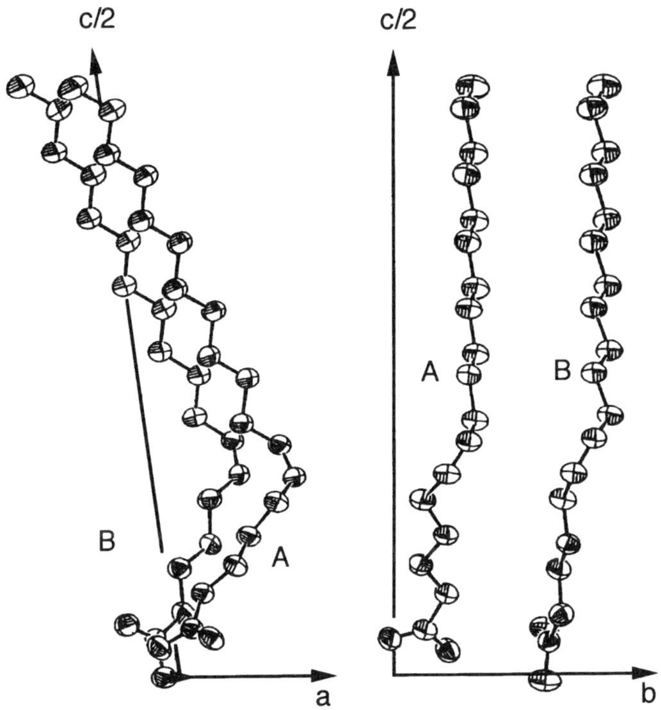

Fig. 11 Crystal structure of the high-melting phase of petroselinic acid.

cally independent molecules in an asymmetric unit. The molecular layer is a unique interdigitated lamellar structure, in which methyl and carboxyl groups make up a lamellar interface, keeping juxtaposition of *cis*-C=C groups. The conformation of the C—C=C—C portion is almost *trans-cis-trans* (T-C-T) for both molecules, and the acyl chains form $T_{//}$ subcell. The polymethylene segments incline toward the basal plane by 45°.

For this interdigitated structure, the location of *cis*-C=C bond is of great importance. If the *cis*-C=C bond is apart from the center of acyl chain, a vacancy arises at either the methyl or the carboxyl terminal, which significantly reduces the cohesive energy. Infrared spectral data suggest that the β_2 phase forms a structure similar to that of the β_1 phase with a subtle difference in the carboxyl group [49].

2. *cis-Polyunsaturated Fatty Acids*

Only the crystal structure of linoleic acid was reported, which has common characteristics with the γ phase of *cis*-monounsaturated fatty acids [55]. The unit cell

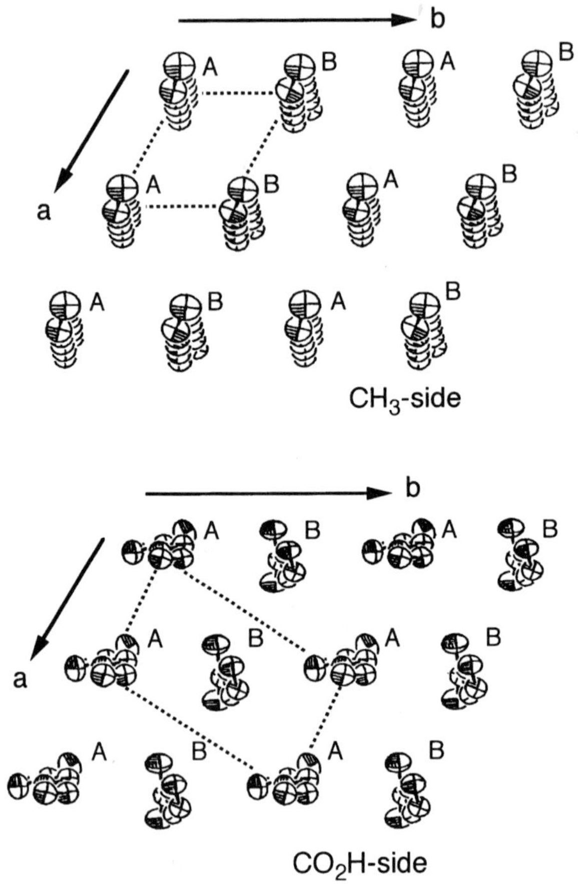

Fig. 12 Lateral packings of the methyl-sided and carboxyl-sided chains in the HM phase of petroselinic acid.

belongs to a monoclinic system of symmetry $P2_1/a$ and consists of two dimers. The conformation of C—C=C—C—C=C—C is S′—C—S—S—C—S′, which results in the $O'_{//}$ subcell. On the basis of unit cell parameters, it is considered that α-linolenic acid and arachidonic acid form similar crystal structures, which are characterized by the S—C—S′ type conformation: a S′—C—S—S—C—S′—S′—C—S conformation for α-linolenic acid and a S′—C—S—S—C—S′—S′—C—S—S—C—S′ one for arachidonic acid.

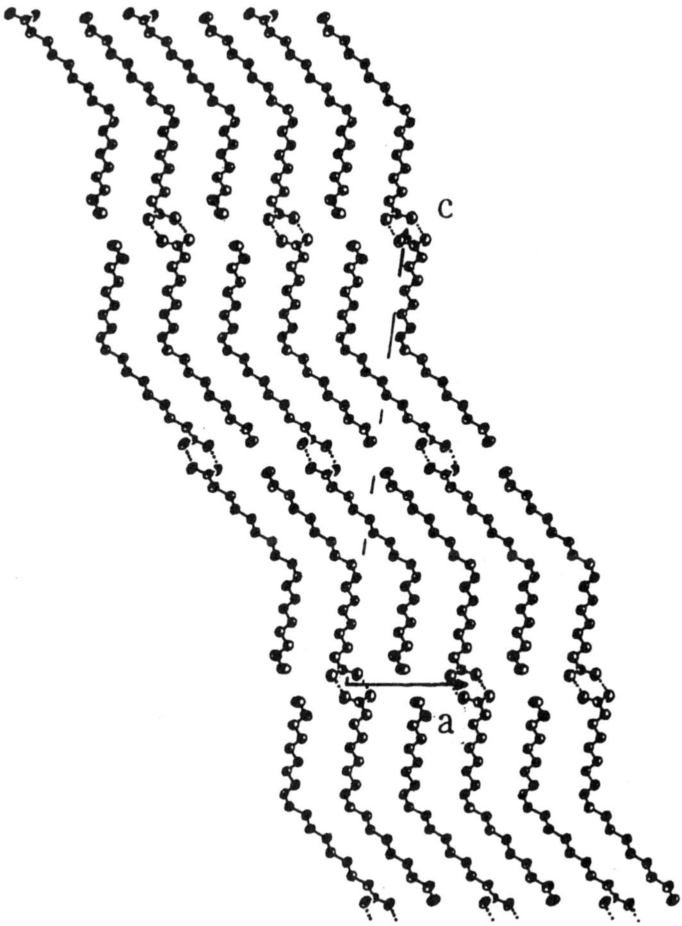

Fig. 13 Crystal structure of the β_1 phase of oleic acid.

3. trans-Monounsaturated Fatty Acid

So far, detailed polymorphic studies have been done only on elaidic acid [50–52]. There are two polymorphs: high-melting and low-melting phases. The crystal structure of the low-melting phase was fully analyzed [52], as shown in Fig. 14. The unit cell belonged to a monoclinic system of $A2/a$ and contained eight molecules. There were two bimolecular layers in a repeating unit along the c axis. As for the conformation around *trans*-C=C bond, the molecule approximated to S—T—S′ conformation, which made the skeletal planes of the methyl-sided and

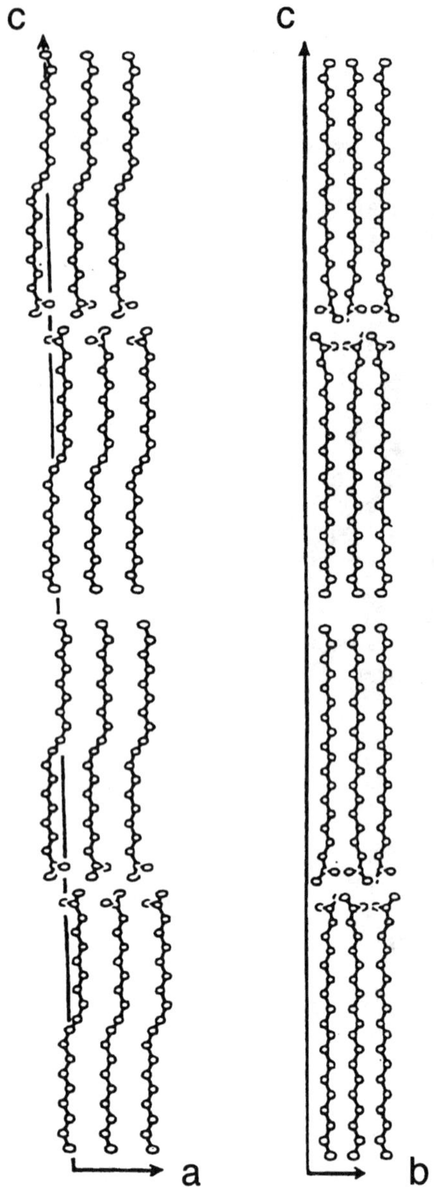

Fig. 14 Crystal structure of the low-melting phase of elaidic acid.

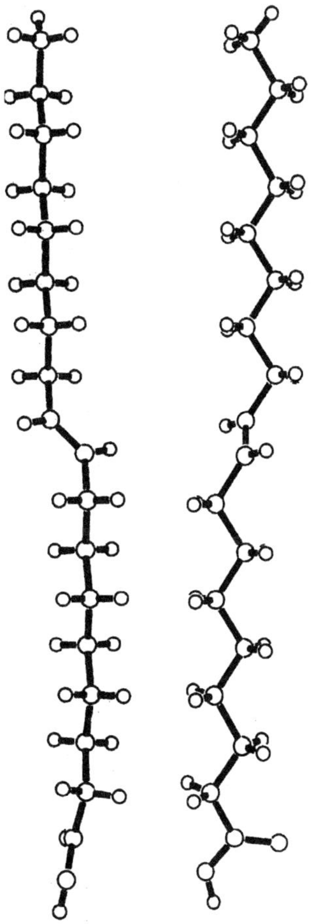

Fig. 15 Molecular form of elaidic acid in the low-melting phase.

carboxyl-sided chains parallel to each other (Fig. 15). Polymethylene segments forming the $O_{\perp}$ subcell were arranged perpendicular to the lamellar surfaces, as well as odd-numbered *n*-alkanes of orthorhombic lattice [1,53,54].

C. Polytypism of Fatty Acids

Polytypism has been found in *n*-alkanes, *n*-alcohols, and fatty acids [12–15,56]. As described above, there are two polytypes in fatty acids: ordinary single-layered structure (Mon) and double-layered structure (Orth II). The polymorphs that exhibit polytypism have the same structural feature; polymethylene chains are accommodated in the $O_\perp$ subcell whose (0 1 1) plane is parallel to the lamellar interface. This suggests that the intermolecular interaction between the methyl terminals is a significant factor for polytypism.

For both polytypes, one layer has the monoclinic symmetry $P2_1/a$. Its unit cell contains two dimers that are related to each other by twofold screw axes and a glide plane. In the Orth II polytype, two layers in a repeating period are related by a twofold screw axis which is perpendicular to the lamellar interface. Consequently, the chain direction of inclination is inverted at every lamellar stacking, and the unit cell has a higher orthorhombic symmetry, *Pbca*.

It has been clarified that polytypic structures have a significant influence on thermodynamic and mechanical properties [5,57].

D. Acylglycerols

Numerous studies have been done on the polymorphism of acylglycerols. Due to space limitations, only the results of recent X-ray crystal structural analyses using single-crystal specimens on mix-chain acylglycerols are described. For previous and other noteworthy studies, see Refs. 1 and 6 and the note at the end of the manuscript.

1. 1,2-Dipalmitoyl-3-acetyl-sn-glycerol (PP2)

Two molecules are present in the monoclinic unit cell of space group $P2_1$ [58]. The unit cell forms a trilayer structure, as shown in Fig. 16: an interdigitated monolayer of acetyl groups is sandwiched with two layers of palmitoyl chains, which adopt the $T_{//}$ subcell and tilt toward the basal plane by 27°. With respect to the glycerol conformation, PP2 is markedly different from saturated monoacid triacylglycerols [59–61] and from 1,2-diacyl-*sn*-glycerols [62,63] but very similar to the A conformer of 1,2-dimyristoyl-*sn*-glycero-3-phosphocholine [64]; the glycerol backbone and acetyl group extend roughly linearly from the *sn*-1 chain, and the *sn*-2 chain takes a *gauche* conformation at the C_2—C_3 bond.

2. 1-Stearoyl-3-oleoylglycerol (1,3-SODG)

Four molecules are present in the monoclinic unit cell of space group *Cc* [65]. The unit cell forms a four-chain-length structure consisting of two double-layer leaflets, as shown in Fig. 17. The molecule forms an extended V-shaped conformation with the oleoyl and stearoyl chains. The stearoyl and oleoyl chains pack separately in individual layers. The stearoyl chain is roughly straight and packed in the $T_{//}$ subcell. Both methyl- and carboxyl-sided chains of oleoyl groups adopt the $T_{//}$ subcell. The torsion angle sequence along the oleoyl chain from C_7 to C_{13} is 173°, −152°, −17°, −157°, −163°, and −178°, which is different from that

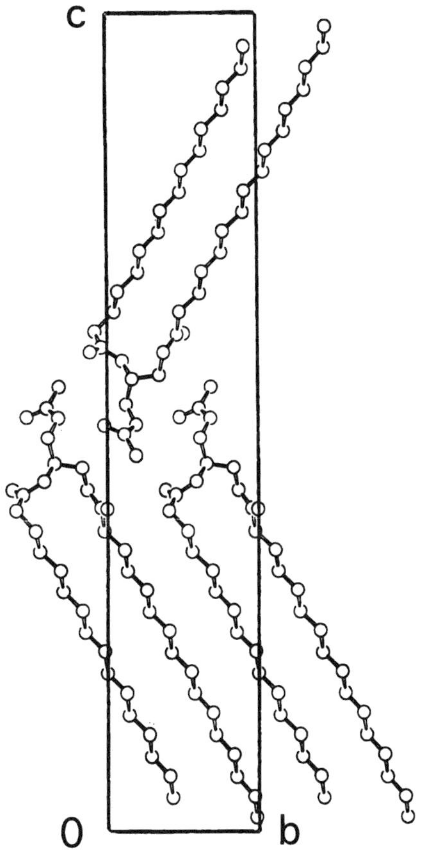

Fig. 16 Crystal structure of 1,2-dipalmitoyl-3-acetyl-*sn*-glycerol.

in the polymorphs of *cis*-monounsaturated fatty acids. In the glycerol region, the molecules are linked by hydrogen bonds 2.78 Å long, between the free hydroxyl group at the second position of the glycerol group and the carbonyl oxygen on the oleoyl chain. As for the conformation of the glycerol ester part, 1,3-SODG is rather similar to the A conformer of *rac*-monolaurin β form [66] and to 1,3-bromo-diundecanoin [67].

3. β′-2 Polymorph of 1,2-Dipalmitoyl-3-Myristoyl-sn-Glycerol

A crystal structure of β′ polymorph of triacylglycerols was analyzed for the first time at the atomic level on β′-2, one of the two β′ forms of PPM [68]. There are eight molecules in a monoclinic unit cell with space group *C*2. The asymmet-

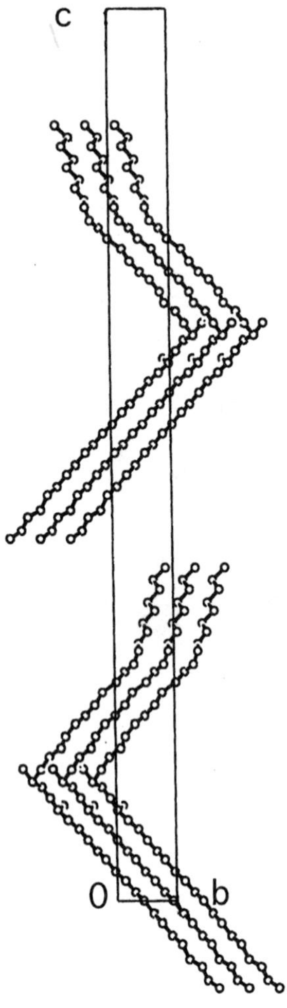

Fig. 17 Crystal structure of 1-stearoyl-3-oleoyl-glycerol.

ric unit contains two crystallographically independent molecules, A and B, and the two molecules reveal different conformations of acyl chains in the vicinity of glycerol group. In A, the *sn*-1 and *sn*-2 palmitoyl chains take all-*trans* conformation, while the *sn*-3 myristoyl chain has a *gauche* bond near its ester bond. In B, the *sn*-1 chain takes a *gauche* conformation and the other chains take all-*trans* conformation. The two types of molecules have the same glycerol structure similar to that of tricaprine [59,60].

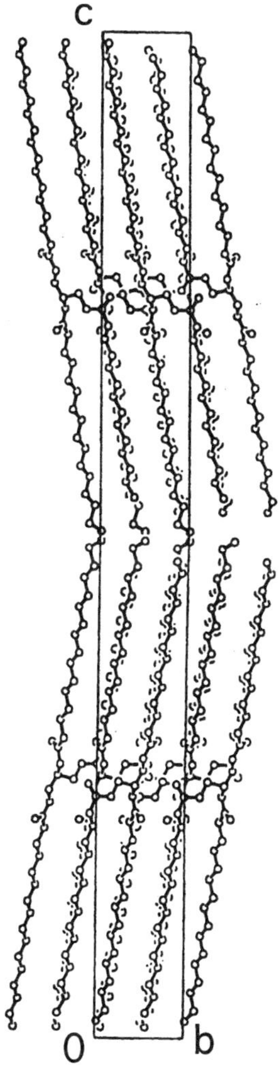

Fig. 18 Crystal structure of β′-2 polymorph of 1,2-palmitoyl-3-myristoyl-*sn*-glycerol.

The unit cell shows four-chain-length structure consisting of two double-layer leaflets which are related by the twofold axis along the *b* axis, as shown in Fig. 18. The two double-layer leaflets are combined end by end, which results in two types of interfaces for the double-layer leaflets. One interface consists of only palmitoyl chains: the *sn*-1 and *sn*-2 chains of A and the *sn*-2 chain of B. The other one consists of 1:2 mixture of palmitoyl chains and myristoyl chains:

the *sn*-3 chain of A and the *sn*-1 and *sn*-2 chains of B. The acyl chains incline against the interfaces, and exhibit alternative chain inclination of every double-layer leaflet, in the projection of the *bc* plane.

The polymethylene chains are packed in a new type of hybrid subcell arrangements (HS3) of a disordered orthorhombic perpendicular type [1,6,69]. The orthorhombic subcell parameters are $a_S = 15.0$, $b_S = 7.5$, and $c_S = 2.55$ Å. The a_S axis is three times as long as that of the usual simple orthorhombic perpendicular subcell.

III. SOLID-STATE PHASE TRANSITIONS

Fatty acids exhibit diversified solid-state phase transitions [1,5,6]. A great variety of molecular motions are activated at these transition points. The phase transition mechanism is greatly dependent on molecular structure. In this section, the phase transition mechanisms clarified by recent studies are dealt with.

A. E → C and B → C Transitions of Saturated Fatty Acids

Since the C form of the *n*-saturated fatty acids becomes stable in the vicinity of the melting point, the B and E forms irreversibly transform to the C form on heating [70–72]. For stearic acid, the B → C transition occurs around 50°C, and the E → C transition around 45°C. As described above, there is a structural analogy between the B and E forms, and the two transitions have characteristics in common.

On these transitions, the appearance of single crystal is kept unchanged on slow heating. However, the structural studies using electron microscope, X-ray diffraction, and IR spectroscopy showed that significantly large displacements take place cooperatively over the entire region of a single crystal. Furthermore, the systematic differences in transition behavior between the Mon and Orth II polytypes have been confirmed with oblique IR transmission method [73].

Figure 19 shows the experimental procedure of the oblique transmission method. The inclination direction of the acyl chain can be estimated with spectral changes by sample rotation about the θ and ϕ axes. First, with normal incident rays, the orientation of starting E (or B) crystal is adjusted by changing θ until the strongest intensity of the b_S component of CH_2 rocking $r_b(CH_2)$ mode at 720 cm^{-1} is obtained. At this angle, there are only two possible inclination directions: upward or downward. By observing the intensity change due to the change of ϕ, the inclination direction can be determined. For the resultant C crystal, the inclination direction can be determined in a similar way, using the a_S component of the CH_2 rocking mode $r_a(CH_2)$ at 730 cm^{-1}.

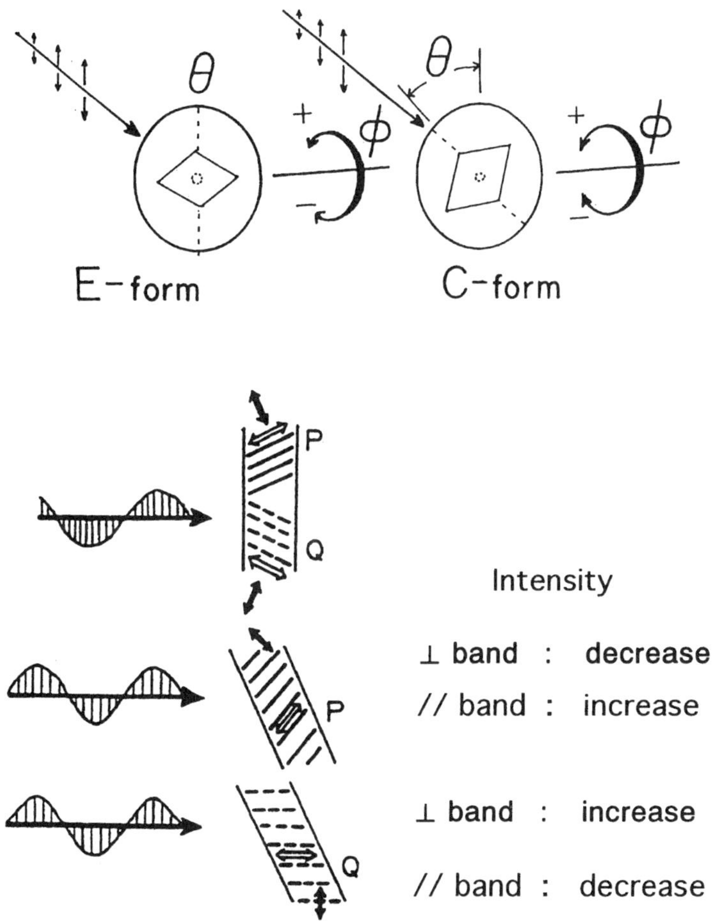

Fig. 19 Experimental procedure of the oblique transmission method.

Figure 20 shows an example of spectral changes during the E(Mon) → C transition. In this specimen, the angle of θ is found to be rotated by only 10°. However, the order of the intensity change in $r(CH_2)$, O—H out-of-plane δ(O—H), and O—C═O deformation δ(O—C═O) bands is inverted, which means that during this phase transition the inclination direction of acyl chains changes collectively by 170° around the normal of the basal plane. Such a large precessional motion has been confirmed as to the B(Mon) → C and E(Mon) → C transitions. The displacements observed in IR experiment were divided into

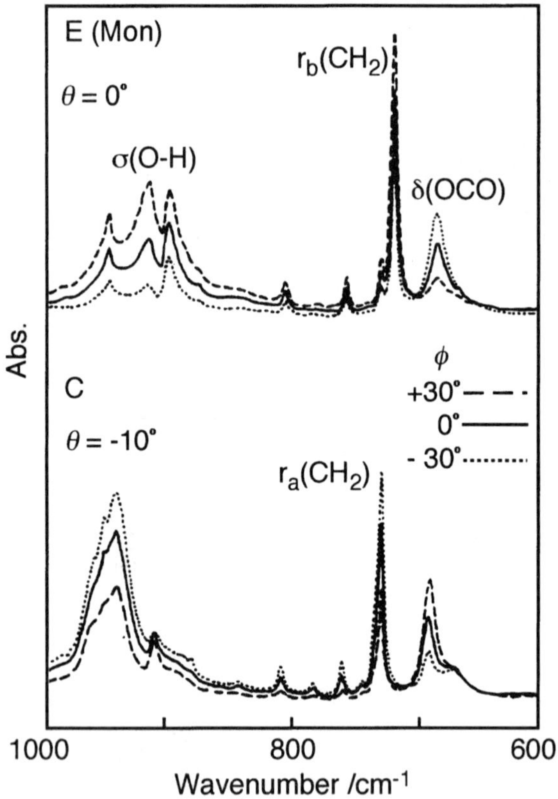

Fig. 20 Spectral changes during the E(Mon) → C transition.

two cases; in case I, the inclination direction of acyl chains changes by about 150°–180°, while in case II it changes by 60°–80°. The frequency of occurrence for case I is somewhat larger than that for case II.

As for single crystals of Orth II type, a polymorphic and polytypic composite transition takes place. Figure 21 shows the spectral changes on the E(Orth II) → C transition. The starting E(Orth II) crystal shows no clear spectral changes with the alternation of ϕ since acyl chains incline in opposite directions, taking the same probability. The resultant C crystal showed marked changes in intensity as an ordinary crystal. This spectral change means that the E crystal of double-layered structure transforms to the C crystal of single-layered structure. In this case, θ varies by only 10°, which suggests that two kinds of displacements take place alternately; in one layer the chain tilt direction is almost inverted and

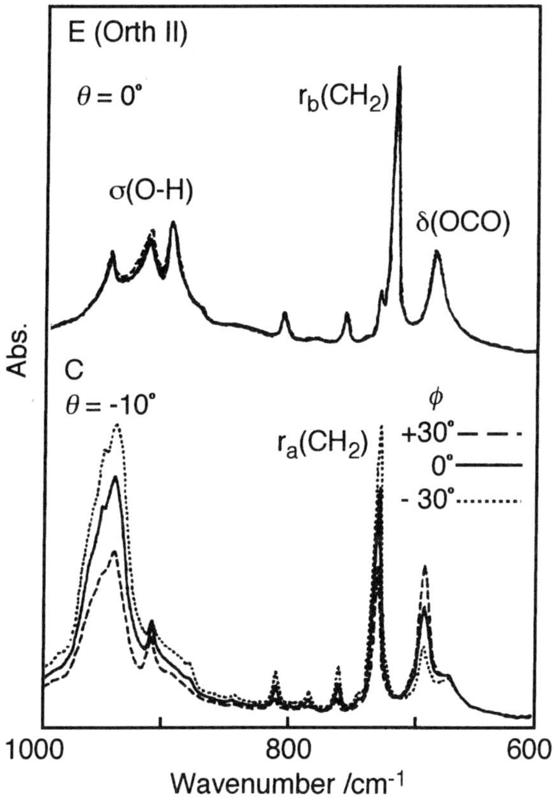

Fig. 21 Spectral changes during the E(Orth II) → C transition.

changes by only 10° in the neighboring layers (Fig. 22). A similar behavior has been observed for the B(Orth II) → C transition.

B. Reversible Phase Transitions of *cis*-Monounsaturated Fatty Acids

So far two kinds of reversible phase transitions have been found in *cis*-monounsaturated fatty acids. One is the $\gamma \rightarrow \alpha$ transition, which has been found in oleic, palmitoleic, erucic, gondoic, and asclepic acids, and the other is the $\gamma_1 \rightarrow \alpha_1$ transition of erucic acid. Although it has been confirmed that the higher-temperature phases α and α_1 contain conformational disorders in the methyl-sided chains [74–76], a recent study on erucic acid showed that there is a large difference in

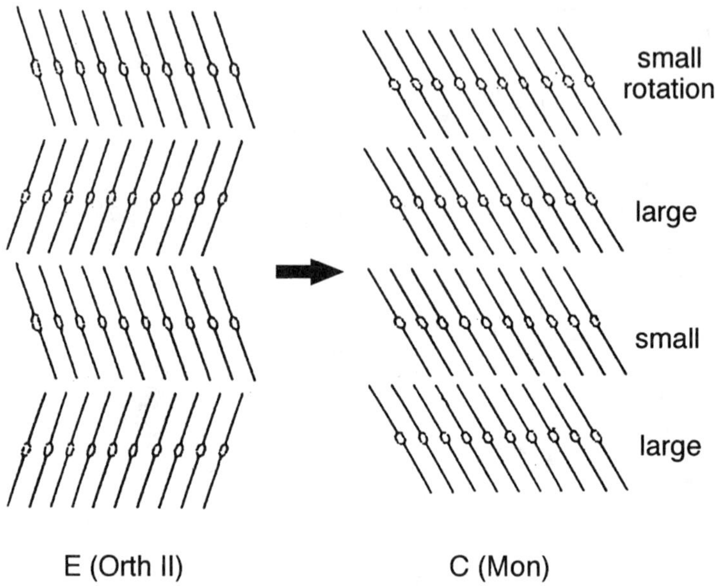

Fig. 22 Polymorphic and polytypic composite structural change during the E(Orth II) → C transition.

the transition behavior between the $\gamma \rightarrow \alpha$ and $\gamma_1 \rightarrow \alpha_1$ transitions [77]. Contrary to the abrupt occurrence of conformational disorders at the $\gamma \rightarrow \alpha$ phase transition, the disordering starts in the lower-temperature phase γ_1 about 30°C below the $\gamma_1 \rightarrow \alpha_1$ transition point without striking stepwise increase of disorders at this temperature, as shown in Fig. 23. The amount of disordered chains is much larger in the α_1 phase than in the α phase.

On the $\gamma \rightarrow \alpha$ transition, a large conformational change takes place at *cis*-olefin group, as shown in Fig. 24. The S–C–S′ type conformation of the γ phase changes to S–C–T(*trans*) type conformation, which results in the selective structural change of the methyl-sided chains. The $O'_{//}$ subcell transforms to a looser packing similar to the $O_{\perp}$ subcell; thus, the mobility of the methyl-sided chain is increased.

On the other hand, the results of structure analysis of the α_1 phase (Fig. 25) show that no conspicuous conformational change occurs on the $\gamma_1 \rightarrow \alpha_1$ phase transition point. The S–C–S type conformation of *cis*-olefin group and the $T_{//}$ type subcell of the methyl-sided and carboxyl-sided chains are kept during the transition. On the contrary, the inclination of both chains changes at the transition point; the extent of the inclination becomes somewhat smaller in the α_1 phase.

There is a close relationship between the conformational disordering and

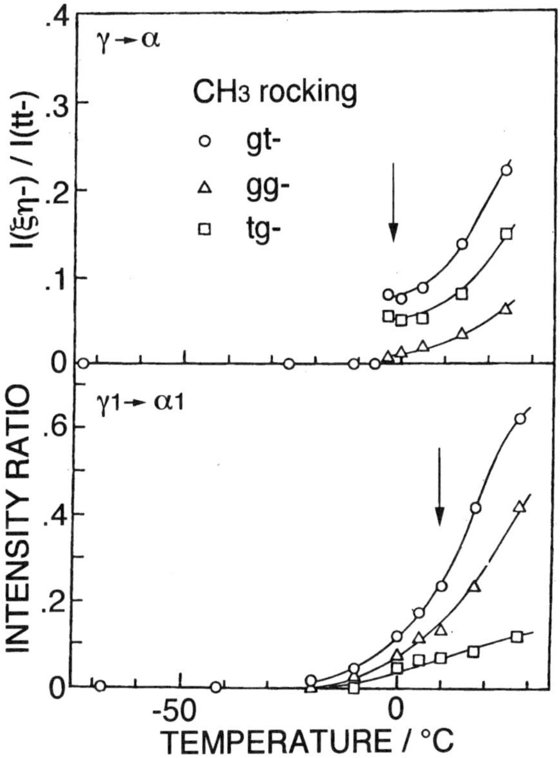

Fig. 23 Temperature dependence of disorder bands intensity in erucic acid.

the arrangement of methyl terminals at the lamellar interfaces. The methyl terminals are tightly packed and well stabilized in the γ phase. However, a large rearrangement of the methyl terminals is accompanied by the subcell structural change of the methyl-sided chains at the $\gamma \rightarrow \alpha$ transition, which leads to the unstable packing of methyl terminals in the α phase. This would be the main cause for the abrupt occurrence of conformational disorders.

The important feature of the γ_1 phase is the large inclination of the methyl-sided chains; the inclination angle of 42° is quite large compared with that of any other solid modifications of *n*-saturated fatty acids and *cis*-monounsaturated fatty acids, except for the β phase of oleic acid. With this inclination, the methyl terminals in the γ_1 phase are packed loosely compared with those in the γ phase, and the stabilization due to cohesive energy becomes smaller. Therefore, the conformational disorders at the methyl terminals can be induced by thermal agitation even at a relatively low temperature. At the $\gamma_1 \rightarrow \alpha_1$ transition point, the

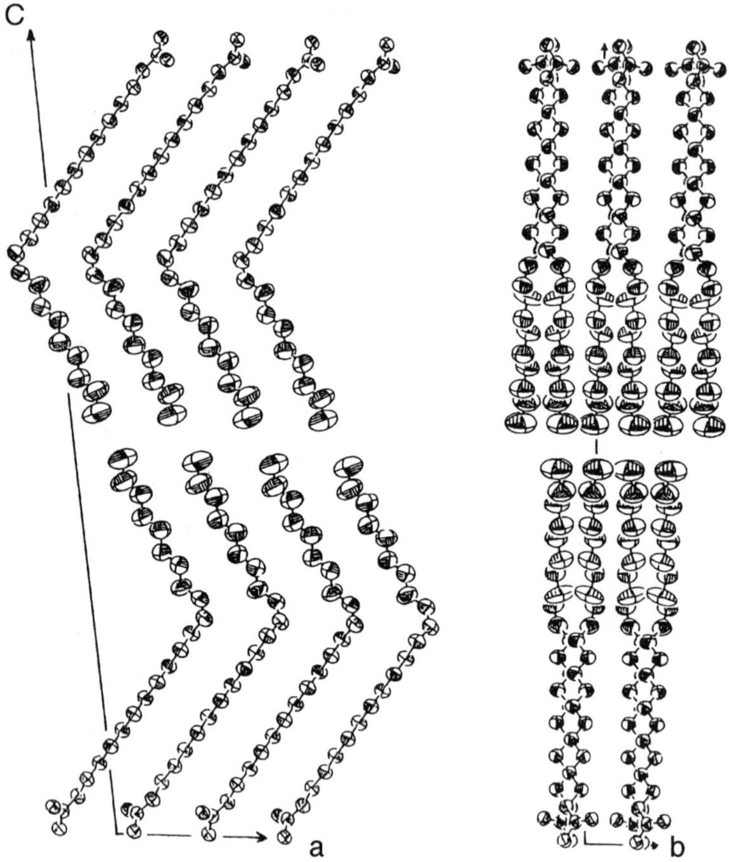

Fig. 24 Crystal structure changes of erucic acid α.

methyl terminals as well as the subcell structure show no large structural changes. As a result, no stepwise conformational disordering occurs at the transition point.

C. Polymorphic and Polytypic Phase Transitions during Crystallization

It has been found that specific stable phases of long-chain compounds need metastable phases for their occurrence. A typical example is the crystal growth mechanism of the B form of stearic acid [78,79]. Although the crystallization of stearic acid has been studied in detail [15,80–83], the existence of the E form was disre-

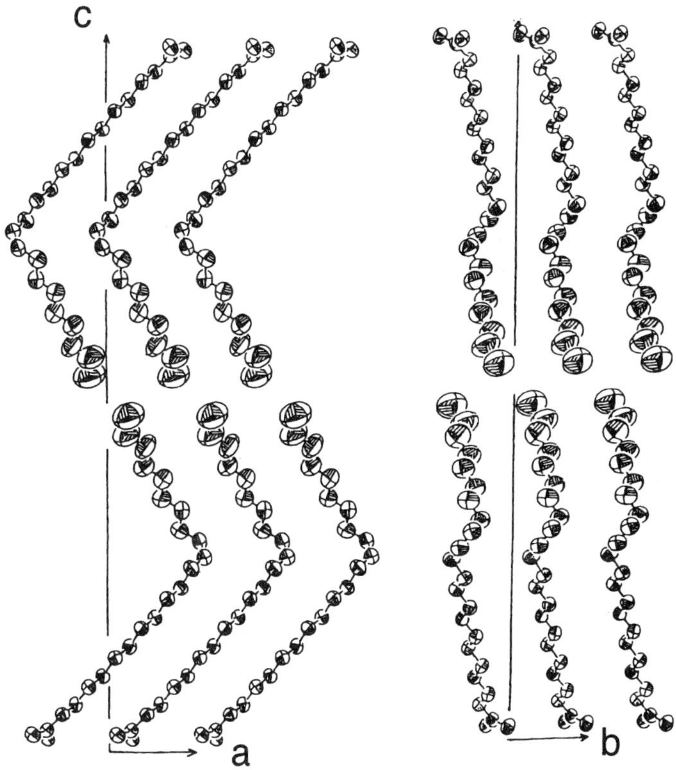

Fig. 25 Crystal structure of erucic acid α_1.

garded, owing to the lack of information about its occurrence condition and crystal structure, which was unveiled rather recently [27]. Because of their structural similarity, the E and B forms of stearic acid exhibit markedly intense X-ray reflections at almost the same 2θ angles showing the same crystal morphology, a lozenge plate with an acute interedge angle of 74°. Therefore, the identification of the E form by optical observation and X-ray diffraction studies is impossible. In the previous study, it might be possible to consider E and B as one polymorph. However, the difference in the conformation around carboxyl group is definitely reflected in IR spectra. For example, the O—C=O in-plane deformation band appears at 648 cm^{-1} in B and 688 cm^{-1} in E.

A careful examination was made to differentiate E and B forms for the single crystals grown from solution by cooling the supersaturated solution and keeping it at the same temperature after the occurrence of single crystal to induce solvent-mediated transformations. The crystals obtained within one hour after the

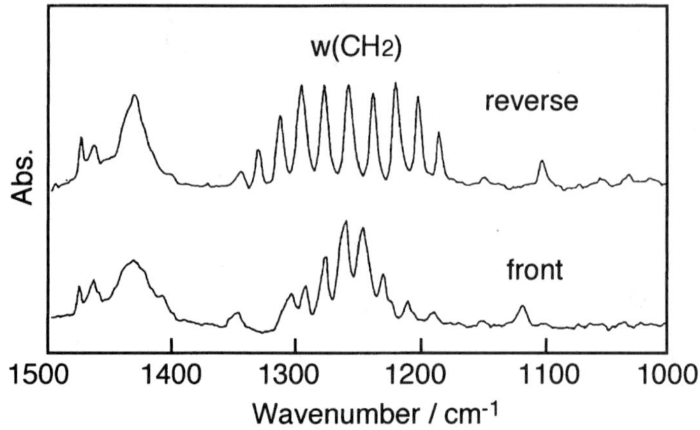

Fig. 26 *p*-Polarized ATR spectra taken on the {001} faces of a single crystal of stearic acid.

start of crystallization exhibited the IR spectral feature of E. Some single crystals began to show the spectral feature of B, which are referred to here as BE type. After the occurrence of the BE-type crystals, single crystals of the B form increased in number in the solution, and after one day all single crystals present in the solution were of the B form.

The ATR-IR spectra taken at the {001} faces of the BE-type crystal showed the layered structure of the B and E forms, as shown in Fig. 26. The front face exhibited the spectral features of B, whereas the reverse face shows those of E. The BE-type crystal gradually transformed to the B form through the solid-state phase transition. However, pure single crystals of the E form did not transform spontaneously, which means that the heterogeneous nucleation of the B form on the {001} faces of the E crystal was prerequisite to the E $\rightarrow$ B solid-state phase transition. Once the B domain nucleated on the single crystal of E in solution, the transition proceeded irrespective of the ambience; the single crystal of the BE type transformed to the B form both in and out of solution.

It is inferred that the E $\rightarrow$ B transition proceeds in a bimolecular layer through the succession of the structural change from an all-*trans* to a bent conformation at the phase boundary between B and E forms (Fig. 27). However, the interlamellar interactions are so weak compared with the intralamellar interactions that the structural change on the crystal surface may not induce the phase transition in the interior. Actually, not all single crystals transform to the B form; in particular, most crystals generated at relatively moderate conditions remain as the E form.

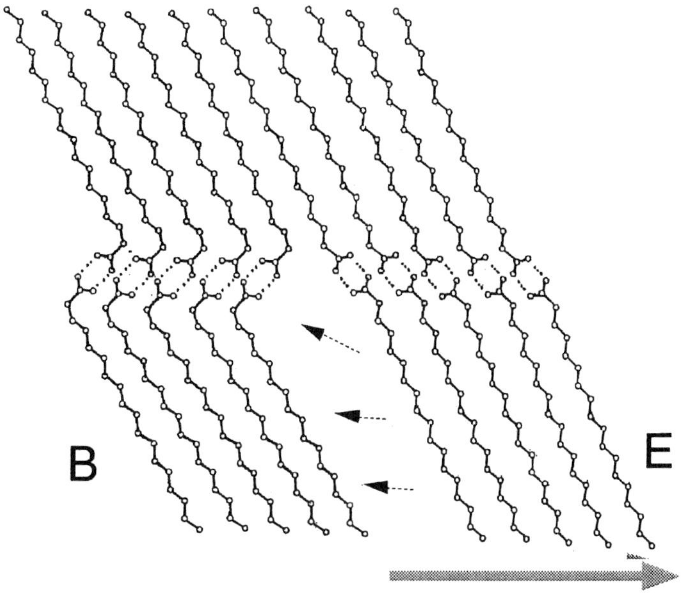

Fig. 27 Model for the structural changes at the phase boundary between B and E.

Chemical etching experiment shed light on this problem. The single crystals, which have speedily transformed to B, exhibit many lozenge etch pits on the (001) plane, some of which reach the substrata of the crystal, while very few pits appear in the crystals that have not converted to B. Considering the fact that the growth of the {001} faces of *n*-fatty acids is principally controlled by the spiral growth mechanism, one may explain the striking contrast in chemical etching to the E → B transition as follows: (1) Growth spiral steps due to screw dislocations are extremely long, durable, and catalyze the heterogeneous nucleation of the B form. (2) Since bimolecular layers form a spiral structure due to screw dislocations, the E → B transition is able to travel from a {001} face to the interior without interruption by driving on the ramp in a circle about the dislocation line.

In addition to this polymorphic transition, the polytypic transition from single-layered to double-layered structure has been confirmed. Figure 28 shows the spectral change during the polytypic transformation of E form. In this experiment, the crystallization is initiated by stirring a supersaturated solution, and then the solution is kept quiescent for hours. The $\delta_s(CH_3)$ and $\delta_a(CH_3)$ bands due to the Orth II type structure are not clearly observed at the inception of crystallization, but gradually increase in intensity. Contrary to this spectral change, the

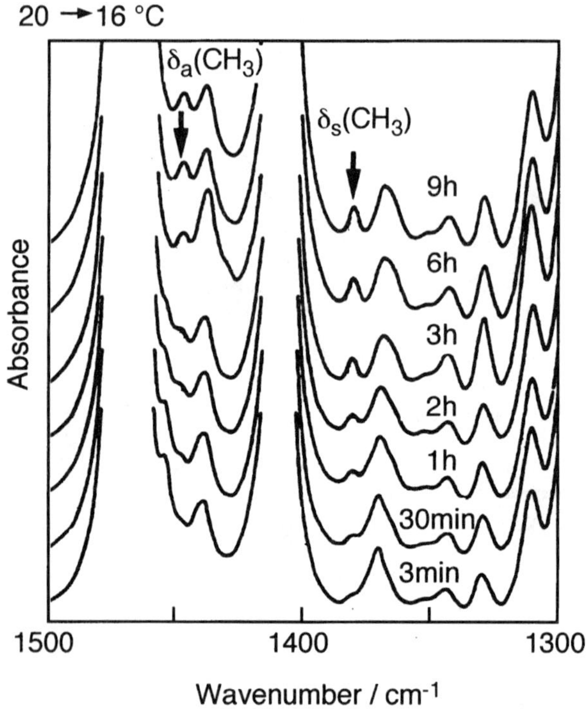

Fig. 28 An increase of the IR bands due to the Orth II type structure growing during the crystallization of stearic acid E form. The two arrows indicate the IR bands assigned to the antisymmetric and symmetric CH_3 deformation modes of E(Orth II) bands, respectively.

spectral features of the E form in the other spectral regions do not change. Namely, the first occurring single crystals are of the Mon type, which gradually change to the Orth II ones. This transformation is due to the relative stability between polytypes; Orth II is thermodynamically more stable than Mon. At a high supersaturation the metastable Mon occurs for kinetic reasons. As the growth rate of Mon falls with decreasing solute concentration, the growth rate of Orth II surpasses that of Mon near saturation with respect to Mon. Accordingly, single crystals begin to grow with Orth II–type stacking.

A polymorphic transformation during the crystal growth was also observed in the β phase of oleic acid [48,49], and the Mon → Orth II polytypic transition was also confirmed in the LM phase of petroselinic acid [45].

D. Martensitic Transformations of Petroselinic Acid and Pentadecanoic Acid

1. Petroselinic Acid

As described in the previous section, the LM phase of petroselinic acid shows polytypism. The polytypic structures of the LM phase control not only the thermodynamic properties but also the solid-state phase transition behavior [45].

The HM phase is the stable phase above 18.7°C. On slow heating, a solid-state phase transition from the LM phase occurs. There is another method to initiate the LM → HM phase transition. When a mechanical stress is placed on a point on a single crystal of LM(Mon), e.g., a press with the end of a needle, a part of a single crystal transforms to the HM phase in a moment (Fig. 29a). The growth of the product phase is so fast that it cannot be followed with the naked eye. There is a certain relationship in the direction of the crystallographic axes between the mother and product phases. The transformed region often displays complex zigzag patterns as shown in Fig. 29b, but each transformed domain is highly oriented.

A first-order solid-state phase transition is divided into two stages: nucleation and growth. The growth mechanism of the product phase can be classified into two types, namely molecular jumps and cooperative displacement [84]. The latter is called martensitic transformation and has been found in metal alloys, chemical compounds, and macromolecules. In the former type, the rearrangement of molecules at the growing face of the product phase takes place through thermally activated molecular jumps, while a martensitic transformation is accompanied by cooperative displacements of numerous molecules at the growth face. Thus, the growth of the product phase in a martensitic transition is usually significantly fast compared with the solid-state transition of the former type. The transition of the LM(Mon) phase induced by mechanical stress can be regarded as a martensitic phase transition.

On the other hand, the Orth II type of the LM phase does not exhibit any transition with mechanical stress, even with a stress enough to crack it. Usually a martensitic transformation is not accompanied by large displacements of atoms or molecules. However, large molecular rearrangements from a double- to single-layered structure are required for the solid-state phase transition from the LM(Orth II) phase to the HM phase. Under this condition, cooperative displacements of numerous molecules seems significantly difficult. This phenomenon suggests a possibility that new physical properties of a polymorph can be deduced by controlling polytype structure.

The polytypic structures have a significant influence also on the transition behavior by heating. When a specimen of LM(Mon) is heated rapidly (more than 10 K/min) to a temperature above the melting point of the LM phase, it trans-

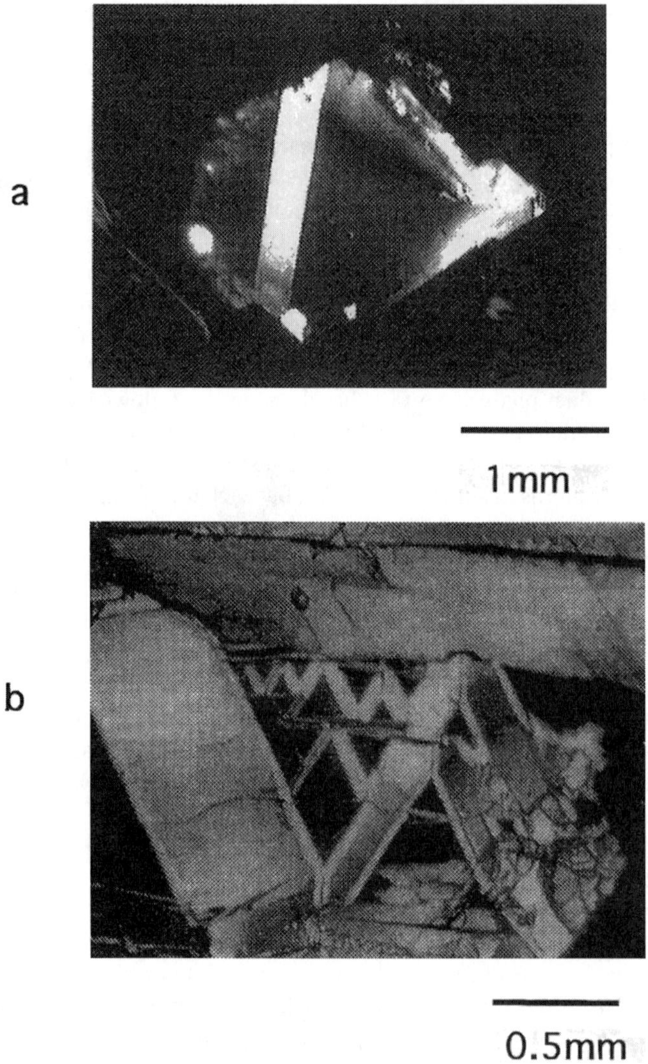

Fig. 29 Resultant transformed regions of the martensitic transition from the LM phase to the HM phase in petroselinic acid.

forms to the HM phase instantaneously. For a single crystal of Orth II, rapid heating results in fusion of its greater part, and a few infusible domains grow as the HM phase in the melt.

2. Pentadecanoic Acid

Similar rapid phase transitions involving cooperative molecular displacements have been found for pentadecanoic acid [85,86]. There are three polymorphs, A′, B′, and C′. The A′ form can be divided into A'_h ($T > 23°C$) and A'_l ($T < 23°C$). The C′ form exists only in the vicinity of the melting point. The stable phase changes successively to B′, A'_h, and A'_l, as temperature decreases. On heating, the A'_h phase transforms to the B′ phase irreversibly.

The transition from B′ to A'_h occurs when a mechanical stress is imposed on a crystal, and its reverse transition can also be induced by the same way. These transitions are accomplished in a moment. The transformed regions are highly oriented and have some specific orientations against the mother phase just like the LM(Mon) $\rightarrow$ HM transition of petroselinic acid.

There is a reversible solid-state transition between the A'_h and A'_l phases. For a single-crystal specimen, an overall morphological change takes place on these transitions, keeping its single-crystal state during this phase transition. The transition behavior depends significantly on the condition of specimens. However, a fairly good single crystal exhibits a rapid morphological change. Figure 30 shows an example. The transition starts at a long edge and propagates toward the opposite side. In this crystal, both the $A'_h \rightarrow A'_l$ and $A'_l \rightarrow A'_h$ transitions finish within 0.1 sec. Furthermore, it has been confirmed that a mechanical stress can initiate both transitions.

It is inferred that the molecular packing of odd-numbered fatty acids is an important factor for these martensitic transitions. Odd-numbered fatty acids tend to possess large molecular volume compared with even-numbered ones, which results in slightly lower melting points [1]. This relatively loose packing would make it possible for molecules to slide cooperatively.

E. Solid-State Phase Transitions of Mixed-Acid Triacylglycerol

Most complex lipids in nature contain *cis*-unsaturated acyl chains. It was confirmed that *cis*-unsaturation adds complexity to the polymorphic behavior of triacylglycerols [1,3,87–89]. This phenomenon is important, in connection with the saturated chain–unsaturated chain interaction in biological system [90]. Polymorphic transformations in two saturated-unsaturated mixed acid triacylglycerols, SOS (*sn*-1,3-distearoyl-2-oleoylglycerol) and OSO (*sn*-1,3-dioleoyl-2-stearoylglycerol), have been studied by FT-IR spectroscopy using deuterated specimens in which stearoyl chains were fully deuterated [91].

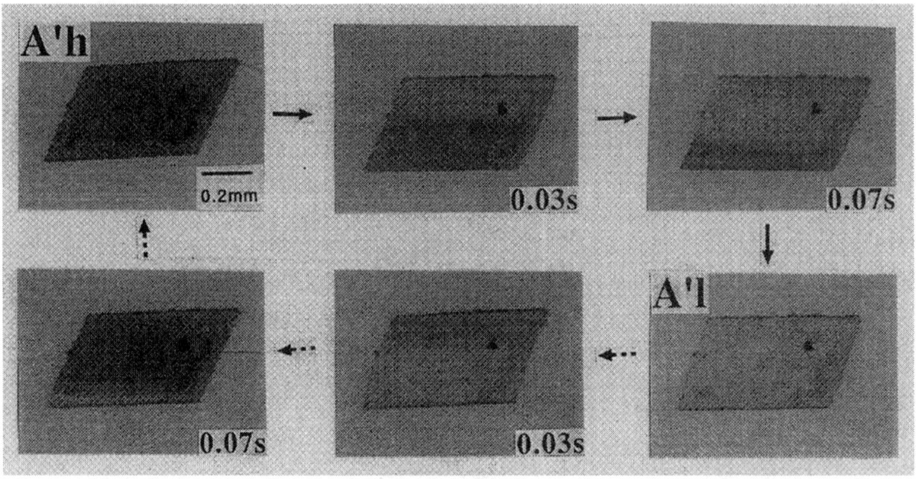

Fig. 30 Rapid morphological change on the $A'_l \leftrightarrow A'_h$ transition of pentadecanoic acid induced by a temperature change ($A'_l \rightarrow A'_h$: heating process, and $A'_h \rightarrow A'_l$: cooling process).

There are five polymorphs for SOS (sub-α, α, γ, β′, and β) and four polymorphs (sub-α, α, β′, and β) for OSO. In both SOS and OSO, the α phase transforms to the sub-α phase reversibly by cooling, and the α phase finally transforms to the stable β phase with a series of irreversible transitions; $\alpha \rightarrow \gamma \rightarrow \beta' \rightarrow \beta$ for SOS and $\alpha \rightarrow \beta' \rightarrow \beta$ for OSO.

Deuteration of stearoyl chain enables one to evaluate the conformation and lateral packing of stearoyl and oleoyl chains separately. Figure 31 shows the IR spectra of deuterated SOS in $\delta(CH_2)$ and $\delta(CD_2)$ regions. A change in lateral packing mode of hydrocarbon chains results in frequency shift and/or band splittings of the $\delta(CH_2)$ and $\delta(CD_2)$ bands, and the lateral packing mode of oleoyl chain is reflected in the $\delta(CH_2)$ region and that of stearoyl chain is reflected in the $\delta(CD_2)$ region. For example, the $\delta(CD_2)$ band due to deuterated stearoyl chains splits into a doublet on the α ↔ sub-α phase transition, while the $\delta(CH_2)$ band due to oleoyl chains remains unchanged. These spectral data suggest that only stearoyl chains change their lateral packing into a perpendicular packing.

For molecular conformation, the progression bands due to the ν_3 branch [CH_2 wagging: $w(CH_2)$] modes of polymethylene segments can be used as a probe for the conformational state of acyl chains [92,93]; the frequencies of these bands are sensitive to the length of all-*trans* segments. Figure 32 shows the IR spectra in $w(CH_2)$ region of SOS and deuterated SOS. In SOS, the conformational state of stearoyl chains mainly determines the profile of this region, while oleoyl chain

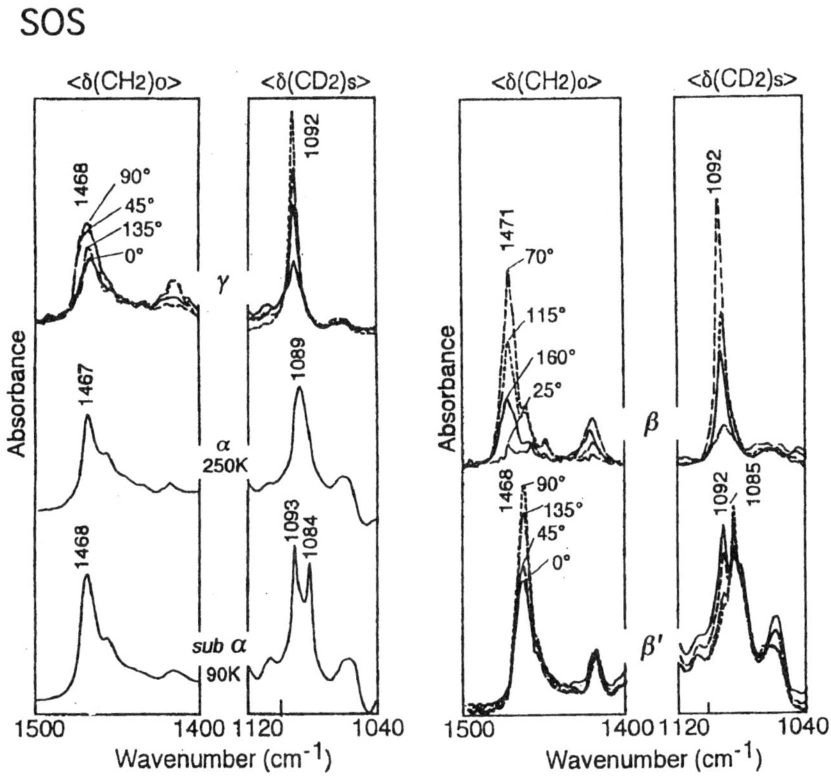

Fig. 31 The δ(CH_2) and δ(CD_2) bands of deuterated SOS.

has a dominant influence in deuterated SOS. Taking the α phase, for instance, the *w*(CH_2) progression bands appear in SOS, whereas only the broad bands are observed in deuterated OSO; these spectral features mean that the stearoyl chains take all-*trans* conformation, but the oleoyl chain is conformationally disordered.

Using the above characteristics of IR spectroscopy and X-ray diffraction, a detailed investigation has been done about the structures of crystalline states of SOS and OSO. The results obtained are summarized in Table 3, which is explained as follows:

1. The α → sub-α reversible transition of SOS is principally due to the orientational change of stearoyl chains in the lateral directions from the hexagonal subcell to a perpendicularly packed one, as schematically depicted in Fig. 33. In α and sub-α, the greater part of stearoyl chains

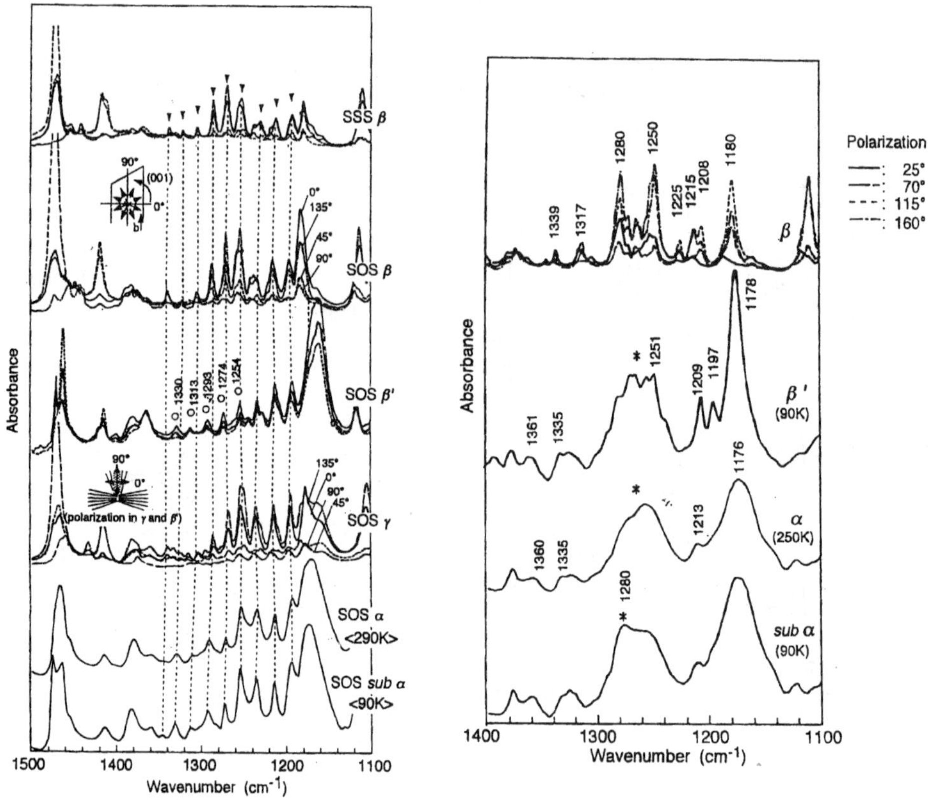

Fig. 32 Transmission IR spectra in $w(CH_2)$ region of SOS and deuterated SOS. The spectra of γ, β′, and β of SOS and deuterated SOS were measured with polarized radiations. Open circles and wedges show the ν_3 bands of the *sn*-1,2 straight and *sn*-3 bent stearoyl chains, respectively.

take all-*trans* conformation, while there are many conformational disorders in the oleoyl chains.

2. As the first stage of a series of irreversible transitions from α to β in SOS, the bilayer structure transforms to the trilayer structure on the α → γ transition, and the segregation of unsaturated chain from the saturated chains takes place; an oleoyl leaflet is held between stearoyl leaflets. Two stearoyl chains take more ordered conformation; one chain takes all-*trans* conformation, whereas the other has a bent conformation in the vicinity of the ester bond. The trilayer structure and this conformation of stearoyl chains are kept until the β phase fuses. The

Table 3 Chain-Length Structure and Subcell Packing

TAG	Form	Layered structure (nm)		Subcell structure: Stearoyl-leaflet	Subcell structure: Oleoyl-leaflet
SOS	α	Double	4.83	Hexagonal	
	γ	Triple	7.05	//-Type	Hexagonal
	β′	Triple	7.00	$O_{\perp}$	Hexagonal
	β	Triple	6.60	$T_{//}$	$T_{//}$
OSO	α	Double	5.20	Hexagonal	
	β′	Double	4.50	$O_{\perp}$ (partially)	
	β	Triple	6.50	$T_{//}$	$T_{//}$

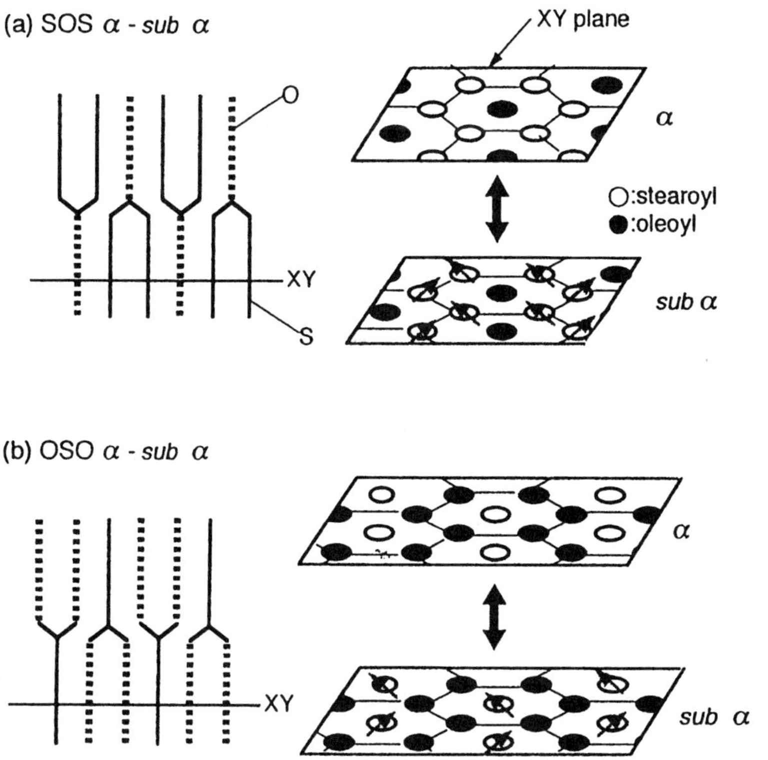

Fig. 33 Schematic representation of sub-α ↔ α transition in SOS and OSO. (Left) Bilayer structure in which the stearoyl and oleoyl chains are indicated by solid and dotted lines, respectively. (Right) Chain arrangement in lamella. Open circles and closed circles show the stearoyl and oleoyl chain positions in the XY plane indicated in the left side.

lateral packing of stearoyl chains changes to a parallel type, while oleoyl chains are accommodated in the hexagonal packing.

3. On the $\gamma \rightarrow \beta'$ phase transition, the stearoyl chains transform their packing into the $O_\perp$ subcell, but the oleoyl chain still retains its hexagonal packing and contains conformational disorder at a high level.
4. Finally, the stearoyl and oleoyl chains form the $T_{//}$ subcell during the $\beta' \rightarrow \beta$ phase transition. The oleoyl chain becomes conformationally ordered, and its *cis*-olefin group takes S-C-S′ conformation.
5. In the metastable phases of OSO, molecules form a bilayer structure. Consequently, the stearoyl and oleoyl chains are mixed in the same leaflet in the α and β' phases. The $\alpha \rightarrow \beta'$ transition corresponds to the stabilization of the stearoyl groups. In the β' phase, the stearoyl chain takes all-*trans* conformation, whereas the oleoyl chains still contain a large amount of conformational disorders. The stearoyl and oleoyl groups together make a loosely packed $O_\perp$ subcell, owing to the steric hindrance between them.
6. The bilayer to trilayer transformation with the segregation of saturated and unsaturated chains takes place during the $\beta' \rightarrow \beta$ transition of OSO. In the β phase a stearoyl leaflet is sandwiched between a couple of oleoyl leaflets, and both the stearoyl and oleoyl chains form $T_{//}$-type subcell structure.

For the polymorphism of SOS and OSO, the steric hindrance between saturated and unsaturated acyl chains has an important influence. The hindrance results in the occurrence of various metastable forms and complicated polymorphic transformations as described above. At the first stage, the ordering as to conformation and lateral packing takes place in the saturated chains, and then the stabilization of the unsaturated moieties is followed.

NOTE

Recently, several noteworthy X-ray crystal structure analyses have been done on several triacylglycerols. It has been confirmed that the β phase of tripalmitin is isomorphous to that of trilaurin [94]. It has been clarified that the β phase of trielaidin has a structure similar to that of saturated TAGs; the *sn*-3 chain takes a *gauche* conformation at the C_2—C_3 bond. As for the conformation around *trans*-C=C bond, all acyl chains approximate skew-trans-skew′ conformation [95]. The crystal structure of the β' phase of $C_nC_{n+2}C_n$-type TAGs has been determined for 1,3-didecanoyl-2-dodecanoylglycerol [96]. The molecule has a bend at the glycerol moiety; the sn-2 chain is adjacent to the sn-3 chain, and the sn-1 chain makes an angle of about 130° with the sn-2 and sn-3 chains.

REFERENCES

1. *The Physical Chemistry of Lipids*, D. M. Small, ed. (Plenum, New York, 1986).
2. B. L. Silver, *The Physical Chemistry of Membranes* (Solomon, New York, 1985).
3. J. M. Hagemann, in *Crystallization and Polymorphism of Fats and Fatty Acids*, N. Garti and K. Sato, eds. (Marcel Dekker, New York, 1988), pp. 9–95.
4. E. von Sydow, *Ark. Kemi, 9*: 231 (1956).
5. M. Kobayashi, in *Crystallization and Polymorphism of Fats and Fatty Acids*, N. Garti and K. Sato, eds. (Marcel Dekker, New York, 1988), pp. 139–187.
6. L. Hernqvist, in *Crystallization and Polymorphism of Fats and Fatty Acids*, N. Garti and K. Sato, eds. (Marcel Dekker, New York, 1988), pp. 97–137.
7. A. R. Verma and P. Krishna, *Polymorphism and Polytypism in Crystals* (Wiley, New York, 1966).
8. E. von Sydow, *Acta Chem. Scand., 10*: 1 (1956).
9. M. Goto and E. Asada, *Bull. Chem. Soc. Jpn., 51*: 70 (1978).
10. T. R. Lomer, *Acta Crystallogr., 16*: 984 (1963).
11. T. Kobayashi, M. Kobayashi, and H. Tadokoro, *Mol. Cryst. Liq. Cryst., 104*: 193 (1984).
12. S. Amelinckx, *Acta Crystallogr., 8*: 530 (1955); *9*: 16, 217 (1956).
13. M. Kobayashi, T. Kobayashi, Y. Itoh, and K. Sato, *J. Phys. Chem., 91*: 2273 (1987).
14. K. Sato, M. Kobayashi, and H. Morishita, *J. Crystal Growth, 87*: 236 (1988).
15. K. Sato and M. Kobayashi, *Crystals*, Vol. 13, N. Karl, ed. (Springer, Berlin and Heidelberg, 1991), pp. 65–108.
16. M. Goto and E. Asada, *Bull. Chem. Soc. Jpn., 51*: 2456 (1978).
17. F. Kaneko, H. Sakashita, M. Kobayashi, Y. Kitagawa, and Y. Matsuura, *Acta Crystallogr., C50*: 245 (1994).
18. C. W. Bunn, *Trans. Fraday Soc., 35*: 482 (1936).
19. V. Vand, W. M. Morley, and T. R. Lomer, *Acta Crystallogr., 5*: 324 (1951).
20. V. Marta, G. Celotti, R. Zanetti, and A. F. Martelli, *J. Chem. Soc. B*: 548 (1971).
21. S. Hayashi and J. Umemura, *J. Chem. Phys., 63*: 1732 (1975).
22. J. Umemura, *J. Chem. Phys., 68*: 42 (1978).
23. A. J. Horsewill, A. Heidemann, and S. Hayashi, *Z. Phys. B,90*: 319 (1993).
24. R. F. Holland and J. R. Nielsen, *J. Mol. Spectrosc., 9*: 436 (1962).
25. R. F. Holland and J. R. Nielsen, *Acta Crystallogr., 16*: 902 (1963).
26. M. Kobayashi, T. Kobayashi, Y. Cho, and F. Kaneko, *Makromol. Chem. Macromol. Symp., 5*: 1 (1986).
27. F. Kaneko, M. Kobayashi, Y. Kitagawa, and Y. Matsuura, *Acta Crystallogr., C46*: 1490 (1990).
28. F. Kaneko, H. Sakashita, M. Kobayashi, Y. Kitagawa, and Y. Matsuura, *Acta Crystallogr., C50*: 247 (1994).
29. F. Kaneko, T. Simofuku, H. Miyamoto, M. Kobayashi, and M. Suzuki, *J. Phys. Chem., 96*: 10554 (1992).
30. M. Goto and E. Asada, *Bull. Chem. Soc. Jpn., 53*: 2111 (1980).
31. E. von Sydow, *Acta Crystallogr., 7*: 529 (1954); *8*: 845 (1955).
32. E. von Sydow, *Acta Crystallogr., 7*: 823 (1954).

33. M. Goto and E. Asada, *Bull. Chem. Soc. Jpn.*, *57*: 1145 (1984).
34. E. von Sydow, *Acta Crystallogr.*, *8*: 810 (1955).
35. K. Sato, in *Advances in Applied Lipids Research*, Vol. 2, F. Padley, ed. (JAI Press, New York, 1996), pp. 213–268.
36. F. Kaneko, J. Yano, and K. Sato, *Curr. Opin. Struc. Biol.*, *8*: 328 (1998).
37. M. Suzuki, T. Ogaki, and K. Sato, *J. Am. Oil Chem. Soc.*, *62*: 1600 (1985).
38. M. Suzuki, K. Sato, N. Yoshimoto, and M. Kobayashi, *J. Am. Oil Chem. Soc.*, *65*: 1942 (1988).
39. K. Sato, J. Yano, I. Kawada, M. Kawano, F. Kaneko, and M. Suzuki, *J. Am. Oil Chem. Soc.*, *74*: 1153 (1997).
40. K. Sato, in *Crystallization and Polymorphism of Fats and Fatty Acids*, N. Garti and K. Sato, eds. (Marcel Dekker, New York, 1988), pp. 227–263.
41. S. Abrahamsson and I. Ryderstadt-Nahringbauer, *Acta Crystallogr.*, *15*: 1261 (1962).
42. F. Kaneko, K. Yamazaki, and M. Kobayashi, *Acta Crystallogr.*, *C49*: 123 (1993).
43. F. Kaneko, M. Kobayashi, Y. Kitagawa, Y. Matsuura, K. Sato, and M. Suzuki, *Acta Crystallogr.*, *C48*: 1060 (1992).
44. K. Sato, N. Yoshimoto, M. Suzuki, M. Kobayashi, and F. Kaneko, *J. Phys. Chem.*, *94*: 523180 (1990).
45. F. Kaneko, M. Kobayashi, K. Sato, and M. Suzuki, *J. Phys. Chem.*, *101*: 285 (1997).
46. F. Kaneko, M. Kobayashi, Y. Kitagawa, Y. Matsuura, K. Sato, and M. Suzuki, *Acta Crystallogr.*, *C48*: 1054 (1992).
47. F. Kaneko, M. Kobayashi, Y. Kitagawa, Y. Matsuura, K. Sato, and M. Suzuki, *Acta Crystallogr.*, *C48*: 1057 (1992).
48. F. Kaneko, K. Yamazaki, K. Kitagawa, T. Kikyo, M. Kobayashi, Y. Kitagawa, Y. Matsuura, K. Sato, and M. Suzuki, *J. Phys. Chem.*, *101*: 1803 (1997).
49. F. Kaneko, K. Tashiro, and M. Kobayashi, *J. Crystal Growth*, *198/199*: 1352 (1999).
50. R. R. Mad, J. A. Harris, and E. L. Skau, *J. Chem. Eng. Data*, *13*: 115 (1968).
51. S. Ueno, T. Suetaka, J. Yano, M. Suzuki, and K. Sato, *Chem. Phys. Lipids*, *72*: 27 (1994).
52. F. Kaneko, K. Sumiya, and K. Tashiro, to be published.
53. A. E. Smith, *J. Chem. Phys.*, *21*: 2229 (1953).
54. A. A. Schaerer, C. J. Busso, A. E. Smith, and L. B. Skinner, *J. Am. Chem. Soc.*, *77*: 2017 (1955).
55. J. Ernst, W. S. Sheldick, J.-H. Fuhrhop, *Z. Naturforsch.*, *346*: 706 (1979).
56. R. Boistelle, B. Simon, and G. Pepe, *Acta Crystallogr.*, *B32*: 1240 (1976).
57. Y. Itoh and M. Kobayashi, *J. Phys. Chem.*, *95*: 1794 (1991).
58. M. Goto, D. R. Kodali, D. M. Small, K. Honda, K. Kozawa, and T. Uchida, *Proc. Natl. Acad. Sci. USA*, *89*: 8083 (1992).
59. L. H. Jensen and A. J. Mabis, *Nature(London)*, *197*: 681 (1963).
60. L. H. Jensen and A. J. Mabis, *Acta Crystallogr.*, *21*: 770 (1966).
61. K. Larsson, *Ark. Kemi*, *23*: 1 (1964).
62. I. Pascher, S. Sundell, and H. Hauser, *J. Mol. Biol.* *153*: 791 (1981).
63. D. L. Dorset and W. A. Pangborn, *Chem. Phys. Lipids*, *48*: 19 (1988).
64. K. Halos, H. Eibl, I. Pascher, and S. Sundell, *Chem. Phys. Lipids*, *34*: 115 (1984).
65. M. Goto, K. Honda, L. Di, and D. M. Small, *J. Lipid Res.*, *36*: 2185 (1995).
66. M. Goto and T. Takiguchi, *Bull. Chem. Soc. Jpn.*, *58*: 1319 (1985).

67. A. Hybl and D. Dorset, *Acta Crystallogr., B27*: 977 (1971).
68. K. Sato, M. Goto, J. Yano, K. Honda, D. R. Kodali, and D. M. Small, *J. Lipid Res., 42*: 338 (2001).
69. S. Abrahamsson, B. Dahlén, H. Löfgren, and I. Pascher, *Prog. Chem. Fats Other Lipids, 16*: 125 (1978).
70. E. Stenhagen and E. von Sydow, *Ark. Kemi., 6*: 309 (1953).
71. K. Inaoka, M. Kobayashi, M. Okada, and K. Sato, *J. Cryst. Growth, 87*: 243 (1988).
72. F. Kaneko, M. Kobayashi, Y. Kitagawa, and Y. Matsuura, *J. Phys. Chem., 96*: 7104 (1992).
73. F. Kaneko, O. Shirai, H. Miyamoto, and M. Kobayashi, *J. Phys. Chem., 98*: 2185 (1994).
74. M. Kobayashi, F. Kaneko, K. Sato, and M. Suzuki, *J. Phys. Chem., 90*: 6371 (1986).
75. Y. Kim, H. Strauss, and R. G. Snyder, *J. Phys. Chem., 93*: 485 (1989).
76. F. Kaneko, K. Yamazaki, and M. Kobayashi, *Spectrochim. Acta, 50A*: 1589 (1994).
77. F. Kaneko, K. Yamazaki, K. Kitagawa, T. Kikyo, M. Kobayashi, Y. Kitagawa, Y. Matsuura, K. Sato, and M. Suzuki, *J. Phys. Chem., 101*: 1803 (1997).
78. F. Kaneko, T. Simofuku, H. Miyamoto, M. Kobayashi, and M. Suzuki, *J. Phys. Chem., 96*: 10554 (1992).
79. F. Kaneko, H. Sakashita, M. Kobayashi, and M. Suzuki, *J. Phys. Chem., 98*: 3801 (1994).
80. K. Sato, M. Suzuki, M. Okada, and N. Garti, *J. Crystal Growth, 74*: 236 (1985).
81. N. Garti, in *Crystallization and Polymorphism of Fats and Fatty Acids*, N. Garti and K. Sato, eds. (Marcel Dekker, New York, 1988), pp. 267–303.
82. W. Beckmann, R. Boistelle, and K. Sato, *J. Chem. Eng. Data, 29*: 215 (1984).
83. W. Beckmann and R. Boistelle, *J. Crystal Growth, 67*: 271 (1984).
84. V. Raghavan and M. Cohen, in *Treatise on Solid State Chemistry*, Vol. 5, *Changes of State*, N. B. Hannay, ed. (Plenum, New York, 1975), pp. 67–127.
85. F. Kaneko, J. Yano, H. Tsujiuchi, and K. Tashiro, *J. Phys. Chem., 102*: 320 (1998).
86. F. Kaneko, H. Tsujiuchi, J. Yano, and K. Tashiro, *J. Phys. Chem., 102*: 6187 (1998).
87. D. R. Kodali, D. Atkinson, T. G. Redgrave, and D. M. Small, *J. Lipid Res., 28*: 403 (1987).
88. K. Sato, T. Arishima, Z. H. Wang, K. Ojima, N. Sagi, and H. Mori, *J. Am. Oil Chem. Soc., 66*: 664 (1989).
89. J. Yano, S. Ueno, K. Sato, T. Arishima, N. Sagi, F. Kaneko, and M. Kobayashi, *J. Phys. Chem., 97*: 12967 (1993).
90. D. M. Small, *J. Lipid Res., 32*: 1635 (1991).
91. J. Yano, F. Kaneko, K. Sato, D. R. Kodali, and D. M. Small, *J. Lipid Res., 40*, 140 (1999).
92. J. Yano, F. Kaneko, M. Kobayashi, and K. Sato, *J. Phys. Chem., 101*: 8112 (1997).
93. J. Yano, F. Kaneko, M. Kobayashi, D. R. Kodali, D. M. Small, and K. Sato, *J. Phys. Chem., 101*: 8120 (1997).
94. A. Van Langevelde, K. Van Malssen, F. Hollander, R. Peschar, and H. Schenk, *Acta Crystallogr., B55*: 114 (1999).
95. C. Culot, B. Norberg, G. Evrard, and F. Durant, *Acta Crystallogr., B56*: 317 (1999).
96. A. Van Langevelde, K. van Malssen, R. Driessen, K. Goubitz, F. Hollander, R. Peschar, P. Zwart, and H. Schenk, *Acta Crystallogr., B56*: 1103 (1999).

3

Morphological Connected Net–Roughening Transition Theory

Application to β-2 Crystals of Triacylglycerols

P. Bennema, F. F. A. Hollander, S. X. M. Boerrigter, R. F. P. Grimbergen,* J. van de Streek, and H. Meekes
University of Nijmegen, Nijmegen, The Netherlands

I. INTRODUCTION

The habit or morphology of crystals plays a very important role in crystallization processes. Often crystals grow faceted with flat, well-defined faces, but in many cases some of the faces appear as rounded-off roughened faces. In these cases the growth velocity of such faces is much higher than the flat ones. In extreme cases even dendritic, or cellular, growth occurs. An intermediate case is the one for which spherulitic crystals grow. This case is often encountered for fat crystals. The crystals then appear as spherulites made up of many faceted crystals growing out of an ill-defined nucleus. In this chapter we analyze the morphology of the individual faceted crystals of β-2 crystals of fats. Although we treat the case of faceted crystals, it will turn out that the morphology of these crystals, to a large extent, is determined by the tendency of some of their faces to become rough easily. Closely connected to this tendency is the roughening temperature of flat face above which they appear as rounded-off. This effect is known as thermal roughening. The growth velocity of these faces becomes high below this critical temperature. Besides, as a result of its roughening temperature a flat face can become rough beyond a critical supersaturation, known as kinetical roughening, resulting in comparable high growth velocities. These roughening effects can

* Current affiliation: DSM Research, Geleen, The Netherlands.

explain the elongated morphology of fat crystals, as explained in this chapter, and, in particular, the morphology of the top facets of fat crystals of the β-2 structure will be explained on the basis of the roughening theory of crystal faces.

In order to demonstrate how the phenomena of thermal and kinetical roughening in reality can be observed on crystal facets (*hkl*) of growing crystals, we first present the pictures of the experimentally observed crystals of Figs. 1 and 2, respectively [1,2]. It can be seen from Fig. 1a that lozenge-shaped paraffin crystals are bounded by two large parallel faces of the form {001} and four thin side faces of the form {110}. One face $(00\bar{1})$ of the form {001} consisting of two parallel faces is lying on the bottom of the crystallization cell. The four thin side faces, which are more or less perpendicular to the bottom of the crystallization cell are observed as straight boundaries of the paraffin crystal. It can also be seen from Fig. 1b that when the saturation temperature of the solution increases, a critical temperature occurs for the faces {110}, beyond which the side faces do not grow, even at low supersaturations, as flat faces but as rough rounded-off faces without any crystallographic orientation. For about 15 years the observed critical temperature has been interpreted as the macroscopic manifestation of the thermal roughening phase transition. This is a thermodynamic surface phase tran-

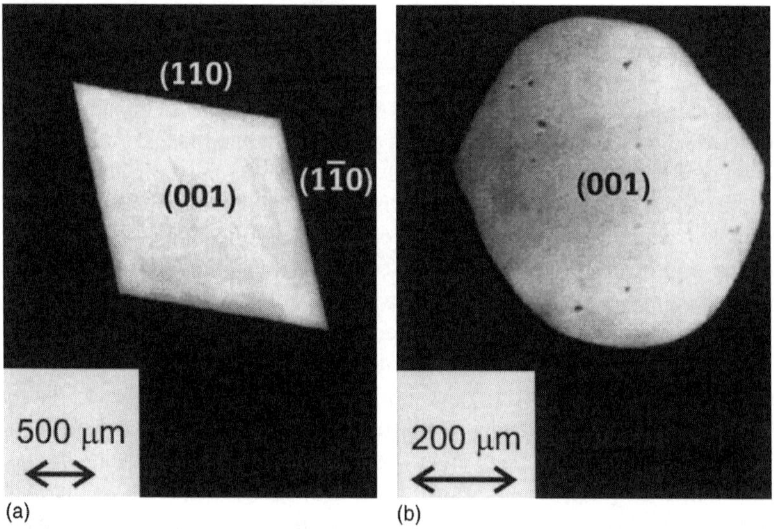

Fig. 1 (a) Paraffin crystal growing from a hexane solution below the roughening temperature of about 11°C at 6.14°C in a closed vessel at a supersaturation corresponding to 0.01°C; the faces {110} are straight. (b) Same as (a) but the paraffin crystal was growing above the roughening temperature at 19.36°C at a supersaturation of $\Delta T = 0.01$°C. Now the faces {110} are rounded off.

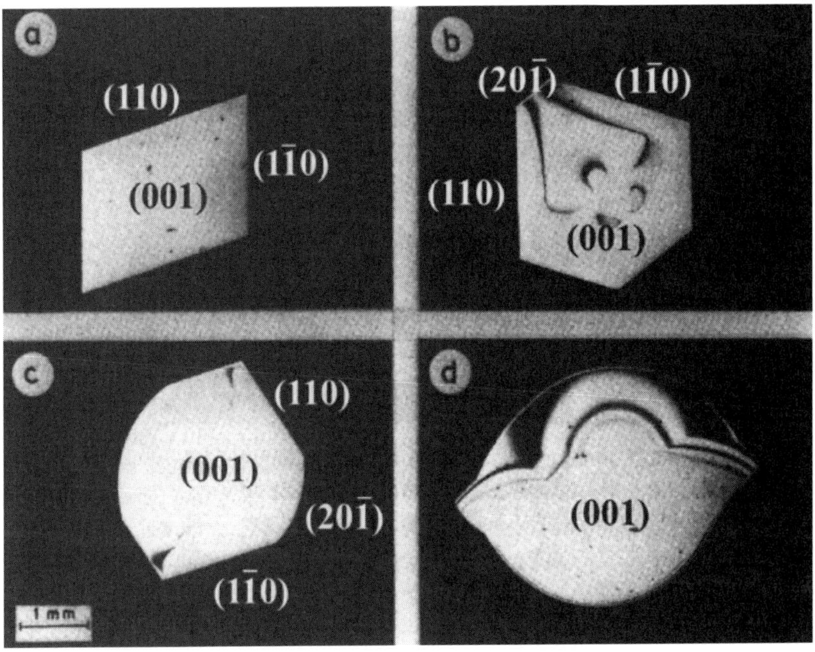

Fig. 2 (a) Picture of a crystal of naphthalene growing from a toluene solution. The crystal is lying on one (001) plane. The four {110} faces can clearly be seen. The relative supersaturation is 0.88%. (b) Identical to (a), but two faces {20$\bar{1}$} can be seen, because the relative supersaturation is lower: 0.32%. (c). Identical to (b). The relative supersaturation is higher: 1.14%. Due to kinetical roughening first the {20$\bar{1}$} faces become rounded-off faces. (d) Identical to (b). The relative supersaturation is higher: 1.47%. Now also the {110} faces become kinetically rough.

sition. This concept of roughening transition has for about 20 years been strongly rooted in statistical mechanical models of crystal surfaces.

Moreover, it is demonstrated in Fig. 2 that similarly to naphthalene, the two faces of the form {20$\bar{1}$} and the four faces of the form {110}, respectively, become rounded. In this case, however, the critical parameter is not the temperature but the driving force for crystallization, i.e., the relative supersaturation. It can be seen from Figs. 2b–d that beyond a certain critical supersaturation, which differs for the faces of the form {110} as compared to faces of the form {20$\bar{1}$}, the flat growing faces are transformed into rounded-off faces again. This phenomenon is called kinetical roughening. It is now well known that the phenomena of thermal and kinetical roughening are determined by the concept of edge free energy of steps occurring on surfaces: the lower the edge free energy the lower the critical roughening temperature and the lower the supersaturation beyond which kinetical roughening occurs.

The implications of the concept of roughening transition to explain crystal growth phenomena together with crystal morphology are very profound. Rough faces have, generally speaking, higher growth rates than flat faces, and in most cases these faces do not appear on the crystal morphology.

It will be shown in Sec. III that most aspects of modern theories of crystal growth and morphology, including the newest developments, are needed to explain the ''flat needle'' or ''plank'' like habit of fat crystals. In this explanation, the parallel relation between roughening temperature and morphological importance of faces of form $\{hkl\}$ will play an essential role. It is shown that recent extensions of the Hartman-Perdok theory and the concept of roughening transition can be integrated on the basis of the concept of connected net. These extensions will be applied to fat crystals of the β-2 structure. Contrary to the results of a paper published in 1992, in which some crucial discrepancies between predicted and observed morphologies were reported, according to a paper published in 1999 these discrepancies vanish. In Sec. II we review the developments in the morphological connected net–roughening transition theories. For experimental and theoretical data on roughening we refer to the survey paper of Bennema, corresponding to Part 7 of the *Handbook of Crystal Growth*, Vol. 1 [3].

II. GENERAL PRINCIPLES OF THEORIES ON CRYSTAL GROWTH AND MORPHOLOGY

A. The BCF Step Edge-Growth Spiral Picture at the End of the Second Millennium

Half a century ago, Burton, Cabrera, and Frank (BCF) developed the famous spiral growth theory [4], after Frank ''invented'' the originally purely theoretical concept of growth spiral [5]. In the famous BCF paper, rightly called the ''bible of the science of crystal growth,'' surface and step models, expressed in the statistical mechanical language of Ising models, were applied to the (001) face of a simple primitive cubic crystal or Kossel crystal. These theories were logically integrated with 2D nucleation theories and screw dislocations, giving rise to growth spirals.

In statistical mechanical Ising models for the interface between crystal and mother phase, the content of solid cells corresponds to ''growth units'' (GUs), i.e., atoms, ions, molecules, complexes, etc., which are the building blocks, which together form the growing crystal. An important parameter in these Ising models is the interaction energy between neighboring cells, which depends on the actual properties of the two cells. These energies can be different for the type of neighbors, e.g., nearest or next-nearest neighbors. Note that in the BCF theory, Onsager's statistical mechanical theory of the order-disorder phase transition in a 2D Ising model played an essential role [6]. Further, BCF showed that a step, which is a 1D system, contrary to a 2D surface, does not show a phase transition down to 0 K. This implies that steps are always rough.

From a modern point of view statistical mechanical 2D Ising models are, in principle, good models to describe crystal surfaces. Therefore, Onsager's 2D order-disorder phase transition can be interpreted as the roughening transition, which is defined as follows:

$$\begin{aligned} \theta < \theta^c \rightarrow \gamma > 0 \\ \theta \geq \theta^c \rightarrow \gamma = 0 \end{aligned} \tag{1}$$

Here θ is a dimensionless temperature, and θ^c is a critical dimensionless temperature which can, in principle, be considered as the order-disorder phase transition temperature of a given interface between crystal and mother phase. This critical temperature depends on the energy of broken bonds at the interface, the topology of the bond structure of the surface (*hkl*), its interaction with the mother phase, and the edge free energies γ of steps at its surface. The dimensionless temperatures are given by

$$\theta = \frac{2kT}{\phi} \tag{2a}$$

where ϕ is a generalized bond energy having the form

$$\phi = \phi^{sf} - \tfrac{1}{2}(\phi^{ss} + \phi^{ff}) \tag{2b}$$

In Eq. (2b), sf refers to the solid-fluid (mother phase) bonds, ss to the solid-solid bonds within the crystal, and ff to the fluid-fluid bonds. Note that these three bond energies are usually defined in reference to vacuum, where growth units are infinitely far apart from each other. Conventionally θ and θ^c are expressed in reference to the strongest bond or highest ϕ of the crystal (or crystal graph), according to Eq. (2).

We note that very often it is assumed that the bond energies ϕ and ϕ^{ss} are proportional; see Sec. II.C. The concept of crystal graph will be discussed in Sec. IIC. See further the survey papers of Bennema [1,7].

About 15 years ago it turned out to be possible to calculate the Ising temperatures θ^c even for complex 2D Ising lattices consisting of many different cells [8,9]. Based on the concept of roughening transition two basically different crystal growth mechanisms can be distinguished: Below the roughening temperature a layer growth mechanism occurs, either a spiral growth or a 2D nucleation mechanism, or a combination of both. Above the roughening temperature a continuous growth mechanism leading to a rough face occurs.

In the first case, a surface grows macroscopically flat, with a well-defined orientation (*hkl*), in the second case as a microscopically rough and macroscopically unstable rounded-off face, without any crystallographic orientation. The most probable reason for the observation that just beyond the roughening temperature a face becomes rounded is that the edge free energy vanishes and a step growth mechanism changes into a continuous one. According to Chernov and Nishinaga [10], a step growth mechanism shows a high resistance against instabilities, i.e., deviations from growth as a flat face, caused by variations in supersaturation, occurring at different positions at the growing surface. Experimentally this transition shows up macroscopically as a rather abrupt change from a growing facet to a rounded surface, over a small temperature range of the crystallization temperature (Fig. 1).

As explained before, kinetical roughening occurs at a surface growing below the roughening transition temperature as a flat face. When for relatively weak faces, having a low edge free energy and roughening temperature, supersaturation increases, at relatively low supersaturations the size of 2D critical nuclei becomes so small that it corresponds to a few growth units. Note that the critical nucleus is inversely proportional to the supersaturation. If this happens, again no step growth mechanism occurs and the same instabilities occur as for thermal roughening, leading to nonstable rounded-off surfaces (Fig. 2).

B. Computer Simulations and Recent Experimental Observations

A widely used computer simulation of crystal surfaces, using Ising-like models, is the Monte Carlo model. In such a simulation, a crystal surface is represented by a 2D grid of columns that represents the local height. In Fig. 3 results of such simulations, carried out less than 30 years ago by Gilmer and Bennema [11] and de Haan et al. [12], are presented. Here the dimensionless growth rate is plotted versus the dimensionless driving force for crystallization $\beta = \Delta\mu/kT$, being roughly equal to the relative supersaturation. Each curve corresponds to a value of $\alpha = 4\phi/kT$, which corresponds to a dimensionless reciprocal temperature. It can be seen that the higher is α or the lower is the temperature, for constant ϕ, the lower is the rate of growth R. This is understandable because the lower the temperature of the crystal surface, the less rough the surface will be at the atomic scale. Furthermore, the flatter the surface, the smaller the fraction of growth units arriving at the surface that will actually stick as a result of the lack of sufficient kink sites. The most important finding of the simulations was the change from linear to curved R versus β curves. It is now generally accepted that this change marks the roughening transition. These results from computer simulations inspired theoreticians like Leamy and Gilmer [13], van Beijeren [14], van der Eerden and Knops [15], Weeks and Gilmer [16], and other (also see references in Ref. 1) to develop a more sophisticated roughening transition theory for a solid-

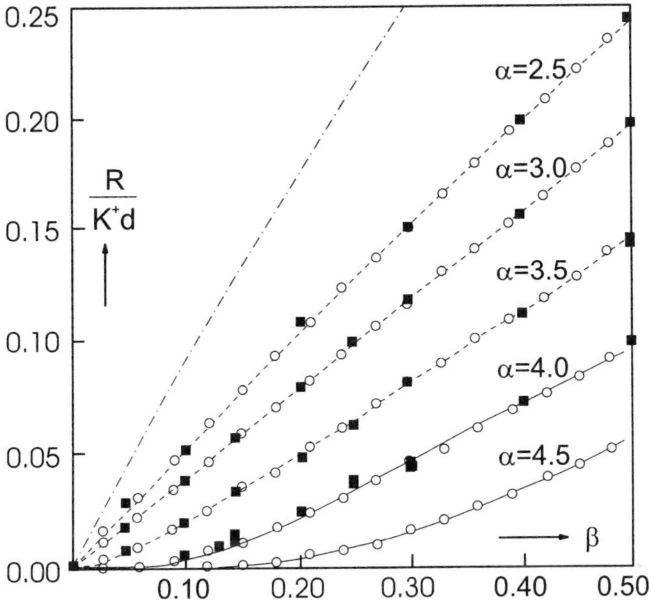

Fig. 3 Results from Monte Carlo simulations for several values of the reciprocal temperature $\alpha = 4\phi/kT$. Plotted is the dimensionless growth rate R/k^+d versus $\beta = \Delta\mu/kT$; full squares are results from a general computer [11]; open circles are special-purpose computer results [12]; solid lines represent theoretical two-dimensional birth-and-spread nucleation curves; dashed lines empirical fitting curves. The uppermost line represents the maximal rate of growth (of a kinked or rough surface). This is the so-called Wilson-Frenkel law.

on-solid (SOS) model of the (001) surface of the Ising-Kossel model and the body-centered SOS (BCSOS) model.

In the last decade new theories leading to new types of surface phases and new critical surface phase transitions were developed. We just mention one example. After the pioneering paper of van Beijeren of 1977 [14] in 1995, Mazzeo, Carlon, and van Beijeren [17] showed the existence of a new surface phase. This is the DOF (disordered flat) phase of the (001) surface of the SOS CsCl structure. This DOF phase is between the flat phase and the rough phase of the (001) faces of, respectively, the SOS Kossel and BCSOS model crystals. This example shows that by simply replacing an atom in the center of a BCC model crystal by a different atom, implying that a BCC structure is replaced by a CsCl structure, a much richer surface phase diagram is obtained. Recently, these ideas were, after some modifications, applied to explain the most peculiar behavior of the faces {011} of naphthalene crystals growing from the vapor [18]. We will not discuss these developments in theories of roughening further, since we will not apply these new ideas to explain the morphology of fat crystals of the β-2 structure. It may be important, however, for future research on growth and morphology of crystals to be aware of these developments.

The uninterrupted curves of Fig. 3 represent fits to analytical 2D nucleation birth and spread curves, implying that 2D nuclei are formed by statistical fluctuations, that these nuclei grow or spread, form terraces, and that on top of these terraces new nuclei are formed, etc. According to relation (1), above the roughening temperature the edge free energy vanishes, which explains the linear rate versus supersaturation curves in Fig. 3, which resulted from the simulations.

Since the introduction of the BCF theory, surface Ising models proved to be useful for the development of theories on crystal surfaces and crystal growth mechanisms. However, Ising models suffer from the ad hoc approximation that growth units at the surface have only two properties. They are either "solid," i.e., in a completely well defined crystalline state, determined by the crystal structure, or "fluid," i.e., without any translational order. In order to overcome the limitations of Ising models, simulations of crystal surfaces have been carried out using molecular dynamics (MD). Recently, Huitema, and coworkers [19] obtained rate versus supersaturation curves from MD studies, applying methods to obtain relevant results within a reasonable time. Growth rates of (111) faces of FCC crystals growing from supersaturated solutions resulted, for the first time, in curved and linear R versus $\Delta\mu/kT$ curves. These curves were qualitatively, and to some extent quantitatively, the same as the curves resulting from Monte Carlo simulations of the (001) face of an Ising SOS Kossel crystal surface, presented in Fig. 3.

The integrated BCF–Ising–growth spiral–2D nucleation theory has now been, directly as well as indirectly, to a very high extent confirmed for a large range of crystals. This range extents from very simple FCC helium and argon crystals, growing from the melt at very low temperatures, to crystals of huge complex molecules, such as molecules of fats and extremely large molecules of proteins, growing from solution. During the last half-century, first ex situ observations with advanced optical and electron optical techniques were made. Later came the in situ observations of, for example, rotating spirals, which could be performed using advanced optical techniques. In the last 10 years both in situ and ex situ observations have been made with additional modern techniques, especially scanning probe microscopy (SPM).

For example, Tsukamoto et al. [20] apply highly advanced optical techniques to observe in situ the movement of monomolecular trains of steps on crystal surfaces growing from a supersaturated solution. Rotating growth spirals could be seen in situ and, simultaneously rate versus supersaturation curves could be measured on growing (100) surfaces of barium nitrate crystals with screw dislocations, also growing from solution.

C. Hartman-Perdok Theory, Connected Nets, and Roughening Transition

In order to predict the morphology of crystals from the crystal structure and to understand crystal growth mechanisms taking place at various surfaces with different orientations (*hkl*), the following connected net analysis is carried out. This analysis is based on Hartman-Perdok theory [21–23]. Note that in case of polymorphism one always considers a single-crystal structure corresponding to a certain polymorph.

1. From a certain crystal structure, the growth units (GUs) are defined. Note that GUs depend on the mother phase; for one mother phase GUs may be, for example, monomers, for another mother phase dimers, etc.
2. The interaction energies between GUs are calculated. These energies are referred to as bonds.
3. The GUs are reduced to centers of gravity or centers of geometry and a selection is made of the strongest bonds (often first nearest-neighbor bonds). The bonds and the centers of gravity together result in the corresponding crystal graph.
4. The connected nets are determined using the crystal graph. These connected nets define the possible faces {*hkl*} that may appear on the morphology.
5. Using connected nets, crystallographic directions [*uvw*] within a connected net are identified along which steps may occur during the crystal growth process. As will be explained, actual edge energies for these steps are calculated. Using these edge energies relative Ising temperatures θ^c, which are a good measure for the relative roughening temperatures, can then be calculated.
6. The final step of the prediction procedure is the construction of the morphology using the information obtained.

In the following, we take a closer look at steps 1 to 4 and 6. In the next section, we discuss the implications of step 5. In that section the newest developments in the connected net-roughening theory will be discussed, which will be applied to fat crystals of the β-2 structure in Sec. III.

Steps 1 to 3 can be carried out in many different ways. Note that for crystals consisting of complex molecules, such as fat molecules, overall bonds between molecules or GUs generally are the result of numerous atom-atom interactions and interactions between charged areas of molecules (of GUs). We note that, as discussed briefly in Sec. III.C, overall bond energies of bonds calculated between GUs are the bond energies ϕ^{ss} referenced to vacuum. In order to estimate actual bond energies, often the ad hoc proportionality condition is introduced, which implies that the ratio of bond energies referenced to vacuum is the same as the corresponding bond energies for growth from solution. So all generalized bond energies have the shape of Eq. (2) and for all bonds i the following relation holds:

$$\phi_i = \kappa\phi_i^{ss} \tag{3}$$

where κ is the proportionality constant. This relation implies that for each connected net found, its interaction with the mother phase is assumed to be identical.

We now first define the concept of connected net. We start to define a connected net in a sloppy fashion as a 2D object, having translational symmetry in two dimensions. It consists of the GUs, i.e., their corresponding centers of gravity, in the crystal graph, all of which are connected to each other by the defined bonds of the crystal graph. The relation between a connected net and the crystal graph can be explained as follows. Assume that we make a ''cheap'' 2D cut in the crystal graph along an orientation (hkl) in such a way that the broken bond surface energy corresponds to a relative minimum out of possible alternative cuts. The cut is chosen to be a complete cut of the crystal graph so that the crystal graph can be cleaved into two halves. Then, due to the periodicity of the crystal graph, at a distance d_{hkl}, exactly the same cut can be made. Here d_{hkl} is defined as the interplanar distance, corrected for the selection rules of the space group. We note that quite often, several different connected nets all having the same thickness d_{hkl}, can be constructed for one orientation (hkl).

The importance of the concept of the connected net as defined is that a connection can be made between real crystal structures and statistical mechanical ''Ising surfaces.'' In all cases, the integrating principle of crystal surfaces and crystal growth processes is based on the important concept of thermal roughening transition, in which the order parameter edge free energy plays an essential role. From relation (1) this can be seen at once, because this relation can be interpreted in a reversed order as follows: If in all crystallographic direction $[uvw]$ coplanar with the orientation (hkl) edge energies are larger than zero, then also the corresponding edge free energies will be larger than zero below a certain temperature, and the corresponding roughening temperature will be larger than 0 K. If in at least one direction $[uvw]$, the edge energy is zero, the corresponding edge free energy will be zero for all temperatures down to 0 K. This implies that a face (hkl) which is parallel to such a ''disconnected net'' will be rough at all temperatures down to 0 K.

The new version of the Hartman-Perdok theory will be called connected net theory. This theory is related to classical Hartman-Perdok theory in which the classification of faces in F(lat), S(tepped), and K(inked) faces, together with the concept of periodic bond chain (PBC), plays a key role. Without going into details a PBC is defined as an uninterrupted chain of bonds, selected from the crystal graph, having the periodicity [*uvw*] of the crystal structure and crystal graph. F faces are defined as faces parallel to at least two different, nonparallel PBCs. S faces contain only one PBC, and K faces are not parallel to any PBC at all. From a modern point of view an F face is defined as a face with an orientation (*hkl*) in which at least one connected net can be identified. As mentioned and defined, such a connected net has an overall thickness d_{hkl}. Note that the boundaries of a connected net need not be flat; these boundaries may be ''wavy.'' It is obvious that S and K faces are not parallel to a connected net and have a roughening temperature of 0 K (see Refs. 1 and 7).

Nowadays, using software developed by our group, called FACELIFT [24], recently integrated into Cerius^2 software [25], all connected nets of a certain crystal graph can be determined and visualized. This software is based on the method developed by Strom [26]. According to this method all possible PBCs of a given crystal graph are generated, and from these all connected nets are generated by checking which combinations of PBCs give rise to connected nets.

In order to discuss step 6 we first introduce the concepts of attachment energy E^{att} and slice energy E^{slice}, which did play an essential role in classical Hartman-Perdok theory. According to this theory we have to distinguish E^{att} of a connected net (*hkl*), cut by the boundary of the connected net, and the complementary energy E^{slice} of the unbroken bonds of the corresponding growth layer. The sum of these complementary quantities corresponding to the crystallization energy per unit cell or stoichiometric unit of the crystal graph, is the crystallization energy E^{cr}. This is a bulk property, independent of the orientation. Therefore, the following relation holds for a given orientation (*hkl*):

$$E^{att}_{(hkl)i} + E^{slice}_{(hkl)i} = E^{cr} \tag{4}$$

All bonds belonging to the bonds cut by the boundary lead to E^{att}, depending on the connected net labeled i, and all bonds within the corresponding growth layer lead to E^{slice}. These bonds, in principle, have the form of Eq. (2).

We note that without knowledge of connected nets in the past the morphology was predicted using the well-known attachment energy (E^{att}) method, which, for example, can be found in the morphology module of the Cerius^2 software [25]. By taking the distance from a face to a common origin proportional to E^{att} the growth form is constructed using the Wulff construction. In this case, the boundary is chosen ad hoc to be a flat plane and the connectedness of the net underneath this boundary is not checked. Using, however, knowledge of connected nets the calculation of E^{att} can be refined and a better prediction of the morphology will result.

Looking back to the procedure, consisting of six points, the three following aspects can be distinguished:

1. A chemical aspect, concerning the choice of GUs and the bonds of the crystal graph
2. A crystallographic-topological aspect, concerning structures of connected nets and the way the nets for a certain orientation (*hkl*) are embedded within a crystal graph
3. A statistical mechanical aspect, concerning among others, the calculation of the Ising temperature as a measure for the roughening temperature

D. Physical Implications of the Topology of Connected Nets with Oblique Bonds

For one orientation (*hkl*) of a given crystal graph, quite often more than one connected net can be identified. Using modern software such as FACELIFT, it is not uncommon that, say 10 to a few tens of connected nets for one orientation (*hkl*) can be identified and constructed. Then the question is how are such cases of multiple connected nets dealt with. Until a few years ago we assumed that in case several connected nets occur for one orientation (*hkl*), the strongest connected net, having for example the highest E^{slice}, would be the surviving connected net and dominate during the growth process of the face (*hkl*) [1]. Recently, however, it was discovered that from a physical point of view, for certain configurations for a given orientation (*hkl*), with one or more connected nets, these nets cannot be treated as independent connected nets. In fact, the topology of the connected nets turns out to be of crucial importance for the stability of the corresponding crystal surface, even for single connected nets when oblique bonds are present. For details see Grimbergen et al. [27,29] and Meekes et al. [28].

It will be demonstrated with a simple model crystal graph that in such cases a substitute net can be constructed. For such a substitute net new step edge energies and, therefore, a new relative Ising temperature θ^c can be calculated. Having a correct picture of an interface with orientation (*hkl*) is of course essential to calculate the relative Ising temperature θ^c or the roughening temperature θ^R in order to understand the crystal growth mechanisms of a face (*hkl*) and the resulting relative growth rate. This information will in turn lead to an adequate prediction of the morphological importance for the face (*hkl*). This will now be demonstrated using the two connected nets for the (001) orientation in a very simple model graph, as depicted in Fig. 4. In Sec. III this principle will be applied to the morphology of fat crystals.

It can be seen from Fig. 4 that in the direction [100], seen end on, two PBCs consisting only of *c* bonds occur. These PBCs do not play an explicit role

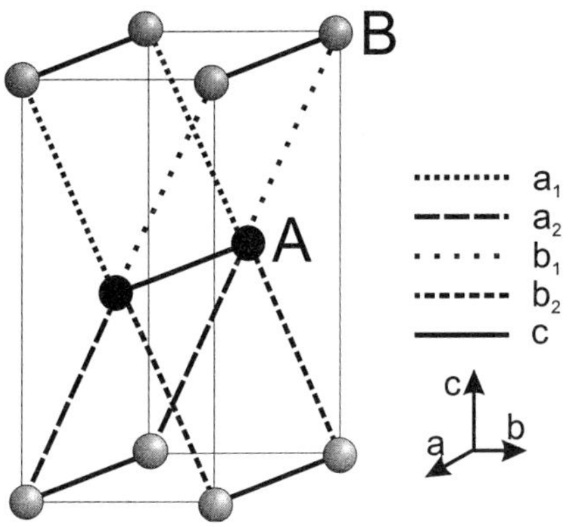

Fig. 4a Unit cell of a model crystal graph with two growth units *A* and *B*.

in the following argument, but will only make sure that for the orientation (001) genuine connected nets occur. Four bonds consisting of two pairs of oblique bonds $a_{1,2}$ and $b_{1,2}$, respectively, connect via first-nearest-neighbor interactions the two different GUs *A* and *B* of the crystal graph in the unit cell. The corresponding positive generalized broken bond energies, having the form of Eq. (2), will be indicated in the following as $\phi_{a1,2}$ and $\phi_{b1,2}$, respectively. It is demonstrated in Figs. 4b,c that together with the *c* bonds these four bonds give rise to two

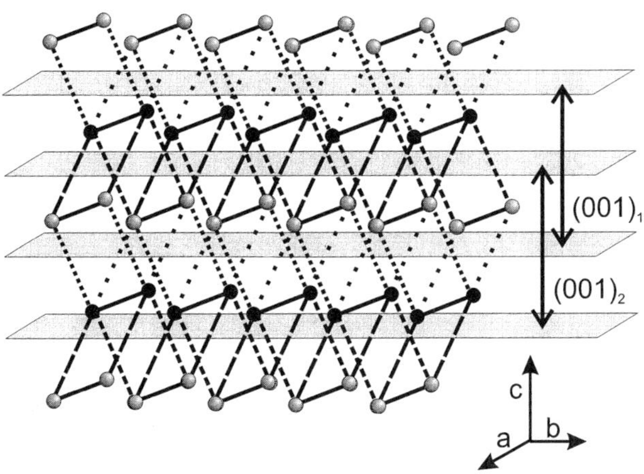

Fig. 4b The two connected nets for the (001) orientation: $(001)_1$ and $(001)_2$.

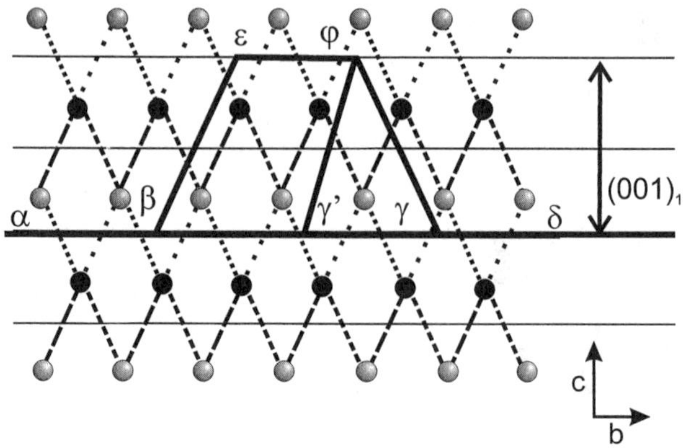

Fig. 4c The two connected nets of Fig. 4b seen in the [100] projection, that is, along the *c*-bonds. The profiles of two alternative 2D nuclei are indicated as $\alpha\beta\varepsilon\phi\gamma\delta$ and $\alpha\beta\varepsilon\phi\gamma'\delta$, respectively.

alternative connected nets $(001)_1$ and $(001)_2$ for the (001) orientation. These connected nets have a rooflike pattern.

It is shown in Fig. 4b that the two connected nets $(001)_1$ and $(001)_2$ can be constructed with the proper interplanar thickness d_{001}. Such nets are shifted $\frac{1}{2}\, d_{001}$ in reference to each other. The core of the $(001)_1$ connected net consists of bonds a_2 and b_2 and the core of the $(001)_2$ connected net of bonds a_j and b_1, respectively. In the following we assume that the connected net $(001)_1$ is the most stable one and that $\phi_{a1} < \phi_{a2}$ and $\phi_{b1} < \phi_{b2}$. Moreover, only edge energies ϵ are considered instead of edge free energies γ. For the general case we refer to Ref. 27. Furthermore, we only consider the 2D nucleation mechanism as the relevant growth mechanism. In Sec. III we argue that this mechanism is the relevant growth mechanism for the top faces of flat needle-shaped fat crystals.

The key problem for the topology of connected nets having oblique bonds is the calculation of the lowest possible edge energy for a step. Inspired by van Beijeren and Nolden [30], we show how to find the lowest possible activation barrier to form a 2D nucleus on top of a surface by a statistical fluctuation, bounded by the most stable connected net $(001)_1$. The shape of the 2D nucleus is indicated in Fig. 4c. The lowest possible edge energy for a 2D nucleus is found to be the difference in cut energy between the profile $\alpha\beta\epsilon\phi\gamma\delta$ and the straight profile $\alpha\beta\gamma\delta$ in the [010] direction, along the boundaries of the connected nets $(001)_1$. This difference is given by

$$\varepsilon_{[100]1} = (\phi_{a2} - \phi_{a1}) + (\phi_{b2} - \phi_{b1}) = \Delta\phi_a + \Delta\phi_b \tag{5}$$

Now, a substitute connected net can be constructed that has the same 2D nucleation edge energy as the connected net $(001)_1$, for which the bond energies ϕ_{a1} and ϕ_{b1} are replaced by $\Delta\phi_a$ and $\Delta\phi_b$, respectively. This net is a rectangular net with bond energies ϕ_c and $\epsilon_{[100]1}$, respectively. The Ising temperature of such a net can be determined as an approximate value for the roughening temperature of the original net. Note that the Ising temperature of a connected net usually is determined by the strengths of the bonds making up the net. Since $\Delta\phi_a + \Delta\phi_b$ can be much smaller than $\phi_{a2} + \phi_{b2}$, the Ising temperature of the substitute net can be much lower than that of the original net. This case was treated in detail in Refs. 27–29.

As a special case we can distinguish a subcase implying that one bond of the set of four bonds $a_{1,2}$ and $b_{1,2}$ becomes zero. Let us take as an example the case

$$\phi_{a1} < \phi_{a2}, \quad \phi_{b1} < \phi_{b2}, \quad \text{and} \quad \phi_{b1} = 0 \tag{6}$$

From Fig. 4, it can be seen that since the bond b_1 is no longer a part of the crystal graph, the original connected net $(001)_2$ is no longer a connected net. Now, the whole argument to replace the connected net (001), in principle, remains valid, taking $\Delta\phi_b = \phi_{b2}$. So the absence of the bond b_1 will make the surviving single connected net $(001)_1$ stronger. Nevertheless, it will still be weaker compared to the original connected net $(001)_1$, due to the bond a_1. In fact, this case is encountered for the top faces of needle-shaped fat crystals, as explained in Sec. III.

In an even more extreme case, for which $\phi_{a1} = \phi_{a2}$ and $\phi_{b1} = \phi_{b2}$ and, thus, $\Delta\phi_a = \Delta\phi_b = 0$, the orientation (001) will already be rough at 0 K. This is a case of symmetry roughening [28].

In the following, a general procedure for finding the "cheapest 2D nucleus" is presented, which has to be applied to all possible faces $\{hkl\}$ of the crystal graph. This 2D nucleus can be used to calculate the Ising temperature for the face $\{hkl\}$. To start this generalization we first define, analogously to the definition of the classical distinction between E^{att}_{hkl} and E^{slice}_{hkl}, for an orientation $\{hkl\}$ two sets of bonds within the crystal graph, which we call $F^{\text{att}}_{hkl,i} = \{\phi_1, \phi_2, \ldots, \phi_n\}_i$ and $F^{\text{slice}}_{hkl,i} = \{\psi_1, \psi_2, \ldots, \psi_m\}_i$, where i labels the connected net under consideration. In case of the connected net $(001)_1$ in Fig. 4c, $F^{\text{slice}} = \{a_2, b_2, c\}$ and $F^{\text{att}} = \{a_1, b_1\}$. Together these sets make up the crystal graph, which can be represented in a similar way by F^{cr}, analogous to Eq. (4). Note that F^{slice} contains all the bonds that make up the growth layer for the orientation $\{hkl\}$ under consideration. The flat profile of a surface bounded by a connected net without a 2D nucleus cuts only F^{att} bonds. The profile $\alpha\beta\gamma\delta$ in Fig. 4c is an example of such a case. The surface with a 2D nucleus cuts additional bonds. The island profile $\alpha\beta\varepsilon\varphi\gamma'\delta$ cuts, for example, bonds also from F^{slice}. The differences of broken bonds of both profiles result in the edge energy $\varepsilon_{[100]}$. For the profile $\alpha\beta\varepsilon\varphi\gamma'\delta$ this edge energy is determined only by the bonds a_2 and thus only by F^{slice} bonds.

Alternatively, the profile αβεφγδ has an edge energy determined by F^{slice} and F^{att} bonds according to Eq. (5). Note that from all alternative profiles that can be created, the one that results in the lowest broken bond energy is the one that is most important for the growth. This exercise results in all edge energies for all possible directions parallel to the face $\{hkl\}$. Using the edge energies obtained a substitute connected net can be constructed that can be used for the estimation of the roughening temperature.

The concepts outlined illustrate the extensions of the classical Hartman-Perdok theory. The essential difference is that in the latest approach the actual growth mechanism, in the present case being 2D nucleation, is taken into account. The determination of the relevant step energies is essential, and for this a distinction between Kossel-like topologies and non-Kossel-like topologies is necessary.

For Kossel-like topologies, the substitute net for the corresponding surface has an edge energy that is totally determined by bonds within F^{slice}. Therefore, the bonds that make up the growth layer determine the edge free energy, and the bonds within F^{att} are not involved. For non-Kossel-like topologies the substitute connected net contains at least one edge energy that is made up using bonds of F^{att}. Therefore, the (non)-Kossel-like character is fully determined by the profile of the 2D nucleus under consideration. For some connected nets only Kossel-like profiles are possible, and we call such nets Kossel-like connected nets. For other nets both types of profile can be constructed; we call these non-Kossel-like connected nets.

The classical Hartman-Perdok theory can be considered as a theory based on a generalized Kossel crystal, with the assumption that bonds cut by the boundary and uncut bonds within the connected net can clearly be distinguished. Cutting these latter bonds yields unambiguous edge energies, for different crystallographic directions, consisting of bonds occurring within the connected net. The new approach allows for step energies that also include attachment bonds often resulting in relatively low step energies and therefore low Ising temperatures and, thus, relatively high growth rates compared to the classical approach.

E. Construction of the Crystal Morphology

As result of the considerations of the previous section, we conclude that the classical growth law according to the attachment energy method that assumes a parallel relationship between the growth rate R of an orientation (hkl) and the attachment energy of its strongest connected net,

$$R \propto E^{\text{att}} \tag{7}$$

is no longer generally valid. In order to obtain a qualitative morphological model for the crystal we will, inspired by the recent developments in crystal growth theories, outlined above, introduce for the moment a very simple ‘‘minimal hypothesis.’’ This implies that there is an antiparallel relation between the Ising temperature θ^c of the face $\{hkl\}$ and the growth rate R of that face at a certain supersaturation. Therefore, there is also an antiparallel relation between the morphological importance (MI) of this face. Note that the MI of a face (hkl) is defined as some qualitative statistical measure for the relative frequency of occurrence of that face on a sample of crystals under consideration and/or the relative size of that face. This antiparallel relation implies

$$(\theta^c_{hkl} > \theta^c_{h'k'l'}) \rightarrow (R_{hkl} < R_{h'k'l'}) \rightarrow (MI_{hkl} < MI_{h'k'l'}) \tag{8}$$

These ideas have recently been applied to the $\{011\}$ faces of naphthalene crystals growing from the vapor (Grimbergen et al. [31,18]) to the growth of paraffin crystals (Grimbergen et al. [32]) as well as to the morphology and growth behavior of triacylglycerols crystals (fats) (Hollander et al. [33]).

This last application will be discussed in Sec. III. It will then be shown that only by taking the implications of the character of Kossel or non-Kossel connected nets of the crystal graph of the β-2 phase of fat crystals explicitly into account, can a satisfactory interpretation of the observed morphology of the side but especially of the top faces of fat crystals, determined by a large variety of competing orientations, be given.

III. APPLICATION TO MONOTRIACYLGLYCEROL CRYSTALS OF THE β-2 STRUCTURE

A. Introduction

In the following we will label triacylglycerol molecules as TAG molecules and fat crystals of mono β-2 triacylglycerol molecules as β-2 TAG crystals, or sometimes simply as TAG crystals. Following Sec. II.D and the paper on the morphology of β-2 TAG crystals of Hollander et al. [33], it will be shown that discrepancies between predicted and observed morphology of β-2 TAG crystals, reported earlier by Bennema, Vogels, and de Jong [34], vanish to a very high extent. The reason is that Ref. 34 is based on the old principle of choosing the strongest net out of the set of connected nets for one orientation (hkl) instead of using the recently developed concept of substitute net. The program of six points—crystal structure, bonds, crystal graph, connected nets, roughening temperature, and morphology construction—as outlined in Sec. II.C and II.D will therefore also be used as a guideline in this chapter.

B. Crystal Structure

1. Crystal Structure of β-2 TAGs

We will closely follow de Jong [35] and de Jong and van Soest [36] in their analysis and description of the structure of β-2 TAG crystals. Moreover, following Bennema et al. [34] and Hollander et al. [33], we limit ourselves to crystals of the mono β-2 TAG structures, which implies that the alkane chains have equal length, to be explained shortly. For additional information on alternative thermodynamically stable and unstable structures of triacylglycerol crystals with varying chain lengths, see de Jong [35] and Wesdorp [37]. In this part the same crystallographic setting will be used as in Ref. 33. According to the conventions used in Refs. 33–36, TAG molecules will be labeled $R_1 \cdot R_2 \cdot R_3$, where R_i is the length of the ith alkane chain. In the following, both theoretical and experimental studies on the morphology of β-2 TAG crystals, having three even saturated alkane chains R with equal lengths of $n = 10, \ldots, 22$ carbon atoms, will be presented.

Until now, experimental coordinates were available only for the 10.10.10, 12.12.12, and 16.16.16 TAG crystals. In order to obtain reliable structure data from X-ray diffraction data of β-2 TAG crystals, crystals of good quality and sizes of approximately 100–1000 μm in each dimension must be grown. The size of the required crystals, however, may go down considerably if high-intensity X-ray synchrotron radiation is used. Unfortunately, obviously due to their structure and weak van der Waals interactions, fat crystals tend to grow in a needle-like habit and disorder and twinning easily occur. In addition, spherulites are easily formed if attempts to grow single crystals of TAG crystals are carried out. Note that growing a large enough single TAG crystal and successfully elucidating the crystal structure is difficult; only three structures have been reported up to now.

The first study of TAG crystals by the X-ray single method was carried out by Vand and Bell, who analyzed the β-2 form of 12.12.12 TAG crystals [38]. Their crystal structure determination, however, was incomplete and the details of the molecular arrangements remained unknown. Later, Vand found a more detailed structure, but he did not publish it (see Ref. 39). The work of Larson, who studied the solid state of several TAG structures mainly by X-ray methods, was a leap forward. He was the first to publish a fully detailed TAG structure of 12.12.12 [40]. Shortly thereafter, Jensen and Mabis published an accurate crystal structure of the homologue 10.10.10 [41]. Recently the structure of 16.16.16 was solved by van Langevelde et al. [42], showing that this series of $n.n.n$ TAGs is indeed homologous. This simply implies that all structures of these TAGs within this series are isostructural. This confirmed the derivation of the remaining structures of the unknown TAGs within the series already done by de Jong [35] and de Jong and van Soest [36], who used extrapolation rules to obtain the other structures not yet elucidated. One simple example of the homologous series is demonstrated in Table 1, which shows that the lengths of the a- and b-axes of the unit cells do not depend on the chain length n, but that only the length of the c-axis increases linearly with n.

Many different crystallographic settings have been used in the literature concerning β-2 TAG crystals. The following transformations have to be applied to transform the coordinates mentioned in the cited references.

Table 1 Enthalpy of Fusion, Melting Temperature, and Crystallographic Data for the Homologous Series of β-2 *n.n.n* TAGs

n	ΔH_β [kcal/mol]	T_m [°C]	a [Å]	b [Å]	c [Å]	α [°]	β [°]	γ [°]
10.10.10	95	31.6	14.26	5.488	33.54	115.7	125.7	57.1
12.12.12	122	45.7	14.27	5.44	38.87	114.4	125.0	58.7
14.14.14	147	57.1	14.19	5.49	43.78	113.6	124.2	57.0
16.16.16	171	65.9	14.19	5.46	48.81	113.0	123.8	57.7
18.18.18	194	72.5	14.06	5.45	53.69	111.9	122.9	57.3
20.20.20	220	77.8	14.05	5.45	58.81	111.0	122.3	57.38
22.22.22	245	81.7	14.02	5.45	63.83	110.1	121.6	57.32

Crystallographic Studies. We refer to Van Langevelde et al. [42], where various transformation matrices can be found to crystallographic data published in the 20th century. To transform their setting into ours, the following matrix has to be used: [a′ → −a + b, b′ → 5a−b−c, c′ → b].

Morphological Studies. To transform the settings used in various morphological studies to the one used in this study, one uses

1. Skoda and van den Tempel [54] and Skoda et al. [55]: (left-handed) [a′ → −a, b′ → b, c′ → −a + 2b + c]
2. Bennema, Vogels, and de Jong [34]: (left-handed) [a′ → −a, b′ → b, c′ → c − b]

An example of the β-2 TAG structure of 16.16.16 seen in the direction of the *b*-axis is shown in Fig. 5a. A detail of the structure of the glycerol base is depicted in Fig. 5b and shows the 1,3 or ‘‘tuning fork’’ configuration. This implies that in order to describe the configuration of a molecule in the mono β-2 TAG structure in reference to a molecule as drawn above, the middle alkane chain R_2 has to change its orientation in reference to the two outer alkane chains $R_{1,3}$. Another possible TAG conformation is the 1,2 or ‘‘chair’’ conformation, which was suggested for the structure of *n.n* + 2.*n* TAGs of the β′-2 form. A packing concept for these structures was suggested by van Langevelde et al. [43]. Van de Streek et al. [44] suggested a different structure using molecular mechanics and compared this with X-ray powder diffraction data.

It can be seen in Fig. 5a that two TAG molecules occur per unit cell. The space group is *P*-1. The two molecules, situated within the unit cell, are transformed into each other by centers of symmetry, located exactly in the middle between adjacent molecules. It can be observed from Fig. 5a that for this β-2 TAG structure the three alkane chains of a molecule are extended as far as possible. In order to describe the shape of the molecules, within the unit cell of the

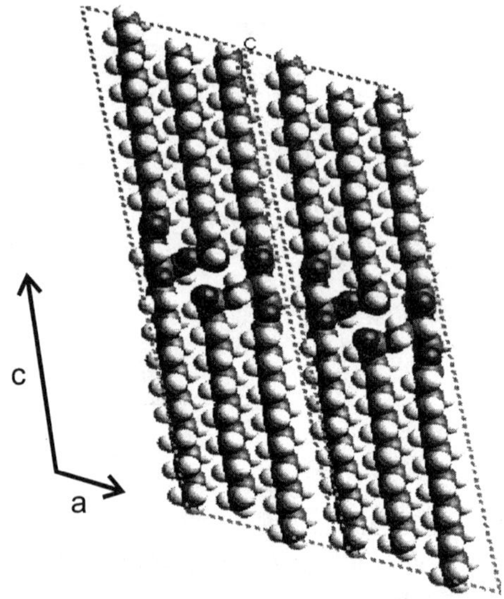

Fig. 5a Projection of the *n.n.n* TAG crystal structure for $n = 16$, seen in the [010] projection. The glycerol base and the alkane chains can be clearly recognized.

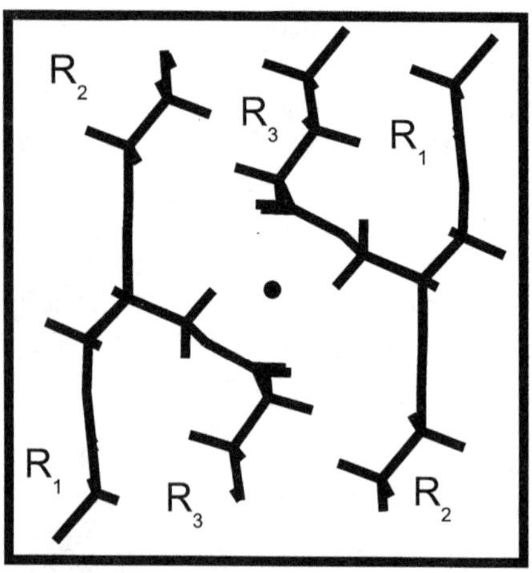

Fig. 5b Closeup of the glycerol base. The typical tuning fork conformation is shown.

n.n.n β-2 TAG structure, presented in Fig. 5a, we will use the picture of a 2-D ''tuning fork''–like molecule. This implies that a molecule has two ''alkane legs,'' one ''alkane back,'' and a ''glycerol base,'' to be called, respectively, legs, back, and base of a molecule.

2. *Analogy between TAG and Paraffin Structures*

In Fig. 6a a projection is presented of the same structure as that of Fig. 5a, but now seen more or less in the direction of the *c*-axis. The packing of the zigzagging alkane chains *R*, seen end-on in the direction of the chains, can now be recognized. According to the principle of close packing, as described by Kitaigorodskii [45], two types of closest packing of alkane chains with an even number of C atoms occur in TAGs: namely a triclinic $T_{//}$ and an orthorhombic $O_{\perp}$ subcell. In the structure in Figs. 5a and 6a,b, the alkane chains have a triclinic stacking. It can be seen that the planes of both the alkane chains of the legs and the backs of neighboring molecules are parallel to each other. The glycerol base stands more or less perpendicular to the planes of the alkane chains. In contrast to an orthorhombic stacking, in which alkane molecules have two different symmetrically related orientations, in a triclinic stacking the planes of the alkane chains are parallel. It is interesting to note that triclinic paraffin and TAG crystals have the same triclinic $T_{//}$ subcell, which is an identical stacking of alkane chains. This implies that in this case the ''perturbation'' of the glycerol base of the TAG

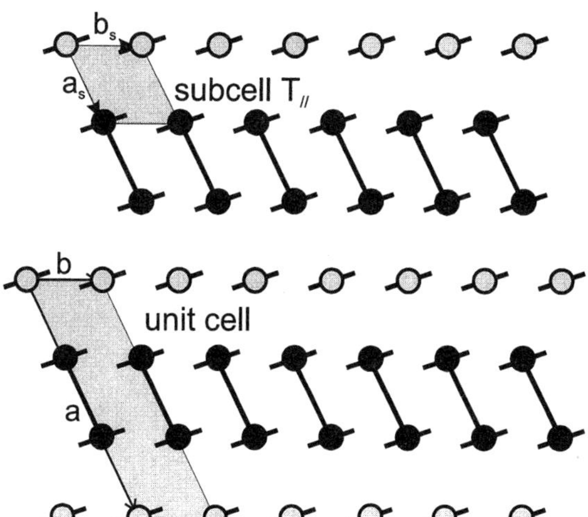

Fig. 6a [001] projection of the TAG crystal structure. The alkane chains of the TAG molecules are represented by circles. Full circles represent two legs of a single TAG molecule. Open circles denote the back of the neighboring molecules.

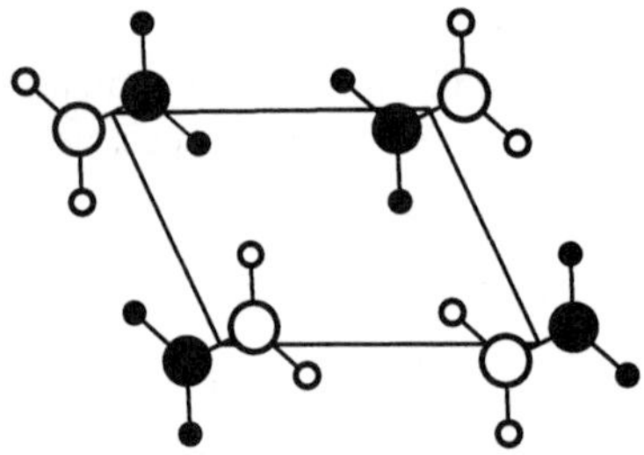

Fig. 6b Projection of the triclinic subcell $T_{//}$ along the *c*-axis. The carbon and hydrogen atoms of the alkane chains are reduced to large and small circles, respectively.

molecules is less important for the crystal packing. The height of this subcell, roughly parallel to the *c*-axis, corresponds to the length of a C_2H_4 group within an alkane chain (Figs. 6a,b). For a more detailed description of the stacking of alkane chains in mono β-2 crystals see de Jong [35], de Jong and van Soest [36], and Bennema et al. [34]. For discussions of triclinic and orthorhombic stackings of alkane chains in paraffin crystals, see Nyburg and Potworowski [46], Smith [47], Boistelle et al. [48], and Mnykh [49]. According to Broadhurst [50], information on stackings of paraffin chains in different kinds of paraffin crystals can be summarized as follows: The *n*-alkanes C_nH_{2n+2} ($n > 6$) crystallize in one of the four systems:

(i) Triclinic n (even) < 26
(ii) Monoclinic $26 < n$ (even) < 44
(iii) Orthorhombic $22 < n$ (even) < 44 and not quite pure
(iv) Orthorhombic $11 < n$ (odd) < 33

We note that seen from above in the direction of the *c*-axis, the planar group of the triclinic subcell has symmetry *p*2 and the monoclinic and orthorhombic subcell (for alkane chains with an even or odd number of C atoms) has symmetry *pgg*. In the same way as even *n*-alkanes, β-2 TAG crystals with an even number of C atoms in the alkane chains crystallize, roughly speaking, below a value of $n = 24$ (provided the alkane is reasonably pure) in a triclinic structure and above this value in an orthorhombic structure. This last case corresponds to the β′ structures, and these structures correspond to the alternative orthorhombic stacking of alkane chains.

C. Bonds and Bond Energies

1. *Calculation of Bond Energies for β-2 TAG Crystals*

In order to calculate overall bond energies between two molecules, the coordinates of all C, O, and H atoms of molecules within a unit cell must be known with a reasonable degree of accuracy. From these data accurate mutual distances between all C, O, and H atoms occurring within the two molecules have to be calculated. Next all mutual atom-atom interactions of the three types of atoms, which all belong to the two TAG molecules under consideration, can be calculated using proper

1. Short-range repulsive interaction
2. London–van der Waals attractive interactions
3. Electrostatic interactions

The sum of all atom-atom interactions of all atoms belonging to the two TAG molecules under consideration, respectively, together with the electrostatic interactions between molecules corresponds to the overall bond energy between two molecules. In Bennema et al. [34] the interaction energies were calculated from empirical parameters, which were taken for the C—C and H—H interactions from Kitaigorodskii et al. [51] and for the O—O interactions from Pople and Beveridge [52], respectively. For the interactions between different atoms, de Jong used averages over the empirical constants [35,36], mentioned above. In Hollander et al. [33] a different approach was used to calculate bond energies. There the consistent valence force field (CVFF), developed by Dauber-Osguthorpe and coworkers [53], was used to calculate the same interatomic interactions.

In order to make sure that the proper electrostatic interaction potentials were used, especially for the glycerol part of the TAG molecules the partial atomic charges for 10.10.10 were calculated quantum mechanically. It turned out that the results of the CVFF force field and the results of this calculation did agree quite well and justified the use of this force field for the other TAGs within the homologous series.

For all structures, the crystal energy was minimized, using Cerius^2 molecular mechanics software. However, during energy minimization the glycerol backbone deformed as a result of intramolecular stress. This was especially due to a wrong parameterization of ester groups of the glycerol part. Since we were only interested in intermolecular interactions for the construction of crystal graphs, it was decided to fix the carbon and oxygen atoms of the glycerol backbone during energy minimization.

2. *Topological Recognition of Bonds*

According to the procedure outlined above, we first identify important strong bonds between neighboring TAG molecules followed by the weaker next-nearest-neighbor bonds. Anticipating the presentation and detailed discussion of precise results of the calculation of overall bond energies between neighboring molecules, we first start to identify bonds between β-2 TAG molecules within the β-2 TAG structure, topologically. Following Refs. 33 and 34, the various bonds are labeled p, q_i, r_i, s_i, and t_i. It is easy to first reduce the alkane chains of legs and

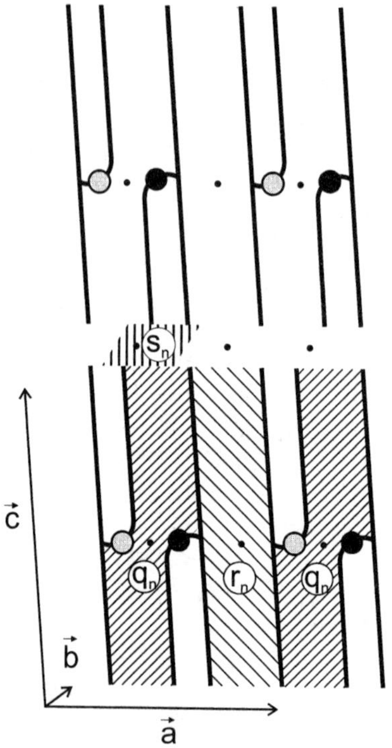

Fig. 7a Topological recognition of various bonds. The interaction areas are hatched. The small dots represent the inversion centers.

back of TAG molecules to sticks and the glycerol bases to curved lines (compare Figs. 5a, 7a). Now three types of areas of interaction between adjacent molecules can be distinguished, namely,

1. Base, leg and back-base, leg and back areas indicated with $q_{1,2}$
2. Back and leg-back and leg areas, indicated with $r_{1,2}$
3. Head-tail contacts, indicated with $s_{1,2}$ areas, corresponding to mutual contacts between ends of backs and legs of molecules located in adjacent (001) layers (Figs. 5a and 7a)

We refer here to Fig. 7a, where alkane chains are reduced to sticks and the complete molecules to a kind of tuning fork.

Now in Figs. 5a and 7a, TAG molecules, located in adjacent unit cells, are stacked atop each other in the direction of the *b*-axis, seen end on. Two neighboring TAG molecules lying atop each other show a close contact of two legs and one back. These close contacts altogether yield the strongest overall bond p of the crystal graph to be discussed below. This bond is parallel to the b-axis and is the only bond of the crystal graph, which corresponds to one full translational distance b.

For the areas $q_{1,2}$ and $r_{1,2}$, we note that any TAG molecule is in reference to an adjacent TAG molecule, lying upside down, due to the center of symmetry, exactly in the middle of the two TAG molecules (Fig. 7a). In the direction of the b-axis, seen end on, two adjacent molecules are, roughly speaking, situated at heights which differ roughly by $\frac{1}{2}\,b$, i.e., half of the unit cell, seen in the b direction. This implies that a reference TAG molecule "feels" an overall first nearest neighbor q and r bond upward and an overall first nearest neighbor q and r bond downward, respectively. These four bonds, together with two p bonds, i.e. $+b$ and $-b$, make a sixfold coordination of a single TAG molecule.

If the possibility of next-nearest-neighbor bonds is taken into consideration, then at distances of about 1.5b two next-nearest-neighbor bonds—$r_{0,3}$ upward and also $q_{0,3}$ downward—can be distinguished. In order to have a complete crystal graph from which all kinds of orientations (hkl) can be constructed, relatively weak overall head-tail bonds between molecules of adjacent layers need to be introduced, namely bonds s_1 and s_2, respectively (Fig. 7a). These bonds are all much smaller than the five nearest-neighbor bonds.

3. *Vector Representation of Bonds*

In order to describe the (overall) bonds between TAG molecules, we label, following the usual convention (see also Ref. 34), the centers of gravity of the TAG molecules by the translational distance, represented as a vector in reference to the selected reference molecule within the unit cell: 1[000] (Fig. 7b). The bonds are then indicated as vectors spanning the distance between the reference center of gravity 1[000] and the center of gravity of molecule $j[uvw]$, indicated simply as

$$1.j\text{-}[uvw] \tag{9}$$

With u, v, w integer, $[uvw]$ represents the translation vector $\mathbf{t}$:

$$\mathbf{t} = u\mathbf{a} + v\mathbf{b} + w\mathbf{c} \tag{10}$$

in reference to the point 1([000]). For each TAG structure, a list of intermolecular interaction energies was calculated by subtracting the energy of the molecules at infinite distance from the energy of the two molecules at the distance $[uvw]$. The interaction energies are presented in Table 2 and will be discussed here.

In Table 2 the bond labels are given in the first column. In the second column the labels according to (9), and in the other columns the calculated bond energies in kcal/mol, for the range of $n.n.n$ of β-2 TAG structures, with $10 \leq n$

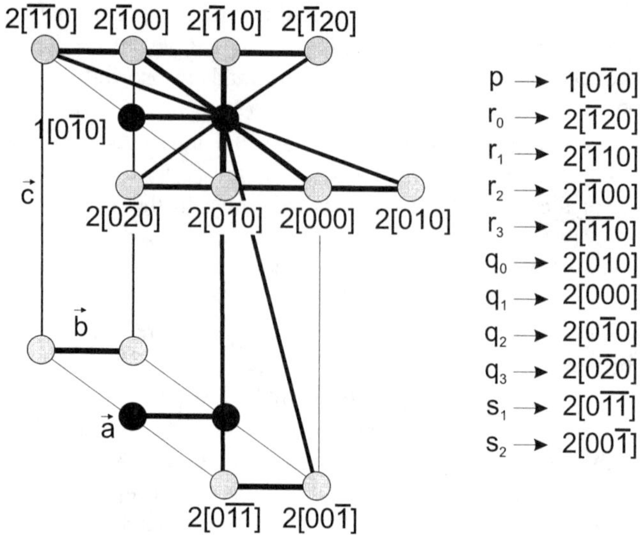

Fig. 7b Crystal graph 2. The bond notation is explained in the text.

≤ 22, can be found. All bond energies, presented as positive broken bond energies depending on n, are plotted in Fig. 8. The order of bond energies and to a high extent the absolute values of the β-2 TAG structure energies, presented in Table 2 and Fig. 8, agree quite well with the order of bond energies published seven years earlier [34]. As mentioned, these values were calculated in a classical way, using semiempirical potentials.

Note that only bonds of types p, q, and r depend on the chain length n, while the s bonds are independent (Table 2 and Fig. 8). This can easily be understood because, when the TAG molecules are elongated in Fig. 7a, the interaction areas corresponding to p, q, and r increase while the area s remains constant. The n-dependent bonds are the overall body-body bonds between TAG molecules situated within the strong layers, parallel to the (001) faces. Since these bonds are coplanar with the (001) orientation, the last index $w = 0$. These bonds are to a high extent caused by the mutual interaction of the alkane chains. This explains that these bond energies increase almost exactly linearly with chain length n of the $n.n.n$ TAG molecules.

The bonds independent of n are the bonds s_i and t_i, which belong to the interactions between a TAG molecule in a given (001) layer and an adjacent TAG molecule in an adjacent (001) layer. This explains the last index $w = \pm 1$. These bond energies are caused primarily by "head-tail contacts," due to van der Waals interactions between CH_3 groups, situated at the ends of legs and backs

Table 2 Different Bond Strengths in kcal/mol for the Various TAGs[a]

	$CM_1.CM_2$-[uvw]	10.10.10	12.12.12	14.14.14	16.16.16	18.18.18	20.20.20	22.22.22
p	1.1-$[0\bar{1}0]$	−21.7	−25.4	−29.0	−32.6	−36.3	−39.8	−43.4
q_0	1.2-$[010]$	−0.6	−0.8	−1.0	−1.2	−1.3	−1.5	−1.7
q_1	1.2-$[000]$	−14.4	−17.6	−20.8	−24.0	−27.1	−30.1	−33.5
q_2	1.2-$[0\bar{1}0]$	−20.7	−23.3	−25.9	−28.6	−31.3	−34.0	−36.8
q_3	1.2-$[0\bar{2}0]$	−1.1	−1.3	−1.5	−1.7	−1.8	−2.0	−2.2
r_0	1.2-$[\bar{1}20]$	−1.1	−1.2	−1.4	−1.6	−1.7	−1.9	−2.1
r_1	1.2-$[\bar{1}10]$	−16.1	−18.8	−21.5	−24.3	−27.1	−29.9	−32.8
r_2	1.2-$[\bar{1}00]$	−19.8	−22.9	−26.0	−29.1	−32.2	−35.3	−38.5
r_3	1.2-$[\bar{1}\bar{1}0]$	−1.2	−1.4	−1.6	−1.7	−1.9	−2.1	−2.3
s_1	1.2-$[0\bar{1}\bar{1}]$	−2.5	−2.5	−2.5	−2.5	−2.5	−2.5	−2.6
s_2	1.2-$[00\bar{1}]$	−2.9	−2.9	−2.9	−2.8	−2.8	−2.8	−2.8
q_4	1.1-$[100]$	−1.1	−1.3	−1.6	−1.8	−2.0	−2.3	−2.5
t_1	1.2-$[\bar{1}11]$	−1.1	−1.0	−1.0	−1.0	−1.1	−1.1	−1.1
q_5	1.1-$[1\bar{1}0]$	−0.4	−0.5	−0.6	−0.7	−0.8	−0.9	−0.9
q_6	1.1-$[110]$	−0.4	−0.5	−0.6	−0.6	−0.8	−0.9	−0.9
s_3	1.1-$[10\bar{1}]$	−0.7	−0.7	−0.7	−0.7	−0.7	−0.7	−0.7
s_4	1.1-$[00\bar{1}]$	−0.6	−0.6	−0.6	−0.6	−0.6	−0.6	−0.6
t_2	1.2-$[\bar{1}01]$	−0.6	−0.6	−0.6	−0.6	−0.6	−0.6	−0.6

[a] Bond strengths calculated using the CVFF force field [53].

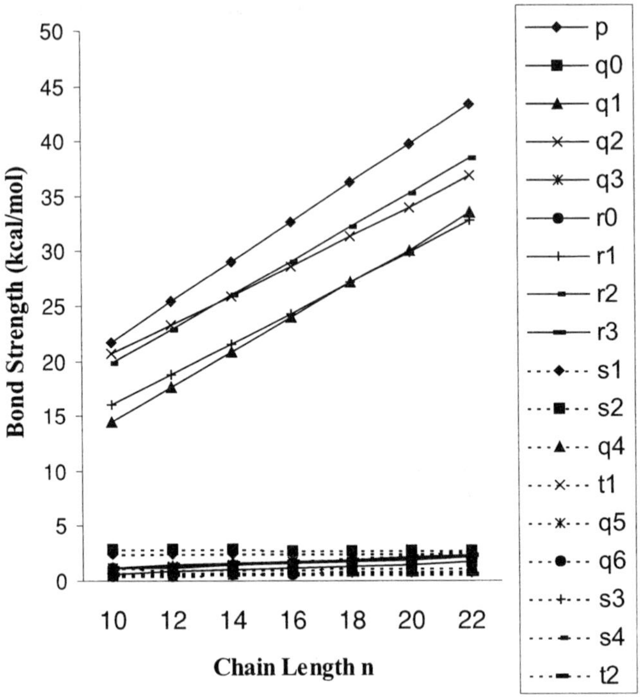

Fig. 8 Bond strength as a function of the chain length n for all bonds of crystal graph 3.

with the same CH_3 groups of neighboring molecules in adjacent (001) layers. It can be seen from Table 2 and Fig. 7b that the bonds $s_{1,2}$ can be considered as a kind of first-nearest-neighbor vertical bonds and the bonds $s_{3,4}$ and $t_{1,2}$ as next-nearest-neighbor bonds. The latter bonds have not been drawn in Fig. 7b.

D. Choice of Bonds and Alternative Crystal Graphs

On our way to the final crystal graph, we next reduce the chairlike TAG molecules to centers of mass or gravity. In addition, the "bond areas" p, q, r, and s are reduced to strings connecting two centers of gravity. We thus obtain the crystal graph, seen in perspective, presented in Fig. 7b. From the crystal graph we construct the connected nets. Note that the precise location of the centers of mass or gravity of a crystal graph is not critical in the search for connected nets. Within this context one can always go back and forth from representations of connected nets, limiting a certain crystal surface with an orientation (hkl), of centers of gravity to representations of connected nets with all kinds of molecules. To calculate Ising temperatures θ^c for a connected net or corresponding substitute net for an orientation *(hkl)*, the actual bond energies are always used.

From Table 2 and Fig. 7b we propose the following three alternative crystal graphs, characterized by different sets of bonds:

1. A crystal graph only consisting of the set of first-nearest-neighbor bonds: p, $q_{1,2}$, $r_{1,2}$, and $s_{1,2}$
2. A crystal graph like the preceding one but also consisting of the next-nearest-neighbor lateral bonds r and q, so the set of bonds becomes p, $q_{0,1,2,3}$, $r_{0,1,2,3}$, and $s_{1,2}$
3. A crystal graph with the set of all bonds of Table 2: p, $q_{0,1,2,3,4,5,6}$, $r_{0,1,2,3}$, $s_{1,2,3,4}$, and $t_{1,2}$

Sometimes the subscript i is used in the following to indicate that the whole family of bonds is meant, which depends on the crystal graph considered; i.e., q_i denotes $\{q_0, q_1, q_2, q_3\}$ for crystal graph 2 and $\{q_1, q_2\}$ for the small crystal graph 1. Although in general the problem of the choice of bonds for a given crystal graph is not discussed explicitly, this choice offers in fact a problem. Mostly a lower bound for the set of bond energies is chosen at say the value kT, where T is the crystallization temperature. Some rules of thumb are given below that helped us to choose which lower bound to take.

The bond energies, calculated with Cerius^2, are referenced to vacuum and refer to the bond energies ϕ^{ss} of Eqs. (2) and (3). The actual broken bond energy ϕ results not only from ϕ^{ss} but also from the ''compensating'' solid-fluid interaction energy ϕ^{sf} and the fluid-fluid interaction energy ϕ^{ff}. This makes the broken bond energy ϕ lower than the calculated bond energy ϕ^{ss}.

As in previous papers [1,33,34] we use the ad hoc recipe of the proportionality hypotheses. Using the proportionality hypothesis implies that we assume we know the ratio of bond energies but that absolute values of the bond energies are not known. Together with the previous point, this implies that we assume that the interaction of the fluid with the crystal face is isotropic, i.e., identical for each face $\{hkl\}$. In our case, we will relax this condition by applying it only to certain sets of crystal faces.

In more or less complex connected nets a few extra weak bonds may still influence the ''connectedness'' of a connected net and, hence, the resulting relative Ising or roughening temperature. However, note that the smaller the bond energy, the smaller will be the influence on the Ising temperature.

In our analysis, it turned out that the extra information of the full crystal graph compared to the second crystal graph was negligible. We therefore only analyze the implications of crystal graphs 1 and 2, respectively. Occasionally, however, we look back to some implications of the full crystal graph 3.

It will be shown that to consider crystal graph 1 as a special case of the more encompassing crystal graph 2, is quite fruitful, both from a topological point of view and from a statistical mechanical point of view.

E. Determination of Connected Nets

We will apply three methods to determine the connected nets of crystal graphs 1 and 2 to analyze their mutual relations of the logically derived results from these graphs. These methods are

1. Visual inspection, leading to shapes of connected nets and their mutual relations in case more connected nets occur for one orientation (*hkl*)
2. Search for combinatorial principles, which will allow us to calculate a priori the number of connected nets and check the mutual relations of a set of connected nets
3. Determination of all connected nets by the FACELIFT software [24]

All connected nets will be categorized in two different zones, the [010] zone and other zones. Considering the needle-like habit of the fat molecules, this corresponds to the large basal face and elongated side faces, on one hand, and the small top faces, on the other hand.

1. *Connected Nets of the Zone [010]*

In Fig. 7b crystal graph 2 is presented and in Fig. 9a the connected net (001) seen from above is shown. First, the PBC [010], consisting of the strongest bond *p* of the crystal graphs having the length of the translation *b*, can be recognized.

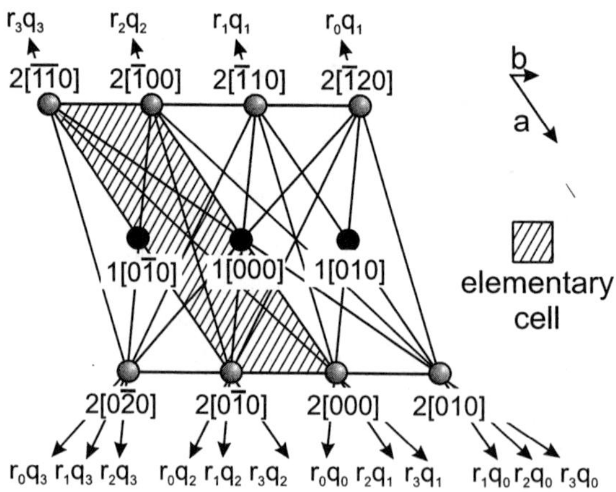

Fig. 9a [001] projection of crystal graph 2. All PBCs [*uv0*] are indicated.

Next all q and r PBCs $[uv0]$ which are situated in the connected net (001) are indicated. In Fig. 9a, 16 different PBCs can be observed, which are formed by a combination of one q and one r bond from crystal graph 2. Note that although 16 different PBCs can be constructed ($4q \times 4r = 16$), only seven different directions ($[1v0]$, where $-4 \leq v \leq 2$) can be made.

Figure 9a corresponds to three planar unit cells of this connected net (001), seen from above. Due to the first-nearest-neighbor strongest bonds p, q_i, and r_i, the connected net (001) will be the connected net with the highest slice energy. It turns out also to have the highest Ising temperature of all connected nets. Hence, the two faces of the form {001} will be dominant on the growth form of β-2 TAG crystals. This is indeed the case, as will be shown.

In order to have a closer look at the strongest connected nets parallel to the strongest bond p in the [010] direction, we first present a projection in the [010] direction in Fig. 9b. In this figure, the horizontal lines, referring to the connected net (001), are now seen from aside in the [010] direction. The q_i and r_i areas of crystal graph 2 lying within the connected net (001) can be seen from above in Fig. 9a.

Comparing Fig. 9b with Figs. 7a,b, it can be observed that the "vertical bonds" of Fig. 9b represent the $s_{1,2}$ bonds. So in summary it can be stated that the horizontal lines of Fig. 9b correspond respectively to the "planes" of the bonds q_i and r_i and the vertical lines to the "planes" of the bonds s_i.

Looking at the bricklayer like projection obtained in Fig. 9b, it can be observed that this structure can be divided into only three sets of different mutu-

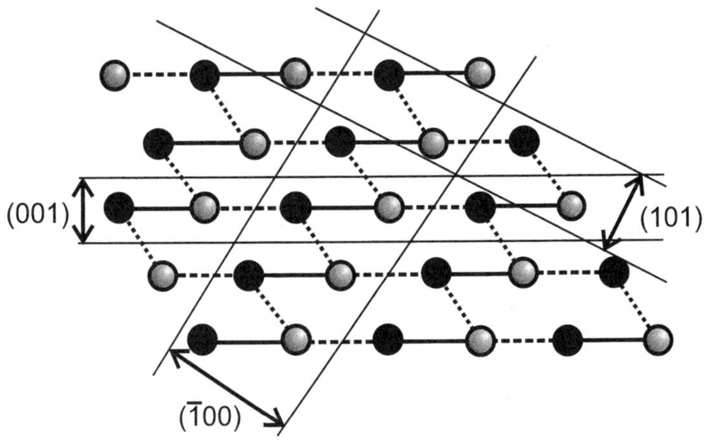

Fig. 9b [010] projection of crystal graph 2. The three connected nets in zone [010] are indicated.

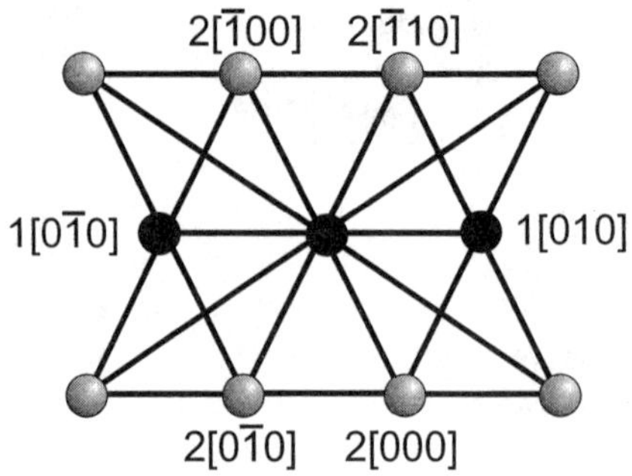

Fig. 9c Top view of connected net (001).

ally equal parallel connected nets, with the proper thickness d_{hkl} corresponding to the connected nets (001), (100), and (101), respectively. The connected net (001) consists, apart from the p bond, seen end on, of the two half-planes of q and r bonds. The connected nets (100) and (101) consist of a strong half of the connected net (001) corresponding respectively to the bonds q_i and the bonds r_i and a weak half consisting of the vertical s_i bonds.

In order to have another look at the three connected nets (001), (100), and (101), these nets seen from above are presented in Figs. 9c,d,e, respectively. Comparing Figs. 7a and 9c, the strong lower half of q bonds and strong upper half of r bonds within the connected net (001) can be seen. Comparing Figs. 7a and 9d,e within the connected net (100), the strong half of q bonds and a weak half of s bonds and within the connected net (101) the strong half with r bonds and the weak half with s bonds can be recognized, respectively.

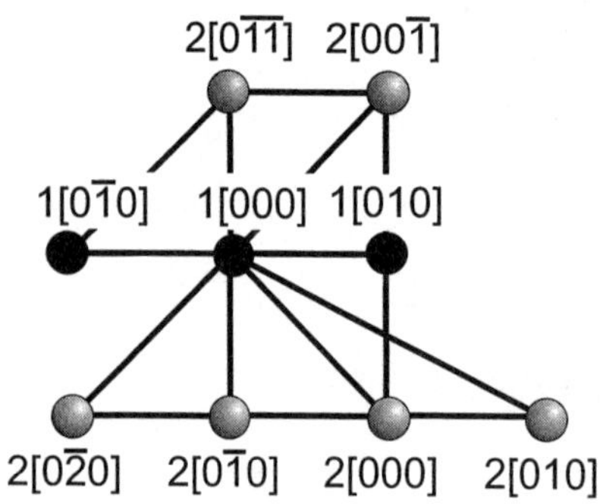

Fig. 9d Top view of connected net (100).

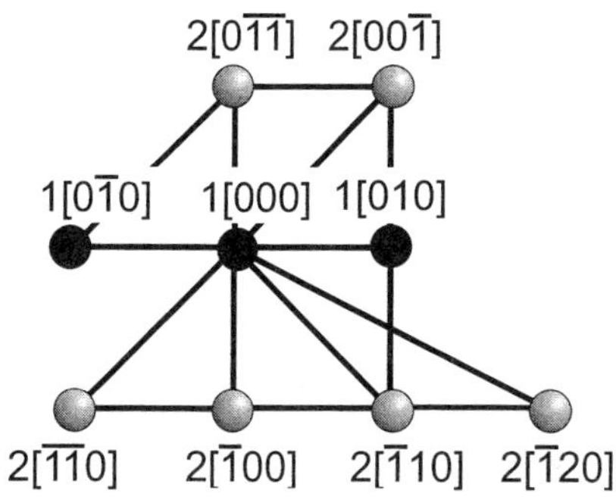

Fig. 9e Top view of connected net (101).

Anticipating a comparison of observed crystal forms and predicted crystal forms, below, note that the combination of strong slowly growing large {001} faces and the faster-growing weaker, and therefore thinner, {100} or {101} faces are indeed observed. This combination leads to the typical ''flat needle'' habit of β-2 TAG crystals.

It is most important, in order to get a complete understanding of growth and morphology, of β-2 TAG crystals, to study the symmetrically equivalent relatively fast-growing faces, occurring at the top and bottom of β-2 TAG crystals. We therefore investigate the topology of connected nets of these faces in the following section.

2. Connected Nets of Faces Not Parallel to the [010] Direction

In order to derive all remaining connected nets of crystal graph 2, we follow the procedure described in Ref. 34. This procedure is based on the principle of combining the 16 lateral PBCs, presented in Fig. 9a, with the vertical bonds $s_{1,2}$, which will lead to $2s \times 4q \times 4r = 32$ different connected nets. The procedure and the resulting connected nets are demonstrated using Table 3.

We will use in the following description of the 32 connected nets the possibility of describing bonds as vectors. Two vectors, corresponding to bonds, are combined to form a new vector (i.e., the vector sum), having a period [*uvw*] of the crystal graph. Such a vector sum corresponds for the present case to a PBC, sometimes called a ''PBC vector.'' It can be seen from Tables 3 and 4 that in case for one orientation (*hkl*) two connected nets can be constructed; these nets are different because they consist of different *q*, *r*, and *s* bonds. Hence, these two connected nets are not related by symmetry. Therefore, these pairs are never symmetry roughened [28], although their roughening temperatures will turn out to be rather low.

Table 3 Combinations of Various PBCs That Show All Possible Top Faces for Crystal Graph 2

	Horizontal PBC		Vertical PBCs				Face
No.	Bonds	$[uvw]_i$	Bonds	$[uvw]_i$	Bonds	$[uvw]_k$	(hkl)
1	q_0r_0	$[1\bar{1}0]_1$	s_1r_0	$[\bar{1}31]$	s_1q_0	$[021]$	$(11\bar{2})$
2	q_0r_0	$[1\bar{1}0]_1$	s_2r_0	$[\bar{1}21]_1$	s_2q_0	$[011]_1$	$(11\bar{1})_1$
3	q_0r_1	$[100]_1$	s_1r_1	$[\bar{1}21]_2$	s_1q_0	$[021]$	$(01\bar{2})$
4	q_0r_1	$[100]_1$	s_2r_1	$[\bar{1}11]_1$	s_2q_0	$[011]_1$	$(01\bar{1})_1$
5	q_0r_2	$[110]_1$	s_1r_2	$[\bar{1}11]_2$	s_1q_0	$[021]$	$(\bar{1}1\bar{2})$
6	q_0r_2	$[110]_1$	s_2r_2	$[\bar{1}01]_1$	s_2q_0	$[011]_1$	$(\bar{1}1\bar{1})_1$
7	q_0r_3	$[120]$	s_1r_3	$[\bar{1}01]_2$	s_1q_0	$[021]$	$(\bar{2}1\bar{2})$
8	q_0r_3	$[120]$	s_2r_3	$[\bar{1}11]$	s_2q_0	$[011]_1$	$(\bar{2}1\bar{1})$
9	q_1r_0	$[1\bar{2}0]_1$	s_1r_0	$[\bar{1}31]$	s_1q_1	$[011]_2$	$(21\bar{1})$
10	q_1r_0	$[1\bar{2}0]_1$	s_2r_0	$[\bar{1}21]_1$	s_2q_1	$[001]_1$	$(210)_1$
11	q_1r_1	$[1\bar{1}0]_2$	s_1r_1	$[\bar{1}21]_2$	s_1q_1	$[011]_2$	$(11\bar{1})_2$
12	q_1r_1	$[1\bar{1}0]_2$	s_2r_1	$[\bar{1}11]_1$	s_2q_1	$[001]_1$	$(110)_1$
13	q_1r_2	$[100]_2$	s_1r_2	$[\bar{1}11]_2$	s_1q_1	$[011]_2$	$(01\bar{1})_2$
14	q_1r_2	$[100]_2$	s_2r_2	$[\bar{1}01]_1$	s_2q_1	$[001]_1$	$(010)_1$
15	q_1r_3	$[110]_2$	s_1r_3	$[\bar{1}01]_2$	s_1q_1	$[011]_2$	$(\bar{1}1\bar{1})_2$
16	q_1r_3	$[110]_2$	s_2r_3	$[\bar{1}11]$	s_2q_1	$[001]_1$	$(\bar{1}10)$
17	q_2r_0	$[1\bar{3}0]_1$	s_1r_0	$[\bar{1}31]$	s_1q_2	$[0\bar{0}1]_2$	(310)
18	q_2r_0	$[1\bar{3}0]_1$	s_2r_0	$[\bar{1}21]_1$	s_2q_2	$[0\bar{1}1]_1$	$(311)_1$
19	q_2r_1	$[1\bar{2}0]_2$	s_1r_1	$[\bar{1}21]_2$	s_1q_2	$[001]_2$	$(210)_2$
20	q_2r_1	$[1\bar{2}0]_2$	s_2r_1	$[\bar{1}11]_1$	s_2q_2	$[0\bar{1}1]_1$	$(211)_1$
21	q_2r_2	$[1\bar{1}0]_3$	s_1r_2	$[\bar{1}11]_2$	s_1q_2	$[001]_2$	$(110)_2$
22	q_2r_2	$[1\bar{1}0]_3$	s_2r_2	$[\bar{1}01]_1$	s_2q_2	$[0\bar{1}1]_1$	$(111)_1$
23	q_2r_3	$[100]_2$	s_1r_3	$[\bar{1}01]_2$	s_1q_2	$[001]_2$	$(010)_2$
24	q_2r_3	$[100]_2$	s_2r_3	$[\bar{1}11]$	s_2q_2	$[0\bar{1}1]_1$	(011)
25	q_3r_0	$[1\bar{4}0]$	s_1r_0	$[\bar{1}31]$	s_1q_3	$[0\bar{1}1]_2$	(411)
26	q_3r_0	$[1\bar{4}0]$	s_2r_0	$[\bar{1}21]_1$	s_2q_3	$[0\bar{2}1]$	(412)
27	q_3r_1	$[1\bar{3}0]_2$	s_1r_1	$[\bar{1}21]_2$	s_1q_3	$[0\bar{1}1]_2$	$(311)_2$
28	q_3r_1	$[1\bar{3}0]_2$	s_2r_1	$[\bar{1}11]_1$	s_2q_3	$[0\bar{2}1]$	(312)
29	q_3r_2	$[1\bar{2}0]_2$	s_1r_2	$[\bar{1}11]_2$	s_1q_3	$[0\bar{1}1]_2$	(212)
30	q_3r_2	$[1\bar{2}0]_2$	s_2r_2	$[\bar{1}01]_1$	s_2q_3	$[0\bar{2}1]$	$(211)_2$
31	q_3r_3	$[1\bar{1}0]_4$	s_1r_3	$[\bar{1}01]_2$	s_1q_3	$[0\bar{1}1]_2$	(112)
32	q_3r_3	$[1\bar{1}0]_4$	s_2r_3	$[\bar{1}11]$	s_2q_3	$[0\bar{2}1]$	$(111)_2$

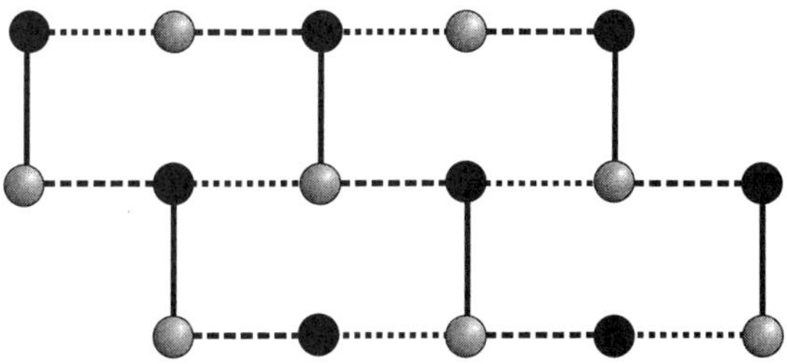

Fig. 9f Brick-layer-like structure of the 32 connected nets of the top faces.

In Fig. 9f the characteristic structure of all 32 connected nets, resulting from 4*q*, 4*r*, and 2*s* bonds, is presented. It can be seen from this schematized presentation, seen from above, that again a brick-layer-like structure results. Seen in perspective, the 32 connected nets, summarized in Tables 3 and 4, have the shape of a tripod (Fig. 9g). In case two connected nets $(hkl)_{1,2}$ occur for one orientation (*hkl*) these together have the shape of a tripod and a different upside down contratripod. This is also demonstrated in Fig. 9g.

In passing we want to stress that in this case of the rather simple crystal graph of the β-2 TAG structure as shown in Fig. 7b, all 35 connected nets belonging to graph 2 discussed above can be derived simply by visual inspection and applying combinatorial principles (see also Ref. 34). Previous tedious ''manual work of visual inspection'' to determine the connected nets of the crystal graph

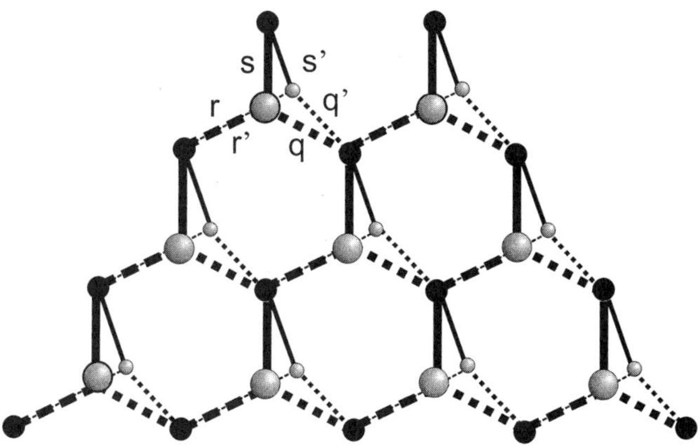

Fig. 9g Tripod-antitripod configuration of a pair of connected nets for a top face.

of the β-2 TAG structure, thanks to the FACELIFT software [24], recently implemented in the Cerius^2 software [25], can now be done using a computer.

3. *Classification*

Looking at Fig. 9b and Table 3, it can be seen that for the crystal graph 2 no other connected nets occur than the three connected nets parallel to the [010] direction and the 32 not parallel to this direction. According to the needle-like habit, we can now make the following classification in three subsets of 35 connected nets (repeating the classification above):

1. Three connected nets containing the [010] PBC
   - One connected net concerning the large basal face, i.e., (001)
   - Two connected nets concerning the side faces, i.e., (100) and (101)
2. Thirty-two connected nets not parallel to the [010] PBC
   - Fourteen connected nets (*hkl*) occurring as single nets for the orientation (*hkl*)
   - Eighteen connected nets (*hkl*) occurring in nine pairs of connected nets $(hkl)_1$ and $(hkl)_2$ for the same orientation (*hkl*)

Analogous to the situation introduced in Sec. II.D, each of the nine pairs of connected nets will have to be replaced by one substitute net. This leads to a reduction of the set of 32 connected nets to $32 - 9 = 23$ different faces $\{hkl\}$ which can be found in Table 4. It is also interesting to look at the connected nets, which belong to the crystal graph 1, having, as defined above, the next-nearest-neighbor bonds $q_{0,3}$ and $r_{0,3}$ omitted. Looking at Table 4, it is clear that now from the original set of 32 connected nets, only $2q \times 2r \times 2s = 8$ connected nets survive. This subset of connected nets is printed in bold in Table 4. Within this set only one combination forms a pair of two connected nets for the orientation (110). Therefore, this crystal graph leads to seven different faces $\{hkl\}$ outside the zone [010].

F. Calculation of Ising Temperatures

On our way to a calculation of physically relevant Ising temperatures, we now discuss the physical implications of the topological analyses leading to the four types of connected nets. The calculation of Ising temperatures is the last station on our way to predicting the morphology of β-2 TAG crystals.

1. *Ising Temperature of Connected Net (001)*

From Fig. 9a it can be seen that the simplest way to calculate the Ising temperature θ^c of the (001) connected net is to omit the weakest next-nearest-neighbor bonds $q_{0,3}$ and $r_{0,3}$. This results in a hexagonal-like connected net. Applying Onsager's original approach to a hexagonal net with a given ratio of bond energies, an exact (relative) Ising temperature can be calculated [6]. For an exact calculation of Ising temperatures, see Ref. 9, and for a general method for all kinds of connected nets see Refs. 8 and 1. Note that by omitting the bonds $q_{0,3}$ and $r_{0,3}$, we consider the connected net (001) as a connected net belonging to crystal graph 1, discussed in Sec. III.E. This procedure was applied in Ref. 34. We note that there the way to calculate the Ising temperature of this face was based only on Kossel-like edge energies. This implies that no small edge energies were found that include bonds from F^{att} (= $\{s_1, s_2\}$). However, looking now at the [010] projection of the crystal graph in Fig. 9b, within the connected net (001) in exactly the same way as discussed in Sec. II.D, Kossel-like paths and non-Kossel-like paths can be distinguished, respectively. Therefore, this approach was not totally correct, because also for this topology an edge energy can be constructed in the [100] direction consisting of broken q_i and r_i bonds reduced with the broken s_i bonds. However, as can be seen from the relative energies in Table 2, the Ising temperature of, e.g., the (001) face of 16.16.16 decreases from 15,900 to 14,600 K if these difference bonds are included. For the time being this justifies the classical way of calculating the relative Ising temperature for the (001) connected net. Note that the Ising temperatures are expressed in kelvins, referenced to vacuum. The actual roughening temperature will, however, be lower, due to the neglected effect of the solvent.

2. *Ising Temperatures of Connected Nets (100) and (101)*

Looking now again at the [010] projection of Fig. 9b, it can be seen that for the (100) and (101) connected nets in essence the same situation occurs as for the two alternative connected nets $(001)_{1,2}$ described in Sec. II.D. There it was shown that in order to obtain the brick-layer-like projection as in Fig. 9b one bond must vanish. In our case this is the weak t_i bond. The formalism mentioned can directly be applied to the configuration of Fig. 9b. It can be seen that for the connected net (100) the edge energy for steps parallel to the [010] direction for crystal graphs 1 and 2 is given by

$$\varepsilon_{[100]} = (\phi_{qi} - \phi_{ri} + \phi_{si}) \tag{11a}$$

and for the connected net (101) the corresponding edge energy is

$$\varepsilon_{[100]} = (\phi_{ri} - \phi_{qi} + \phi_{si}) \tag{11b}$$

Similar to the (001) face, in Ref. 34 the effect of difference bonds was not taken into account. Again, the Ising temperatures drop, now, from 5470 to 3366 K for {101} and from 5480 to 4041 K for {100}, again calculated for 16.16.16. The drop in temperature is not extreme because the edge energy in the [010] direction remains the same and ϕ_p mainly determines the Ising temperature. Note that in this case the Ising temperature is sensitive for the crystal graph used. For the small crystal graph, the Ising temperatures for {100} and {101} are, respectively, 3324 K and 4041 K, while for the third crystal graph these are 4041 K and 3366 K, respectively, (again for 16.16.16). Note that the order is switched between these two faces.

3. *Ising Temperatures of the 14 Single and 9 Double Connected Nets (hkl)*

Recall that the 32 connected nets not parallel to the [010] zone consisted of a set of 14 single connected nets and 9 substitute connected nets as depicted in Table 4. In case a tripod and an antitripod occur for one orientation (*hkl*), the rules for the construction of the substitute net appear to be straightforward, again analogous to the case of Sec. II.D. Again we will, according to Ref. 34, apply the principle of defining differences of bond energies, which were so far applied to one crystallographic direction or better PBC only, to each of the three pairs of PBCs belonging to the pairs of tripods. Note that the difference interactions needed for the top faces are always made up from identical types, i.e., s, q, or r. However, the picture of only a tripod and an antitripod is not sufficient, because we have neglected the effect of oblique bonds until now. Consider now each tripod-antitripod projected from above as a star with three radii (an s-, r-, and q-radius), as depicted in Fig. 9g. In this projection we can distinguish three different PBC directions, made up from two star radii, i.e., s-r, r-q, and s-q. For each top face and connected nets these PBCs are given in Table 4. Now the oblique bonds are the cases when, in addition to the PBCs that make up the connected nets, an alternative PBC can be found. For example, q_1-r_1 and q_2-r_2 are such alternatives for the direction [110].

To calculate the Ising temperatures of the top face {*hkl*}, the island with the smallest step free energy is chosen. The calculation is done by performing four steps. Because all top faces have a similar topology of tripod-antitripod, we calculate the Ising temperature of the {110} face of 16.16.16 as an example. We use the smallest data set of only seven bonds.

1. *Determination of the PBCs parallel to the face {hkl}*. This implies that for each PBC the condition $[uvw] \cdot \{hkl\} = 0$ must be fulfilled. From Table 3 the following six PBCs are found: q_1r_1, q_2r_2 in the $[1\bar{1}0]$ direction, s_2r_1, s_1r_2 in the $[\bar{1}11]$ direction, and s_2q_1, s_1q_2 in the [001] direction.

2. *Constructing the graph in the most probable configuration*. For our example, the bonds found in Table 3 make up a complete tripod-antitripod topology as depicted in Fig. 9g. Here the lower "slice" tripod is formed by the following set of bonds: $F^{\text{slice}} = \{s_1, q_2, r_2\}$, while the antitripod is given by $F^{\text{att}} = \{s_2, q_1, r_1\}$. As can be seen in Table 4, the former tripod has the highest slice energy

and is therefore assumed to be the connected net that makes up the slice from a classical viewpoint.

3. *Determining the profiles from the graph.* In Fig. 9f a brick-layered pattern is shown, from which the various steps that make up the 2D nuclei can be found. By inspection, we found that the cheapest island is a small square-like island made up from the two cheapest nonparallel steps. Constructing different steps, which do not have to be straight, six possible candidates are found: cutting only identical types of bonds *q-q*, *r-r*, and *s-s* or an alternating sequence of two different bond types: *q-r*, *s-r*, and *s-q*. Note that for the first three profiles, one difference bond and a complete bond are cut. So we have to find the cheapest two profiles, or

$$\{\Phi_1, \Phi_2\} = \tfrac{1}{2}\min[(s + \Delta s), (r + \Delta r), (q + \Delta q), (\Delta s + \Delta q), (\Delta s + \Delta r), (\Delta r + \Delta q)] \quad (12)$$

This leads in our example to $\Phi_1 = \tfrac{1}{2}(s + \Delta s) = 5.4$ kJ/mol and $\Phi_2 = \tfrac{1}{2}(\Delta s + \Delta q) = 20.5$ kJ/mol.

4. *Calculating the Ising temperature.* Considering this rectangular island, we can use the following equation used by Onsager [6] and Rijpkema [8] half a century ago:

$$\sinh\left[\frac{\Phi_1}{RT_1}\right]\sinh\left[\frac{\Phi_2}{RT_1}\right] = 1 \quad (13)$$

Solving this equation, we find $T_1 = 780$ K.

For all triangular substitute connected nets the Ising temperatures dependent on chain length *n* of the alkanes within the TAG were calculated. In Table 4, column 7, 23 calculated Ising temperatures (referenced to vacuum) corresponding to 14 original single connected nets and 9 substitute nets, respectively, are presented for the case of 16.16.16. So far we discussed Ising temperatures of crystal graph 2. For the simpler crystal graph 1 only seven faces $\{hkl\}$ survive, and they are boldface in Table 4.

4. *Ising Temperatures of 26 Connected Nets Dependent on n*

In Fig. 10a the Ising temperatures calculated for the various faces considering the crystal graph 2 are plotted as a function of *n*, the alkane chain length of the TAGs. In Fig. 10b an identical plot is presented for the small crystal graph. Considering Fig. 10a, it can be seen that within the set of the 23 top faces, essentially two different subsets of connected nets can be distinguished. Namely, faces of which the Ising temperature strongly depends on *n* and faces of which the Ising temperature hardly depends on *n*. In the following, these two types of faces

Table 4 Bond Strengths and Ising Temperatures Calculated for the Various Top Faces Found Using All Bonds in Table 2[a]

Zone	Conn. net	Bonds	E^{att} (16.16.16) [kcal/mol]	Ranking E^{att}	T-Ising (16.16.16) [K]	Ranking T-Ising	T-Ising dependence (W/S)
$[1\bar{4}0]$	{412}	$s_2q_3r_0$	89.7	25	1332	13	W
	{411}	$s_1q_3r_0$	89.9	28	1677	6	W
$[1\bar{3}0]$	{312}	$s_2q_3r_1$	78.4	17	573	16	S
	{311}	$s_1q_3r_1$	78.5	18	433	19	S
		$s_2q_2r_0$	76.3	13	—		
	{310}	$s_1q_2r_0$	76.4	14	350	22	S
$[1\bar{2}0]$	{212}	$s_2q_3r_2$	76.0	9	1630	7	S
	{211}	$\mathbf{s_2q_2r_1}$	**64.9**	**4**	**1781**	**4**	S
		$s_1q_3r_2$	76.1	10	—		
	{210}	$s_2q_1r_0$	65.1	6	1507	10	S
		$\mathbf{s_1q_2r_1}$	**78.6**	**20**	—		
	$\{21\bar{1}\}$	$s_1q_1r_0$	78.7	22	1874	1	S
$[1\bar{1}0]$	{112}	$s_2q_3r_3$	89.7	25	355	21	W
	{111}	$\mathbf{s_2q_2r_2}$	**62.5**	**1**	**886**	**14**	W
		$s_1q_3r_3$	89.8	27	—		

	{110}	$\mathbf{s_2q_1r_1}$	**62.7**	**2**	**780**	**15**	*W*
		$\mathbf{s_1q_2r_2}$	**67.2**	**7**	—		
	{$11\bar{1}$}	$s_2q_0r_0$	90.0	30	1499	11	*W*
		$\mathbf{s_1q_1r_1}$	**67.4**	**14**	—		
	{$11\bar{2}$}	$s_1q_0r_0$	90.1	32	394	20	*W*
[100]	{011}	$s_2q_2r_3$	76.2	11	1603	8	*S*
	{010}	$\mathbf{s_2q_1r_2}$	**76.4**	**14**	**1797**	**3**	*S*
		$s_1q_2r_3$	64.8	3	—		
	{$01\bar{1}$}	$s_2q_0r_1$	65.0	5	1545	9	*S*
		$\mathbf{s_1q_1r_2}$	**78.6**	**20**	—		
	{$01\bar{2}$}	$s_1q_0r_1$	78.8	24	1848	2	*S*
[110]	{$\bar{1}10$}	$s_2q_1r_3$	78.5	18	559	17	*S*
	{$\bar{1}1\bar{1}$}	$s_2q_0r_2$	78.7	22	462	18	*S*
		$s_1q_1r_3$	76.2	11	—		
	{$\bar{1}1\bar{2}$}	$s_1q_0r_2$	76.4	14	334	23	*S*
[120]	{$\bar{2}1\bar{1}$}	$s_2q_0r_3$	89.9	28	1724	5	*W*
	{$\bar{2}1\bar{2}$}	$s_1q_0r_3$	90.1	32	1453	12	*W*

[a] Boldface entries are those that also appear in the small data set, using only the seven strongest bonds.

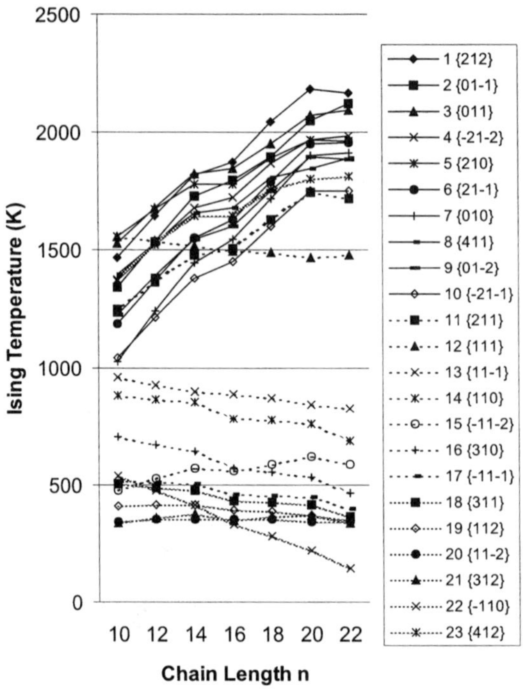

Fig. 10a Ising temperatures as a function of the chain length n for all top faces of crystal graph 2.

will be labeled as faces having the property S(strong dependence) or W(weak dependence), respectively. Note that this distinction is different from the previous one in single and double connected nets.

The conditions for the occurrence of the two sets of S or W faces, respectively, can be explained by the construction of the substitute connected nets. Because for W faces, the substitute nets must also be independent of the chain length n. So we can take a closer look at these substitute nets and derive some general rules for them. First, we look again at the values of the bond energies plotted as a function of n in Fig. 8. It can clearly be seen that the bond energies of the five strongest bonds p, $q_{1,2}$, and $r_{1,2}$ strongly depend on n. The reason is that, as pointed out in Sec. III.C, these nearest-neighbor interactions are mainly due to the alkane chains. The n dependence of the bond energies of the weaker bonds, such as the next nearest bonds $q_{0,3}$ and $r_{0,3}$, is obviously less strong. Note that the bonds s and t, which describe the interactions between {001} layers, are independent of n. Second, the slope of the five strongest bonds is almost identical,

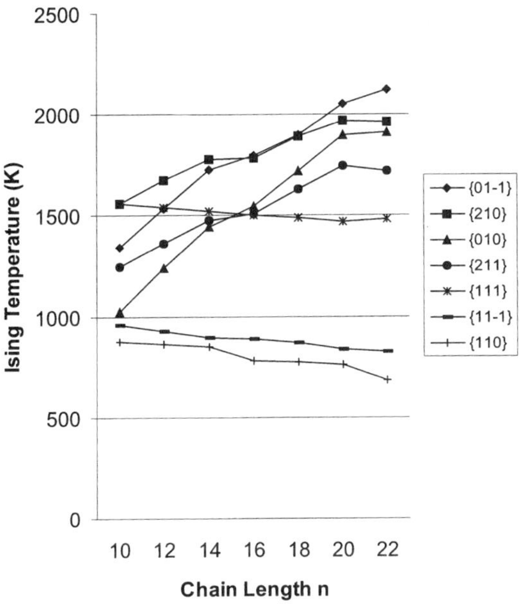

Fig. 10b Ising temperatures as a function of the chain length n for all top faces of crystal graph 1.

which implies that all difference interactions between these bonds are also independent of n. Therefore, substitute nets that are made up from only difference interactions of the five strongest bonds and s and t bonds will hardly change as a function of n. This automatically implies that the Ising temperatures of such faces are also independent of n. We note that these independencies of the chain length only occur for some top faces, because the substitute nets of the faces {001}, {100}, and {101} always include the p interaction, which depends strongly on n. A nice example of the independence of n is that all faces $\{11l\}$ always contain the PBCs q_2-r_2 and q_1-r_1 in the [110] direction. This implies that the substitute net includes the difference interactions $\Delta q = \phi_{q2} - \phi_{q1}$ and $\Delta_r = \phi_{r2} - \phi_{r1}$ which are independent of n. Because all corresponding s interactions that complete the substitute net are also independent of n, all faces $\{11l\}$ will be independent of n.

The results of our theoretical considerations for crystal graph 2 are summarized in Table 4. In addition to the Ising temperatures for the different faces for 16.16.16, the attachment energy E^{att} for each connected net is given. For both the Ising temperature and the attachment energy the result of a ranking, according to increasing Ising temperature or decreasing E^{att}, is given. In the next section a tentative theoretical growth form of β-2 TAG crystals, based especially on data of Table 4, will be presented, and this growth form will be compared to observed growth forms of β-2 TAG crystals.

G. Comparison of Predicted Qualitative Growth Form and Observed Growth Forms

Here we on the basis of the results of the theoretical considerations obtained so far, derive in a qualitative logical way a theoretical growth form of β-2 TAG crystals. We only use the parallel relation [8] between θ_{hkl}^{c} and MI_{hkl}, and we only describe growth forms in a qualitative way. For the concepts we refer to Refs. 1 and 9, and for the application of this approach to β-2 TAG crystals see Refs. 32 and 33.

1. Qualitative Theoretical Growth Form

We are primarily interested in the morphology of the top and bottom of β-2 TAG crystals. Since, however, the most important aspect of the morphology is related to the overall morphology, we first focus on the morphology caused by the three different zones of the flat needle-like habit. We first note that the two faces of the form {001}, both from a more classical attachment energy approach like Ref. 34 and a for a more modern approach including Ising temperatures, the connected net (001) will remain the strongest connected net. Second, the connected nets {100} and {101} will be the next strongest connected nets. Third, the faces of the top and bottom are far less important. From these considerations it can be concluded that β-2 TAG crystals will have the shape of a quite long plank, bounded by two large parallel faces of the form {001} and two thin parallel faces {100} and/or {101}. The edges of the length direction of the plank will be parallel to the strongest PBC {010}, consisting of the strongest *p* bond.

In order to predict the morphology of β-2 TAG crystals for the top faces we will make full use of the information of Table 4. It can be seen that faces lying in zones $[1\bar{2}0]$ and [100], considering the Ising temperatures, will dominate the top of the crystal growth form of β-2 TAG crystals. It can be observed that for zone $[1\bar{2}0]$ the faces of the form $\{21\bar{1}\}$ would be dominant, immediately followed by faces of the form {211}. In the same way it can be seen that for the zone [100] the faces of the form $\{01\bar{2}\}$ will be dominant, immediately followed by the faces of the form {010}.

We have to consider that the Ising temperatures presented in column 7 of Table 4 are calculated for 16.16.16. Figure 10a,b shows that Ising temperatures of faces in zone $[1\bar{1}0]$ remain almost constant with decreasing *n*, while the Ising temperatures of the faces of zones $[1\bar{2}0]$ and [100] decrease rapidly with decreasing *n*. Note that for the smallest TAG, i.e., 10.10.10, the Ising temperatures of the three zones are almost identical. Applying now rule [8] this would imply that for the smallest TAGs also crystal faces of zone $[1\ \bar{1}\ 0]$ may show up on growth forms of β-2 TAG crystals. After these purely theoretical qualitative predictions for the morphology of β-2 TAG crystals, we now turn to the experimental observed habits.

2. *Observed Habits of β-2 TAG Crystals*

Anticipating the discussion of experimentally observed growth forms of β-2 TAG crystals as presented in Fig. 11, we first discuss problems concerning the accuracy of measured Miller indices (*hkl*) of faces on β)-2 TAG crystals. The largest crystals obtained have approximate dimensions of 5 mm × 700 μm × 80 μm. Knowing that the largest face is the {001} face, the angles with the side faces {100} or {101} are rather easy and accurate to obtain with an optical goniometer. Contrary to the three faces of the [010] zone, unfortunately the size of the faces of the forms {*hkl*} of the top or bottom faces are much smaller than the three faces just mentioned. In addition, the quality of these faces is worse compared to the quality of the faces of the [010] zone. This may be caused by the fact that, due

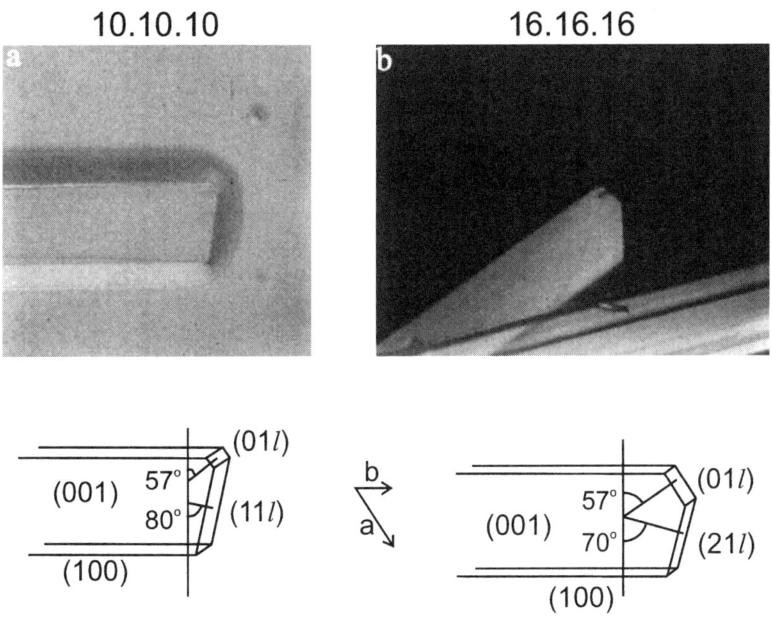

Fig. 11 Experimental growth form for 10.10.10 TAG and 16.16.16 TAG crystals. The faces are indexed in the drawing below.

to strong interaction of the mother phase with these rather weak faces, the separation of mother liquid from the crystal is not possible for the top faces. This is called the shutoff effect. Another problem arises from the fact that fat crystals easily bend. This makes an accurate measurement of, especially, angles of top faces questionable. Note that the different *l*-indices are closely situated to each other, because of the relatively long *c*-axis.

We note from Fig. 11 that the lateral angles can be measured with quite a reasonable degree of accuracy, not only with the help of a goniometer but also with an optical microscope. This makes it possible to determine the indices *h* and *k* of top faces unambiguously. All these problems show that the inaccuracy of the *l*-index is at least $\pm$ 1.

In Fig. 11b, the most general habit of a typical 16.16.16 crystal, grown from a dodecane solution, is presented. The needle-like shape is always present. However, there is a variety in the top faces of the crystals. Here a crystal is presented with both $\{21l\}$ and $\{01l\}$ top faces. The *MI* of these top faces may, however, vary for different crystals and circumstances (low and high supersaturation). For TAGs with long alkane chains similar habits were observed. For these crystals, however, newer faces within the zone [110] were found, which were actually only found for the shorter 10.10.10 crystals. In Fig. 11a, a 10.10.10 crystal is depicted including a $\{11l\}$ top face. This crystal was grown at low supersaturation by slow evaporation from an acetone solution. Note that for these short TAGs $\{01l\}$ and $\{21l\}$ faces have also been reported [34].

3. Comparing the Experimentally Observed and Predicted Faces

We first note that the theoretically expected dominant plank shape can be clearly recognized from both TAGs in Fig. 11. These observations are to a high extent in agreement with our expectations, derived from the theory presented. Even the fact that according to the Ising temperature the $\{100\}$ face would be dominant to the $\{101\}$ face is verified by experiments. Note that although both faces were observed [33], the $\{100\}$ face is observed more frequently; see, for example, Skoda and van den Tempel [54], who never observed this face. The crystallographic setting used in Ref. 54 can be transformed into the one used here, by applying the transformation $\boldsymbol{a} \rightarrow \boldsymbol{a}'$, $\boldsymbol{b} \rightarrow -\boldsymbol{b}'$, and $\boldsymbol{c} \rightarrow \boldsymbol{c}' - \boldsymbol{b}'$. The observed morphology of the TAG crystals can be summarized by the occurrence of $\{01l\}$ and $\{21l\}$ faces for all chain lengths *n*, while for the shorter TAGs sometimes the $\{11l\}$ faces are also observed. This is in agreement with the expectations of the theory as can be seen from Figs. 10a, b. Particularly for the smallest crystal graph it can be seen that for 10.10.10 the Ising temperatures for $\{210\}$, $\{01\bar{1}\}$, and $\{11\bar{1}\}$ are relatively high. Because $\{111\}$ is independent of *n*, only $\{210\}$ and $\{01\bar{1}\}$ remain important for TAGs with a higher chain length *n*. Taking into consideration the error bar $\Delta l = \pm 1$ of the l index and looking at the calculated values of Ising temperatures of Table 4, we conclude that the dominant top and bottom faces are probably {010} and $\{01\bar{1}\}$ for zones [100] and {211} and perhaps $\{21\bar{1}\}$ for zone $\{1\bar{2}0\}$.

In Skoda and van den Tempel [54], Skoda et al. [55], and Hollander et al. [33] it was reported that for β-2 TAG crystals were grown from a dodecane or trioleate solutions at low supersaturations, faces of the form {311} lying in zone [1–30] were quite often observed. More experimental data of the morphology of β-2 TAGs can be found in Skoda and van den Tempel [54], Skoda et al. [55], and Albon and Parker [56]. The crystallographic setting used in Ref. 55 can be transformed into the one used here by applying the transformation $\boldsymbol{a} \rightarrow \boldsymbol{a}'$, $\boldsymbol{b} \rightarrow -\boldsymbol{b}'$ and $\boldsymbol{c} \rightarrow \boldsymbol{c}' - \boldsymbol{b}'$. Because of the experimental uncertainty of the index l, we refer to these faces as $\{31l\}$. According to Table 4 these faces of zone [1–30] could be {312}, {311}, or {310}. However, these faces are weak according to their ranking in Ising temperatures and therefore have a low *MI*. So from a theoretical point of view, these faces have a very small chance of appearing on the growth forms of β-2 TAG crystals.

We note that the experimental fact of the observation of faces of the form {311} is in conflict with the expectations from our qualitative theory, based among others on a parallel relation between the Ising temperature and the morphological importance of a form $\{hkl\}$. Note that also a classical approach based on the attachment energy fails here, according to the ranking in E^{slice} given in Table 4.

In order to explain the occurrence of faces of the form $\{31l\}$, which are only observed for growth from alkane solutions, dodecane and trioleate, respectively, the hypothesis was introduced in Ref. 55 and also used in Ref. 33, that as a result of a special interaction of the alkane (like) solvent molecules with the surface may occur, due to the availability of suited elongated absorption sites for paraffin-like molecules on surfaces of $(31l)$ faces. The removal of absorbed paraffin molecules by the TAG growth units from $(31l)$ faces during growth may be a time-consuming process. Note that according to Skoda and Van den Tempel [54], these faces grow out easily with increasing supersaturation and give rise to spherulites, which supports this hypothesis.

Very recently real monomolecular growth spirals were observed on (001) faces of β-2 TAG crystals, using an AFM; see Hollander et al. [57]. The spirals have a very thin elongated shape, which can be regarded as a reflection of the bulk morphology on this {001} surface. The length direction is, as could be expected, parallel to the strongest p bond or the corresponding [010] direction. The tops of the elongated spiral steps show some degree of polygonization and it is very interesting to note that apart from this very long step two small pieces of steps can also be distinguished, having the crystallographic directions [100] and $[1\bar{2}0]$, respectively. Analogously straight steps in the [110] direction were not observed. It can be shown that for this direction also difference interactions between the PBCs q_1-r_1 and q_2-r_2 prohibit this step from being polygonized, because according to BCF [4] this step will have a low kink density and therefore a high advance velocity. This observation is in agreement with the results of our theory, because it can be seen from Tables 3 and 4 that after the strongest PBC the next strongest PBCs are the q_1-r_2 ([100]) and q_2-r_1 ([$1\bar{2}0$]) PBCs.

IV. CONCLUSIONS, FUTURE RESEARCH, AND TECHNOLOGICAL IMPLICATIONS

A. Conclusion

In the original Hartman-Perdok, which provides a method to predict growth forms of crystals theory, the concepts of E^{att} and E^{slice} play a key role. Assuming that (in principle) F faces will dominate crystal growth forms, theoretical growth forms are constructed by assuming that the rate of growth R of F faces (hkl) is proportional to E^{att}, the energy needed to remove a growth unit from a flat surface (hkl). The parallel relation between R and E^{att} could be justified to a high extent on the basis of all kinds of crystal growth models [58].

In an earlier extension of the Hartman-Perdok theory (Rijpkema et al. [8]), this theory was footed on the statistical mechanical concept of roughening transition of a crystal face (hkl). In order to calculate the relative roughening temperatures of faces (hkl), relative Ising temperatures of faces (hkl) were used. In order to calculate relative Ising temperatures, the concept of connected net was introduced. These connected nets were treated as independent connected nets, and the strongest net was supposed to be the dominant connected net. This method was used to calculate the Ising temperatures of the connected nets of many structures [1,8,9].

Very recently a new development in the connected net–roughening transition theory took place, because of problems which arose for the case that oblique bonds occur for a crystal surface (hkl) [27–29,31–33]. One of the results is that in such cases the connected nets can be replaced by substitute nets. This result has been applied to explain the morphology of β-2 TAG crystals by calculating the Ising temperatures for the substitute nets. On the basis of these Ising temperatures the shape of β-2 TAG crystals, also dependent on the chain length of alkane chains has been derived in a qualitative way. It turned out that predicted and observed growth forms were in very good agreement with each other. We refer here to Hollander et al. [33].

B. Future Research

In the near future research will be carried out on the growth kinetics of the various faces (hkl), characterized by their single connected nets or substitute connected nets, by recently developed software to carry out Monte Carlo simulations for any type of crystal surface. Then the morphology of β-2 TAG crystals dependent on supersaturation can be studied theoretically in detail.

If in future other reliable structures of crystals of TAG molecules of, among others, β′ structures become available, the same approach as described in this part should be applied in order to understand the morphology of these types of crystals. Hopefully the extreme elongated needle-like shape of some of the β′ structures will then be explained. In addition, using AFM techniques, step patterns and growth spirals should be observed. Also using AFM techniques, pictures of molecules situated within connected nets at crystal surfaces can be studied. Such observations give apart from macroscopic morphological data and data of spirals and monomolecular steps and their structures, independent information of molecular structures of interfaces, suited to check the predictions of theories as presented above.

Research as described could be extended to mixtures of two or, perhaps, more different TAG molecules with alkane chains, which may differ mutually a few C_2H_4 fragments of alkane chains. Carrying out this kind of research may lead to a better understanding of what is going on during technical crystallization processes in complex mixtures of TAG molecules.

C. Technological Implications

The survey of research of crystal growth and morphology of β-2 monotriacylglycerol crystals started more than 35 years ago by Skoda and van den Tempel of the Unilever Research Laboratory at Vlaardingen, the Netherlands [54,55]. In parallel, research on growth and morphology of β-2 monotriacylglycerol crystals, developed over the past three decades, first at the University of Groningen and later a cooperation started between Unilever and the University of Nijmegen [34]. This research is now going on in Nijmegen and is an example of pure fundamental research. This fundamental research does, however, in principle have practical implications for technical crystallization processes.

So first of all, understanding crystal growth mechanisms leading to special crystal habits, like the occurrence of elongated planks or needles, is, among other information, crucial for understanding the rheology of melts of TAG molecules and for carrying out effective separation processes.

We further note that during rapid crystallization from the melt at high undercooling, TAG molecules form crystals, which together form spherulites, consisting of many crystals. The density of the spherulites depends strongly on the length-to-width ratio of the grown TAG crystals, and this depends on the habit of the TAG crystals, which is in turn determined by the structure of the crystals, the mother phase from which the crystals are growing, and the driving force for crystallization. Note that the density of spherulites is an important parameter for the efficiency of separation processes with the aim of separating different TAG molecules with varying alkane chain lengths.

It can be stated that by using technical crystallization, aiming to get the most efficient separation of different TAG molecules from mixtures, using crystallization from the melt or solution, respectively, more and more fundamental knowledge on nucleation, growth, and morphology will be required. In order to get relevant information for crystallization processes occurring during technical crystallization processes, in future more fundamental research should be applied to different β and β' TAG structures, to theoretical and experimental research, concerning the influence of varying driving forces on crystal habits of different types of TAG crystals, and to crystallization of TAG crystals from a limited number of well-defined mixtures of TAG molecules.

REFERENCES

1. P. Bennema, X. Y. Liu, K. Lewtas, R. D. Tack, J. J. M. Rijpkema, and K. J. Roberts, *J. Crystal Growth, 121*: 679–696 (1992).
2. L. A. M. J. Jetten, H. J. Human, P. Bennema, and J. P. van der Eerden, *J. Crystal Growth, 68*: 503–516 (1984).
3. P. Bennema, *Handbook of Crystal Growth*, D. T. J. Hurle, ed., Elsevier, Amsterdam, pp. 477–581, 1993.
4. W. K. Burton, N. Cabrera, and F. C. Frank, *Phil. Trans. R. Soc. London, Ser. A, 243*: 299–358 (1951).
5. F. C. Frank, *Disc. Farad. Soc., 5*: 48–54 (1949).
6. L. Onsager, *Phys. Rev., 65*: 117–149 (1944).
7. P. Bennema, *J. Crystal Growth, 166*: 17–28 (1996).
8. J. J. M. Rijpkema, H. J. F. Knops, P. Bennema, and J. P. van der Eerden, *J. Crystal Growth, 61*: 295–306 (1982).
9. P. Bennema and J. P. van der Eerden, *Morphology of Crystals*, Part A, I. Sunagawa, ed., TERRAPUB, Tokyo, pp. 1–75, 1987.
10. A. A. Chernov and T. Nishinaga, *Morphology of Crystals*, Part A, I. Sunagawa, ed., TERRAPUB, Tokyo, pp. 211–267, 1987.
11. G. H. Gilmer and P. Bennema, *J. Appl. Phys., 43*: 1347–1360 (1972).
12. S. W. H. de Haan, V. J. A. Meeuwsen, B. P. Veltman, P. Bennema, C. van Leeuwen, and G. H. Gilmer, *J. Crystal Growth, 24/25*: 491–494 (1974).
13. H. J. Leamy and G. H. Gilmer, *J. Crystal Growth, 24/25*: 499–502 (1974).
14. H. van Beijeren, *Phys. Rev. Lett., 38*: 993–996 (1977).
15. J. P. van der Eerden and H. J. F. Knops, *Phys. Lett., 66A*: 334–336 (1978).
16. J. D. Weeks and G. H. Gilmer, *Adv. Chem. Phys., 40*: 157–228 (1979).
17. G. Mazzeo, E. Carlon, and H. van Beijeren, *Phys. Rev. Lett., 74*: 1391–1394 (1995).
18. R. F. P. Grimbergen, H. Meekes, P. Bennema, H. J. F. Knops, and M. den Nijs, *Phys. Rev. B, 58*: 5258–5265 (1998).

19. H. E. A. Huitema, B. van Hengstum, and J. P. van der Eerden, *J. Chem. Phys.*, *111*: 10248–10260 (1999).
20. K. Tsukamoto, E. Yokoyama, K. Maiwa, K. Shimizu, R. F. Sekerka, T. S. Morita, and S. Yoda, *J. Jpn. Soc. Microgravity Appl.*, *15*: 2–9 (1998).
21. P. Hartman and W. G. Perdok, *Acta Cryst.*, *8*: 49–52, 521–529, (1955).
22. P. Hartman, *Crystal Growth: An Introduction*, P. Hartman, ed., North-Holland, Amsterdam, pp. 387–402, 1973.
23. P. Hartman, *Morphology of Crystals*, Part A, I. Sunagawa, ed., TERRAPUB, Tokyo, pp. 269–319, 1987.
24. R. F. P. Grimbergen, H. Meekes, and S. X. M. Boerrigter, C-program FACELIFT for connected net analysis, Department of Solid State Chemistry, University of Nijmegen, The Netherlands, 1997; further information see hugom@sci.kun.nl.
25. Cerius^2 User Guide, September 1998. San Diego: Molecular Simulations Inc., 1998.
26. C. S. Strom, *Z. Kristallogr.*, *153*: 99–113 (1980); *154*: 31–43 (1981); *172*: 11–24 (1985).
27. R. F. P. Grimbergen, H. Meekes, P. Bennema, C. S. Strom, and L. J. P. Vogels, *Acta Cryst. A*, *54*: 491–500 (1998).
28. H. Meekes, P. Bennema, and R. F. P. Grimbergen, *Acta Cryst. A*, *54*: 501–510 (1998).
29. R. F. P. Grimbergen, P. Bennema, and H. Meekes, *Acta Cryst. A*, *55*: 84–94 (1999).
30. H. van Beijeren and I. M. Nolden, *Topics in Current Physics, Structures and Dynamics of Surfaces*, II43, Springer-Verlag, Berlin, pp. 259–300, 1986.
31. R. F. P. Grimbergen, M. F. Reedijk, H. Meekes, and P. Bennema, *J. Phys. Chem. B*, *102*: 2646–2653 (1998).
32. R. F. P. Grimbergen, P. J. C. M. van Hoof, H. Meekes, and P. Bennema, *J. Cryst. Growth*, *191*: 846–860 (1998).
33. F. F. A. Hollander, S. X. M. Boerrigter, J. van der Streek, R. F. P. Grimbergen, H. Meekes, and P. Bennema, *J. Phys. Chem. B*, *103*: 8301–8309 (1999).
34. P. Bennema, L. J. P. Vogels, and S. de Jong, *J. Crystal Growth*, *123*: 141–162 (1992).
35. S. de Jong, Ph.D. thesis, University of Utrecht, 1980.
36. S. de Jong and T. C. van Soest, *Acta Cryst. B*, *34*: 1570–1583 (1978).
37. L. H. Wesdorp, Ph.D. thesis, Technical University of Delft, 1990.
38. V. Vand and I. P. Bell, *Acta Cryst.*, *4*: 465–469 (1951).
39. D. Chapman, *Chem. Rev.*, *62*: 433–439 (1962).
40. K. Larson, *Ark. Kemi*, *1*: 1–14 (1964).
41. I. H. Jensen and A. J. Mabis, *Acta Cryst.*, *21*: 770–781 (1966).
42. A. van Langevelde, K. van Malssen, F. Hollander, R. Peschar, and H. Schenk, *Acta Cryst. B*, *55*: 114–122 (1999).
43. A. van Langevelde, K. van Malssen, E. Sonneveld, R. Peschar, and H. Schenk, *J. Am. Oil. Chem. Soc.*, *76*: 603–609 (1999).
44. J. van de Streek, P. Verwer, R. de Gelder, and F. Hollander, *J. Am. Oil. Chem. Soc.*, *76*: 1333–1341 (1999).
45. A. I. Kitaigorodski, *Molecular Crystals and Molecules*, Academic Press, New York, 1973; *Molekül Kristalle*, Akedemie-Verlag, Berlin, 1979.

46. S. C. Nyburg and J. A. Potworowski, *Acta Cryst. B*, *29*: 347–352 (1973).
47. A. E. Smith, *J. Chem. Phys.*, *21*: 2229–2231 (1953).
48. R. Boistelle, B. Simon, and G. Pèpe, *Acta Cryst. B*, *32*: 1240–1243 (1976).
49. Y. V. Mnykh, *J. Phys. Chem. Solids*, *24*: 631–640 (1963).
50. M. G. Broadhurst, *J. Res. Nat. Bur. Stds.*, *66A*: 241–248 (1962).
51. A. I. Kitaigorodskii, K. V. Mirskaya, and V. V. Nauchitel, *Sov. Phys. Cryst.*, *14*: 769–781 (1970).
52. J. A. Pople and D. I. Beveridge, *Approximate Molecular Orbital Theory*, McGraw-Hill, New York, 1970.
53. P. Dauber-Osguthorpe, V. A. Roberts, D. J. Osguthorpe, D. Wolff, M. Genest, A. Hagler, and T. Hagler, *Proteins: Struct. Funct. Genet.*, *4*: 31–47 (1988).
54. W. Skoda and M. van den Tempel, *J. Crystal Growth*, *1*: 207–217 (1967).
55. W. Skoda, L. L. Hoekstra, T. C. van Soest, P. Bennema, and M. van den Tempel, *Kolloid Z. Polym.*, *21*: 149–156 (1967).
56. N. Albon and A. Parker, *Nature*, *207*: 87–88 (1965).
57. F. F. A. Hollander, M. Plomp, J. van de Streek, and W. J. P. van Enckevort, submitted.
58. P. Hartman and P. Bennema, *J. Crystal Growth*, *49*: 145–156 (1980).

4

Nucleation and Growth in the Solid-Solid Phase Transitions of *n*-Alkanes

Koji Nozaki
Yamaguchi University, Yamaguchi, Japan

Masamichi Hikosaka
Hiroshima University, Higashi-Hiroshima, Japan

I. INTRODUCTION

Chain molecules show complicated successive first-order phase transitions, such as crystallization or melting between the melt and crystalline phase, ''liquid crystallization'' from the melt to liquid-crystalline mesophase, and various solid-solid phase transitions [1,2]. Crystallization has been well studied [3], whereas understanding of ''liquid crystallization'' and solid-solid phase transitions is still insufficient and the mechanism has been open to question [4].

Typical phase transition from the mesophase, such as ''rotator phase'' (R phase), to ordered phase is a kind of solid-solid phase transition. The transition is usually called ''rotator phase transition'' (R-phase transition) and is interesting because it is closely related to various physical, chemical,and biological properties, which plays important roles in industrial, biological, and food systems.

The fats and lipids systems are composed of hydrophobic and polar components. The former is usually composed of linear chain hydrocarbons, which have high ability to self-organize due to the van der Waals attractive force between linear chains. This plays an important role in the complicated crystallizations and phase transitions of the fats and lipids systems. The simplest linear chain hydrocarbon molecules are *n*-alkane systems. So the study on phase transitions of *n*-alkane systems should give the basic solution to understand the mechanism of the complicated phase transitions of fats and lipid systems [1].

The rotator phase transitions can be seen very widely in the fats and lipids systems. In the rotator phases, the hydrocarbon chains seems to be ''rotated'' in the sense of average for some long time, which results in the hexagonal packing of the subcell. The hexagonal phases are named differently, for example, α form in the case of triacylglycerols. The rotator phase transition can be seen in the phase transition from α to β forms of the triacylglycerols [5].

With regard to the kinetics of the solid-solid phase transition, there have been few studies, and they are mainly theoretical, not only for chain molecules but also for inorganic atomic systems. This is due to the difficulties of observing the changes directly during the solid-solid transition, because the transition is usually completed within a short time. Recently, several studies on the kinetics of the first-order phase transition, such as martensitic transformation, were reported [6]. For ferroelectric substances a microscopic model of the first-order phase transition through an intermediate state was also proposed [7]. Further investigations must be carried out with new experimental techniques in order to clarify the mechanism of the solid-solid phase transitions.

In recent years, it has become possible to investigate the kinetics of the first-order transition in a solid, due to the development of an experimental technique, as reported for dielectric materials [8,9], metal alloys [10,11], and other low-molecular-weight materials [12–16].

In long-chain materials, such as lipids, oils, and polymers, the rate of transition in solids is considered to be lower than that in low-molecular-weight materials. However, there have rarely been any kinetic studies, so far as the authors know, on the solid-solid phase transition in the long-chain materials.

Another interesting problem is the origin and mechanism of ''hysteresis'' in the R phase transition, which means that the transition temperature shows significant difference between heating and cooling processes.

A. Phase Transitions of Long-Chain Molecules

The rate of the solid-solid phase transition in long-chain compounds and polymer materials is considered to be smaller than that in low-molecular-weight materials. Therefore it is possible to observe the changes during the transition in the long-chain compounds and polymers. Furthermore, the hysteresis becomes more remarkable with increase in molecular weight of the long-chain compounds. For example, in crystalline polymers the observed crystallization temperature on slow cooling is about a few ten of degrees lower than the corresponding melting temperature on heating [17]. In the crystal of a normal higher alcohol, significantly large hysteresis is also observed in the order-disorder type transition [18]. Takamizawa et al. [19] reported that the freezing phenomenon of a high-temperature phase was observed in longer *n*-alkanes when the sample was cooled to room temperature. This is an extreme case, in which the transition temperature is depressed below room temperature.

Linear hydrocarbon molecules, such as lipids, oils, and polymers, often show characteristic R-phase transitions between the low-temperature ordered (LO) phase and the R phase [20,21]. The phase diagram is shown in Fig. 1. In the LO phase, *n*-alkane molecules are fully extended taking *all-trans* conformations and are registrated in layers in which the molecular chain axes are parallel with each other, having long-range order with respect to the orientation about their long axes. In the R phase, the *all-trans* conformation of the molecules is partially deformed. The crystal still has three-dimensional long-range positional order crystallographically, but lacks the long-range order in the orientation about their long axes. Recent X-ray investigations have shown the presence of five R phases which have different structures [22–29].

The molecules of *n*-alkane in the R phase behave as if they were in a liquid crystalline state and have high mobility. The long-range diffusion of the

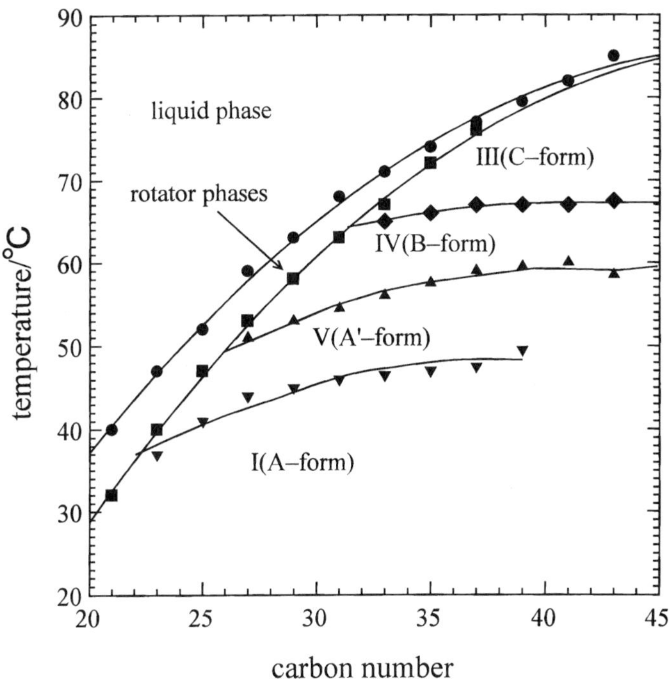

Fig. 1 Phase diagram of odd *n*-alkanes, C_nH_{2n+2}: C*n*. (From Ref. 44)

molecules is observed by means of X-ray powder diffraction [30] and optical microscopy [31,32]. These characteristics are considered to be related to various functions of biological membranes. Hence, in recent years, there has been a growing interest in the R phase.

B. Phase Transitions of *n*-Alkanes

Phase behavior depends on the carbon number, a typical example of which is shown in Fig. 2 on *n*-pentacosane ($C_{25}H_{52}$: C25). The transitions I → V → RV

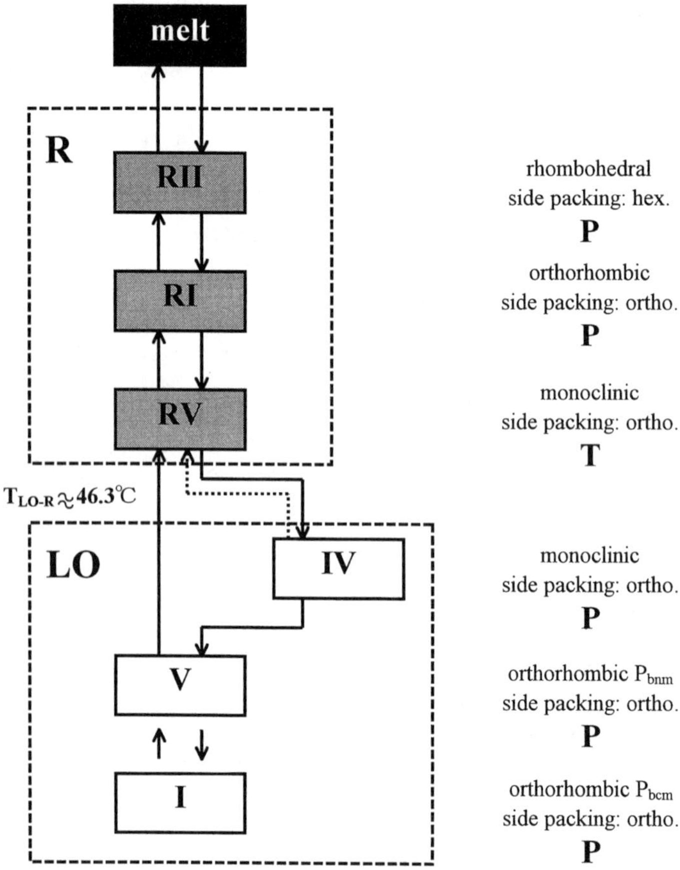

Fig. 2 Schematic of the phase behavior of C25. T and P mean that the molecules are perpendicular to the layer surface and tilt from the normal to the layer surface, respectively. Phases I, V, and IV belong to the LO phases, and the RV, RI, and RII to the R phases. (From Ref. 39)

→ RI → RII and RII → RI → RV → IV → V → I are observed during heating and cooling, respectively [28,33,34], where I, V, and IV are the LO phases, and RV, RI, and RII are the R phases. It is interesting that asymmetric phase behaviors are sometimes observed in chain molecular systems. For example, phase IV appears only on cooling and is not observed on heating. Therefore phase IV should be metastable, although it can be observed for a long time without transforming into the most stable phase V. This is an example of the general fact that transient metastable phases often play an important role in the first-order phase transitions, which gives a significant hint for solving molecular and kinetic mechanisms of the transition. This phenomenon is a kind of "Ostwald's step rule" [35], which will be discussed later.

Another apparently complicated thing is that the phase behaviors are strongly affected by the purity of the sample [36], for example, phase IV of C25 is not seen on heating when the purity of the sample is higher than 99.9% [34], while it is seen when the purity becomes much lower [33].

In all the LO phases of C25, the side packing is orthorhombic (herringbone type), which is the same as that in the orthorhombic crystal of polyethylene, and the molecular chain axes are perpendicular to the layer surface (denoted "P" in Fig. 2), but only the layer stackings are different [33,37]. The RV and RI phases have the same orthorhombic side packing, but the molecules in the RV phase are tilted to the layer surface [28], denoted "T" in Fig. 2. In the RII phase, the side packing is hexagonal, and the chain axes are perpendicular to the layer surface.

C. Effect of Size or Defect on the Phase Transition

Effect of size or defect on the phase transition is significantly strong and important. The finite size affects the transition temperature (T_c) due to the surface effect which results in loss of free energy of the crystal. The T_c of the finite size is given by the Gibbs-Thomson equation, assuming a cubic shape of a crystal [38]:

$$T_c = T_c^0\left(1 - \frac{6V\sigma}{r\Delta H}\right) \tag{1}$$

where T_c^0 is the equilibrium transition temperature from B to A phases defined for the infinite size of phase A, σ is the surface free energy, V is the specific volume, ΔH ($\equiv H_B - H_A$) denotes the enthalpy of the phase transition, ΔH_A and ΔH_B are the enthalpy in phase A and that in phase B, respectively, and r is the size of phase A crystal. Thus the decrease of the transition temperature is proportional to r^{-1}.

The effect of the disordered structure on the phase transition is the same as that of the finite size. The free energy of crystal increases due to the increase in internal energy W. The transition temperature is given by a similar Gibbs-Thomson equation:

$$T_c = T_c^0\left(1 - \frac{W}{\Delta H}\right) \tag{2}$$

When a crystal of chain molecules is heated, defects will be introduced due to significant anisotropic thermal expansion. The orthorhombic lattice of n-alkane with herringbone side packing shows anisotoropic thermal expansion, i.e., expansion along the a axis is larger than that along the b and c directions. Here a, b, and c are defined with respect to the subcell, in which the c axis is parallel to the chain axis, and the a and b axes are perpendicular to the chain axis. The lattice constant a shows a significant increase within the LO phase with increasing temperature and jumps at the R-phase transition [23,26]. The lattice constant b also increases with increasing temperature within the LO phase, but the expansion coefficient is smaller than that of a. At the R phase transition, b decreases discontinuously. It will be discussed in the latter part that the difference in expansion of the lattice is related to the origin of the ''precursor''of a nucleus named ''wrinkles.''

II. THERMAL, STRUCTURAL, AND MORPHOLOGICAL STUDIES

A. DSC and X-ray Diffraction

Figure 3 shows the DSC heating and cooling thermograms of C25. The observed peaks correspond to the phase transitions I → V → RV → RI → RII → melt on heating and melt → RII → RI → RV → IV → V → I on cooling, respectively. The transition temperature does not show significant dependence on the heating and cooling rates, although not shown in the figure. The R-phase transition temperature on heating was obtained to be $T_{\text{LO-R}} = 46.3°\text{C}$. The supercooling of 1°C was observed in the R-phase transition, which shows significant hysteresis.

The X-ray powder diffraction patterns are shown in Fig. 4a, b. The transition temperature $T_{\text{LO-R}} = 46.3°\text{C}$ was confirmed, which agrees well with the results obtained by DSC. Supercooling of 1–1.3°C was observed on cooling in the transition. Phase IV appeared only on cooling, but is not seen on heating, as mentioned in the introduction.

Figure 4c shows the change in lattice constants a and b of C25 on heating, which is consistent with the previous works by Doucet et al. [22] and Ungar [27]. The lattice constant a increases by about 1% with increasing temperature in the LO phase, while the lattice constant b shows a smaller increase of about 0.2%.

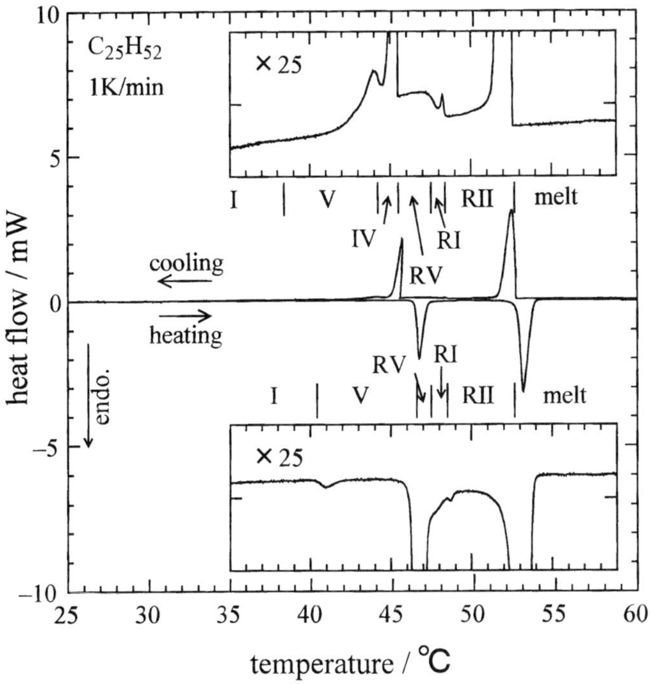

Fig. 3 (a) DSC thermograms of C25 on heating and on cooling. The observed peaks are regarded as corresponding to the I → V → RV → RI → RII → melt transition on heating and melt → RII → RI → RV → IV → V → I on cooling. (b) Transition temperature against various heating and cooling rates determined by DSC measurement. (From Ref. 39)

B. Optical Morphology on Heating

The optically observed transition temperature was determined to be $T_{\text{LO-R}} =$ 46.30°C, which agreed well with that obtained by DSC and X-ray methods. It is assumed that the observed $T_{\text{LO-R}}$ on heating corresponds to the equilibrium transition temperature $T^0_{\text{LO-R}}$. The degree of superheating, $\Delta T \equiv T - T^0_{\text{LO-R}}$, is approximated by $\Delta T \simeq T - T_{\text{LO-R}}$, where T is the observed temperature.

Figure 5 shows a series of optical micrographs on heating. When the crystal in the LO phase is heated to 0.5–1.0°C below $T_{\text{LO-R}}$, thin ''wrinkles'' [39] appeared parallel to the *b* axis (Fig. 5a). The wrinkle will be concluded to be a ''precursor'' of nucleation in the next section.

Figure 5b shows the initial stage of the R-phase transition. First, the R phase grows rapidly along the wrinkle, i.e., along the *bc* plane, and then the growth front advanced normal to the wrinkle along the *a* axis with time. The R-

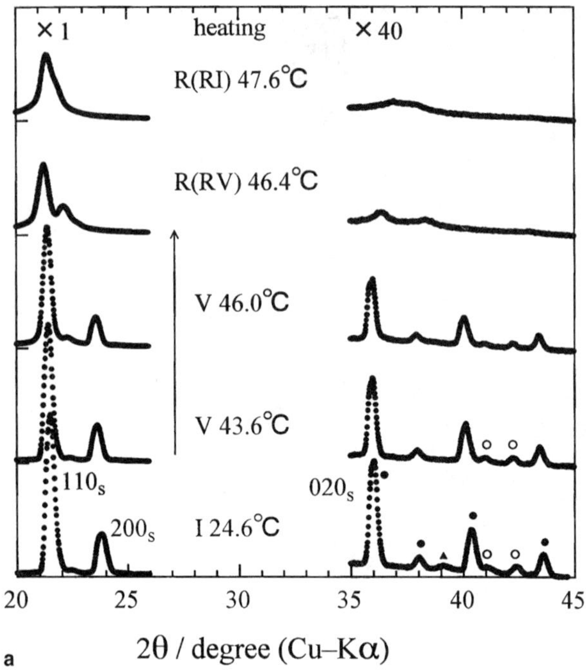
×1
heating
×40
R(RI) 47.6°C
R(RV) 46.4°C
V 46.0°C
V 43.6°C
110s
020s
200s
I 24.6°C
20 25 30 35 40 45
2θ / degree (Cu–Kα)

a

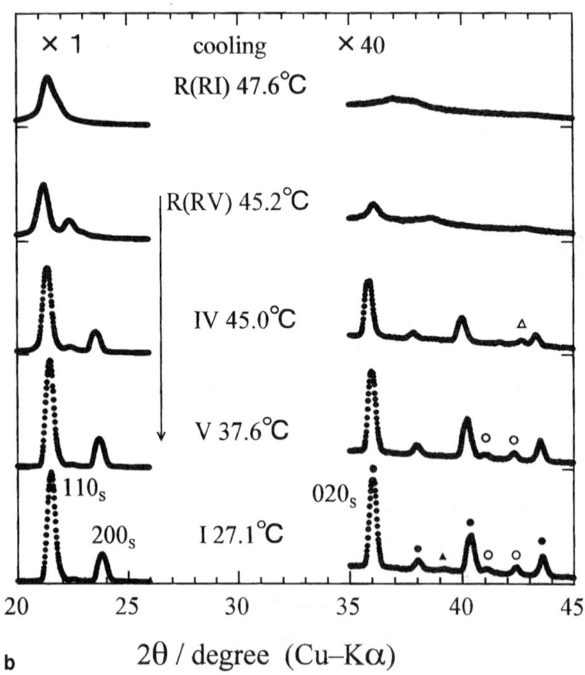
×1
cooling
×40
R(RI) 47.6°C
R(RV) 45.2°C
IV 45.0°C
V 37.6°C
110s
020s
200s
I 27.1°C
20 25 30 35 40 45
2θ / degree (Cu–Kα)

b

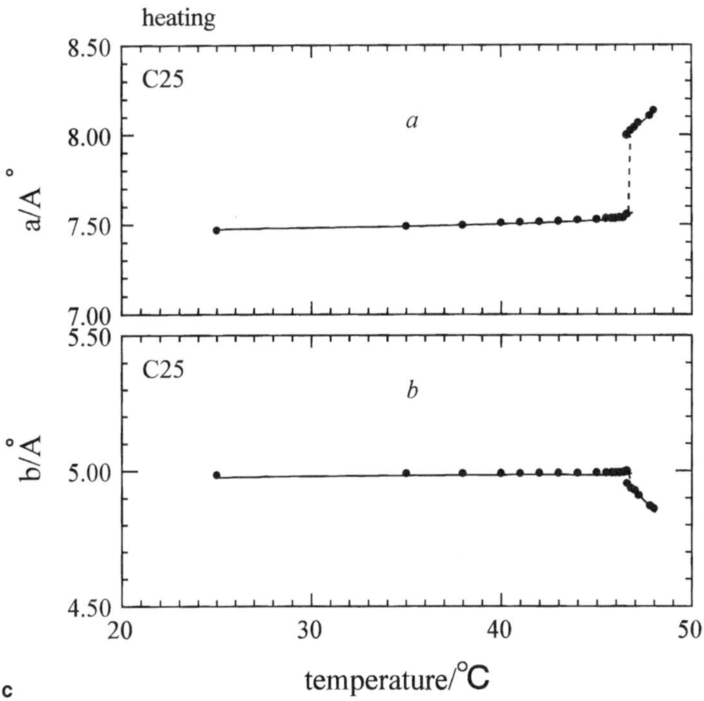

c

Fig. 4 X-ray powder pattern of C25 at various temperatures (a) on heating and (b) on cooling. The 110_s, 200_s, 020_s are the reflections from the side packing. Full circles represent the reflections of LO phase, and open circles those of phases I and V. ▲, reflection of phase I; △, reflection of phase IV. The change in lattice constants *a* and *b* on heating (c). The lattice constants increase with increasing temperature in the LO phase and show discontinuous change at the R phase transition. The change in *a* is larger than that in *b*. (From Ref. 39)

phase region is dark, and the contrast of the front surface is clear. The growth front is nearly straight at first but changes to a slightly curved shape with growth. At small ΔT ($\leq 0.1°C$), the growth front of the R phase stopped advancing before the completion of transition. This may be due to the increase in internal pressure within the transformed region. At large ΔT ($\geq 0.1°C$), the transition is completed; i.e., all regions of the crystal were transformed to the R phase (Fig. 5c).

After the completion of the R-phase transition, new thick wrinkles appeared on the R-phase crystal. This type of wrinkle was reported by Piesczek et al. [37], but is different from the ''precursor wrinkle'' which appeared on the LO phase crystal below $T_{LO\text{-}R}$.

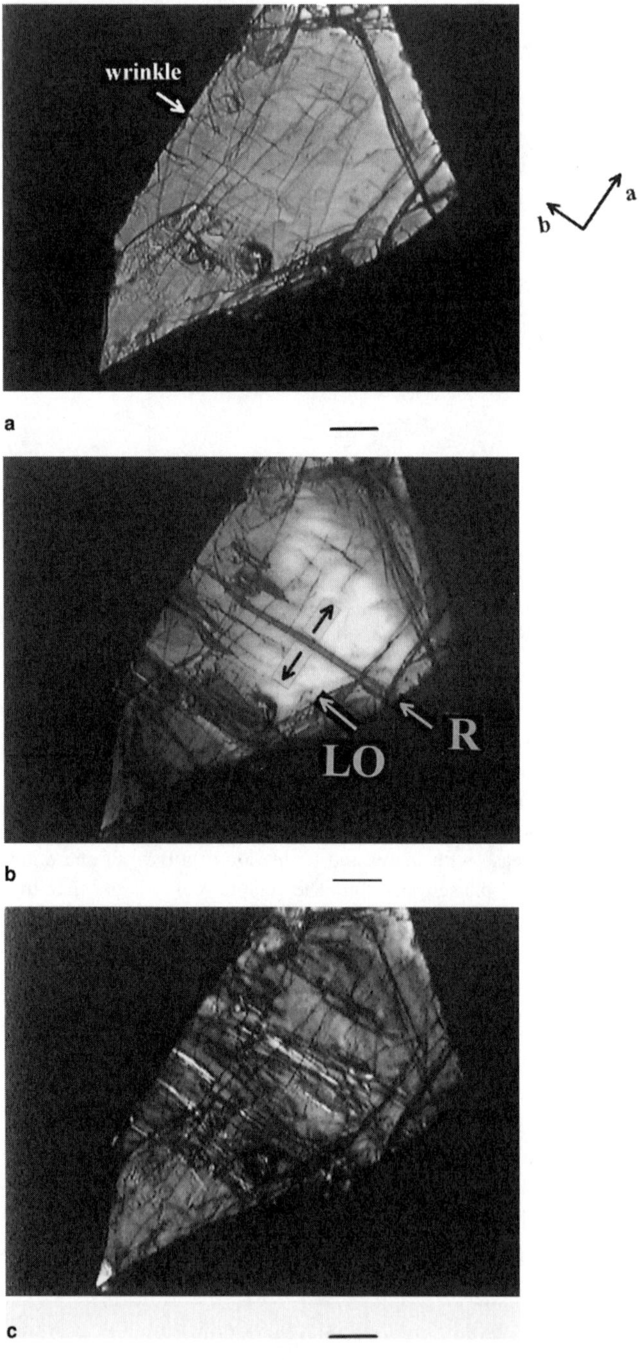
wrinkle
b
a
a
LO
R
b
c

C. "Wrinkle," a Precursor of Nucleus

The morphology of the wrinkle is revealed by means of SEM. Figure 6a shows a SEM image of the wrinkles. The three-dimensional shape of a wrinkle is estimated from the SEM images obtained by changing the tilt angles (θ_{tilt}) of the sample (Fig. 6b), which is schematically shown in Fig. 6c. The typical size of the wrinkle is about 3μm wide, 0.5 μm high, and several tens of micrometers long.

The number density of the wrinkle increases with time and saturates after some time under isothermal condition. The saturated number density increased with increasing temperature (Fig. 7). With increasing temperature up to $T_{LO\text{-}R}$, the saturated number density of the wrinkles increased. When temperature was decreased, most of the wrinkles disappeared, whereas the number of wrinkles increased again when temperature was increased again. The wrinkles always appeared at the same positions. At $T_{LO\text{-}R}$, the increase in the saturated number density of the wrinkle was finally stopped. This phenomenon is closely related to the formation mechanism of the wrinkle (see later).

Three mechanisms for the formation of the wrinkle are considered: (a) onset of the solid-solid transition within LO phases, such as from V → IV phases, (b) onset of the R-phase transition, and (c) a precursor of the R-phase transition.

Mechanism (a) can be easily rejected, because C25 with high purity does not show V → IV phase transition on heating, as shown in this chapter, and because the wrinkle was formed in n-$C_{23}H_{48}$, which does not show phase IV [33]. Mechanism (b) is also denied because no growth of R phase was observed from the wrinkle The third one is the most possible mechanism for the appearance of the wrinkle. Thus it is concluded that the wrinkle is a precursor of the LO-R transition.

Figure 8 shows a mechanism of the formation of the wrinkle. Many defects are formed during the crystallization from the solution. The thermal expansion coefficient β of amorphous phase is usually much larger than that of crystal phase, so β around the defect should be larger than that in the ordered region. Therefore, the defects cause significant lattice expansion within a single crystal, resulting in local accumulation of strain. With increase in temperature, strain increases. When the accumulated strain reaches a critical barrier, a wrinkle forms at the defect to relax the strain. This is the reason why the wrinkles always appear at the same positions. The significant anisotropic expansion makes the wrinkles

---

Fig. 5 A series of optical polarizing micrographs of the R phase transition upon heating and successive cooling: (a) 45.8°C and $\Delta T = -0.5°C$ (showing LO with wrinkles), (b) 46.39°C and $\Delta T = 0.09°C < 0.1°C$ (showing LO → R transition), (c) 46.69°C and $\Delta T = 0.39°C > 0.1°C$ (showing completion of LO → R transition). Scale bar = 100 μm. (From Ref. 45)

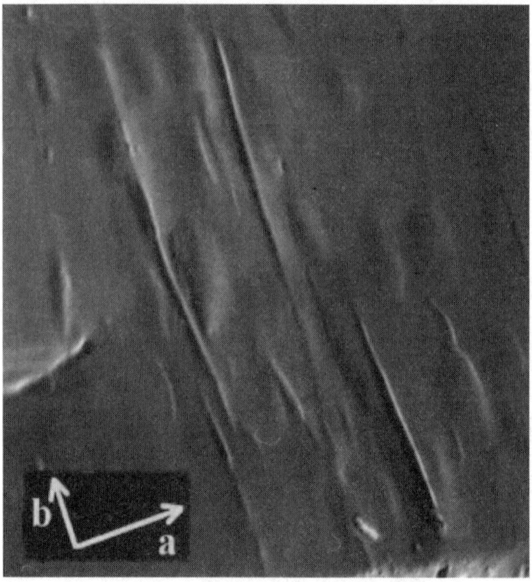

a

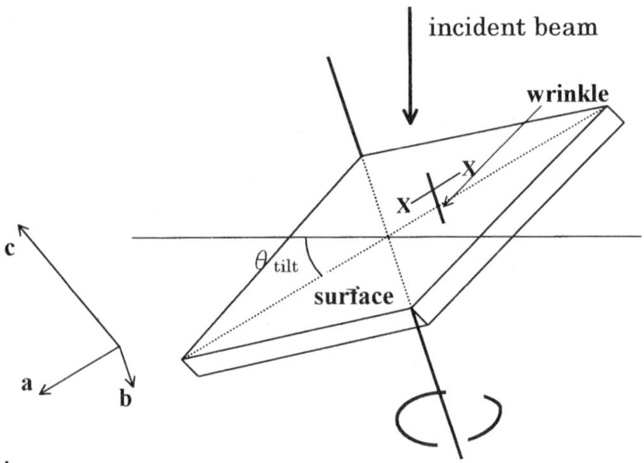

b

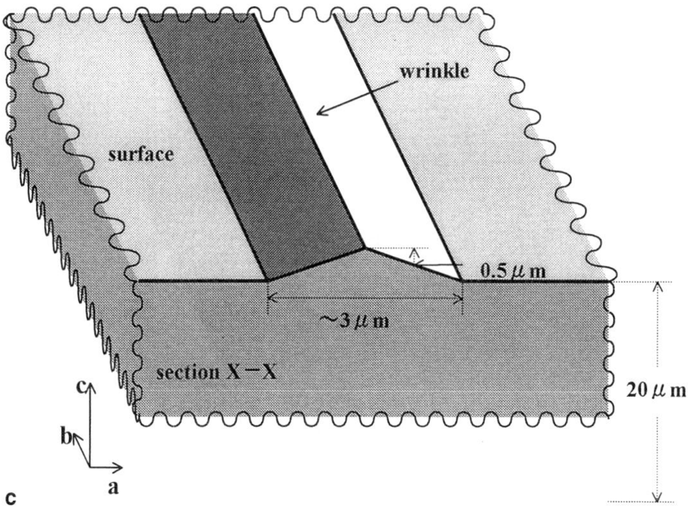

c

Fig. 6 (a) SEM image of the typical wrinkles. Scale bar = 10 μm. (b) Geometry of the SEM to observe the shape of wrinkles. (c) Schematic illustration of the morphology of the typical wrinkle. (From Ref. 39)

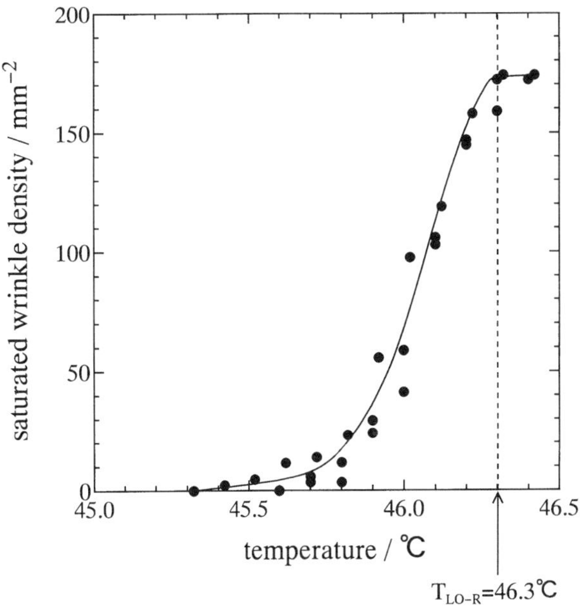

Fig. 7 Temperature dependence of the saturated density of the wrinkle. (From Ref. 39)

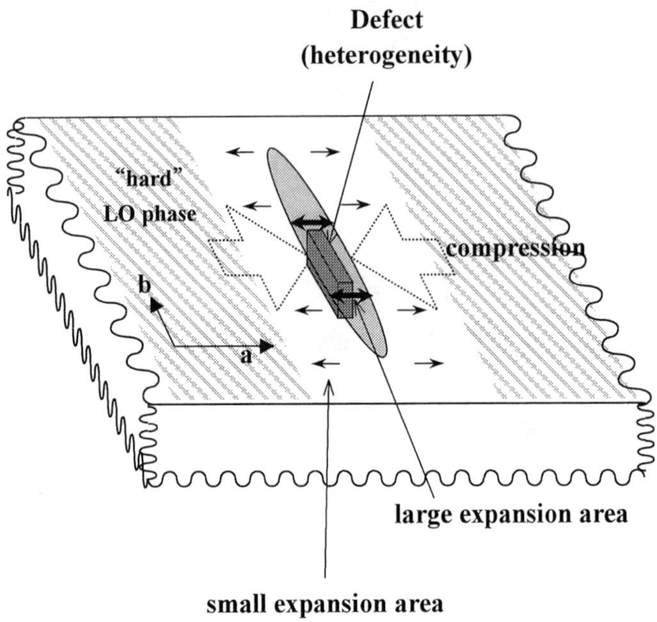

Fig. 8 Model of the origin of the wrinkle prior to the R phase transition in a single crystal of *n*-alkane. (From Ref. 39)

parallel to the *b* axis. This model of the origin of the wrinkle corresponds to a microscopic model proposed by Strobl et al. [40]. They suggested that the longitudinal motion of the molecules along the chain, such as flip-flop screw jump, is activated even in the LO phase.

D. Morphology on Cooling: "Intermediate State"

In the following, the interesting fact that the R phase transforms to the LO phase through a "transient intermediate phase" will be shown. The transient phase is a metastable phase. This is an interesting example of Ostwalds's step rule, which is widely seen and has important significance in the first-order phase transition of the long-chain molecules [38].

When the sample is cooled from the R phase to a temperature slightly below $T_{\text{LO-R}}$, it shows significant supercooling. When the sample is cooled and kept at 44.3°C, where ΔT is significantly large ($\Delta T = 2$°C), it does not transform for 4 ~ 5 sec, suddenly transforms into fully dark phase (named "transient intermediate

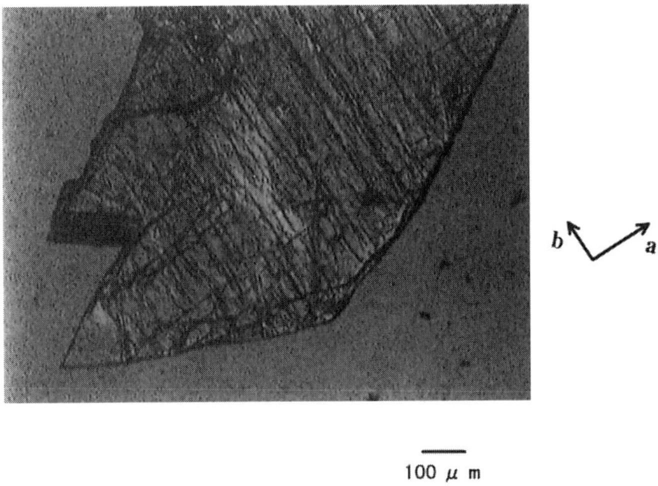

Fig. 9 Optical image of the intermediate state at 44.8°C of the C25 single crystal during the R-phase transition from RV → LO on cooling. (From Ref. 39)

state'') within 1/30 sec (Fig. 9), and finally transforms into the bright LO phase within 1 sec. Figure 10 shows a schematic mechanism of the two-stage transition. The starting R phase is RV, where chains are tilted from the normal direction of the surface of the layer (called the T state). The dark phase suggests that the chains are rearranged from T state into a perpendicular state (P state) with respect to the surface of the layer. The final transformation into LO phase should be the ordering process of the side packing. This mechanism is confirmed by the fact that the two-stage transition is limited to the transition from the T to P states and is not observed in the transition from the P to P states, such as in C21.

The two-stage transformation provides some interesting information on the molecular mechanism of the first-order phase transition of the long-chain system, namely that the tilting process of chains is much faster than that of side chain packing. This result is an important example which shows molecular mechanism of the Ostwald's step rule.

When the sample is cooled from the state where the LO and R phases coexisted, on the other hand, the area of the R phase begins shrinking with relatively small degree of supercooling, which indicates that the process is not mainly controlled by the primary nucleation process but only by the growth process.

The entire morphological changes in the R-phase transition are summarized in Fig. 11.

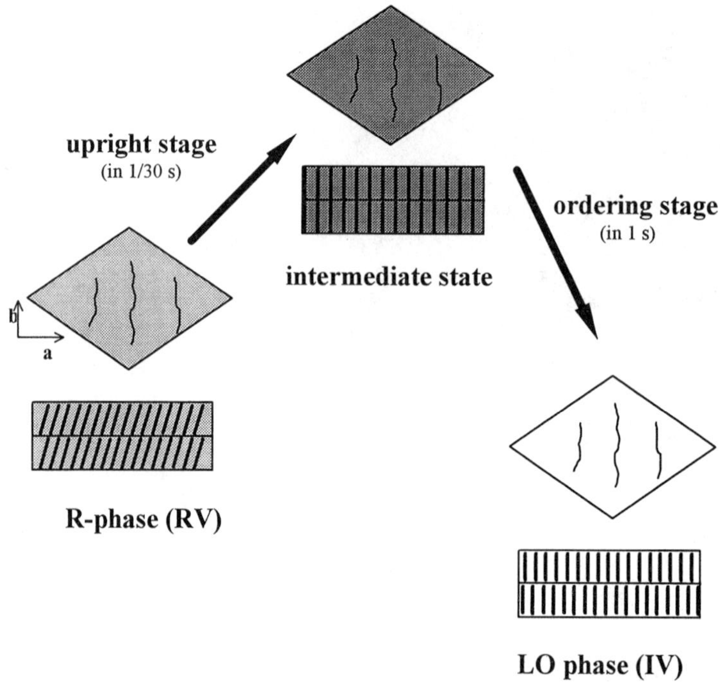

Fig. 10 Structural model of the intermediate state during the T → P phase transition. (From Ref. 46)

E. Hysteresis

The DSC and optical observation showed that transition temperature is $T_{LO\text{-}R}$ = 46.3°C on heating, while that on slow cooling from the R phase is $T_{R\text{-}LO}$ = 44.3°C, which indicates significant hysteresis. On heating, the wrinkles appeared on the crystal surface as a precursor of the transition, and the R phases nucleated and grew from the wrinkles. On slow cooling from the R phase, on the other hand, the transition suddenly started at a large ΔT without any precursor. In another case of cooling from the coexistent state of the LO and R phases, i.e., cooling from the state in which the LO → R transition had not completed, the R → LO transition occurred without any degrees of supercooling. Significant hysteresis was not observed, because primary nucleation of the LO phase is not required. Thus, the origin of the hysteresis should be related to the mechanism of primary nucleation.

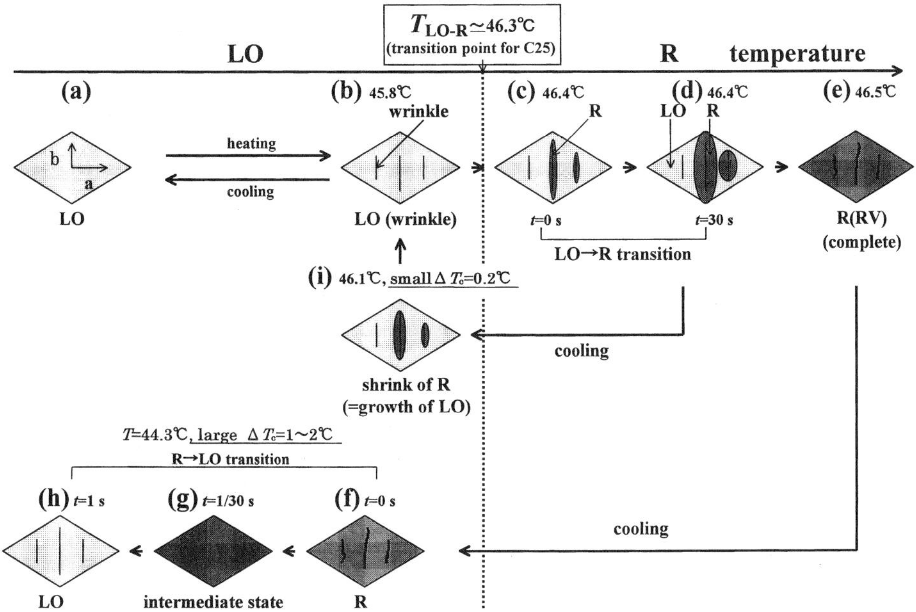

Fig. 11 Illustrations of the observed change in morphology of C25 during the LO → R transition. The lozenge indicates a single crystal of C25, in which the dark region and the vertical line represent the R phase and the wrinkle, respectively. The LO phase, which is phase V or IV in C25, appearing just below the R phase, depends on the sample purity. (From Ref. 39)

III. NUCLEATION AND GROWTH

A. Primary Nucleation

When the R phase starts growing from the wrinkle (precursor), it is interpreted that primary nuclei are formed, which is clearly observed by means of optical polarizing microscopy. The nucleation rate (I) on heating was estimated from the slope of the number density of a primary nucleus (ν) versus time (t):

$$I = \frac{d\nu}{dt} \tag{3}$$

Figure 12 shows log I versus ΔT^{-2} on heating. This gives a straight line; hence we have the experimental formula

$$I = I_0 \exp\left(-\frac{C}{\Delta T^2}\right) \tag{4}$$

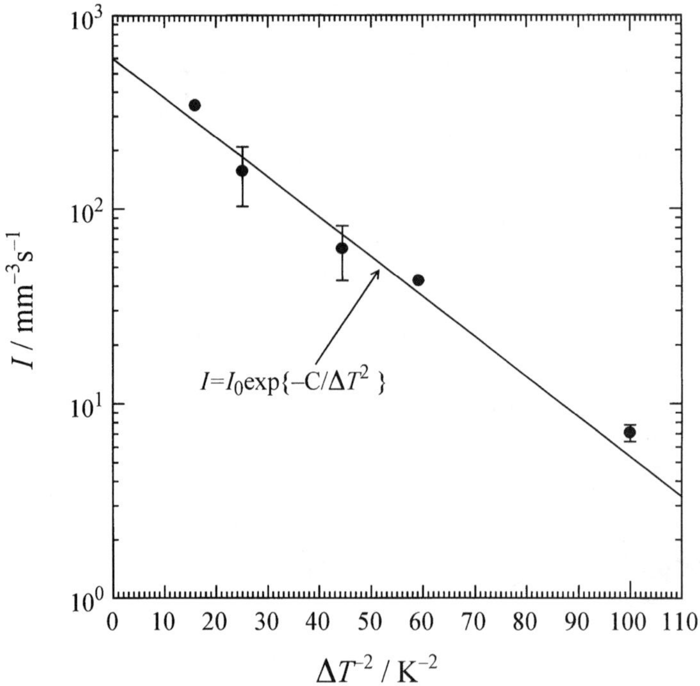

Fig. 12 log I vs. ΔT^{-2}, showing a straight line. (From Ref. 39)

where C is estimated as

$$C_{obs} = 4.7 \times 10^{-2}\ \mathrm{K}^2 \tag{5}$$

This result clearly indicates the important fact that the transition is mainly controlled by a primary three-dimensional (3D) nucleation process [3], which has been confirmed for the first time in solid-solid phase transition of n-alkanes.

Figure 13 illustrates a primary nucleus (3D nucleus) formed within the wrinkle on heating process. Within the wrinkle, the structure is disordered and strain may be accumulated. In this case, the defect may act as an nucleation agent; i.e., the primary nucleation should be a heterogeneous one, which is confirmed by the fact that observed free energy for formation of a 3D critical nucleus, ΔG^*_{3D}(obs), is much smaller than that for a homogeneous nucleus, ΔG^*_{3D}(homo).

The nucleation rate I is given by [41]

$$I \propto \exp\left(-\frac{\Delta G^*_{3D}}{kT}\right) \tag{6}$$

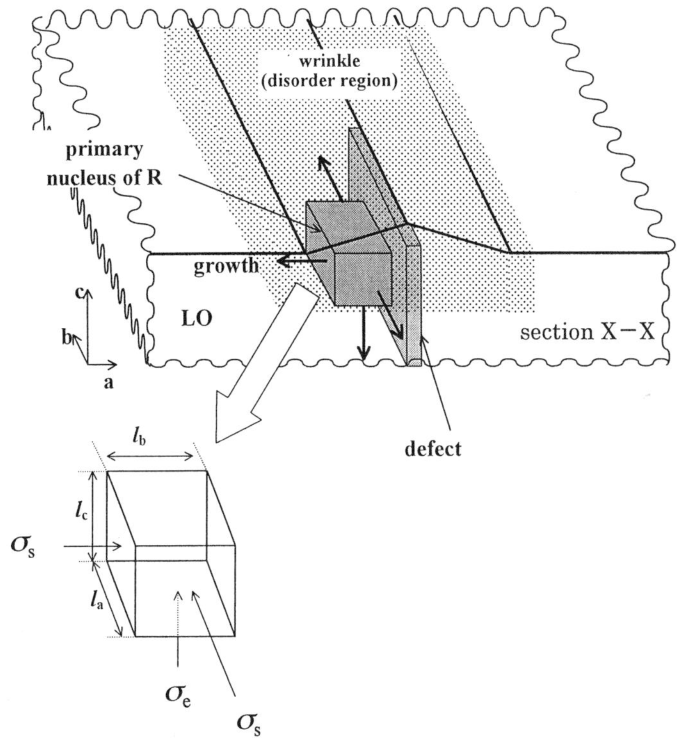

Fig. 13 Schematic of the primary nucleus within the wrinkle. The section X-X corresponds to that in Fig. 6. The defect plays similar role to ''nucleation agent'' in heterogeneous nucleation. (From Ref. 39)

where k is the Boltzmann constant. For homogeneous nucleation C is

$$C_{\text{homo}} = \frac{32T_{\text{LO-R}}\sigma_s^2\sigma_e}{k\Delta h^2} = 7.1 \text{ K}^2 \tag{7}$$

where σ_s and σ_e are surface free energies of the (100) or (010) and (001) boundary surfaces between the LO phase and the R phase and Δh is the enthalpy of the R-phase transition [3]. In this calculation, σ_s and σ_e estimated in the next subsection, and $\Delta h = 6.85 \times 10^7 \text{ J}^2 \cdot \text{m}^{-4}$ [39,42] are used. It is found that $C_{\text{obs}} \ll C_{\text{homo}}$. Thus it is concluded that the observed ΔG^*_{3D} is much smaller than that for a homogeneous nucleus:

$$\Delta G^*_{3D}(\text{obs}) << \Delta G^*_{3D}(\text{homo}) \tag{8}$$

which satisfies the well-known criterion that the nucleation is heterogeneous. This conclusion has a general significance that the onset of the solid-solid phase transition of *n*-alkane systems is mainly controlled by heterogeneous nucleation.

B. Growth

Growth of the R phase was clearly observed by means of optical microscopy. The growth rate V is observed from the equation

$$V = \frac{dx}{dt} \tag{9}$$

where x represents a position of the growth front along the a axis of the R phase. Equation (9) provides important information on molecular mechanism of growth and quantitative estimation of kinetic parameters as surface free energies.

The typical x are plotted against time (t) in Fig. 14. Growth stopped after some time for $\Delta T \leq 0.1°C$, while it does not stop for $\Delta T \geq 0.1°C$. After cessation

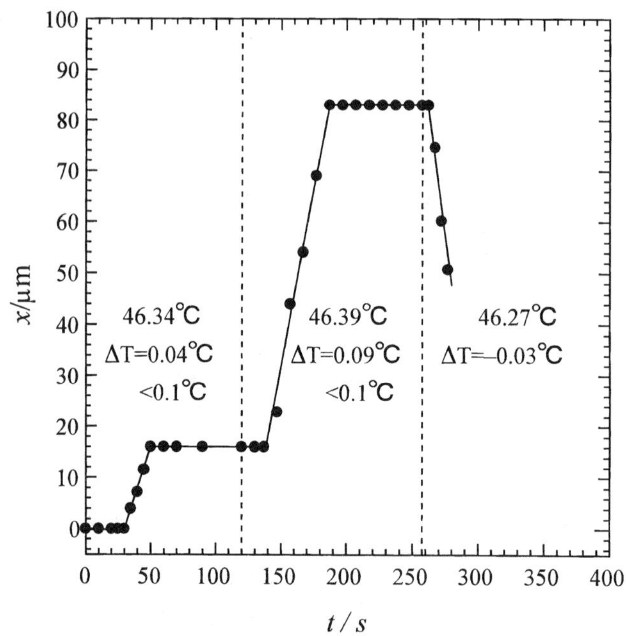

Fig. 14 The distance x vs. time: x increases linearly with time after a few seconds of induction time. At small $\Delta T < 0.1°C$, the increase in x stopped. When the crystal is cooled from the state where the LO and R phases coexist, the R phase starts shrinking immediately; i.e., the R $\rightarrow$ LO transition occurs without nucleation. (From Ref. 45)

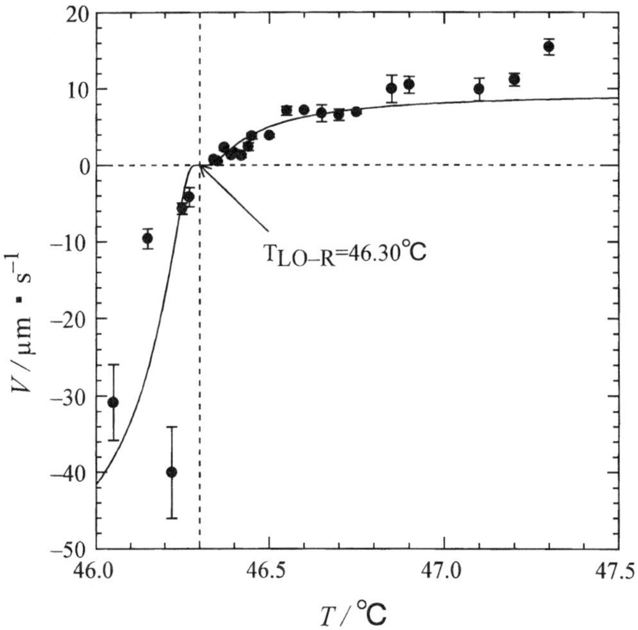

Fig. 15 Temperature dependence of the growth rate of the R phase. The positive and negative growth rates mean the growth of the R phase during the LO → R transition and that of the LO phase during the R → LO transition, respectively. The transition temperature $T_{LO \rightarrow R}$ was determined to be 46.30°C, where $V = 0$. (From Ref. 45)

of growth, when the sample was cooled below the transition temperature ($\Delta T = -0.03$°C), the R phase started shrinking; i.e., the LO phase began growing, from which a negative growth rate is obtained. Figure 15 shows the temperature dependence of the growth rate (V) of the R phase. The positive and negative growth rates mean the growth of the R phase during the LO → R transition and that of the LO phase during the R → LO transition, respectively, from which $T_{LO\text{-}R} =$ 46.30°C is determined correctly. The V versus T curve shows asymmetric shape, the reason for which is not well understood. Similar asymmetric shape has often been seen in the crystallization of polymers [17].

Figure 16 shows log V versus ΔT^{-1}, which shows a straight line given by

$$V \propto \exp\left(-\frac{B}{\Delta T}\right) \tag{10}$$

where B is a constant. This result indicates that growth is mainly controlled by a 2D nucleation process [3]. This is also supported by the fact that the growth front shows a straight plane which corresponds to the typical crystal habit [3].

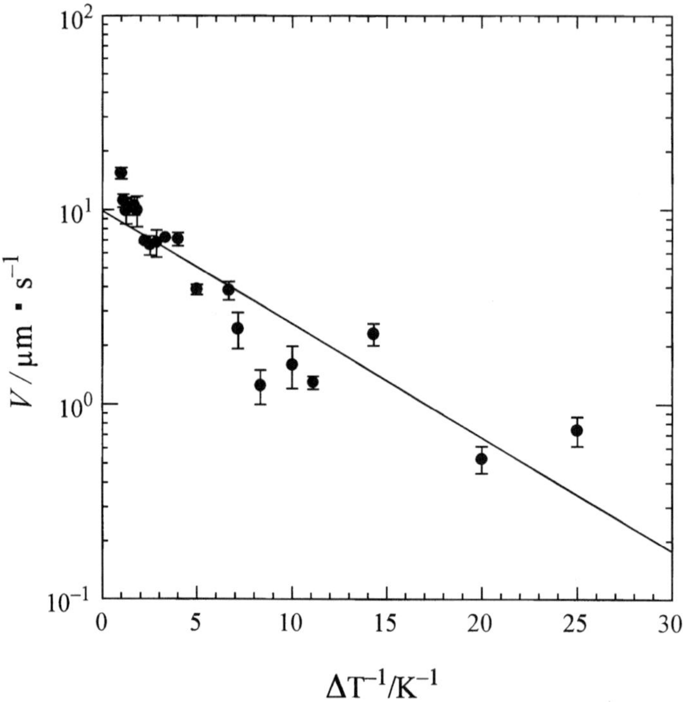

Fig. 16 log V vs. ΔT^{-1} showing a straight line, which suggests 2D nucleation. (From Ref. 45)

The straight growth front slowly changes to a curved one with time, which indicates that 2D nucleation is a multinucleation process. In this case, the growth rate V is proportional to $i^{1/3}$ [43], where i is the nucleation rate of the 2D nucleus, given by

$$i \propto \exp\left(-\frac{\Delta G^*_{2D}}{kT}\right) \tag{11}$$

where ΔG^*_{2D} is the free energy required to form the 2D critical nucleus. Since $i \propto \exp\left(-4\sigma_s\sigma_e/kT\Delta h\Delta T\right)$, V is written as

$$V \propto \exp\left(-\frac{4\sigma_s\sigma_e}{3kT\Delta h\Delta T}\right) \tag{12}$$

$\sigma_s\sigma_e$ is estimated from the slope of the log V versus ΔT^{-1}. Assuming that $\sigma_s \simeq \sigma_e$,

$$\sigma_s \simeq \sigma_e \approx 3.5 \times 10^{-4}\ \mathrm{J \cdot m^2} \tag{13}$$

is obtained, which is significantly smaller than the surface free energy between LO and liquid phases, $\sigma_{s(LO-L)} \approx 9 \times 10^{-3}\ \mathrm{J \cdot m^{-2}}$.

C. Origin of the Hysteresis

Hysteresis is widely observed in first-order phase transitions, but the molecular mechanism is not well understood. It is shown that the primary nucleus of the R phase of an *n*-alkane system is easily formed on heating without significant superheating within the precursor of ''wrinkle.'' In contrast the LO phase is formed on cooling with a large degree of supercooling (a few degrees Celsius), regarded to be ''hysteresis.'' The origin and molecular mechanism of the hysteresis are illustrated in Fig. 17. Within the wrinkle, structure is disordered, so significant thermal expansion is expected with increase of temperature. But the expansion is suppressed by the surrounding LO phase, because the LO phase is more rigid and harder than the R phase, which results in significant accumulation of the strain within the wrinkle. Under this condition, the surface free energies $\Delta\sigma_s$ and σ_e of a nucleus become smaller than those in the usual ordered region

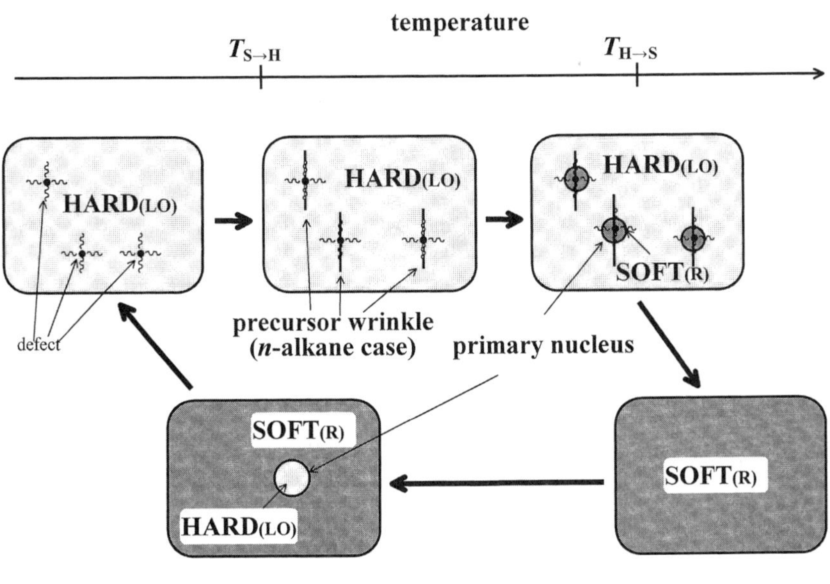

Fig. 17 Schematic of the mechanism of hysteresis in the R phase transition. (From Ref. 39)

of the LO phase, which decreases the critical activation barrier for nucleation, as is well known in heterogeneous nucleation. This is the reason why a nucleus can be formed easily on heating without significant superheating.

On cooling from the R phase, on the other hand, no strain is accumulated around a small nucleus, called the ''embryo'' or impurity, because the embryo or impurity within ''soft'' R phase does not generate significant effects around them. In this case, all the surface free energies are not decreased and large critical activation barrier for nucleation should be required. This is the reason why a nucleus is formed on cooing with significantly large supercooling through homogeneous nucleation.

Thus, in summary the thermal hysteresis in the R-phase transition originates from the different nature of the nucleation: heterogeneous nucleation which minimized superheating and homogeneous nucleation which enhanced supercooling.

Cooling from the coexisting state of the LO and R phases is quite different from the above. The R-LO phase transition is a simple growth process, so no 3D nucleation is necessary. This is the reason why the R-LO phase transition from the coexisting state does not show significant hysteresis.

IV. CONCLUSIONS AND FUTURE RESEARCH

Molecular mechanisms of the solid-solid phase transitions, such as the ''rotator phase transition,'' of *n*-alkanes are quite interesting because they have important relevance to various complicated physical and chemical properties of biological, food, and industrial materials. In was shown in this chapter that the rotator phase transitions of *n*-alkanes are mainly controlled by the primary nucleation and growth processes and that a ''precursor'' named ''wrinkle'' plays an important role in the onset of nucleation. It was also shown that significant hysteresis in transition temperature between the heating and cooling processes is due to accumulation of strain within crystals on heating process which accelerates nucleation. It is shown that the phase sometimes transforms in two steps: in the first, tilt angle of chain axis with respect to the layer surface is changed and then side chain packing is changed. This clearly indicates that tilting and side chain packing are important independent factors in phase transitions. Thus it is concluded that the solid-solid phase transitions of *n*-alkanes are controlled by definite kinetic mechanisms which can be understood by a combination of nature of chain molecules and classical nucleation theory. This conclusion has general significance that the apparently complicated structural changes of *n*-alkanes can be understood by a unified manner.

Much more direct molecular level understanding on the nucleation and formation of the precursor of a nucleus is needed in the future using modern techniques, such as atomic force microscopy and computer simulation. The molecular structure of the transient "intermediate state" seen on cooling process will be clarified using synchrotron radiation, which is often seen in the first-order phase transitions of the long-chain molecular systems.

REFERENCES

1. B. M. Craven, Y. Lange, G. G. Shipley, and J. Steiner, in *The Physical Chemistry of Lipids*, ed. D. M. Small, Plenum Press, New York, 1986, pp. 183–232.
2. G. Goldbeck-Wood, in *Science and Technology of Crystal Growth*, eds. J. P. Eerden and O. S. L. Bruinsma, Kluwer Academic, Boston 1995, pp. 313–328.
3. F. D. Price, in *Nucleation*, ed. A. C. Zettlemoyer, Marcel Dekker, New York, 1969, pp. 405–488.
4. M. Hikosaka and A. Keller, in *Ordering in Macromolecular Systems*, eds. A. Teramoto, M. Kobayashi, and T. Norisuye, Springer-Verlag, Berlin, 1993, pp. 1–15, 89–97.
5. J. W. Hagemann, in *Crystallization and Polymorphism of Fats and Fatty Acids*, eds. N. Garti and K. Sato, Marcel Dekker, New York, 1988, Chap. 2.
6. T. Kakeshita, T. Saburi, K. Kindo, and S. Endo, *Butsuri*, *51*: 498 (1996) [in Japanese].
7. M. Takesada and H. Mashiyama, *J. Phys. Soc. Jpn.*, *63*: 2618 (1994).
8. Y. Yamada, *Ferroelectrics*, *35*: 51 (1981).
9. S. Komori, S. Hayase, and H. Terauchi, *J. Phys. Condens. Matter*, *1*: 3789 (1989).
10. Y. Noda, S. Nishihara, and Y. Yamada, *J. Phys. Soc. Jpn.*, *53*: 4241 (1984).
11. R. F. Shannon, Jr, S. E. Nagler, C. R. Harkless, and R. M. Nicklow, *Phys. Rev.*, *B46*: 40 (1992).
12. N. Hamaya, Y. Yamada, J. D. Axe, D. P. Belanger, and S. M. Shapiro, *Phys. Rev. B33*: 7770 (1986).
13. J. D. Axe and Y. Yamada, *Phys. Rev.*, *B34*: 1599 (1986).
14. H. Iwasaki, Y. Matsuo, K. Ohshima, and S. Hashimoto, *J. Appl. Cryst.*, *23*: 509 (1990).
15. M. Tadakuma, K. Tajima, and G. Masada, *J. Phys. Soc. Jpn.*, *64*: 2074 (1995).
16. T. Kakeshita, K. Kuroiwa, K. Shimizu, T. Ikeda, A. Yamagichi, and M. Date, *Mater. Trans. JIM*, *34*: 423 (1993).
17. B. Wunderlich, in *Macromolecular Physics*, Vol. 3, Academic Press, New York, 1973, Chap. 8.
18. T. Yamamoto, K. Nozaki, and T. Hara, *J. Chem. Phys.*, *92*: 631 (1990).
19. K. Takamizawa, Y. Urabe, J. Fujimoto, H. Ogata, and Y. Ogawa, *Thermochim. Acta*, *267*: 297 (1995).
20. M. G. Broadhurst, *J. Res. Natl. Bur. Stand. Sect. A*, *66*: 241 (1962).
21. A. Müller, *Proc. R. Soc. London Ser. A*, *138*: 514 (1932).
22. J. Doucet, I. Denicolo, and A. Graievich, *J. Chem. Phys.*, *75*: 1523 (1981).
23. J. Doucet, I. Denicolo, A. Graievich, and A. Collet, *J. Chem. Phys.*, *75*: 5125 (1981).

24. I. Denicolo, J. Doucet, and A. F. Graievich, *J. Chem. Phys.*, *78*: 1465 (1983).
25. J. Doucet, I. Denicolo, A. Graievich, and C. Germain, *J. Chem. Phys.*, *80*: 1647 (1984).
26. G. Ungar, *J. Phys. Chem.*, *87*: 689 (1983).
27. G. Ungar and N. Mašić, *J. Phys. Chem.*, *89*: 1036 (1985).
28. E. B. Sirota, H. E. King, Jr., D. M. Singer, and H. H. Shao, *J. Chem. Phys.*, *98*: 5809 (1993).
29. E. B. Sirota and D. M. Singer, *J. Chem. Phys.*, *101*: 10873 (1996).
30. T. Yamamoto and K. Nozaki, *Polymer*, *35*: 3340 (1994).
31. T. Yamamoto and K. Nozaki, *Polymer*, *36*: 2505 (1995).
32. T. Yamamoto, T. Aoki, S. Miyaji, and K. Nozaki, *Polymer*, *38*: 2643 (1997).
33. K. Nozaki, N. Higashitani, T. Yamamoto, and T. Hara, *J. Chem. Phys.*, *103*: 5762 (1995).
34. Y. Urabe and K. Takamizawa, *Technology Reports of Kyusyu University*, *67*: 85 (1994) [in Japanese].
35. W. Ostwald, *Lehrbuch der Allgemeinen Chemie*, Leipzig, 1896.
36. K. Takamizawa, T. Sonada, and Y. Urabe, *Engineering Science Reports of Kyusyu University*, *10*: 363 (1989) [in Japanese].
37. W. Piesczek, G. Strobl, and K. Malzahn, *Acta Crystallogr. Sect. B*, *30*: 1728 (1974).
38. A. Keller, M. Hikosaka, S. Rastogi, A. Toda, P. J. Barham, and G. Goldbeck-Wood, *J. Mater. Sci.*, *29*: 2579 (1994).
39. K. Nozaki and M. Hikosaka, *J. Mater. Sci.*, *35*: 1239 (2000).
40. G. Strobl, B. Ewen, E. W. Fischer, and W. Piesczek, *J. Chem. Phys.*, *61*: 5257 (1974).
41. D. Turnbull and J. C. Fisher, *J. Chem. Phys.*, *17*: 71 (1949).
42. A. Wurflinger and G. M. Schneider, *Ber. Bunsenges. Phys. Chem.*, *77*: 121 (1973).
43. W. B. Hlling, *Acta Met.*, *14*: 1868 (1966).
44. K. Nozaki, T. Yamamoto, and M. Hikosaka, *J. Phys. Soc. Jpn.*, *66*: 3333 (1997).
45. K. Nozaki and M. Hikosaka, *Jpn. J. Appl. Phys.*, *37*: 3450 (1998).
46. K. Nozaki, T. Yamamoto, T. Hara, and M. Hikosaka, *Jpn. J. Appl. Phys.*, *36*: L146 (1997).

5

Molecular Interactions and Phase Behavior of Polymorphic Fats

Kiyotaka Sato and Satoru Ueno
Hiroshima University, Higashi-Hiroshima, Japan

I. INTRODUCTION

Triacylglycerols (TAGs) are the major storage lipids in cells, eggs, and seeds. They are widely used in foods, cosmetics, and pharmaceuticals as nutrients or matrix materials [1,2]. Structural determination of acylglycerol crystals at an atomic level has critical implications in understanding the structure-function relationships of lipids, lipoproteins, fat deposits, and biomembrane lipids, whose numerous structural data were reviewed recently [3–9]. The molecular structures of TAGs are also related to biochemical reactions in absorption, transport, and metabolism of fat [10]. In addition, physical properties such as polymorphism, melting and solidification, density, and molecular flexibility are influenced by the crystalline phases of the TAGs, as first indicated almost a half-century ago [11,12] and summarized in a review [13] and books [14,15]. Recent work has highlighted that the structural properties of the TAG crystals are sensibly influenced by molecular properties of TAGs, such as saturation/unsaturation of the fatty acid moieties, glycerol conformations, symmetry/asymmetry of the fatty acid compositions connected to the glycerol groups, etc. [16,17].

Fats may be represented by TAGs, although diacyl- and monoacylglycerols are categorized as fats. The physical properties of TAGs depend on their fatty acid compositions (Fig. 1a). One may call the TAGs having only one type of acyl chain monoacid TAGs, and those having multiple types of chains mixed-acid TAGs. The physical properties of the mixed-acid TAGs are complicated by the fact that the middle carbon at the *sn*-2 position is potentially asymmetrical as different substitutions at *sn*-1 and *sn*-3 carbons which generate stereoisomers

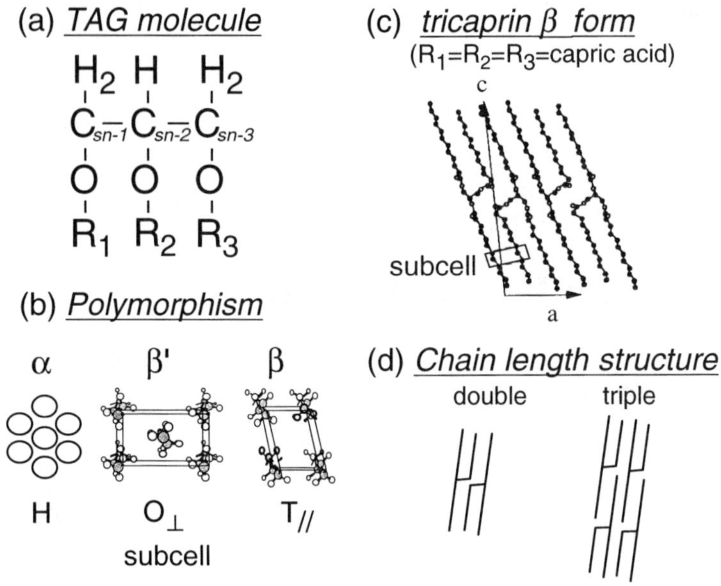

Fig. 1 (a) Chemical structure of a TAG, (b) subcell structures, (c) β form structure of tricaproyl glycerol (tricaprin), and (d) chain length structures.

of the TAGs derivatives. And it is important to note that many naturally occurring TAGs are mixed-acid TAGs (for example, see Chapters 12 and 13 this volume).

So far, fundamental research of polymorphism of fats has been conducted on structures of polymorphic modifications, morphology of fat crystals, thermodynamic stability, phase behavior of fat mixtures, etc. It would be reasonable to define that these studies are categorized as static properties of fats, since the work has been made on rather thermodynamically stable conditions. However, many problems have been left unresolved in such research fields as the kinetic aspects of polymorphic transformation, roles of chain-chain interactions during polymorphic transformations and mixing-separating-agglomerating processes of fats, kinetic phase behavior of metastable forms in both binary and ternary TAG mixtures, dynamic changes in morphology and network of fat crystals forming in dispersed phases, etc. These subjects are categorized as dynamic properties of fats, which are more relevant to real systems of industrial fat productions compared to the static properties. For this purpose, new experimental techniques have recently been applied, in addition to ''conventional'' analytical techniques. In particular, the use of synchrotron radiation (SR) X-ray diffraction [17] and Fourier transform infrared (FT-IR) absorption spectroscopy with multiple facilities

[18] have unveiled new and important phenomena inherent to the molecular interactions and dynamic properties of fats.

This chapter emphasizes molecular interactions and dynamic phase behavior of the polymorphic forms of the mixed-acid TAGs. As a reference, kinetic properties of the monoacid TAGs will be described, since the polymorphic transformation is visualized in fairly simplified manner in these simple TAGs.

II. BASIC CONCEPTS OF FAT POLYMORPHISM

A. Basic Polymorphic Structures

Full details of structural definitions of the polymorphic modifications of TAGs have already been described [1,9,14,16,17]. Here, a brief look is taken at typical characteristics of the polymorphism of the fat crystal as depicted in Fig. 1. The nature and compositions of three fatty acid moieties of a TAG molecule, labeled R_1, R_2, and R_3, determine its chemical property, as shown in Fig. 1a. The three polymorphs α, β′, and β are named in accordance with subcell structures [19], which correspond to cross-sectional packing modes of the zigzag aliphatic chain (Fig. 1b). Figure 1c illustrates a single-crystal structure of tricaproylglycerol (CCC) β form [20,21]. The chain length structure produces a repetitive sequence of the acyl chains involved in a unit cell lamella along the long-chain axis [22] (Fig. 1d). A double-chain-length structure is formed when the chemical properties of the three acid moieties are the same or very similar, as shown in CCC β form. When the chemical properties, however, of one or two of the three chain moieties are largely different from the others, a triple-chain-length structure is formed because of chain sorting [23; Chapter 2 this volume]. The chain length structure plays critical roles in the mixing phase behavior of different types of TAGs in solid phases. The basic three polymorphic forms are modified, depending on the fatty acid compositions, in such a manner that no β is present, multiple β′ or β forms occurs, or other new forms occur, etc. Such diversity in polymorphism of TAGs have been extensively examined in the past 10 years, as reviewed here.

B. Polymorphic Transformation Pathways

Among the three typical polymorphic forms in Fig. 1b, the β form is most stable, β′ is metastable, and α is the least stable. The transformation pathways among the polymorphic crystals and liquid are determined by thermodynamic stability relationships, as expressed in Gibbs free energy (G) values versus temperature (Fig. 2a) [24]. Due to the monotropic nature of polymorphism of TAGs, the G values are largest for α, intermediate for β′, and smallest for β, and each polymorphic form has its own melting temperature, T_m, as depicted in Fig. 2a. This G–

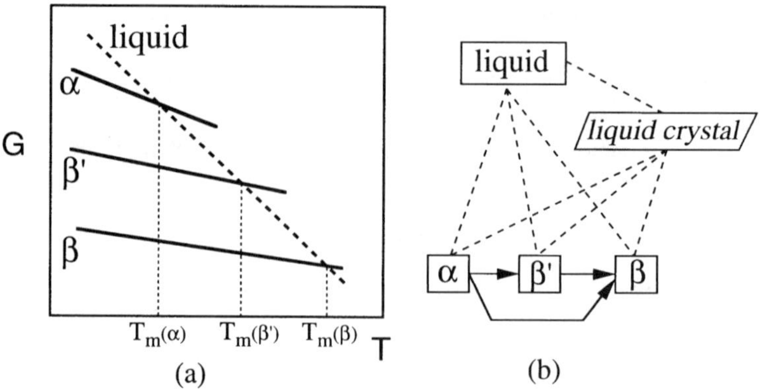

Fig. 2 (a) Gibbs energy (G) and temperature (T) relations of three polymorphs of TAG and (b) transformation pathways.

T relationship explains the following polymorphic crystallization and transformation:

1. Crystallization of the three polymorphs from liquid is possible, yet their rates are highly dependent on the polymorphic nature itself and supercooling.
2. Transformation occurs in a solid state or through melt mediation, the latter being a recrystallization of the more stable forms after the melting of the less stable forms. For example, the melt-mediated transformation easily occurs from α to β through the melting of α.
3. A recent study uncovered the presence of thermotropic liquid crystalline phase, which occurs before the crystallization of the polymorphic crystals or during the melt-mediated transformation as a metastable phase (see Sec. IV). Therefore, the transformation pathway becomes complicated by the presence of the liquid crystal phase, as shown in Fig. 2b; solid-solid, liquid-solid, liquid-liquid crystal transformation, and solid-liquid or liquid crystal-solid.

III. POLYMORPHISM IN MIXED-ACID TRIACYLGLYCEROLS

In many biologically important fats and lipids, combinations of the fatty acid moieties are heterogeneous, e.g., mixed-acid. The heterogeneity is exhibited in such different combinations of the fatty acid moieties as saturated/unsaturated acids, unsaturated fatty acids with *cis*-double bonds placed at different positions, short- and long-chain acids, even and odd carbon-numbered fatty acids, etc. In these specimens, aliphatic chain-chain interactions are critical factors determining their physical properties [25]. In the case of the mixed-acid TAGs, the molecular interactions through the main body of the aliphatic chains, methyl end packing, and glycerol groups are modified compared to those of the monoacid TAG. Therefore, multiplicity and relative stability of the polymorphic forms and their lattice energies are modified. Recent systematic studies have been conducted on structural analyses of the mixed-acid type TAGs, providing physical backgrounds of the materials designation of structured fats.

A. Saturated Mixed-Acid Triacylglycerols

A systematic study has recently been made on polymorphism of a group of TAGs, PP*n*. [23,26–32]. In PP*n* the carbon number (*n* even) of the *sn*-3 fatty acid chain was varied from 0 (1,2-dipalmitoyl-*sn*-glycerol) to 16 (tripalmitoyl glycerol, PPP), while the fatty acids at the *sn*-1 and *sn*-2 positions were palmitic acid. PP*n* samples were synthesized chemically [26]. Their polymorphic transformations and structural information were studied with X-ray diffraction (XRD), differential scanning calorimetry (DSC), Fourier transformed infrared (FT-IR) spectroscopy, and XRD single crystal analysis. This work highlighted a remarkable diversity in the occurrence of polymorphism of asymmetric mixed-acid TAGs composed of saturated fatty acid moieties.

Table 1 shows the occurrence of polymorphic forms of PP*n*, in which *n* was varied from 2 to 14. Quite interestingly, the number of polymorphic forms, relative stability of β' and β, and the chain length structure varied from one TAG to others, singly by converting the acyl chain length at the *sn*-3 position. For example, the single-chain structure was observed in the α forms of PP4, PP6, and PP8, six-chain-length structure and four-chain-length structure were seen in β'_1 of PP10 and β'_2 of PP14, respectively. Interestingly, these unusual chain length structures were revealed in the metastable forms in each TAG. Figure 3a shows the variation in the melting points and long spacing values of the most stable forms of eight types of PP*n*. It is obvious that the chain length structure converted from double (PP2, PP4), triple (PP6, PP8, and PP10), and double (PP12, PP14) again. Correspondingly, the melting point decreased from dipalmitoyl glycerol to PP4, stabilized around 42°C for PP4 through PP10, and increased for PP12, PP14, and PPP. As to the occurrence of the most stable form, β was for PP2, PP4, PP10, and PP12, and β' was for PP6, PP8, and PP14. The transformation behavior involving stable and metastable forms in each substance of the PP*n*-type TAGs was fully summarized elsewhere [29]. Single-crystal X-ray structural

Table 1 Polymorphic Occurrence of PP*n*

n	2			4			6, 8		10					12				14		
Form	α	β'	β	α	β'	β	α	β'	α	β'_3	β'_2	β'_1	β	α	β'_2	β'_1	β	α	β'_2	β'_1
Chainlength	2	3	2	1	3	2	1	3	2	2	2	6	3	2	2	2	2	2	4	2

Source: From Ref. 29.

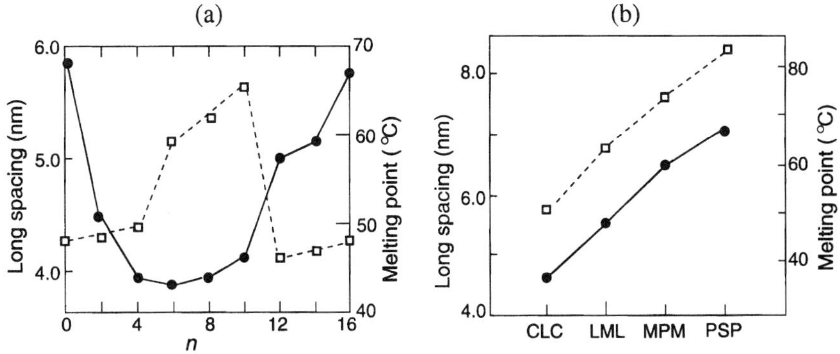

Fig. 3 Long spacing values (open squares) and melting points (closed circles) of (a) PP*n* TAGs [27] and (b) $C_nC_{n+2}C_n$ TAGs [34].

determination proved that PP2 β was of the double-chain-length structure with the $T_{//}$ subcell [23] and PP14 β'_2 was of the four-chain-length structure with the hybrid orthorhombic subcell [32; Chapter 2 of this volume].

So far, no clarification has been completed on the mechanisms operating in the occurrence of the complicated polymorphism in PP*n*. Apparently, some kind of ''molecular magic force'' would be working, indicating that specific chain-chain interactions of the methyl end packing, aliphatic chains, and glycerol groups are modified by changing the length of the *sn*-3 moiety, as elaborated below. This assumption may be rationalized by comparing the PP*n* with P*n*P [33], where the chain length of the *sn*-2 position was varied, while the *sn*-1 and *sn*-3 positions were palmitoyl chains. The P*n*P-type TAGs exhibited the common polymorphic form of β, whose melting points linearly increased with increasing length of the *sn*-2 fatty acid chains. As for the stabilization of the β′ form, it is interesting to refer to polymorphism of a homologous series of $C_nC_{n+2}C_n$, in which *n* was even-numbered carbon atoms ranging from 10 to 16 [34,35]. In these TAGs, the most stable form is β′, and no β form was observed. The long spacing values and melting points of the β′ forms of the four TAGs linearly increased with increasing length of the acyl chains, as shown in Fig. 3b [34]. This means, together with the results of PP*n*, that heterogeneity in the chain length of the three acyl chains is a prerequisite for the stabilization of the β′ form, although its mechanistic reason is still unclear.

Molecular-level structural analyses have been made on two stable β′ forms of PP14 (1,2-dipalmitoyl-3-myristoyl-*sn*-glycerol, abbreviated PPM herewith) using single crystals. Figure 4 shows the transformation pathways of the three forms of PPM and neat liquid [30]. The two β′ forms occur in a delicate manner:

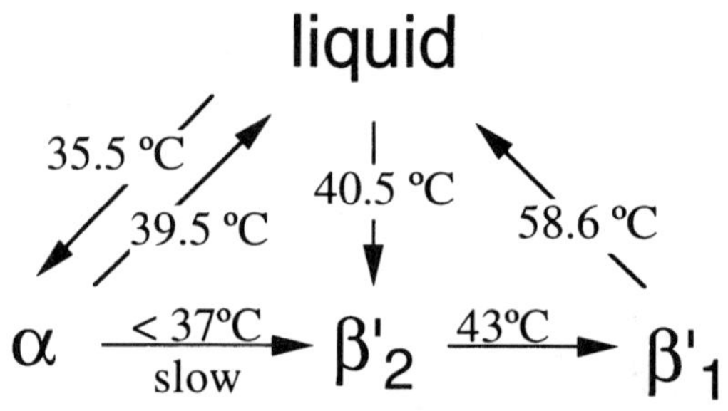

Fig. 4 Transformation pathways of three forms of PPM and liquid [30].

β'_2 is a bit less stable than β'_1, which was formed through the solid-state transformation from β'_2. However, solvent crystallization can form the two β' forms by very slow crystallization. By using the solvent-crystallized single crystals, a basic FT-IR spectral band information of a series of saturated monoacid TAGs [36] has been applied to β'_1 in order to examine the presence of $O_\perp$ subcell, conformation of the three acyl chains with respect to the glycerol group, and the degrees of inclination of the acyl chains with respect to the lamellar interface [18,31]. As for β'_2, single-crystal XRD analysis has been successful in providing atomic coordinates of the orthorhombic crystal structure having hybrid-type $O_\perp$ subcell, four-chain-length structure, and anomalous methyl end packing modes [32]. Since the details of β'_2 structure are described in Chapter 2, typical characteristics of the two β' forms of PPM are discussed in reference to β'_2.

The $O_\perp$ subcell of β'_1 was revealed in clear dichroism of CH_2 scissoring $\delta(CH_2)$ and CH_2 rocking $r(CH_2)$ bands in FT-IR absorption spectra taken for the single crystals of β'_1 (Fig. 5a) [31]. The projections of the a_s- and b_s-axes of the $O_\perp$ subcell are almost perpendicular and parallel to the long axis of the single crystal, respectively. FT-IR analysis with oblique transmission also showed that the acyl chain axis tilts against both the a_s- and b_s-axes of the $O_\perp$ subcell, yet the degree of inclination against the b_s-axis is slightly larger than that against the a_s-axis (Fig. 5b). Furthermore, the a_s-axis is inclined one directionally, while the b_s-axis is alternately inclined along the successive layer. This finding is related to a paradox concerning the β' structure: whether the chain tilt direction alternates at the glycerol groups [Fig. 5c(i)] or at the methyl end terraces [Fig. 5c(ii)] [37–41]. Single-crystal XRD analysis on β'_2 of PPM has shown that chain alternation occurs at the methyl end terraces in this form [32]. To verify it, the c-axis lattice parameter corresponded to four-chain-length structure [32]. In the case of β'_1 of PPM, however, the long spacing value corresponded to the double-chain-length structure. Therefore, the possible structure of β'_2 of PPM in regard to the chain alternation would be either the model depicted in Fig. 5c(i) or that shown in Fig.

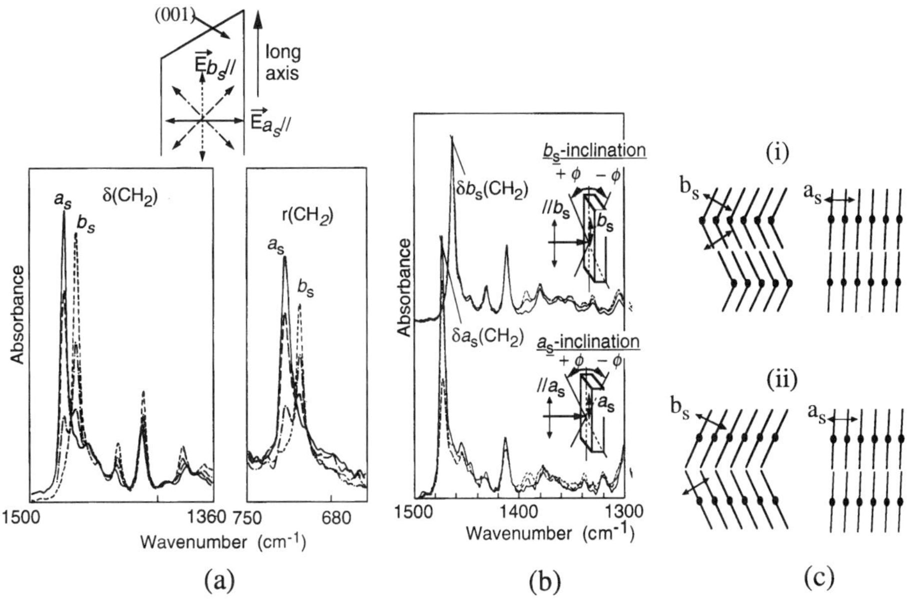

Fig. 5 FT-IR absorption spectra of PPM β'_1 form by (a) polarized microprobe and (b) oblique methods, and (c) two possible structure models for β'_1 [36].

5c(ii) having the nature of polytypism [42]. The structure in Fig. 5c(i) was proposed previously [37,38] and recently reinforced to explain the β' form revealed in $C_nC_{n+2}C_n$, based on both powder crystal XRD spectra and calculation [34]. The final decision must await single-crystal XRD analysis at an atomic level. In any case, the structures proposed in Fig. 5 may shed new light on diversity in the β' structures in view of chain-chain interactions which favor β' over β as the stable structure.

As for the diversity in the polymorphic occurrence in PP*n*, many properties have still not been resolved; e.g., the details of single-chain (PP4, PP6, and PP8) and six-chain-length structure (PP10), three β' forms (PP10), conversion in the chain length structure of double-triple-double (PP2), irreversible transformations involving solid-state and melt mediation, etc. Note that the complexity in polymorphism of PP*n* is symbolic of the saturated mixed-acid TAGs which are relevant to natural fats, such as milk fat TAGs, and to structured fats with specific fatty acid components put on selected glycerol carbon positions on purpose. Likewise, further fundamental research is needed to elucidate other types of saturated mixed-acid TAGs.

B. Saturated/Unsaturated Mixed-Acid Triacylglycerols

1. 1,3-Disaturated-2-unsaturated Mixed-Acid TAGs

Structural analysis of mixed-acid TAGs containing saturated fatty acids at the *sn*-1 and *sn*-3 glycerol carbon positions and unsaturated fatty acids at the *sn*-2 position (St-U-St TAGs) is of great importance. This is because these TAGs are the main components of vegetable fats such as palm oil and cocoa butter.

Recent studies have shown that polymorphism in the St-O-St TAGs, where oleic acid was placed at the *sn*-2 position and the even-numbered saturated fatty acids at the *sn*-1 and *sn*-3 positions, is more complicated than that in the saturated monoacid TAGs [43–49]. The St-O-St TAGs so far examined are POP (P, palmitoyl), SOS (S, stearoyl), AOA (A, arachidoyl), POS, and BOB (B, behenoyl). The chain-chain interactions between the saturated and unsaturated fatty acid moieties play critical roles in the complicated polymorphic transformation [50]. The common properties were observed in the St-U-St TAGs where the *sn*-2 acid was replaced by ricinoleic acid (SRS) [51] and linoleic acid (SLS) [52], and in the St-O-St TAG having different saturated acid moieties as well [53]. For the application, analogous studies between the St-O-St TAGs with cocoa butter gave rise to a novel seeding technique of BOB crystals for controlling polymorphic crystallization of cocoa butter [54–60]. With an interest on an early stage of crystallization of confectionery fats, morphological and thermal analyses were made on SOS and POS [61,62]. The formation mechanism of granular crystals in palm-oil blended fats was interpreted in relation to the crystallization of stable forms of POP present in palm oil [63].

Figure 6 illustrates a typical feature of polymorphic transformation from α to β_1 forms through γ, β', and β_2 of St-O-St TAGs, their long spacing values and melting points. Although POS has a single β form and no γ form, the occurrence of the γ form and two β forms is unique in this group of TAGs [46]. Similar results were obtained for SOA [53]. The difference in melting points between two β forms is 1.5–2.0°C. The observed deviation in POS from the other St-O-St TAGs might be ascribed to its racemic nature. Another uniqueness seen is that the chain length structure converted from double to triple, and the subcell structures changed in different manners between oleic and saturated acid moieties (Table 2). For example, α is double chain length in all TAGs, and the other stable forms, except for POP β', are of triple chain length. Correspondingly, the subcell structures, aliphatic chain conformation, and olefinic conformation converted in such manners as specific to each polymorphic form. This complicated transformation behavior is caused by the steric hindrance of stearic and oleic acid chains, as well as by the structural stabilization of the aliphatic chains and glycerol groups altogether.

Here elaborated are the structural properties and chain-chain interactions in the polymorphic transformations in SOS, which were examined with XRD, DSC [44], FT-IR [47,49], and ^{13}C nuclear magnetic resonance (NMR) [48].

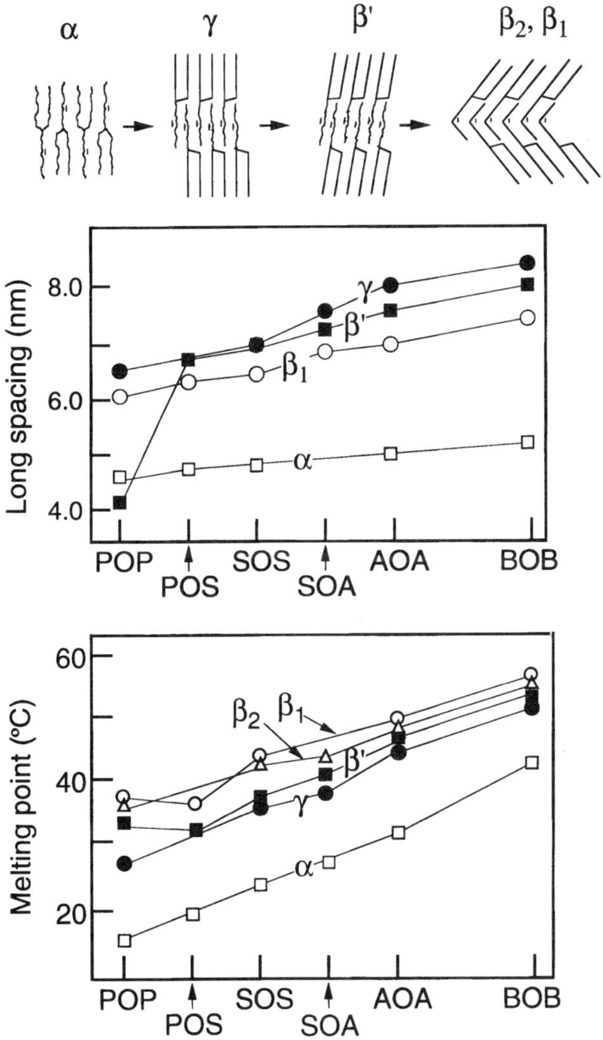

Fig. 6 Structure model, long spacing values, and melting point values of five polymorphs of St-O-St TAGs [44].

Table 2 Occurrence and Polymorphic Structures of Triacylglycerols

TAG[a]						Polymorphic structure				
						subcell		conformation		olefinic
SSS	SOS	POP	POS	SRS	SLS	saturated	oleoyl	saturated	oleoyl	group
α-2	α-2	α-2	α-2	α-2	α-2	H	H	disorder	disorder	n.s.[b]
—	γ-3	γ-3	(δ-3)	γ-3	γ-3	//-type	H	disorder	disorder	n.s.
				β'_2-3						
β′-2	β′-3	β′-2	β′-3	β'_1-3	—	$O_\perp$	H	order	disorder	n.s.
—	β_2-3	β_2-3	—	—	—	T//	T// or O′//	order	order	SCS′[c]
β-2	β_1-3	β_1-3	β-3	—	—	T//	T//	order	order	SCS′

[a] S = stearoyl, O = oleoyl, P = palmitoyl, R = ricinoleoyl, L = linoleoyl
[b] not specified
[c] *skew-cis-skew′*

(a) α *form.* The double-chain-length structure assumes the coexistence of the stearoyl and oleoyl moieties in the same leaflets. The hexagonal subcell was shown in the XRD and FT-IR spectra, leading to a disordered aliphatic conformation. No specific structure was shown for the olefinic conformation.

(b) γ *form.* In triple-chain-length structure, the oleoyl and stearoyl leaflets are separated through the chain sorting during the α-γ transformation. The stearoyl leaflet assumed a specific parallel packing and the hexagonal subcell structure still remained in the oleoyl leaflets, as verified by FT-IR spectral bands of SOS containing fully deuterated stearoyl and hydrogenated oleoyl chains (see below). The increase in the long spacing value per one CH_2 unit, $\Delta L = 0.255$ nm/CH_2, measured for POP, SOS, AOA, and BOB, indicates that the long chains are arranged normal to the lamellar interface.

(c) β' *form.* In the triple-chain-length structure, the stearoyl leaflet assumed the $O_\perp$ subcell and the hexagonal subcell structure still remained in the oleoyl leaflets, as shown in FT-IR spectral bands. The NMR spectra showed clearer differences between the two carbons adjacent to the *cis* double bond and the three glycerol carbons. For β' the value of $\Delta L = 0.242$ nm/CH_2 assumes that the long chains are inclined against the lamellar interface by about 70°.

(d) *Two* β *forms.* The long spacing values of the triple-chain-length structure were 6.75 nm for β_2 and 6.60 nm for β_1, and the value of $\Delta L = 0.20$ nm/CH_2 assumes that the long chains are inclined against the lamellar interface by about 52°. The subcell structure of the stearoyl and oleoyl leaflets was $T_{\parallel}$ in β_1. The subcell structure in β_2 was very close to $T_{//}$ for the two leaflets, but very subtle differences were detectable between the two β forms (see below). The other properties for the two β forms seemed identical.

As described, the importance of olefinic chain conformation is critical in the formation of the triple-chain-length structure of the TAGs containing oleic acid moiety [64]. Here described are the polarized FT-IR assessment of the acyl chain conformations and the differentiation of the two β forms of SOS.

Infrared CH_2 scissoring and CH_2 rocking regions are good indicators of lateral packing, i.e., subcell packing [65]. Since the bands of oleoyl groups overlap those of stearoyl groups for the usual hydrogenated specimens [47], partial deuteration has been attempted in SOS so that stearoyl chains are deuterated and oleoyl chains are hydrogenated, e.g., $S_DO_HS_D$ [49]. Deuterated acyl groups show the CD_2 scissoring $\delta(CD_2)$ and CD_2 rocking $r(CD_2)$ bands around 1090 and 525 cm^{-1}, instead of the $\delta(CH_2)$ and $r(CH_2)$ bands around 1460 and 720 cm^{-1}, respectively.

Figure 7 shows the spectral changes of the $\delta(CH_2)$ and $\delta(CD_2)$ regions, where $\delta(CH_2)_o$ and $\delta(CD_2)_s$ mean the $\delta(CH_2)$ mode of hydrogenated oleoyl chains and the $\delta(CD_2)$ of deuterated stearoyl chains in the $S_DO_HS_D$ specimens [49]. In the α form of $S_DO_HS_D$, a single band was observed at 1467 cm^{-1} for $\delta(CH_2)_o$ and at 1089 cm^{-1} for $\delta(CD_2)_s$, both of which correspond to the hexagonal

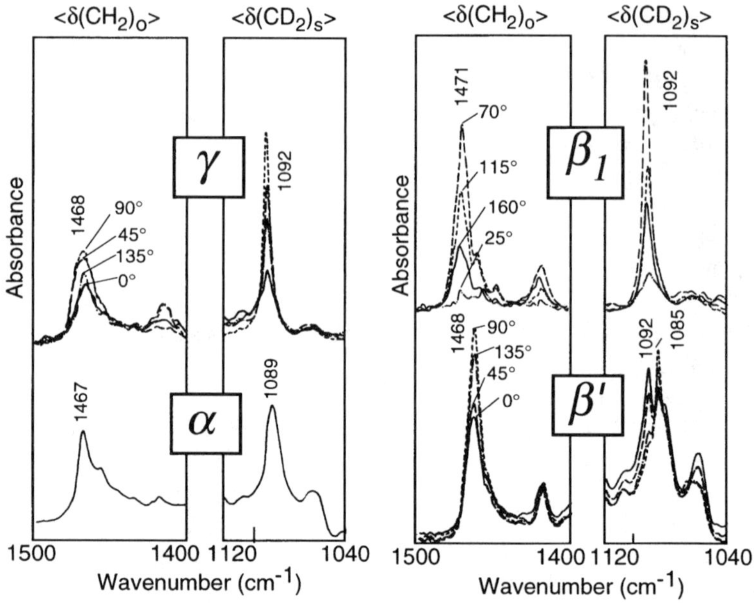

Fig. 7 FT-IR spectra of four polymorphs in partially deuterated SOS, $S_DO_HS_D$. Polarized spectra were taken for γ, β′, and β_1 forms [49].

subcell. In γ a sharp single band of $\delta(CD_2)_s$ appeared at 1092 cm^{-1}, while a broad band of $\delta(CH_2)_o$ was observed at 1468 cm^{-1}. It is suggested that the stearoyl groups form a parallel packing and the oleoyl moiety packs in the hexagonal subcell. In the β′ form, the $\delta(CH_2)$ band had two components, 1473 and 1465 cm^{-1} in the all-hydrogenated SOS, whereas the $\delta(CD_2)_s$ mode split into two components at 1021 and 1085 cm^{-1} in $S_DO_HS_D$. Further, the $\delta(CD_2)_s$ bands showed dichroism in such a manner that the bands at 1092 and 1085 cm^{-1} are maximum at the 0° and 90° polarization directions, respectively. These spectral data indicate the $O_\perp$ subcell of the stearoyl leaflets. This splitting of the $\delta(CD_2)_s$ bands was also observed in deuterated *n*-alkanes taking the $O_\perp$ subcell [66]. As to the oleoyl leaflet, the $\delta(CH_2)_o$ mode became slightly sharper in β′ than in γ, but its frequency was still 1468 cm^{-1}, suggesting a lateral packing similar to the hexagonal subcell in α. In the β_1 form of the all-hydrogenated SOS, the single $\delta(CH_2)$ band appeared at 1470 cm^{-1} with a clear polarization nearly parallel to the *b* axis. In contrast, the $\delta(CH_2)_o$ and $\delta(CD_2)_s$ bands of β_1 of $S_DO_HS_D$ were also observed as single bands at 1471 and 1092 cm^{-1} with a clear polarization. The $\delta(CH_2)_o$ bands appeared at a higher frequency than in the β′ form. These results suggested that both the stearoyl and oleoyl moieties were packed in the $T_{//}$ subcell.

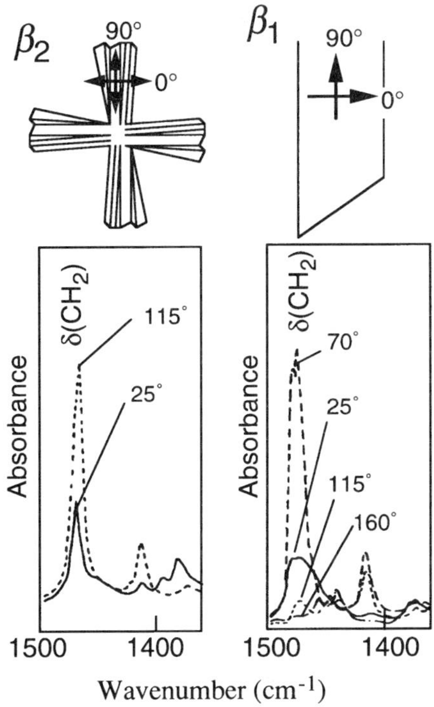

Fig. 8 Polarized microprobe FT-IR spectra of two β forms of SOS [47].

Figure 8 shows the polarized IR spectra of polymethylene scissoring $\delta(CH_2)$ mode taken for single crystals of two β forms of SOS. A complete polarization was observed for β_1, while imperfect polarization was observed for β_2. This indicates the difference in the subcell structures of the oleic acid chains between the two β forms, assuming $O'_{//}$ for β_2 and $T_{//}$ for β_1. However, the FT-IR spectra does not singly differentiate the $T_{//}$ and $O'_{//}$. With ^{13}C NMR spectroscopy, similar results were obtained, again except for the ambiguity in the subcell structures of the oleoyl leaflet [48]. From these data, it is highly suggested that the difference in the molecular structures of the two β forms of SOS must be revealed in the polymethylene chains of the oleic acid moiety. Further clarification, most powerfully with single-crystal X-ray analysis, is needed on this problem.

In regard to the polymorphism in the St-U-St TAGs, the polymorphism in SRS (R, ricinoleoyl) and SLS (L, linoleoyl) is basically the same as that revealed in St-O-St, and that the differences are seen in the absence of the more stable forms (Table 2). Namely, SRS has no β form [51],and β′ and β forms are absent

Table 3 Thermal Data of SSS, SOS [44], SRS [51], and SLS [52] Polymorphs

	Polymorph	Melting point (°C)	ΔH (kJ/mol)	ΔS(J/mol·K)
	α	55.0	109.3	333.08
SSS	β'	61.64	142.8	426.54
	β	73.0	188.4	544.27
	α	23.5	47.7	160.79
	γ	35.4	98.5	319.23
SOS	β'	36.5	104.8	338.47
	β_2	41.0	143.0	455.19
	β_1	43.0	151.0	477.62
	α	25.8	58.1	194.35
SRS	γ	40.6	119.64	381.32
	β'_2	44.3	171.19	539.29
	β'_1	48.0	184.76	575.31
	α	20.8	40.9	139.2
SLS	γ	34.5	137.4	448.7

in SLS [52]. Thermal data of the polymorphic forms of SRS and SLS are shown in Table 3 together with those of tristearoylglycerol (SSS) and SOS. The α and γ forms present in the two TAGs showed the same molecular structures as those in St-O-St. This feature indicates the importance of the chain-chain interactions of the unsaturated fatty acid moieties in the St-U-St TAGs revealing metastable forms. As for the stable forms, hydrogen bonding in the ricinoleoyl chains is so tight that the $O_\perp$ subcell is stabilized through the glycerol groups, making β' the most stable in SRS [51]. Hydrogen bonding in SRS makes the enthalpy and entropy values for melting the β' forms much higher than β' forms of SSS and SOS and even comparable to β of SSS. In SLS the interactions among the linoleoyl chains at the *sn*-2 position, each of which has two *cis* double bonds, may stabilize the γ form, prohibiting the transformation into more stable forms of β' or β. For this reason, the enthalpy and entropy values for melting the γ form of SLS are much larger than those of SOS and SRS. The transformation from γ to β' or β is associated with an inclined chain arrangement with respect to the lamellar interface, which might be prohibited by the linoleoyl chain-chain interactions in SLS [52].

In the polymorphic transformations of St-U-St TAGs depicted in Fig. 6, chain sorting from double- to triple-chain-length structures occurs in α-γ and β'-β in POP. The mechanistic process of the chain sorting is still unclear, although computer modeling calculation was attempted [67].

Table 4 Polymorphic Occurrence of OSO, OEO, and OVO

	Melting point (°C)			Long spacings (nm)		
TAG	α	β′	β	α	β′	β
OSO	−6	—[a]	25	5.2	4.5	6.5
OEO	—	−0.5	9	—	4.5	6.5
OVO	—	1	4	—	4.5	6.5

[a] Not detectable.
Source: Ref. 68.

2. *1,3-Diunsaturated-2-saturated Mixed-Acid TAGs*

A systematic study was carried out on the polymorphic behavior of symmetric mixed-acid TAGs, in which the *sn*-1 and *sn*-3 acids were oleoyl and the *sn*-2 was stearoyl (OSO), elaidoyl (OEO), and vaccinoyl (OVO) chains [68]. Similarly to St-O-St TAGs, conversion in the chain length structure occurred in the polymorphic transformations in the mixed-acid TAGs (Table 4, Fig. 9). On quenching from the liquid, OEO and OVO formed the β′ forms, yet the α form was formed in OSO, all being stacked in the double-chain-length structure. In OSO, α transformed to β′, keeping the double-chain-length structure unchanged. By incubation of the β′ form, it transformed to the triple-chain-length β forms. It is assumed that the driving force to form the triple-chain-length β forms of the three TAGs shown in Table 4 is the inability of the saturated or *trans*-unsaturated acyl chains along with the bent (*cis*) oleoyl chains upon the structure stabilization. This mechanism is essentially the same as that in the St-O-St TAGs, discussed previously.

Other types of saturated-unsaturated mixed-acid TAGs involving *trans*-unsaturated acid have been examined, as summarized in Table 5 [69,70]. Note that, even in the symmetric mixed-acid TAGs, β′ is most stable in PEP (1,3-dipalmi-

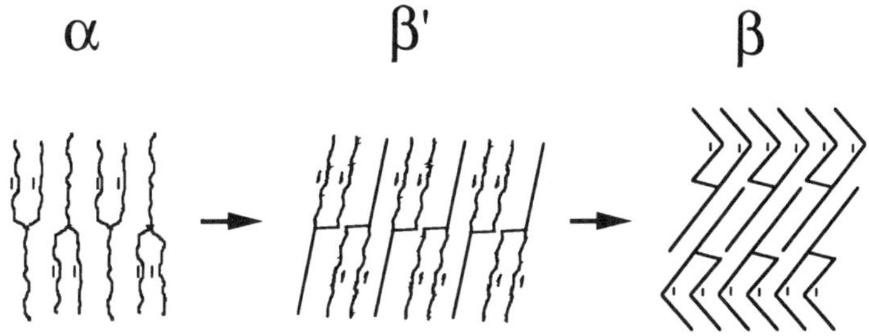

Fig. 9 Structure model of three polymorphs of OSO [68].

Table 5 Polymorphic Occurrence of PEP, EPP, PEE, SES, ESS, and SEE

TAGs	Polymorphic forms			
PEP	α	β'_2	β'_1	—[a]
EPP	α	β'_2	β'_1	—
PEE	α	β'_2	β'_1	—
SES	α	β'_2	β'_1	β
ESS	α	β'_2	β'_1	β
SEE	α	—	—	β

[a] Not detectable
Source: Ref. 70.

toyl-2-elaidoyl-*sn*-glycerol), SES (1,3-distearoyl-2-elaidoyl-*sn*-glycerol) has the most stable form of β [70]. Despite its importance, the mechanisms for the stabilization of the β′ forms in PEP, EPP, and PEE are unknown, in contrast to the stabilization of β in SES, ESS, and SEE. It seems that the chain end packing mode can be a key factor, which must be further clarified by an atomic level structural determination of β′ and β structures for the TAGs listed in Table 5.

In regard to St-U-St TAGs, one must refer to polymorphism of monounsaturated TAGs such as trioleoylglycerol (OOO) and trielaidoylglycerol (EEE), which was examined in single and mixture phases [71]. The structure and intersolubility analyses have shown that EEE has an intermediate polymorphic behavior between SSS and OOO.

IV. DYNAMIC ASPECTS IN POLYMORPHIC TRANSFORMATIONS OF PRINCIPAL TRIACYLGLYCEROLS

Based on the thermodynamic and structural knowledge of polymorphism of TAGs, dynamic aspects in the polymorphic transformations have been highlighted quite recently, mostly by using synchrotron radiation (SR) X-ray beam. The SR X-ray beam has superior advantages over traditional X-ray beam generators of a laboratory scale in its power, coherency, monochromatic wavelength, linearity, etc. The TAGs so far examined with SR X-ray diffraction (SR-XRD) are tripalmitoylglycerol (PPP) [72], tristearoylglycerol (SSS) [73], binary mixtures of PPP and SSS [74], 1,3-distearoyl-2-oleoyl-*sn*-glycerol (SOS) [75,76], binary mixtures of 1,3-dipalmitoyl-2-oleoyl-*sn*-glycerol (POP)/PPP [77], POP/1,2-dipalmitoyl-3-oleoyl-*sn*-glycerol (PPO) [78], POP/1,3-dioleoyl-2-palmitoyl-*sn*-glycerol (OPO) [79], and cocoa butter [80,81]. The newest trial was successful in combining the SR-XRD and DSC in a simultaneous way [82]. All of these studies have given fruitful information on molecular-level understanding of the dynamic processes occurring in the polymorphic crystallization and transformations in single and mixed phases of principal TAGs. SR-XRD studies of two principal TAGs of PPP and SOS are reviewed here. The main results of the binary mixtures will be shown in the next section, and Chapter 12 describes the results on cocoa butter.

A. Monoacid Triacylglycerols

As a typical example of the monoacid TAGs, PPP has been examined by a Belgian group [72–74]. The transformation processes of three forms of PPP are revealed by DSC cooling and heating thermopeaks, as shown in Fig. 10a [83]. After the formation of the α and β′ forms, the DSC heating process produced the melting of α and β′ and successive crystallization of the β form, which melted on further heating. Quite precise and convincing information on the molecular

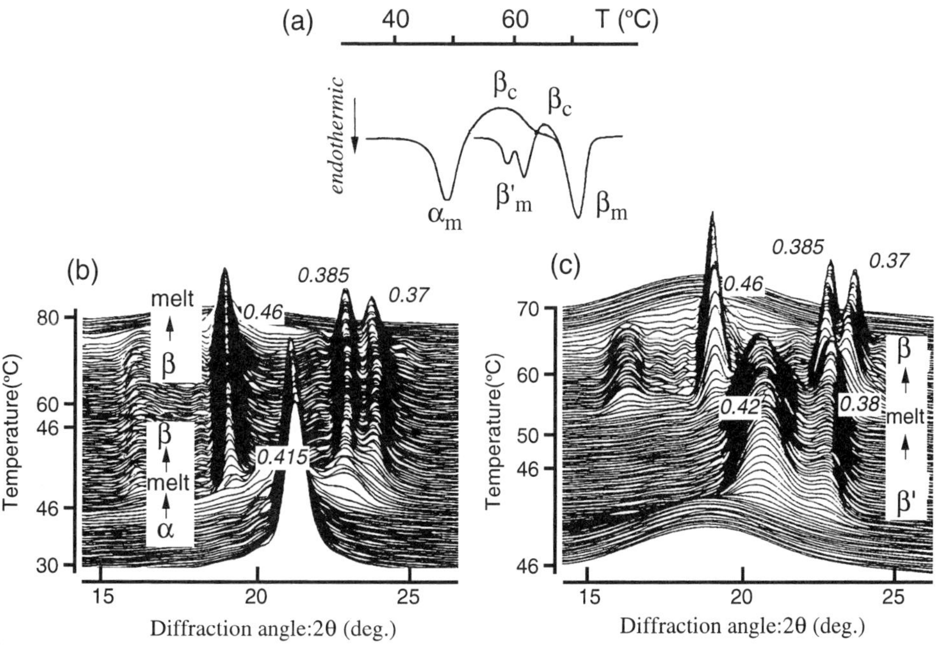

Fig. 10 Synchrotron radiation X-ray diffraction spectra of α-melt-mediated transformation into β of PPP [72].

mechanisms of the transformation of PPP was available with the SR-XRD [72,73]. Figure 10b shows the SR-XRD spectra taken during the heating process from 30 to 80°C, where rapid heating from 30 to 46°C and from 46 to 80°C and isothermal treatment at 46°C over 15 min were applied. The time interval for recording individual SR-XRD spectrum was 15 sec. Similarly to the DSC heating thermopeaks, the α form having a single short spacing spectrum of 0.415 nm melted around 46°C and followed by the crystallization of the β form having three short spacing spectra, which melted around 70°C.

In addition to this overall thermal behavior, an interesting feature was shown by the time-resolved SR-XRD analyses in that the occurrence of the three XRD short spacing spectra of β was from 0.46 nm to 0.385 nm to 0.37 nm, which correspond to the subcell lattice parameters of b_s, $\langle 110 \rangle_s$, and a_s, respectively. A time lag between the occurrence of the b_s- and a_s-axes was about 5 min, meaning the sequential ordering in the subcell arrangement of the $T_{//}$ subcell. Furthermore, it was observed that the β′ form appeared for a very short time during the melt-mediated transformation from α to β, but its stability was minimized. The similar SR-XRD experiment of the β′-β melt-mediated transformation showed that b_s occurred faster than a_s, and that there was a continuation between the spectra of a_s of the $O_{\perp}$ subcell in β′ and $\langle 110 \rangle_s$ of the $T_{//}$ subcell in β during the melt-mediated transformation (Fig. 10c).

As described above, the SR-XRD data highlighted a molecular-level process of the polymorphic transformation of PPP, which was not available by the other techniques. We further note that (a) nucleation/growth processes of the β form crystals occurred during the α-melt–β transformation; (b) no liquid crystalline phase was observed after the melting of α; (c) the SR-XRD long and short spacing spectra disappeared at the same time at the melting of β′, and no lamellar structure was formed after the melting of β [73].

B. Mixed-Acid Triacylglycerols

We review the SR-XRD analysis of the polymorphic transformations in SOS, which was performed by two modes of melt crystallization: simple cooling and melt mediation [75]. The time interval for detection of individual SR-XRD spectrum was 10 sec. The main results are summarized in the following.

(a) *Occurrence of liquid crystal phases.* A very rapid α-melt mediation without thermal annealing unveiled the formation of two types of liquid crystalline structures, called LC1 and LC2 phases, having long spacing values of 5.1 nm and 4.6 nm, respectively. The crystallization of the β′ form was made after the occurrence of the LC1 phase, which eventually disappeared in accordance with the growth of the β′ form after the melting of the α form. A quenching crystallization from neat liquid (46°C) to 25°C also revealed the occurrence of the LC1 phase. It was confirmed that the LC1 phase has its ''melting'' temperature around 32°C.

(b) *Effects of thermal annealing.* Thermal annealing induced the formation of the γ form after the α-melt mediation, since seed crystals of γ were generated in the solid phase of α.

(c) *Ordering sequence.* The formation of lamellar ordering occurred more rapidly than that of subcell packing, as exhibited in the earlier occurrence of SR-XRD long spacing spectra in comparison to the short spacing spectra. Time lags between the occurrence of the long and short spacing became longer for the more stable forms, such as β′, than for the metastable forms, such as γ.

Figure 11a shows the occurrence of the LC1 phase and β′ polymorph during a temperature jump from 10 to 30°C after the rapid cooling of α from liquid. α showed two long spacing spectra, 5.3 and 4.4 nm, soon after the crystallization. After the temperature jump, α melted as indicated by the disappearance of the corresponding long and short spacing spectra. Then a long spacing spectrum of 5.1 nm appeared and lasted over 8 min after the melting of α. However, no corresponding short spacing spectra were detectable during the period when the long spacing spectra of 5.1 nm was maintained. This feature is typical of the occurrence of a liquid crystalline phase. The XRD spectra of β′ occurred before the disappearance of the LC1 phase, and the initiation of the long spacing spectra of β′ was detectable faster than that of corresponding short spacing spectra. A quantitative calculation of diffraction intensity of the long and short spacing spectra clearly showed this conclusion [76]. This means that the formation of the

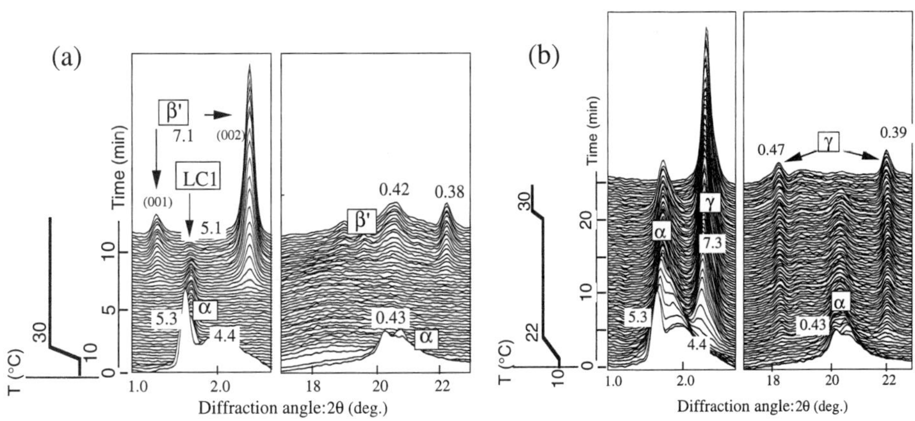

Fig. 11 Synchrotron radiation X-ray diffraction spectra of α-melt-mediated transformation into β′ of SOS [75].

lamellar structure (long spacing) occurred prior to the formation of the subcell packing (short spacing).

In contrast, Fig. 11b shows the time-resolved SR-XRD spectra of α-melt mediation through thermal annealing. The thermal annealing was applied at 22°C, which is a bit below the melting point of α, 23.5°C, for 20 min after the annealing temperature was jumped to 30°C. In this transformation, γ was formed during the thermal annealing process due to an α-γ solid-state transformation as shown in Fig. 11b. The γ crystals thus formed continuously grew at the expense of α. After the melting of α by the temperature jump from 22 to 30°C, the γ form was singly present and the liquid crystalline phase did not occur.

Figure 12 shows the Gibbs free energy (*G*) and temperature (*T*) relationships of the five polymorphs and two liquid crystalline phases of SOS. With regard to the occurrence of the liquid crystalline phase, it is relevant to a long-disputed problem concerning the presence of the ordering structure in the liquid

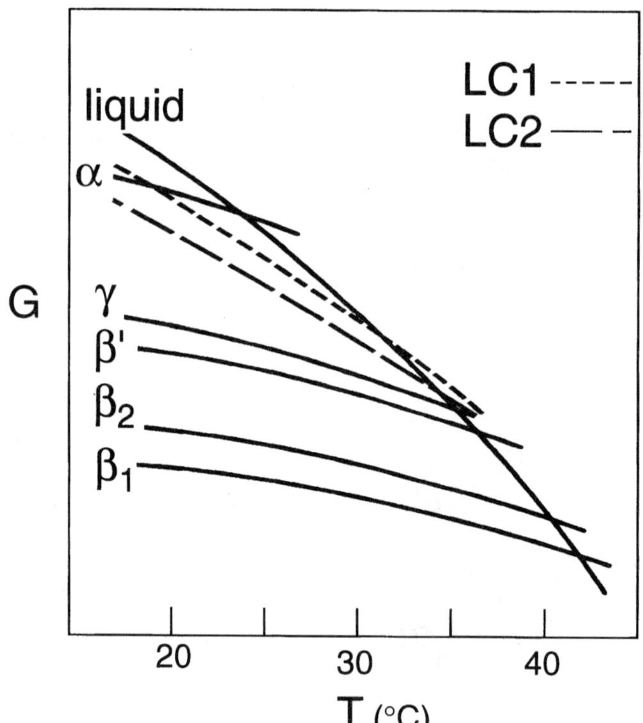

Fig. 12 Gibbs energy (*G*) and temperature (*T*) relations of five polymorphs and two liquid crystalline phases of SOS [75].

state of TAGs. Larsson proposed that the model represented a liquid crystalline state of smectic nature [84]. Callaghan and Jolley supported the existence of the ordering structure in the liquid state of tristearoylglycerol (SSS) by means of a ^{13}C NMR study [85]. Conversely, Cebula et al. [86] performed small-angle neutron scattering experiments of neat liquid of trilauroylglycerol (LLL), claiming that there was no indication that the organized molecular aggregates of the smectic liquid crystals was observed. Instead, they postulated a molecular arrangement such as a nematic phase of liquid crystals [87]. Thereafter, Larsson [88] argued that the ordering structure in liquid of TAGs was like L2 phase, because of the analogy between the XRD low-angle diffraction spectra exhibited by TAGs and L2 phases of polar lipids. The data obtained in SOS with the SR-XRD method by the present authors indicated the presence of the smectic-type liquid crystal.

Referring to a ''memory effect,'' Malssen et al. [89] examined the influence of the thermal history of cocoa butter on its crystallization and melting behavior. They discussed that a tendency of the preferable crystallization of the β-type forms of cocoa butter is due to the presence of the TAGs consisting of high concentrations of stearic acid, which may serve as seed materials. In this respect, the memory effect does not relate to the liquid crystalline phases but to the solid materials. Loisel et al., however, examined the rapid crystallization and transformation of cocoa butter with the simultaneous usage of the SR-XRD and DSC techniques [82], claiming the presence of a liquid crystalline phase in relation to the least stable form of cocoa butter, Form I, defined by Wille and Lutton [90].

To summarize the subject of ''ordering structure in liquid of fats,'' it was obvious that the liquid crystalline phases are present in SOS, as defined by the phases which diffract long spacing spectra without corresponding short spacing spectra, and ''melt'' at defined temperatures. This phase occurred by rapid chilling of liquid and melt-mediated transformation at higher rates. However, the generality of the presence of the liquid crystalline phase in the other TAGs is not established, although some indications have been available in cocoa butter [82] or other fats [91,92]. The SR-XRD analysis on PPP did not unveil the liquid crystalline phases. It is postulated that the appearance of the liquid crystalline phases would be influenced by subtle conditions, due to their lower thermodynamic and structural stability than the polymorphic crystalline forms. Therefore, more dynamic conditions in temperature variations or more sensible monitoring systems may be needed to unveil the feature of liquid crystalline phases of TAGs.

V. THERMODYNAMIC AND KINETIC PHASE PROPERTIES IN BINARY MIXTURES OF TRIACYLGLYCEROLS

Natural fats present in real systems in biotissues or even in food materials are mixtures of different types of TAGs. Therefore, one must elucidate for the complicated behavior of melting, crystallization, and transformations of the natural fats by examining the mixing behavior of binary, ternary, or more multiple phases of specific TAG components. Since a comprehensive review was made by Rossel [93], numerous studies have been made on the mixing behavior of the binary mixtures of principal TAGs.

In general, three typical phases may occur in binary solid mixtures when the two components are miscible in the all proportions in a liquid phase: solid solution phase, eutectic phase, and compound formation [94]. For the TAG mixtures, primary factors determining the phase behavior are the differences in chain length and chemical structures of the fatty acid moieties, as reviewed by Rossel [93] and Small [1]. Further complication is caused by two main reasons: polymorphism and acyl chain compositions attached to the glycerol group. As for the effects of polymorphism, less stable forms such as α or β′ may result in the formation of the miscible phase, because of lesser extent of molecular packing and conformation, even when the eutectic phase is formed in the most stable β form. This property was already reported for the PPP-SSS mixture [93] and more precisely clarified by SR-XRD experiments (see below). The influence of the acyl chain compositions may result in molecular compound formation through specific chain interactions. It was obvious that the molecular compound formation is reflected in turn on its polymorphic properties which are far different from those of the component TAGs.

The phase behavior of the TAG mixtures has critical implications in fat blending and separation of component TAGs from natural fats and oil resources. So, much basic research has recently been carried out, as reviewed in the following.

A. Binary Mixtures of Saturated-Acid Triacylglycerols

In mixtures of saturated monoacid TAGs whose chain length difference is not larger than two carbon atoms, the eutectic phase with a limited region of solid solution is formed for the stable β form, yet the miscible phases are formed in the metastable α and β′ forms. This property was precisely analyzed with a time-resolved SR-XRD study on the PPP-SSS mixtures as shown in Fig. 13 [74]. The 50:50 mixture of PPP:SSS crystallized in α by chilling the mixture liquid. The solid-solution phase of the mixture of α was verified by single long spacing spectrum. Upon heating, the α form transformed to β′ and subsequently to β. The mixture of the β′ form was also formed by cooling the liquid. Irrespective of the formation processes of the β′ form, a miscibility property of PPP and SSS was observed by the SR-XRD analyses. However, the miscibility was broken out when the β′ form transformed to β upon heating, as expressed in splitting of the long spacing spectra (Fig. 13b).

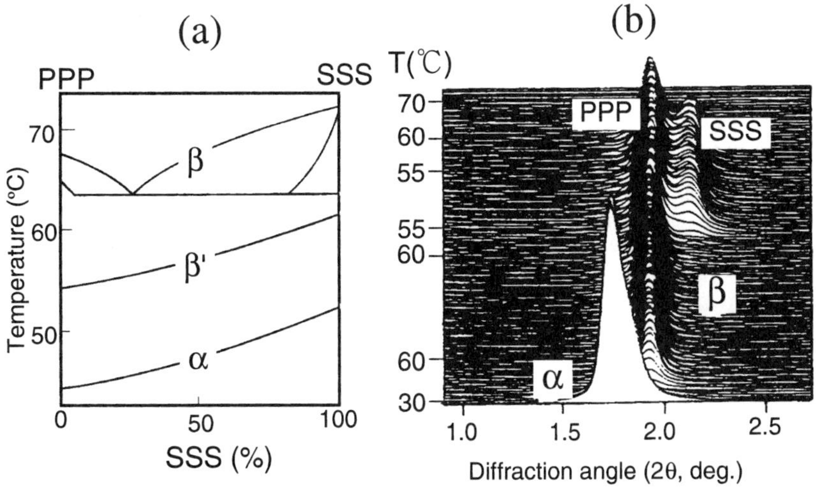

Fig. 13 (a) Phase diagram and (b) synchrotron radiation X-ray diffraction spectra of α-melt-mediated transformation into β of PPP:SSS = 50:50 mixture [74].

B. Binary Mixtures of Saturated-Unsaturated Mixed-Acid Triacylglycerols

In the mixtures between the saturated monoacid TAGs and unsaturated monoacid TAGs, neither miscible nor molecular compound forming systems were observed [93]. However, the formation of the molecular compound crystals was suggested in some binary mixture systems of saturated-unsaturated mixed-acid TAGs [93,95]. In these mixtures, it was assumed that specific chain-chain interactions between the saturated-saturated and unsaturated-unsaturated chains are operating, yet the details have been unveiled. Recent studies have shown that the molecular compound crystals are formed in various combinations of saturated-unsaturated mixed-acid TAGs such as SOS/SSO (1,2-distearoyl-3-oleoyl-*rac*-glycerol) [96], SOS/OSO [97], POP/PPO (1,2-dipalmitoyl-3-oleoyl-*rac*-glycerol) [78], and POP/OPO (1,3-dioleoyl-2-palmitoyl-*sn*-glycerol) [79]. In particular, the combined usage of SR-XRD and DSC has clarified thermodynamic (for the most stable states) and kinetic (for the metastable states) properties of the phase behavior of the molecular compound forming systems. In addition, the molecular structures of the compound crystals of POP/PPO and POP/OPO were assessed with the FT-IR spectroscopic method, indicating that the steric hindrance and glycerol interactions are operating in the formation of the molecular compound crystals [98].

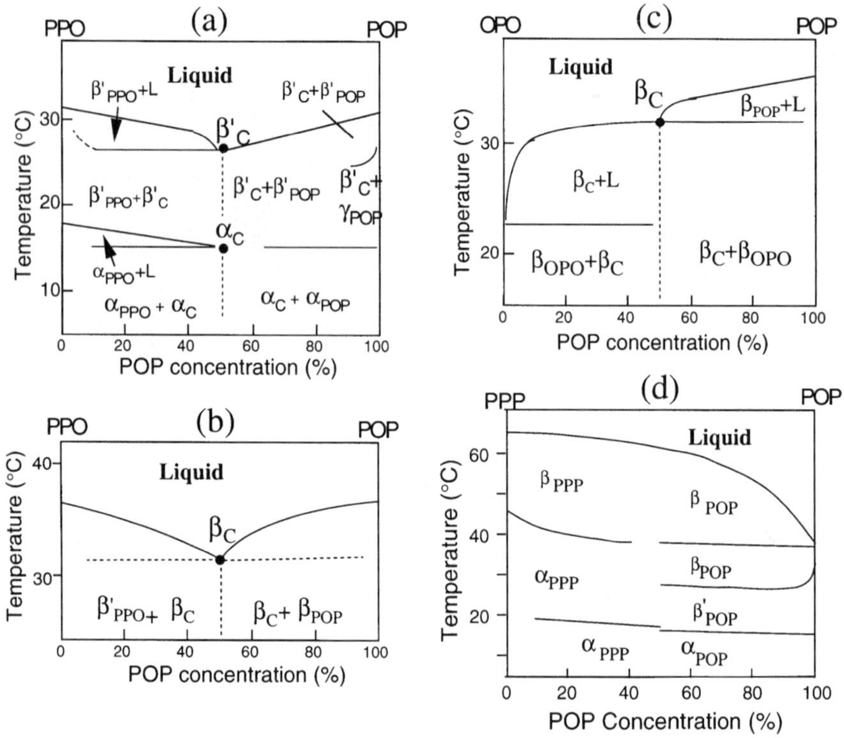

Fig. 14 Phase behavior of (a) metastable forms of PPO/POP mixture, (b) stable forms of PPO/POP mixture [78], (c) stable forms of OPO/POP mixture [79], and (d) metastable and stable forms of PPP/POP mixtures [77].

Figure 14 summarizes the phase diagrams of the binary mixtures between POP and PPO (a, b), OPO (c) and PPP (d). Except for the POP/PPP mixture, the formation of the molecular compound crystals was observed both in the most and metastable polymorphs. In particular, the kinetic phase diagram in Fig. 14a was obtained by a combined study of DSC and time-resolved SR-XRD. The following conclusions can be drawn from these results.

(a) Molecular compounds were formed at the 1:1 concentration ratios in the mixtures of POP/PPO and POP/OPO both in the stable and metastable states. As for the most stable forms, immiscible eutectic or monotectic phases were formed between the component materials and the molecular compounds in a juxtaposed way. For example, in Fig. 14a, α and β' formed the molecular compounds in the metastable forms, and at the same time β of the molecular compound (β_c)

formed the monotectic phases with β of POP (β_{POP}) and β′ of PPO (β'_{PPO}). β'_{PPO} is most stable in PPO.

As a clear evidence, Fig. 15 shows the DSC and SR-XRD data of the transformation in the mixture of POP/PPO = 10/90, in which 20% fraction of the mixture forms the molecular compound and the rest is PPO crystals. Starting

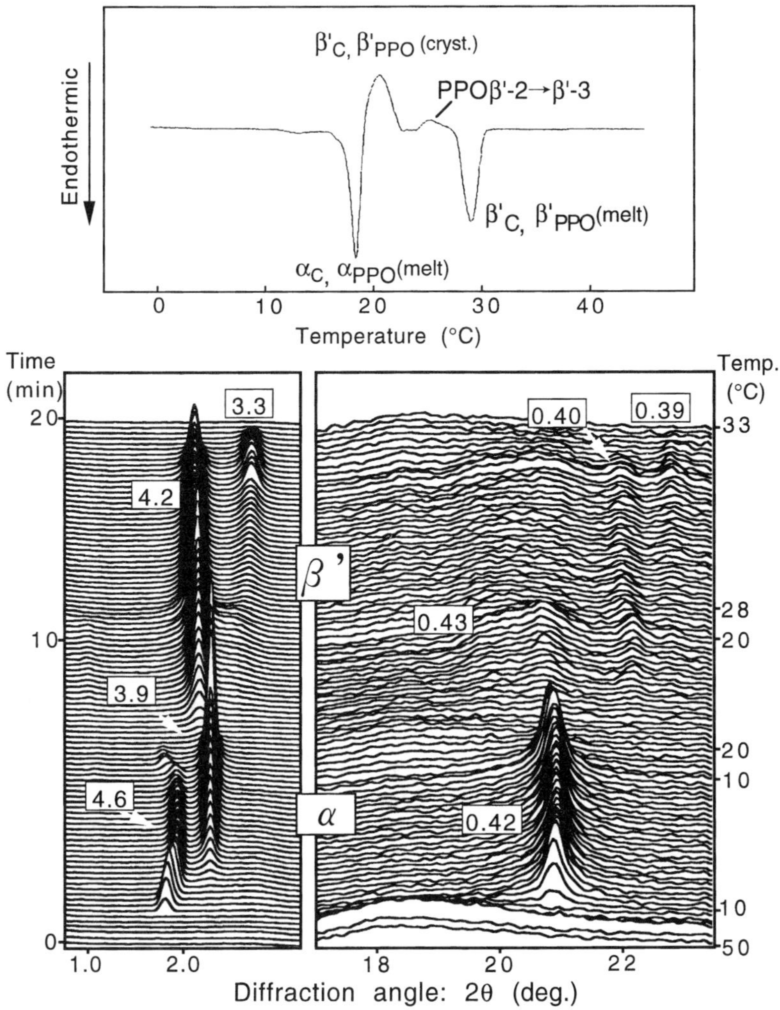

Fig. 15 DSC heating thermopeaks and synchrotron radiation X-ray diffraction spectra of α-melt-mediated transformation into β′ of POP:PPO = 10:90 mixture [78].

from the α forms both of the compound and PPO, the mixture converts into β' through α-melt mediation, and eventually melts after the rapid heating from 28 to 33°C [78]. Without the time-resolved SR-XRD spectra, it was impossible to identify each DSC thermopeak to the crystallization, transformation or melting of the individual crystal fractions of the mixture. However, the SR-XRD small- and wide-angle spectra made the identification of each DSC peak as shown in Fig. 15.

(b) The mixture of POP/PPP did not form the molecular compound crystals, instead immiscible monotectic phases were formed in the three polymorphic forms, as shown in Fig. 14d.

(c) The mixing behavior of the molecular compounds was common between PPO-POP and SSO-SOS, and between POP-OPO and SOS-OSO as well.

(d) In all cases, the molecular compounds are of double chain length, even though the stable forms of the component TAGs are all of triple chain length. This conversion in the chain length structure is primarily caused by molecular interactions through oleic acid chains which are packed in the same leaflets in the double layers, as depicted in Fig. 16.

(e) The subcell structures of the molecular compounds of PPO-POP transformed from H (α_C) to $T_{/\!/}$ (β_C) through $O_{\perp}$ (β'_C) in the same manner as those in the usual types of TAGs. As to molecular conformation of the oleic acid chains in the β_C forms, FT-IR spectroscopic analysis has shown that the olefinic group of β_C (POP-PPO) showed neither S-C-S′ type nor S-C-S type [98]. This means that the molecular compound was formed at the excess of determination energy of the olefinic conformation from the stable S-C-S′ type. This deformation may be caused by steric hindrance between the palmitic or stearic chains and oleic chains. However, this did not occur in β_C(POP-OPO), which revealed the S-C-S′ type.

The structural properties of the molecular compounds are summarized in the following: The $\alpha \rightarrow \beta' \rightarrow \beta$ transformation of the molecular compound crystals occurred in the double-chain-length structure, meaning that the chain segregation of the oleoyl and saturated chains into the two leaflets occurred as a result of steric hindrance. As for the comparison between POP/OPO (or SOS/OSO) and POP/PPO (or SOS/SSO), a complete chain segregation occurred between the saturated and oleic acid chains in the former case. β_C of POP/OPO exhibited the $T_{/\!/}$ subcell and S-C-S′ olefinic conformation. However, the β_C form of POP/PPO exhibited no specific olefinic conformation, neither S-C-S′ nor S-C-S, assuming that the oleoyl chains are distorted from those revealed in the β_C form of POP/OPO and SOS/OSO.

The formation of the molecular compounds in the binary mixtures of the unsaturated-saturated mixed-acid TAGs is related to some applications. For example, the addition of PPO into palm oil retarded the transformation rates from β' to β of palm fats. This may be interpreted by taking into account the molecular

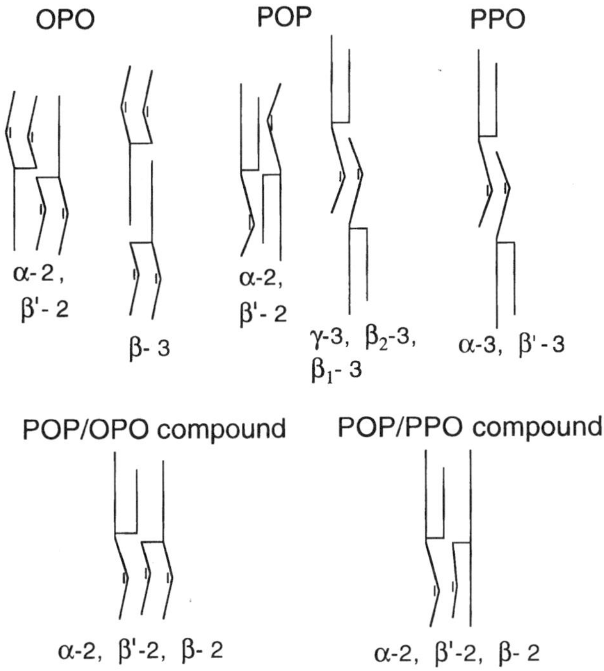

Fig. 16 Structure models of POP, PPO, OPO, POP/OPO compound, and POP/PPO compound crystals.

interactions between the compound of PPO and POP which are the major TAGs of palm oil. The β′ form of palm oil is double chain length, and β is triple chain length. The addition of PPO caused the formation of the molecular compounds in the α, β′, and β forms, all are of double chain length, which may retard the transformation of palm oil from β′ to β. The other outcome of the molecular compound formation is the formation of the β form in the compound, even if the component TAGs have β′ as the stable form. This was seen for the case of POP/PPO and SOS/SSO, in which β′ is the most stable form in PPO and SSO. This may be considered in blending the different fats forming the molecular compound crystals.

V. SUMMARY

This chapter has reviewed the recent work on the molecular interactions and kinetic properties of the polymorphic transformations of the TAGs in the single and mixed states. Progress has recently been made in the molecular-level understanding of the polymorphic transformations of the TAGs. Particularly, the use of the time-resolved X-ray diffraction with synchrotron radiation has provided precise information about the structural changes of the fat crystals at a time scale of 10 sec. Therefore, useful information was obtained on the kinetic and molecular aspects of crystallization and mixing processes of the various types of mixed-acid TAGs, which were not obtained with the traditional thermal and structural techniques because of their complicated structural properties. One may anticipate that, although the experimental sites and machine times are limited, the SR-XRD techniques will be more applied to the fat systems involving the following materials and systems: (a) multicomponent natural fats with and without additives of emulsifiers, proteins, and carbohydrates, (b) fats in dispersed phases such as O/W and W/O emulsions, and (c) crystallization and transformation processes under external influences of hydrostatic pressure, high-power ultrasound stimulation, and shear stress.

The analyses of very complicated mixed-acid TAGs must have critical implication for materials design and physical property control of structured fats. Above all, elucidation for diversity in β' structures of TAGs may be crucial, since it is related to structural deterioration of fat structures in foods and cosmetics, etc. As a first step, the atomic-level structure analysis of β'-stabilized TAGs, as exemplified above in the PP*n*, PEP, and $C_nC_{n+2}C_n$ groups, should be made. For this purpose, collaboration among chemical synthesis and structure analysis is necessary and promising, since it must give rise to fruitful results in exploring the unknown world of functional TAGs.

REFERENCES

1. D. M. Small, Glycerides, in *The Physical Chemistry of Lipids, from Alkanes to Phospholipids*, Handbook of Lipid Research Series, Vol. 4, D. J. Hanahan, ed., Plenum Press, New York, 1986.
2. F. D. Gunstone and F. B. Padley eds., *Lipid Technologies and Applications*, Marcel Dekker, New York, 1997.
3. I. Pascher, S. Sundull, and H. Hauser, *J. Mol. Biol., 153*: 791 (1981).
4. D. L. Dorset and W. A. Pangborn, *Chem. Phys. Lipids, 48*: 19 (1988).
5. H. Hauser, I. Pascher, and S. Sundell, *Biochemistry, 27*: 9166 (1988).
6. L. Di and D. M. Small, *J. Lipid Res., 34*: 1611 (1993).
7. M. Goto, K. Honda, L. Di, and D. M. Small, *J. Lipid Res., 36*: 2185 (1995).
8. I. Pascher, M. Lundmark, P. G. Nyholm, and S. Sundell, *Biochim. Biophys. Acta, 1113*: 339 (1992).
9. I. Pascher, *Curr. Opin. Struct. Biol., 6*: 439 (1996).
10. D. M. Small, *Annu. Rev. Nutr., 11*: 413 (1991).
11. A. E. Bailey, Melting and solidification of pure compounds, in *Melting and Solidification of Fats*, Wiley, New York, 1950, pp. 117–180.

12. E. S. Lutton, *J. Am. Oil Chem. Soc.*, *27*: 276 (1950).
13. R. E. Timms, *Prog. Lipid Res.*, *23*: 1 (1984).
14. N. Garti and K. Sato, eds., *Crystallization and Polymorphism of Fats and Fatty Acids*, Marcel Dekker, New York, 1988.
15. J. M. de Man, Chemical and physical properties of fatty acids, in *Fatty Acids in Their Health Implications*, C. K. Chow, ed., Marcel Dekker, New York, 1992, pp. 17–45.
16. F. Kaneko, J. Yano, and K. Sato, *Curr. Opin. Struct. Biol.*, *8*: 417 (1998).
17. K. Sato, S. Ueno, and J. Yano, *Prog. Lipid Res.*, *38*: 91 (1999).
18. J. Yano and K. Sato, *Food Res. Int.*, *32*: 249 (1999).
19. K. Larsson, *Acta Chem. Scand.*, *20*: 2255 (1966).
20. L. H. Jensen and A. J. Mabis, *Nature (London)*, *197*: 681 (1966).
21. L. H. Jensen and A. J. Mabis, *Acta Crystallogr.*, *21*: 770 (1966).
22. L. Hernqvist, Crystal structures of fats and fatty acids, in Ref. 14, Chap. 3, pp. 97–137.
23. M. Goto, D. R. Kodali, D. M. Small, K. Honda, K. Kozawa, and T. Uchida, *Proc. Natl. Acad. Sci. USA.*, *89*: 8083–8086 (1992).
24. K. Sato, Polymorphism of pure triacylglycerols and natural fats, in *Advances in Applied Lipid Res.*, F. B. Padley, ed., JAI Press, London, 1996, pp. 213–268.
25. D. M. Small, *J. Lipid Res.*, *25*: 1490 (1984).
26. D. R. Kodali, *J. Lipid Res.*, *28*: 464 (1987).
27. D. R. Kodali, D. Atkinson, T. G. Redgrave, and D. M Small, *J. Am. Oil Chem. Soc.*, *61*: 1078 (1984).
28. D. R. Kodali, D. Atkinson, and D. M. Small, *J. Phys. Chem.*, *93*: 4683 (1989).
29. D. R. Kodali, D. Atkinson, and D. M. Small, *J. Disp. Sci. Technol.*, *10*: 393 (1989).
30. D. R. Kodali, D. Atkinson, and D. M. Small, *J. Lipid Res.*, *31*: 1853 (1990).
31. J. Yano, F. Kaneko, M. Kobayashi, D. R. Kodali, D. M. Small, and K. Sato, *J. Phys. Chem. B. 101*: 8120 (1997).
32. K. Sato, M. Goto, K. Honda, J. Yano, D. R. Kodali, and D. M. Small, *J. Lipid Res.*, *42*: 338 (2001).
33. N. V. Lovegren and M. S. Gray, *J. Am. Oil Chem. Soc.*, *55*: 310 (1978).
34. A. J. Van Langevelde, K. van Massen, E. Sonneveld, R. Pechar, and H. Schenk, *J. Am. Oil Chem. Soc.*, *76*: 603 (1999).
35. J. van de Streek, P. Verwer, R. de Gelder, and F. Hollander, *J. Am. Oil Chem. Soc.*, *76*: 1333 (1999).
36. J. Yano, F. Kaneko, M. Kobayashi, and K. Sato, *J. Phys. Chem. B*, *101*: 8112 (1997).
37. U. Riiner, *Lebensm.-Wiss. Technol.*, *4*: 113 (1971).
38. Von D. Precht, E. Frede, and R. Greiff, *Fette Seifen Anstrichm.*, *80*: 344 (1978).
39. L. Hernqvist, and K. Larsson, *Fette Seifen Anstrichm. 84*: 349 (1982).
40. L. Hernqvist, *Fat Sci. Technol.*, *90*: 451 (1988).
41. P. J. M. W. L. Birker, S. de Jong, E. C. Roijers, and T. C. van Soest, *J. Am. Oil Chem. Soc.*, *68*: 895 (1991).
42. K. Sato and M. Kobayashi, *Organic Crystals I. Characterization, Crystals/Growth, Properties and Applications*, vol. 13, N. Karl, ed., Springer-Verlag, Heidelberg, 1991, pp. 65–108.
43. Z. H. Wang, K. Sato, N. Sagi, T. Izumi and H. Mori, *J. Jpn. Oil Chem. Soc. (Yukagaku)*, *36*: 671 (1987).

44. K. Sato, T. Arishima, Z. H. Wang, K. Ojima, N. Sagi, and H. Mori, *J. Am. Oil Chem. Soc.*, *66*: 664 (1989).
45. T. Arishima and K. Sato, *J. Am. Oil Chem. Soc.*, *66*: 1614 (1989).
46. T. Arishima, N. Sagi, H. Mori, and K. Sato, *J. Am. Oil Chem. Soc.*, *66*: 1614 (1989).
47. J. Yano, S. Ueno, K. Sato, T. Arishima, N. Sagi, F. Kaneko, and M. Kobayashi, *J. Phys. Chem.*, *97*: 12967–12973 (1993).
48. T. Arishima, K. Sugimoto, R. Kiwata, H. Mori, and K. Sato, *J. Am. Oil Chem. Soc.*, *73*: 1231–1236 (1996).
49. J. Yano, K. Sato, F. Kaneko, D. M. Small, and D. R. Kodali, *J. Lipid Res.*, *40*: 140–151 (1999).
50. F. Kaneko, J. Yano, and K. Sato, *Curr. Opin. Struct. Biol.*, *8*: 417–425 (1998).
51. K. Boubekri, J. Yano, S. Ueno, and K. Sato, *J. Am. Oil Chem. Soc.*, *77*: 949–955 (1999).
52. M. Takeuchi, S. Ueno, J. Yano, E. Floter, and K. Sato, *J. Am. Oil Chem. Soc.*, to be published.
53. H. Arakawa, T. Kasai, Y. Okumura, and S. Maruzeni, *J. Jpn. Oil Chem. Soc.* (*Yukagaku*), *47*: 19–24 (1998).
54. N. Sagi, T. Arishima, H. Mori, K. Sato, *J. Jpn. Oil Chem. Soc.* (*Yukagaku*), *38*: 306–311 (1989).
55. I. Hachiya, T. Koyano, and K. Sato, *J. Jpn. Oil Chem. Soc.* (*Yukagaku*), *66*: 1757 (1989).
56. I. Hachiya, T. Koyano, and K. Sato, *J. Jpn. Oil Chem. Soc.* (*Yukagaku*), *66*: 1763 (1989).
57. I. Hachiya, T. Koyano, and K. Sato, *J. Am. Oil Chem.*, *66*: 1757 (1989).
58. I. Hachiya, T. Koyano, and K. Sato, *J. Am. Oil Chem.*, *66*: 1763 (1989).
59. I. Hachiya, T. Koyano, and K. Sato, *Food Microstructure*, *8*: 257 (1989).
60. T. Koyano, I. Hachiya, and K. Sato, *Food Structure*, *9*: 231 (1990).
61. P. Rousset and M. Rappaz, *J. Am. Oil Chem.*, *73*: 1051 (1996).
62. P. Rousset and M. Rappaz, *J. Am. Oil Chem.*, *74*: 693 (1997).
63. A. Watanabe, I. Tashima, N. Matsuzaki, J. Kurashige, and K. Sato, *J. Am. Oil Chem. Soc.*, *69*: 1077 (1992).
64. S. de Jong, T. C. Soest, and M. A. van Schaick, *J. Am. Oil Chem. Soc.*, *68*: 371 (1991).
65. S. Abrahamssn, B. Dahlen, H. Lofgren, and I. Pascher, *Prog. Chem. Fats Other Lipids*, *16*: 125 (1978).
66. R. G. Snyder, M. C. Goh, V. J. P. Srivatsavoy, H. J. Strauss, and D. L. Dorset, *J. Phys. Chem.*, *96*: 10008 (1992).
67. J. W. Hagemann and J. A. Rothfus, *J. Am. Oil Chem. Soc.*, *69*: 429 (1992).
68. D. R. Kodali, D. Atkinson, T. G. Redgrave, and D. M. Small, *J. Lipid Res.*, *28*: 403 (1987).
69. P. Elisabettini, A. Desmedt, V. Gibon, and F. Durant, *Fat Sci. Technol.*, *97*: 65 (1995).
70. P. Elisabettini, G. Lognay, A. Desmedt, C. Culot, N. Istasse, E. Deffense, and F. Durant, *J. Am. Oil Chem. Soc.*, *75*: 285 (1998).
71. A. Desmedt, C. Culot, C. Deroanne, F. Durant, and V. Gibon, *J. Am. Oil Chem. Soc.*, *67*: 653 (1990).

72. M. Kellens, W. Meeussen, and C. H. Riekel, *Chem. Phys. Lipids*, *52*: 79 (1990).
73. M. Kellens, W. Meeussen, and G. Gehrke, *Chem. Phys. Lipids*, *58*: 131 (1991).
74. M. Kellens, W. Meeussen, A. Hammersley, and C. H. Riekel, *Chem. Phys. Lipids*, *52*: 79 (1990).
75. S. Ueno, A. Minato, H. Seto, Y. Amemiya, and K. Sato, *J. Phys. Chem. B*, *101*: 6847 (1997).
76. S. Ueno, A. Minato, J. Yano, and K. Sato, *J. Cryst. Growth*, *198/199*: 1326–1329 (1999).
77. A. Minato, S. Ueno, J. Yano, Z. H. Wang, H. Seto, Y. Amemiya, and K. Sato, *J. Am. Oil Chem. Soc.*, *73*: 1567 (1996).
78. A. Minato, S. Ueno, K. Smith, Y. Amemiya, and K. Sato, *J. Phys. Chem. B*, *101*: 3498 (1997).
79. A. Minato, S. Ueno, J. Yano, K. Smith, H. Seto, Y. Amemiya, and K. Sato, *J. Am. Oil Chem. Soc.*, *74*: 1213 (1997).
80. R. N. M. R. van Gelder, N. Hodgson, K. J. Roberts, and A. Rossi, in *Crystal Growth of Organic Materials*, A. S. Myerson, D. A. Green, and P. Meenan, eds., Am. Chem. Soc., New York, 1996, pp. 209–215.
81. S. D. MacMillan K. J. Roberts, A. Rossi, M. Wells, M. Polgree, and I. Smith, in *Proceedings of World Congress on Particle Technology*, Brighton, 1998, pp. 96–103.
82. C. Loisel, G. Keller, G. Leek, C. Bouraux, and M. Ollivon, *J. Am. Oil Chem. Soc.*, *75*: 425 (1998).
83. T. M. Eads, A. E. Blaurock, R. G. Bryant, D. J. Roy, and W. R. Croasman, *J. Am. Oil Chem. Soc.*, *69*: 1057 (1992).
84. K. Larsson, *Fette Seifen. Anstrichm.*, *76*: 136 (1972).
85. P. T. Callaghan and K. W. Jolley, *Chem. Phys. Lipids*, *67*: 4773 (1977).
86. D. J. Cebula, D. J. McClements, and M. J. W. Povey, *J. Am. Oil Chem. Soc.*, *67*: 76 (1990).
87. D. J. Cebula, D. J. McClements, M. J. W. Povey, and P. R. Smith, *J. Am. Oil Chem. Soc.*, *69*: 130 (1992).
88. K. Larsson, *J. Am. Oil Chem. Soc.*, *69*: 835 (1992).
89. K. van Malssen, R. Peschar, C. Brito, and H. Schenk, *J. Am. Oil Chem. Soc.*, *73*: 1225 (1996).
90. R. L. Wille and E. S. Lutton, *J. Am. Oil Chem. Soc.*, *43*: 491 (1966).
91. V. Gibon, F. Durant, and Cl. Deroanne, *J. Am. Oil Chem. Soc. 63*: 1047 (1986).
92. F. Lavigne, and M. Ollivon, *J. Med. Calorim. Anal. Therm.* 237 (1993).
93. J. B. Rossel, *Adv. Lipid Res.*, *5*: 353 (1967).
94. A. I. Kitaigorodsky, *Mixed Crystals*, Springer Series in Solid State Sciences, vol. 33, Springer-Verlag, Berlin, 1984.
95. D. P. Moran, *J. Appl. Chem.*, *13*: 91 (1963).
96. L. J. Engstrom, *Fat Sci. Technol.*, *94*: 173 (1992).
97. T. Koyano, I. Hachiya, and K. Sato, *J. Phys. Chem.*, *96*: 10514 (1992).
98. A. Minato, J. Yano, S. Ueno, K. Smith, and K. Sato, *Chem. Phys. Lipids*, *88*: 63 (1997).

6

The Roles of Emulsifiers in Fat Crystallization

Nissim Garti
The Hebrew University of Jerusalem, Jerusalem, Israel

Junko Yano
Hiroshima University, Higashi-Hiroshima, Japan

I. INTRODUCTION

Long-chain compounds, such as fatty acids and their esters, may occur in different crystal forms, known as polymorphisms [1–4]. Fat polymorphism, therefore, describes phase changes and structural modifications of solid fat phase. The polymorphic fat forms differ in melting points, solubility, crystal morphology and network, rheology, etc. In particular, it is widely acknowledged that the habit of fat crystals is related to the polymorphic state [5]. This property significantly affects the physical and functional properties of food products such as solid-liquid separation, rheology, and particle flow.

As reviewed in Chapter 5, the polymorphism of fats is a complicated phenomenon that is influenced by their composition and many environmental factors. In the most cases, the physical properties of stable polymorphs are not preferable in fat products due to high melting point, slow crystallization, large crystalline sizes, hardening, and so on. In addition, the polymorphic transformation itself can be a source of trouble, since it sometimes causes the phase separation in fat products by destroying the fine-grained network between fat crystals themselves and/or between fats and other ingredients. Therefore, controling the polymorphic transformation of triglycerides and altering the physical properties of fats and their rheological behavior have been the subjects for technologists and scientists.

As a main strategy, the use of emulsifiers has drawn attention for years. The emulsifiers alter the properties of the fat surface and the fat crystallization process, resulting in an altered solid fat content and crystal size. Furthermore, the emulsifiers reduce the oil/water interfacial energy, enabling the formation of water-in-oil (W/O) or oil-in-water (O/W) emulsions. In such confined spaces compartmentalized by the emulsifiers, it has been expected that crystallization phenomena differ from those in bulk systems. As a practical matter, the crystallization of oil phase in O/W emulsion is an important process for coagulation of emulsions occurring during chilled states, deemulsifying process of whipping creams, freezing ice cream, etc. [6–8].

An early review by van den Tempel (1968) [9] and a recent one by Garti [10] have showed that many types of emulsifiers tended to reduce the crystal growth rate of natural fat blends. Since then, further work was performed on the effects of different emulsifiers on fats not only in bulk, but also in emulsion systems. This chapter reviews such recent work carried out under the focus on the effects of emulsifiers on fat systems.

II. CLASSIFICATION OF FATS-EMULSIFIERS INTERACTION

Emulsifiers essentially consist of two parts, hydrophobic and hydrophilic moieties. On the other hand, fats are neutral lipids with minimized hydrophilic nature. Therefore, two different effects are operating for the interactions between the two molecules. Hydrophobic moieties have attractive forces with fat molecules. Depending on homogeneity of the fatty acid moiety, such as chain length and presence of double bond, emulsifiers are physically adsorbed at the fat crystal surface or incorporated in fat crystals. On the other hand, the hydrophilic groups of the emulsifier give rise to repulsive forces, which depend on volume size and chemical formula of the hydrophilic moieties of emulsifiers. Furthermore, emulsifiers reduce the oil/water interfacial energy, which improves the stability of W/O and O/W emulsions.

The following essential cases are considered when the emulsifiers work on fat crystallization in the bulk and emulsion systems (Cases 1–3 in bulk system and Case 4 in emulsion) (Fig. 1).

Case 1. Limited amount of emulsifiers is miscible in fat systems. Small quantity of emulsifiers incorporated in fat crystals acts as impurities and results in imperfections of fat crystals (left side of the phase diagram of Case 1 in Fig. 1). It will promote or retard polymorphic transformation of fat crystals.

Case 2. Fats and emulsifiers are highly miscible. In this case, fats-emulsifiers binary phase shows miscible phase behavior or molecular compound formation.

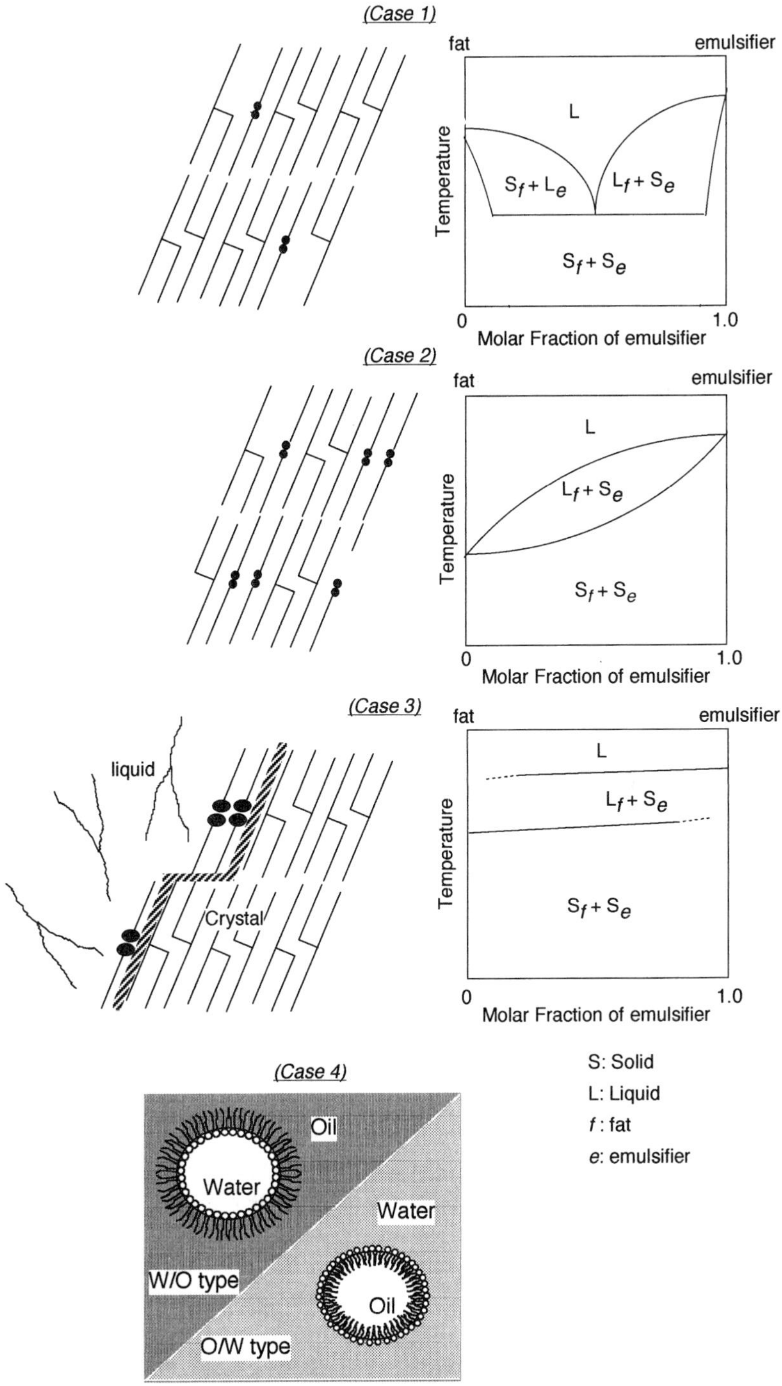

Fig. 1 Classification of fats-emulsifiers interaction. In Cases 1–3, typical phase diagrams of fat-emulsifier binary mixture are also shown. Case 4 shows two types of emulsions, W/O and O/W.

Case 3. Fats and emulsifiers are totally immiscible in a solid state. In nucleation, emulsifiers may work as seeds of fat crystals under special conditions, for example, a template film. In crystal growth, the emulsifiers are adsorbed at steps or kinks of fat crystals at crystal-liquid interfaces and inhibit crystal growth and thereby modify crystal morphology.

Case 4. Emulsion systems. In emulsion, hydrophobic tails of emulsifiers are faced with fat molecules at the water/oil interface, which may induce fat-emulsifier interactions. Then the acceleration of nucleation may occur at the interface of the emulsion, onto which the adsorbed hydrophobic emulsifiers become templates for heterogeneous nucleation.

In this chapter we are mainly concerned with the Cases 2–4.

III. FOOD EMULSIFIERS: STRUCTURES AND PROPERTIES

Food emulsifiers are amphiphilic molecules having long fatty chains and hydrophilic head group. The amphiphiles migrate from bulk water or from lipid/oil/fat systems to the available interface (liquid or solid) spontaneously. It will gain, as a result, some free energy. The list of amphiphiles that is permitted for food applications is rather restricted. Table 1 lists the major food emulsifiers and the levels of use with accordance to the FDA regulations. Most of the food emulsifiers have fatty acyl chains in their structure, such as stearic (in most cases), palmitic (in some cases), and oleic/linoleic as hydrophobic/lipophilic moieties. The fatty emulsifiers consisting of saturated fatty acids are solids at room temperature, while those based on unsaturated fatty acids are waxy or liquids. Those characteristics are important when dealing with the crystallization phenomena.

A. Lecithins

Lecithins (mixtures or purified) are the only natural occurring amphiphiles. They are considered as GRAS (generally recognized as safe) and permitted in foods as additives without any significant restrictions on the added quantities (except bitter flavor that self-restricts the levels of use). Lecithins are extracted from soya and egg. The purification stages include degumming (hydration degumming or acid superdegumming) by chemical treatment, drying, and cooling. The lecithins (Fig. 2) are complex mixtures of phospholipids (ca. 51 wt%), natural lipids, free fatty acids (ca. 38 wt%), glycolipids (ca. 7 wt%), ca. 3 wt% carbohydrates, and ca. 1 wt% water. The lecithins can be further purified by removing the oils/lipids (deoiled powdered lecithin) and/or by fractionating the phospholipids. Commercial lecithins contain 18% phosphatidylcholine (PC), 15% phosphatidyl ethanol-

Table 1 Food Emulsifiers and Their Legal Status

Chemical name	Abbreviation	ADI value[a]	EEC no.	US FDA 21 CFR
Lecithin	—	not limited	E322	§182.1400[b]
Monodiglycerides	MG	not limited	E471	§182.4505[b]
Acetic acid esters of mono-diglycerides	AMG (ACETEM)	not limited	E472a	§172.828
Lactic acid esters of mono-diglycerides	LMG (LACTEM)	not limited	E472b	§172.852
Citric acid esters of mono-diglycerides	CMG (CITREM)	not limited	E472c	—
Diacetyl tartaric acid esters of monoglycerides	DATEM	50	E472e	§182.4101[b]
Succinic acid esters of mono-diglycerides	SMG	—	—	§172.830
Salts of fatty acids (Na,K)	—	not limited	E470	§172.863
Polyglycerol esters of fatty acids	PGE	0–25	E475	§172.856
Propylene glycol esters of fatty acids	PGMS	0–25[c]	E477	§172.856
Sodium stearoyl lactylate	SSL	0–20	E481	§172.846
Calcium stearoyl lactylate	CSL	0–20	E482	§172.844
Sucrose esters of fatty acids	—	0–10	E473	§172.859
Sorbitan monostearate	SMS	0–25	491	§172.842
Polysorbate-60	PS 60	0–25	435	§172.836
Polysorbate-65	PS 65	0–25	436	§172.838
Polysorbate-80	PS 80	0–25	433	§172.840

[a] Acceptable daily intake in mg/kg body weight per day.
[b] Generally recognized as safe (GRAS).
[c] Calculated as propylene glycol.

amine (PE), 11% phospatidyl inositol (PI), 9% polar lipids, and 12% sterol glycosides. Lecithins are major components in many fat applications since they are known as rheology controllers. The activity of lecithin in the bulk fats/oils was believed to control a bulk property behavior. Recent sophisticated work shows that the action is more complex and related to the interference with the fractal network formation of the fat [11]. Its activity in the bulk will be further discussed. However, its effect on polymorphism of fats and polymorphic transformation is not much documented and our knowledge is quite restricted.

Lecithin can be modified chemically and enzymatically. The modifications can be done on the double bond of the unsaturated fatty acid (hydroxylation,

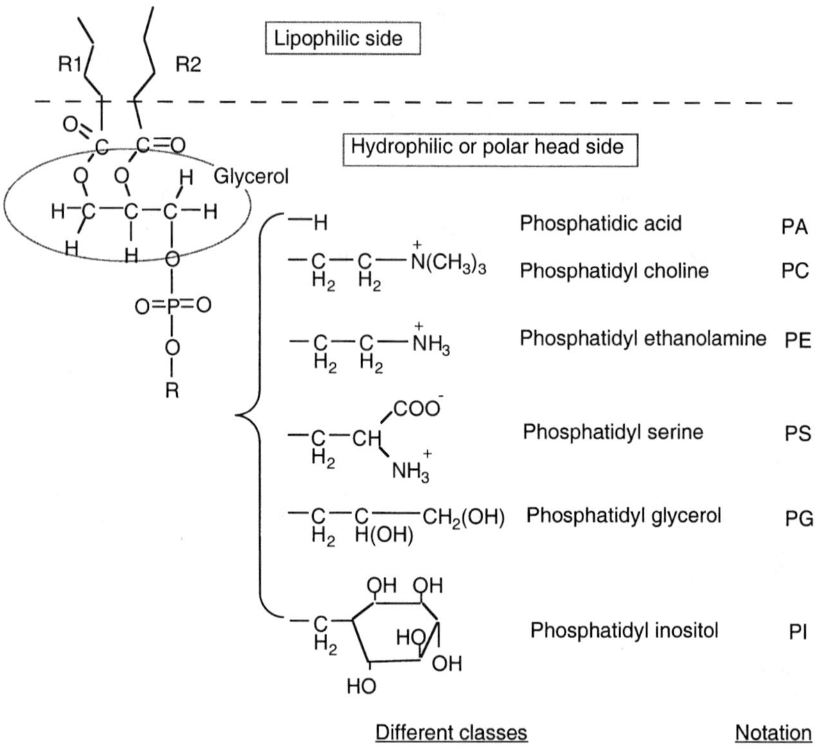

Fig. 2 Six types of glycerophospholipids according to the nature of the substituent, as well as their usual abbreviated notation. The most well known (and often used) is phosphatidylcholine (PC), also called lecithin.

phosphorylation, sulfosuccination, hydrogenation, etc.), on the reactive amino group of the head groups (acylation, methylation), on the glycerol structure (hydrolysis by removal of one fatty acid from position 1 or 2 to form 1-lysolecithin or 2-lysolecithin), by cleavage of the serine, choline, ethanolamine, or the phosphate group, or by transesterification (exchange of fatty acids on the glycerol structure).

The main modified lecithins that have been considered for food applications and fats are lysolecithins (Fig. 3). It has been a good idea to hydrophilize lecithin by attaching its tail chains to hydrophilic functional groups like hydroxyl or epoxy or by hydrolyzing (and removing) one of its tails (fatty acid) to form lysolecithin. Lysolecithin is by far more hydrophilic and can act as a good O/W emulsifier.

Lysolecithin exists in nature but its quantities are very minute and no commercial processes for its direct extraction exist. Attempts have been made to enzymatically hydrolyze lecithin by phospholipase A_2 (PLA_2) to form lysoleci-

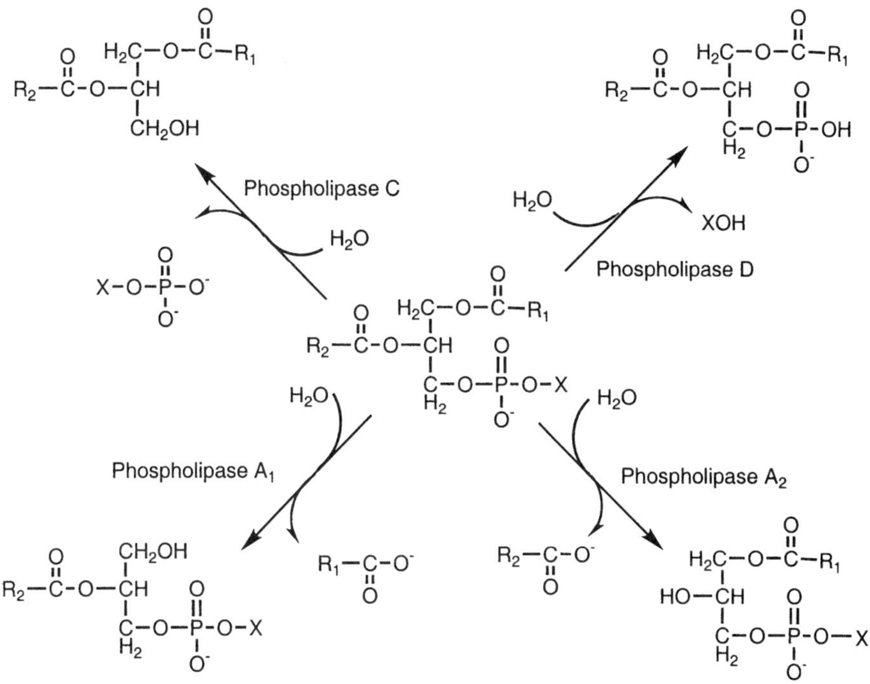

Fig. 3 Cleavage of functional groups by different types of phospholipases.

thin. The product is a good imitation of what nature offers in very small quantities. The partially hydrolyzed lecithins or the fully converted lysolecithins are the subject of recent work conducted to carry the reactions in ''microemulsions as microreactor'' for the PLA_2 enzymatic process [12].

The crystal structures of pure lecithins, such as DLPE·HAc (2,3-dilauroyl-DL-glycerol-l-phosphorylethanolamine acetic acid) [13], DMPC · $2H_2O$ (2,3-dimyristoyl-D-glycerol-1-phosphorylcholine dihydrate) [13], LPPC · H_2O (3-lauroylpropanediol-1-phosphorylcholine monohydrate) [13], and LPPE (3-palmitoyl-DL-glycerol-1-phosphorylethanolamine) [14], were studied some years ago. All the lecithins are arranged in a bilayer structure, and the crystal system is always monoclinic with space group *P*2 and four molecules per unit cell.

A very important amphiphilic ingredient in confectionery products is the so-called YN lecithin, which is essentially ammonium phosphatide. It is prepared from partially hardened rapeseed oil by a series of reactions which include glycerolysis, phosphorylation, neutralization with ammonia, and filtration. The structural formula of the principal component is presented in Fig. 4. Lecithin YN is

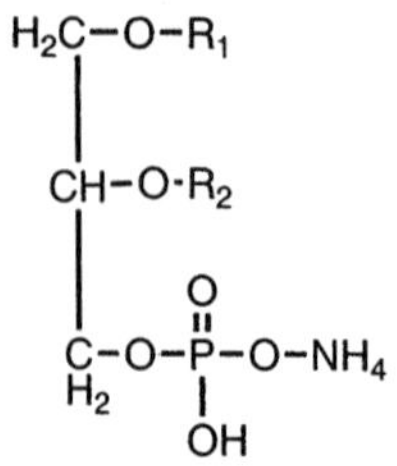

Fig. 4 The principal structural formula of ammonium phosphatide. At lease one of R_1 and R_2 represents a fatty acid moiety and the other may represent a fatty acid moiety or hydrogen.

practically ammonium phosphatide. It has been approved as an additive to cocoa butter and chocolate products under EEC No. 442. The characteristic specifications are 3–3.5% phosphorus, max. 2.5% of insoluble petrol ether, and a pH of 6–8.

B. Mono- and Monodiglycerides

Commercial mono- and monodiglycerides are glycerol fatty acid esters. The products are obtained from direct esterification of glycerol with saturated (iodine value less than 1) or unsaturated fatty acids or from transesterification of fat with glycerol. The products are mixtures of ca. 40% mono (1-glycerol monostearate or palmitate) and ca. 40% distearate (1,3-distearoylglycerol), known as MDGS, with ca. 19% free glycerol, tristearates, and free fatty acids. Upon molecular distillation a significantly purer compound, with a minimum 90% monoesters, is obtained. The product is known as GMS.

GMS and/or MDGS are used as additives to fats in many applications. In molten fats they serve as α tenders (keeping the fat in the final form in the less stable α gel form). The monoglycerides are used also in W/O emulsions (margarine and spreads) as emulsifiers and as stabilizers, and in the ''hydrated form'' or in the ''self-emulsifying form'' (with some soaps added) in various aerated products.

Crystal structures of mono- and monodiglycerides have been extensively studied [15–17]. In the 1,3-diglyceride, the two acyl chains extend to the opposite direction, while in the 2,3-diglyceride the *sn*-3-chain bends back to lie parallel to the *sn*-2-chain. Acetic, lactic, tartaric, and citric acid esters of mono- and diglycerides have been mentioned as possible amphiphiles to retard the polymorphic transformation. However, only few reports exist and the data is scarce and not sufficiently reexamined.

C. Polyglycerol Polyricinoleate

Polyglycerol polyricinoleate (PGPR) is one of the most hydrophobic emulsifiers used in foods (Fig. 5). It has gained much attention since it was recently approved for use in certain confectionery applications, mainly in chocolate-based coatings. Up to 2.64 mg/kg body/day are permitted in the United Kingdom. The product was first introduced by Unilever in patent issued in 1952 [18]. The product is based on a three-step reaction. In the first step glycerol is polymerized at elevated temperatures (ca. 250°C) to form polyglycerol (tri-, tetra-, and pentaglycerol ether). In the second step polycondensed ricinoleic acid is formed from ricinoleic acid [18,19]. An esterification between polyricinoleic acid and polyglycerol is thereafter carried out at lower temperatures to form oligomeric polyglycerol polyricinoleate. Several manufacturers have mastered the process and adjusted the internal composition to its performance in chocolate as a rheology controller. The internal product composition and structure are trade secrets and propriety information of the manufacturers. PGPR has exceptionally good water binding property and this property is of major importance for its effect in chocolate.

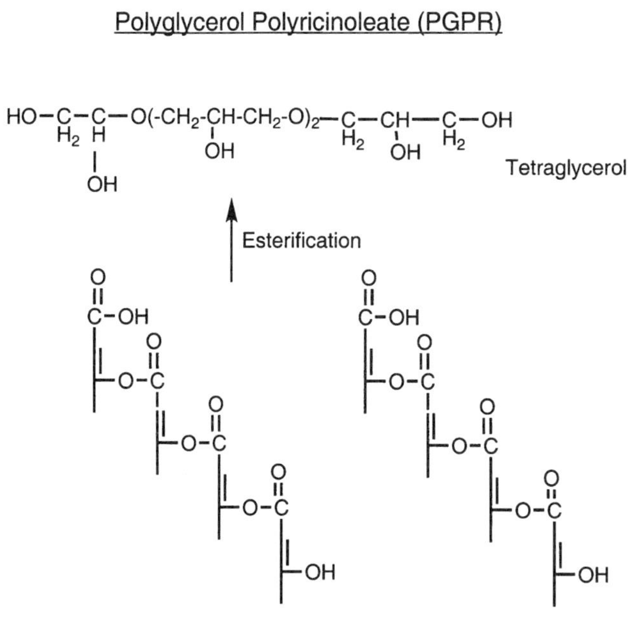

Fig. 5 The principal structural formula of polyglycerol polyricinoleate (PGPR).

It is important to note that PGPR has long been known to reduce the yield value of chocolate mass, whereas the effect on the viscosity is limited [18]. Figure 6 shows the decrease in viscosity, caused by PGPR in the presence of 0.3 wt% of lecithins or ammonium phosphatides (YN). Up to 0.3 wt% lecithins are necessary (as demonstrated from viscosity measurements) to obtain full coverage of the solid particles. Further increase of lecithin levels up to 0.5 wt% of chocolate mass does not have any significant viscosity reduction effect.

Figure 7 shows the flow curve for levels of a typical combination of emulsifiers in a milk chocolate [18]. The synergy between PGPR and lecithin or ammonium phosphatide is clearly shown. There is no difference in the yield value obtained whether lecithin or ammonium phosphatide is used. The slope of the curve, however, shows that ammonium phosphatide is somewhat superior as to reducing viscosity. A dosage of 0.3% PGPR lowers the yield from 26 to 2 Pa. An addition of 0.9% lecithin lowers the yield value only to 24 Pa. No structural or other physical evidence exists to support this claim. By increasing the addition

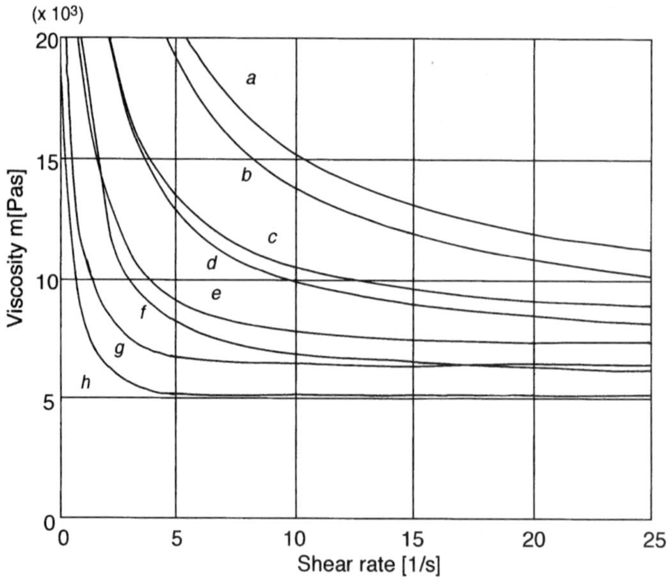

Fig. 6 The viscosity reduction effect of PGPR (P-4125) when lecithin or ammonium phosphatide is present at 0.3%: (a) 0.3% lecithin, (b) 0.3% YN, (c) 0.3% lecithin + 0.1% P-4125, (d) 0.3% YN + 0.1% P-4125, (e) 0.3% lecithin + 0.2% P-4125, (f) 0.3% YN + 0.2% P-4125, (g) 0.3% lecithin + 0.3% P-4125, (h) 0.3% YN + 0.3% P-4125 [18]. The first two curves showing 0.3 addition of lecithin (a) and ammonium phosphatide (b) only are shown as a reference.

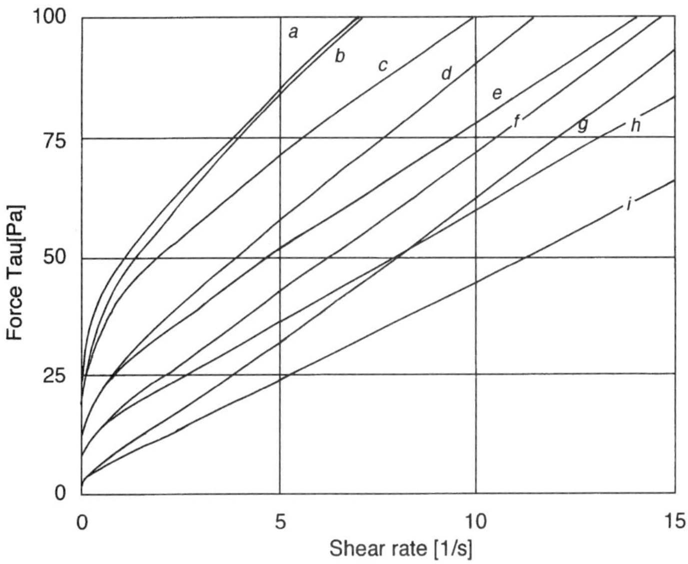

Fig. 7 The flow curves for levels of a typical combination of emulsifiers in a milk chocolate: (a) 0.4% lecithin (yield value, 26), (b) 0.4% YN (22), (c) 0.9% lecithin (24), (d) 0.4% lecithin + 0.1% P-4125 (13), (e) 0.4% YN + 0.1% P-4125 (13), (f) 0.4% lecithin + 0.2% P-4125 (8), (g) 0.4% lecithin + 0.3% P-4125 (2), (h) 0.4% YN + 0.2% P-4125 (8), (i) 0.4% YN + 0.3% P-4125 (2) [18]. As a reference, the flow curves for 0.4% lecithin (a) and ammonium phosphatide (b) are shown.

of PGPR, low viscosity is achieved at low shear rates. Although synergistic effects with lecithin and ammonium phosphatide are known and documented, its mechanism of action is not yet well studied. The manufacturers claim that all three emulsifiers form a monolayer around the nonfat particles, especially around the sugar particles. The sugar particles will thereby be covered by an outer lipophilic layer that binds to the oil (or fat). These surfaces will slide easily against each other when compared with the surfaces of wet sugar particles, which tend to stick to each other. This is a sort of greasing effect that lowers both the plastic viscosity and the yield value. The better ability of PGPR to bind water in comparison with lecithin is explained in terms of its molecular size and sites of binding. In spite of the fact that lecithin and PGPR are believed to primarily and solely bind onto sugar particles, it was recently demonstrated that lecithin, monoglycerides, and other amphiphiles are adsorbed also on fat particles [20]. The competitive adsorption of the emulsifiers on fat and sugars might play an important role in controlling the rheology as well as the polymorphism of the fats. PGPR was not extensively studied, and its adsorption isotherm on fat crystals is unknown. The role of PGPR in the bulk crystallization is also unknown. However, the necessity of PGPR to chocolate producers, ice cream makers, and food technologists is growing. More fundamental work is still needed in view of the reactivity of PGPR in the crystallization process and emulsion formation of fats and cocoa-butter-based products.

D. Sorbitan Esters

Sorbitan esters have been known to the food industry for almost 50 years. They are derived from a simple reaction between sorbitol and fatty acids (stearic, palmitic, oleic, and lauric) to form 1,4- and 1,5-sorbitan monoesters (mostly mono- but di- and triesters also exist). The products are known as Span 60, 40, 80, and 20, respectively. More hydrophilic derivatives are obtained by ethoxylation with ca. 20 moles of ethylene oxide. These complex mixtures are known as Tweens and are widely used in the food industry.

Sorbitan esters are hydrophobic amphiphiles; they can better adsorb onto fat particles and more easily dissolve in molten fat (prior to the cooling process). Of special interest is the sorbitan esters of fatty acids that are solid at room temperatures such as Span 60 and 40. Sorbitan can be esterified with more than one mole of fatty acids to form very lipophilic compounds such as sorbitan tristearate (known also as Span 65). This molecule has three fatty acids extending from the sorbitol head group and has strong tendency to adsorb on fat surfaces from the molten fat during crystallization. It is therefore a good potential molecule for studying polymorphic transition retardation effects.

The activity of sorbitan esters in fat crystallization processes will be further discussed later.

E. Sugar Esters

Sugar esters are new to Western food technologists. The family of amphiphilic molecules has been known since the forties. However, since its production is difficult and expensive, there was not much interest to use or to explore these molecules as crystal structure modifiers of fats. Only recently have sugar esters been approved for food applications, and their prices and availability have improved [21–26].

Sugar esters are products derived from esterification (fatty acids and sucrose) or transesterification (sucrose and fats) (Fig. 8). A large variety of products exist since sugar can be esterified at up to eight hydroxyl positions with as many as eight fatty acids. Monoesters (ca. 70% monoesters and the rest di- and polyesters) are hydrophilic, while di-, tri-, and tetraesters are increasingly more hydrophobic. Esters of stearic, palmitic, oleic, and lauric acids are available. It

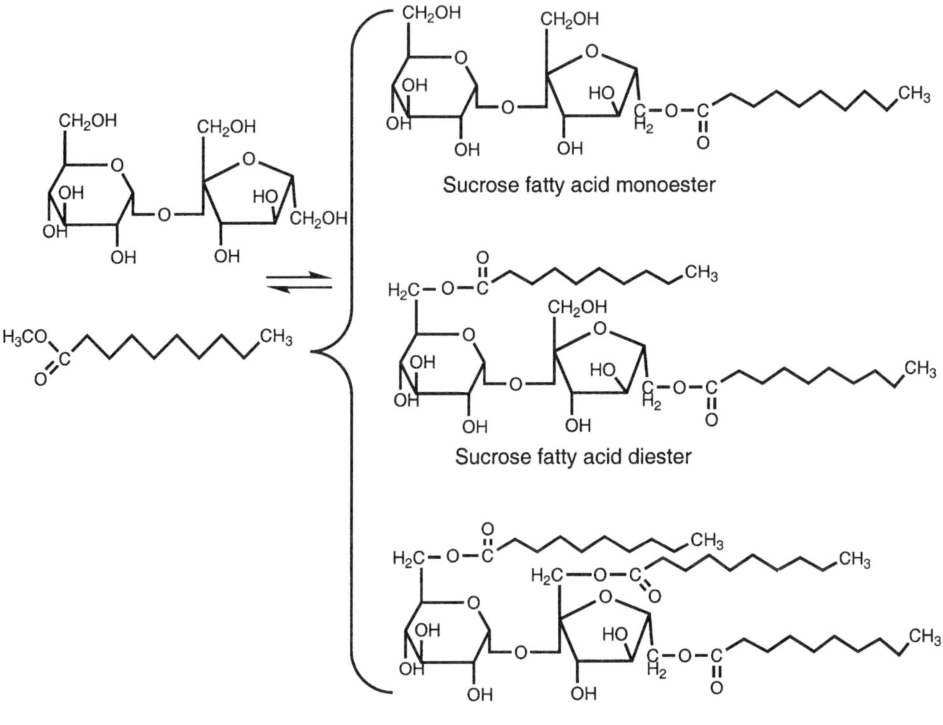

Fig. 8 Synthesis of sucroce esters by transesterification of fatty acid methyl esters.

should be noted that the highly substituted sugar esters are not considered emulsifiers and are not permitted as food additives.

Sugar esters gained the interest of scientists studying fat polymorphism, and some interesting work was carried out. The activity of the sugar esters is further discussed later.

IV. EMULSIFIER EFFECTS ON CRYSTALLIZATION KINETICS

A. Nucleation

In fat systems some emulsifiers act as heteronuclei. Nucleation of fat crystals is accelerated through catalytic actions of such impurities. This effect is rather obvious in emulsion systems [27]. It has been indicated that the acceleration of nucleation occurs at the interface of the emulsion, onto which the adsorbed hydrophobic emulsifiers become templates for heterogeneous nucleation.

As a model system of heterogeneous nucleation, crystallization of oil phase in O/W emulsions has been studied systematically. Figure 9 shows the variation in the ultrasonic velocity values of *n*-hexadecane in the O/W emulsions with and without the addition of sugar esters with stearic moieties (S-170), measured during cooling from 20 to −5°C. On cooling, the ultrasonic velocity values abruptly increased at 2°C in pure microemulsion, due to the crystallization of the *n*-hexadecane phases in the emulsion. The crystallization temperature, T_c, was increased with increasing amount of S-170 (from 0.01 to 1 wt%) from 2 to 12°C. This type of acceleration with added hydrophobic emulsifiers on nucleation was not observed in a bulk system. Therefore, it is considered that the fatty acid chains of S-170, which are adsorbed at the oil-water interface at the molecular level, are solidified on cooling and thereby play the role of a catalytic template film for heterogeneous nucleation of *n*-hexadecane.

Figure 10 shows the T_c variations of *n*-hexadecane in O/W emulsion systems with the additives of four sugar esters having stearic (S-170), palmitic (P-170), lauric (L-195), and oleic (O-170) acids. The addition of S-170 and P-170 showed similar profiles of increasing T_c of *n*-hexadecane through the two stages, depending on their concentrations. Although in a rather moderate manner, the increase in T_c due to L-195 showed basically the same two-stage processes [27].

The two-stage processes can be discussed by taking into account the adsorption of the sucrose oligoester molecules at the oil-water interface and the formation of a reversed micelle phase (Fig. 11). The sugar esters employed in the present study are rather lipophilic and, therefore, are strongly adsorbed onto

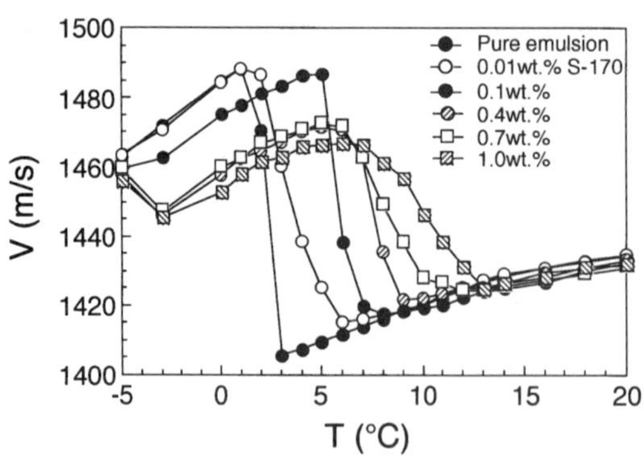

Fig. 9 Temperature variation of ultrasonic velocity (V) of *n*-hexadecane/water emulsion with the additives of S-170 during cooling processes [27].

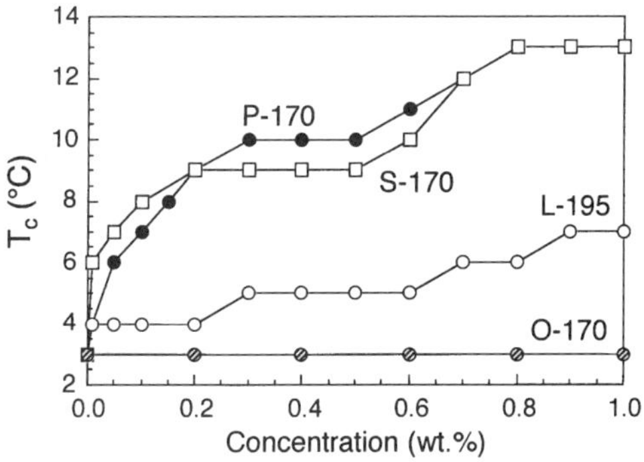

Fig. 10 Variations in crystal growth rate of *n*-hexadecane from bulk liquid at T_c = 17.1°C with P-170, S-170, L-195, and O-170 (1.0 wt% added) [27].

the oil-water interface in the surfactant/water/oil systems [27,28]. Even in the surfactant/oil systems, sugar esters are not solubilized into the oil phase and form molecular aggregates such as reversed micelles in the oil phase [29,30]. Accordingly, when the concentrations of the additives of sugar esters in *n*-hexadecane in the O/W emulsion systems are low, all of them would be adsorbed at the water/oil interface. S-170 and P-170 are adsorbed at the interface and may accelerate the heterogeneous nucleation of the oil phase at the surface of droplets when the O/W emulsion was cooled. The DSC measurement on the crystallization behavior of *n*-hexadecane with addition of P-170 suggested the formation of the molecular aggregations with Tween 20 on the oil-water interface [31].

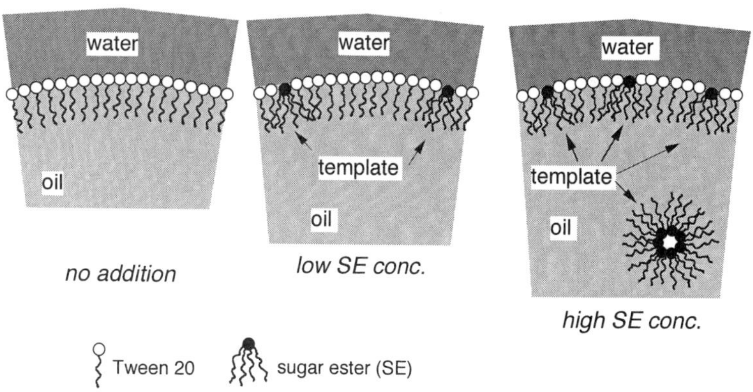

Fig. 11 A model of interface heterogeneous nucleation in O/W emulsions [27].

With higher concentration of the additives, the adsorption at the interface is saturated and excess molecules of the sugar esters may form reversed micelles. Such molecular aggregates might be solidified in the liquid oil phase on the cooling process and accelerate the nucleation in the oil phase [31]. Therefore, as to the two stages of the acceleration of nucleation shown in Fig. 10, the first rapid increase may correspond to the heterogeneous nucleation at the interface, and the second increase is ascribed to the formation of the reversed micelles.

B. Crystal Growth

The transformation of triglycerides from unstable or metastable forms to stable polymorphs occurs spontaneously and irreversibly, at the expense of the unstable one. Figure 12 shows the Gibbs free energy–temperature relationships of three polymorphs of monoacid triglycerides. By cooling the neat liquids below the melting point T_α, the α form crystallized at first. Then, during aging or heating, this form transforms to the most stable β-form directly without melting, in the case of short-chain triglycerides, or after melting and recrystallization in the case of long-chain triglycerides.

Among the polymorphs of fats, the β form is desirable in salad dressings because its physical dimensions prevent the crystals from settling. However, in most cases, the β′-type phase is functional in the fat products due to its small crystal size, about 1 μm long having thin needle-shaped morphology. This rela-

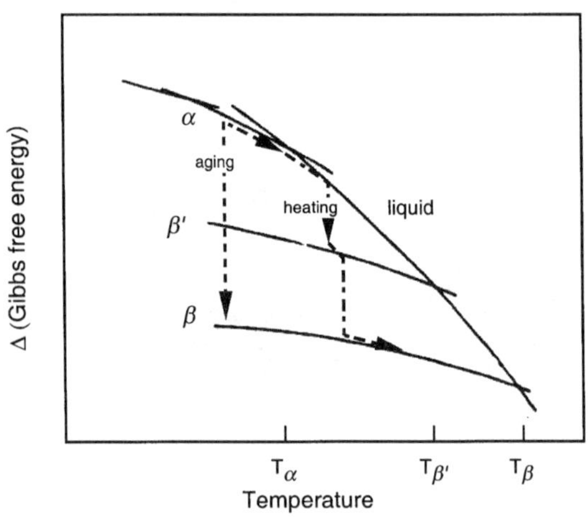

Fig. 12 Gibbs free energy–temperature relationship of monoacid triglyceride polymorphs.

tively small crystal results in good plasticity and gives proper softness to the fat products, such as margarine and shortening. In addition, β′ crystals can make a complicated mixture with other components such as liquid oils, surfactants, and water. Therefore, the β′-to-β transition of fat crystals results in deterioration of the end products.

Emulsifiers sensitively modify the rates of crystal growth and polymorphic transition of fats through the preferred adsorption at or inclusion in fat crystals [32–37]. The retardation or acceleration of the polymorphic transformation is influenced by the hydrophobic moiety structure. Figure 13 shows the effect of emulsifiers on the α-to-β transformation of tristearin during aging at room temperature [10]. Among solid emulsifiers, sorbitan monostearate (SMS) and triglycerol-1-stearate (3G1S), which have in common a particular feature, delayed the transformation; the α-form was stabilized by the presence of these emulsifiers. On the other hand, glycerol-1-stearate (GMS) and lactate glycerol-1-stearate (LGMS) enhanced it. Any liquid surfactant such as sorbitan monolaurate (SML) significantly accelerates the transformation from α to β. The results suggested that the retardation or acceleration of the polymorphic transformation by the additive on the solid state is also connected to the hydrophobic moiety structure.

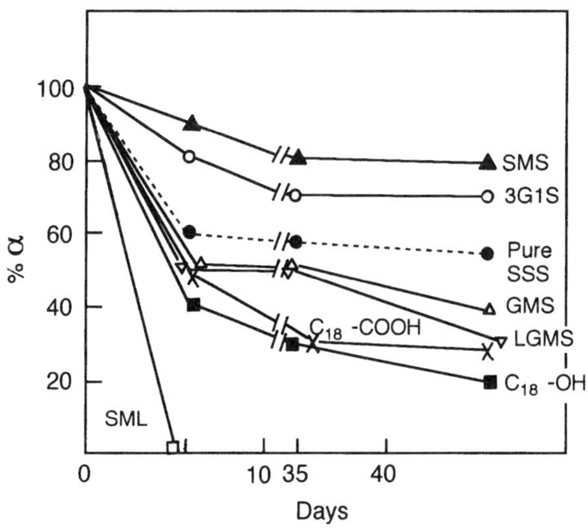

Fig. 13 Aging of tristearin at room temperature in the presence of additives 10 wt%. Neat tristearin (●), tristearin + SMS (▲), tristearin + 3G1S (○), tristearin + GMS (△), tristearin + LGMS (▽), tristearin + stearic acid (×), tristearin + stearoyl alcohol (■), and tristearin + SML (□) [10].

To elucidate the effects of hydrophilic and hydrophobic moieties independently, Smith et al. investigated the effects of lauric-based amphiphilic molecules on the crystallization of trilaurin [38–40]. The crystallization was monitored by using temperature gradient microscopy combined with DSC and X-ray diffraction at forced cooling rates between 10 and 100°C per hour. Without additives, thin lath-like β crystals were formed. These crystals increased in size with time, and nearly perfect and large facets were seen. The addition of lauric acid and monolaurin caused a definite morphological change of β forms: the (001) plane was broadened and a reduction in the facet size was observed (Fig. 14). In addition, the twinning density was increased as evidenced by the increase in imperfection of the crystals. The addition of dilaurin caused the facet and crystal sizes of β crystals to be reduced by a considerable amount. Growth distance against time plots are illustrated in Fig. 15. Both lauric acid and monolaurin (not shown here) increased the growth rate of trilaurin, whereas the dilaurates decreased it. 1,3-Dilaurin is particularly effective at retarding the growth rate. In addition, increasing the concentration of the additive leads to greater retardation for the dilaurin-containing sample with limited increases in growth rates for the monolaurin and lauric-acid-containing samples. Namely, the crystallization rate is increased, while facet and crystal size are reduced by the smaller molecules. This is because

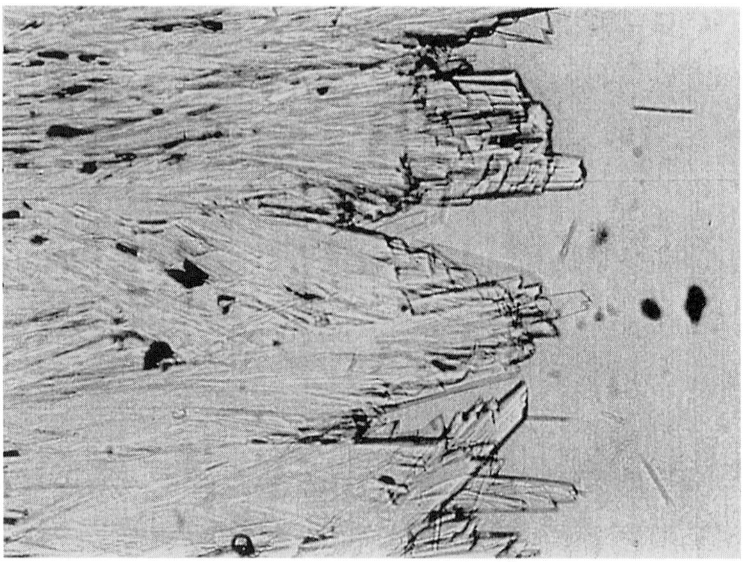

(a)

Fig. 14 Crystals of trilaurin (a), plus 2 wt% lauric acid (b), and plus monolaurin (c) [38].

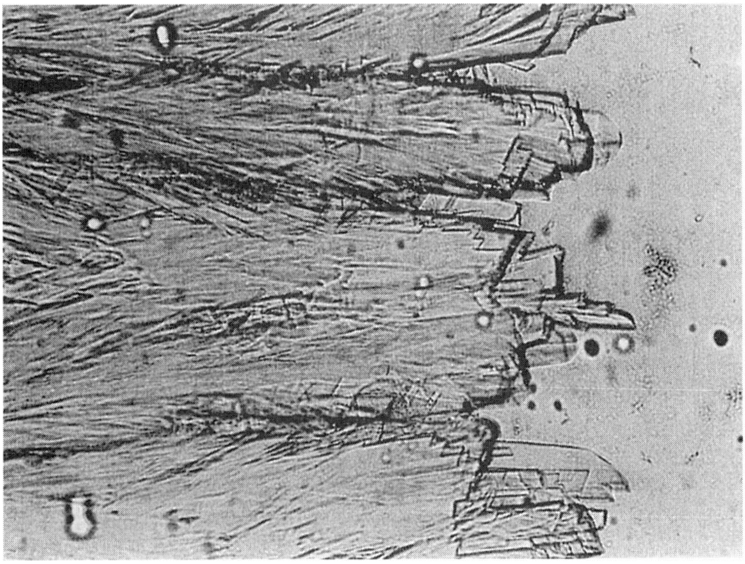

(b)

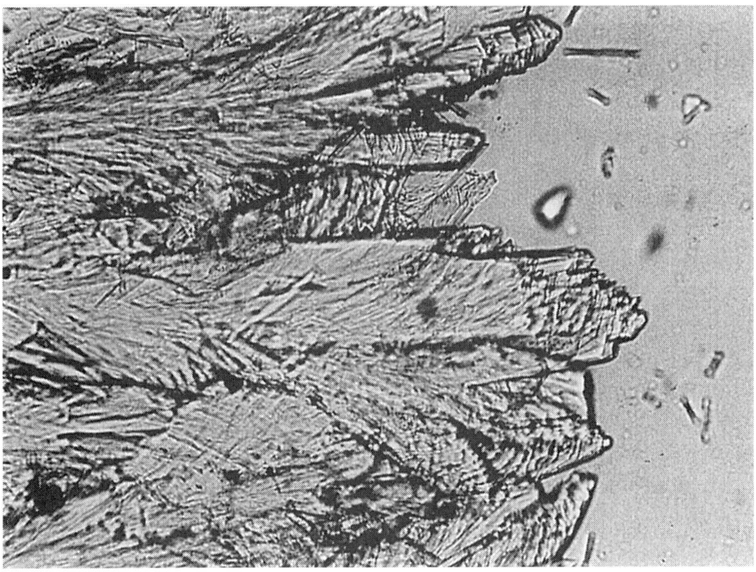

(c)

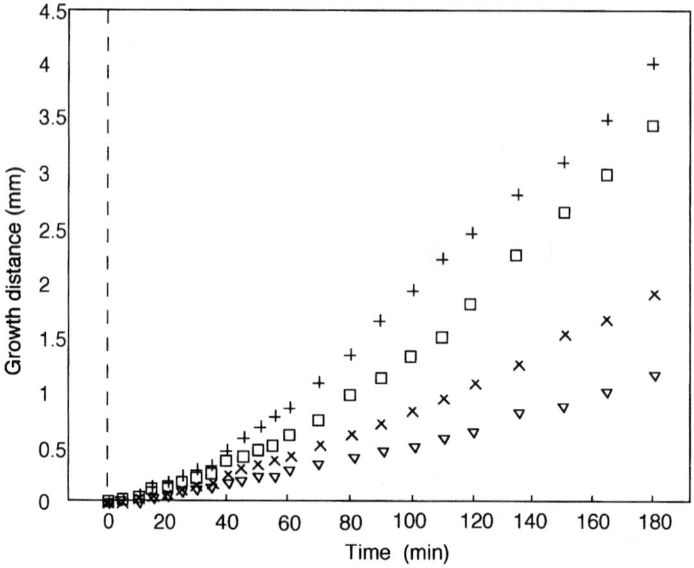

Fig. 15 Growth distance against time for all systems at a slide movement rate of 1 mm per hour (additive level 2% by wt). Trilaurin, □; trilaurin + lauric acid, +; trilaurin + 1,2-dilaurin, ×; trilaurin + 1,3-dilaurin, ▽ [38].

these molecules can easily fit into the trilaurin lattice. The crystal growth rate is slightly increased by these molecules. The defect density increases by the additives. On the other hand, more bulky additives like dilaurin may block the growth sites, which leads to reduction in growth rate. Since 1,3- and 1,2-dilaurin have different shapes and fit into the lattice in different ways, the effect of these two isomers might be different [15–17,38,39].

The effect of hydrocarbon chain length of additives on the crystal growth of trilaurin is also observed by adding caprate (C10)- and myristate (C14)-based amphiphilic molecules [39]. With all of the caprate additives, the growth rate was retarded to some extent. The effect of the myristate additives was the same as that of laurate additives. Consequently, one can obtain the results that the maximal inhibition of crystal growth of trilaurin occurs when the chain length is similar to that of the host and then decreases with increasing difference between host and guest. Among the several additives, 1,3-diglyceride isomer reduced growth rates significantly. Figure 16 shows the effect of side-chain length of 1,3-diglyceride on the growth rate of trilaurin β. From the growth rate dependence on undercooling, a simple incorporation mechanism was proposed for the different additives incorporated into the fat [39].

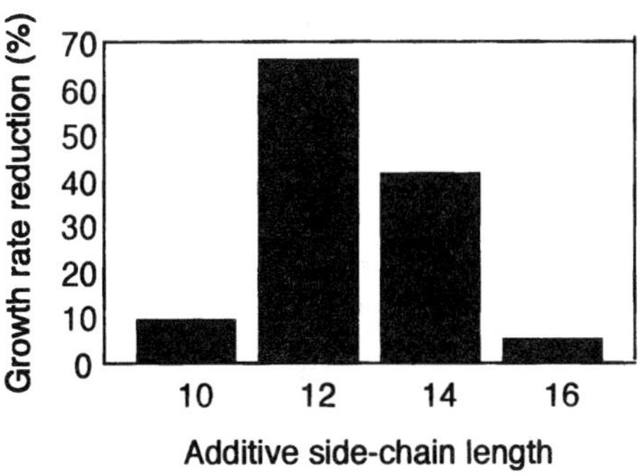

Fig. 16 (a) Effect of 1,3-diglyceride side-chain length on the crystal growth of trilaurin [39].

V. EFFECTS OF EMULSIFIERS IN NATURAL FATS

A. Cocoa Butter

Cocoa butter, the main constituent of chocolate, has six polymorphs, labeled I to VI, which are distinguishable by melting points and X-ray diffraction patterns (see Chapter 12 of this book). Chocolate blooming was explained by two main concepts: (1) phase separation of high- and low-melting triglycerides in cocoa butter, and (2) polymorphic transformation from metastable V form to the most stable VI form. To retard the blooming phenomena, lecithins have been the most favored additives [41]. Also, recent work by Wahnelt et al. showed that diglycerides can retard the crystal growth of cocoa butter [42,43].

Figure 17a shows the effects of added soya lecithin on chocolate viscosity [41]. Maximum thinning is produced by about 0.5% lecithin; above this there is a tendency for the viscosity to increase. Figure 17b states the fat and lecithin contents of chocolate of the same viscosity and indicates the saving in cocoa butter [41]. These two figures show the large viscosity change and the saving in cocoa butter with the addition of lecithin. Although the reason for this effect is not properly understood, Fig. 17c shows some information concerning the site of action. The addition of lecithin to a suspension of cocoa particles in cocoa butter only gives a slight viscosity reduction (Fig. 17c, upper curve). However, the addition of lecithin to a suspension of ground sugar in cocoa butter has a marked effect on the viscosity change (Fig. 17c, lower curve). Therefore, the major contribution to the viscosity-reducing effects of lecithin in chocolate is by action at the surface of the sugar particles. Figure 18 illustrates the viscosity reduction curves of plain chocolate for soya lecithin with three synthetic surface-active lipids [polyglycerol polyricinoleate (PGPR), sucrose dipalmitate, and

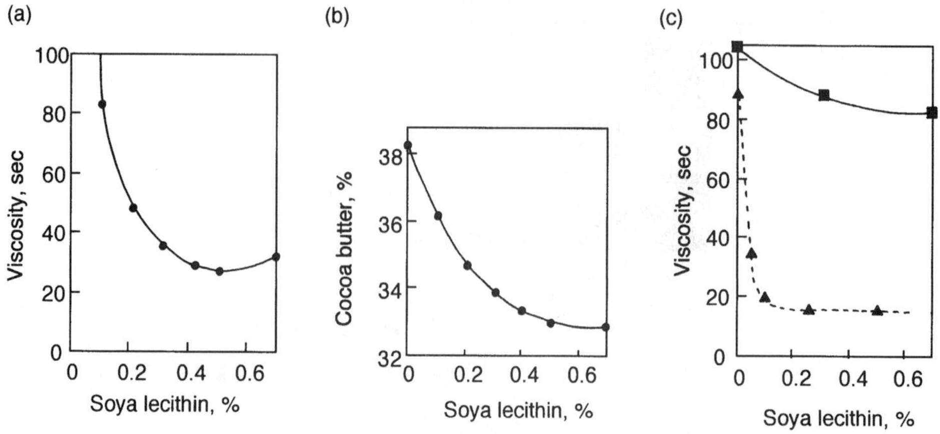

Fig. 17 (a) Effect of adding soya lecithin [initial viscosity of plain chocolate (cocoa butter content, 34.8%), >200 sec] on the viscosity of chocolate. (b) Fat and lecithin contents of chocolate of identical viscosity (42 sec). (c) Effect of adding soya lecithin to suspensions of cocoa particles in cocoa butter (■——■) and ground sugar in cocoa butter (▲——▲) on viscosity. Viscosity was measured at 50°C on Bournville Redwood-type viscometer [41].

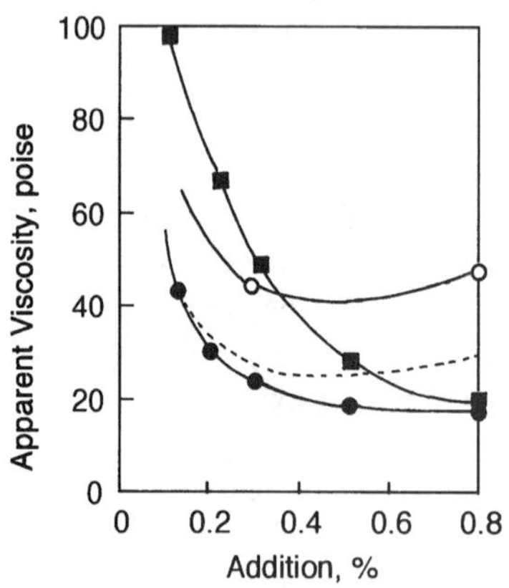

Fig. 18 Reduction of viscosity of plain chocolate by synthetic surface-active lipids. Apparent viscosity was determined on Couette viscometer at 15 sec^{-1} rate of shear at 50°C. Initial viscosity of plain chocolate was 195 poise. With polyglycerol polyricinoleate (■——■); sucrose dipalmitate (○——○); phospholipid YN (●——●); soya lecithin (———) [41].

phospholipid YN]. The sucrose dipalmitate was less effective than any of the other emulsifiers. (Note that since this is an old study, the results of the PGPR do not reflect its real activity. Manufacturers have improved the product and dramatically enhanced its activity.) The flow characteristics of plain chocolate in the presence of the various emulsifiers are summarized in Table 2. Again it can be seen that the sugar ester was found to be inefficient in comparison to the other emulsifiers.

Early studies [44–46] have demonstrated that the more hydrophilic emulsifiers such as ethoxylated sorbitan ester, diacetyltartaric acid esters, and monoglyceride lactate mostly affect polymorphic transformations and have no effect on bulk viscosity. Garti et al. have studied the effect of sorbitan esters and ethoxylated sorbitan esters on the transition among the IV, V, and VI forms [47]. The results suggested that some combinations of those emulsifiers accelerate the transition of form IV into form V by increasing the liquid fraction of the fat prior to its transition. Much additional work is required to reproduce the early findings and to explain (1) the selectivity in adsorption of hydrophilic emulsifiers versus hydrophobic ones on the sugar or fat particles in the bulk after they have been formed, and (2) the effect of these emulsifiers on the crystalline sugar and/or fat particles.

Moran has tried to explore the effect of some emulsifiers on the viscosity of fat/sugar mixtures, using three systems: (1) palm kernel stearin fat (stearin I) + sugar (50 wt% sugar), (2) palm kernel stearin fat (stearin II) + sugar (50 wt% sugar), and (3) chocolate (system 1 or 2 + milk powder and cocoa powder) [48]. The viscosity of the sugar/fat systems was lowered by the incorporation of any of the emulsifiers (sorbitan esters, sucrose esters, lecithin, and polyglycerol stearate). Among the emulsifiers, polyglycerol stearate and sucrose dioleate were more effective than soya lecithin. In addition, the additives retarded the crystallization of the stearin fat at low degree of supercooling. Therefore, the claims made in the past that emulsifiers are adsorbed selectively on sugar particles and do not

Table 2 Flow Characteristics of Plain Chocolate with Added Surface-Active Lipids at 50°C

Addition	Casson plastic viscosity	Casson yield value
0.3% Soya lecithin	6.1	92
0.3% Phospholipid YN	10.3	30
0.3% Polyglycerol polyricinolate	32.5	25
0.3% Sucrose dipalmitate	8.6	166
0.8% Polyglycerol polyricinoleate	20.3	(0)
Cocoa butter to similar plastic viscosity	7.3	72

affect fat particles were not accurate. It also becomes clear that the emulsifiers retard crystallization and slow the polymorphic transformations, and thus affect the rheological properties of the chocolate network. Namely, the emulsifiers are playing a significant role during the crystallization processes and during storage (solid-solid transformation). As the reason for this effect, water and sugar binding capabilities remain to be seen and proven.

B. Sunflower Oil

The effect of sucrose ester (P-170, m.p. 57°C) on the crystallization kinetics of hydrogenated sunflower oil was studied by Herrera and Rocha by means of an optical method [49]. Table 3 shows the effects of the addition of sucrose ester on the induction time of crystallization by cooling to T_C = 30 and 33°C at two cooling rates. At T_C = 30°C, there was only a small difference in induction time at slow and fast cooling rates. At T_C = 33°C, the effect of cooling rate was noticeable and the induction times were shorter at slow crystallization rates. With the addition of sugar ester, the nucleation of β was delayed; the sugar esters

Table 3 Induction Times for Crystallization of Hydrogenated Sunflower Seed Oil

Amount of sucrose esters (%)	T_c (°C)	Cooling rate (°C/min)	Induction time (min)[a]
0	30	2	10.4 ± 0.1
		7	13.6 ± 0.3
	33	2	36.6 ± 0.4
		7	50.7 ± 0.8
0.01	30	2	12.6 ± 0.4
		7	24.1 ± 0.4
	33	2	56.9 ± 1.1
		7	144.8 ± 1.7
0.05	30	2	45.3 ± 1.2
		7	54.6 ± 1.2
	33	2	72.8 ± 1.3
		7	268.4 ± 2.6
0.1	30	2	105.8 ± 2.1
		7	115.5 ± 2.0
	33	2	456.7 ± 5.3
		7	453.5 ± 4.2

[a] Mean ± one standard deviation; T_c = crystallization temperature.

Table 4 Effect of Sucrose Ester of β′-to-β Polymorphic Transition on Hydrogenated Sunflower Seed Oil at 30°C (storage temperature 25°C)

Storage time (d)	Sucrose ester content (%) 0	0.01	0.05	0.1
	(a) Slow crystallization			
0	β′	β′	β′	β′
1	β′	β′	β′	β′
2	β′ ≪ β	β′	β′	β′
3		β′≪β	β′≪β	β′≪β
8	β′ = β	β′ = β		
31			β′ = β	β′ = β
60	β			
76		β		
98			β	β
	(b) Quick crystallization			
0	β′	β′	β′	β′
1	β′	β′	β′	β′
2	β′	β′	β′	β′
3	β′ ≫ β	β′ ≫ β	β′	β′
4			β′ ≫ β	β′ ≫ β
18	β′ = β	β′ = β		
35			β′ = β	β′ = β
76	β			
88		β		
106			β	
120				β

affected the formation of critical nuclei and prolonging induction times. Table 4 indicates the effect of the emulsifier on the β′-to-β transition when the sample was crystallized at 30°C. At all concentrations, the emulsifier delayed the transition by 24 hr. Moreover, long times were needed to complete the transition. The kinetic mode of the transition process was described on the basis of Avrami's equation,

$$1 - X = \exp(-Kt^n)$$

where X is the β fraction, n is the mode of nucleation and growth of β nuclei, t is time, and K is the shape of β. By plotting the transformation curves (β fraction versus time) and fitting the Avrami equation, the n value was obtained approximately as 1 in all the concentrations of additives. This result suggests that only the β′ form could be obtained from the melt. The β form could not be obtained directly from the melt, and the β′-to-β transition is not liquid mediated. It probably occurs from the solid-state transformation.

VI. FAT COLLOIDS AND FAT EMULSIONS

Many food products have complex structures in which fats, proteins, and carbohydrates are structured together in the presence of water. There are only minor compositions of binary or ternary mixtures that can be regarded as true solutions. In most cases, the systems are described as food colloids or colloidal foods. The components will be dispersed one in the other forming microstructures. The simplest example is margarine, a W/O emulsion stabilized by a blend of food emulsifiers, such as lecithin and monoglyceride with fat crystals. In the system, solid fat particles along with liquid water droplets are dispersed in a continuous oil phase. Another example is ice cream in which fat particles are dispersed in a continuous water phase [50,51]. Additional examples are summarized in Table 5. Several other food systems consist of dispersions of particles (fat crystals, sugar crystals, and proteins) in a continuous oil phase. An example of such system is chocolate, which is a dispersion of fat, sugar, and cocoa powder in liquid oil.

In those food products, emulsifiers play an important role in construction of fine microstructures as molecular bridges between various components and molecular wetting agents facilitating such adsorption. However, the adsorption isotherms for the food emulsifiers to different crystals dispersed in oil was not carefully studied until recently.

Table 5 Typical Food Colloids

Food	Type of emulsion[a]	Method of preparation	Mechanism of stabilization
Milk	O/W	Natural product	Protein membrane
Cream	A+O/W	Centrifugation	Protein membrane + particle stabilization of air
Ice cream	A+O/W	Homogenization	Protein membrane + particle stabilization of air + ice network
Butter & margarine	W/O	Churning & in rotator	Fat crystal network
Sauces	O/W	High-speed mixing & homogenization	Protein & polysaccharide
Fabricated meat products	O/W	Low-speed mixing & chopping	Gelled protein matrix
Bakery products	A+O/W	Mixing	Starch & protein network

[a] O = oil, A = air, W = aqueous phase

A. Influence of Emulsifiers on Fat Dispersions in Oil

In an early work by Lucassen-Reynders, it was shown that monoolein can adsorb onto tristearin [52]. Johansson and Bergenstahl carried out a detailed work on model fat systems dispersed in oil [20,53,54]. They studied the adsorption of various emulsifiers to the crystals (fats and sugar) dispersed in oils. The adsorbed amount, the strength of the adsorption, and their relationship to the character of the emulsifiers, crystals, and oils were obtained. Figures 19a, b compare adsorption of monoolein to sugar and fat crystals from different oils (dodecane, decanol, and soybean oil). The adsorption amount (Γ) and the relative change in sedimentation volume (ΔV) strongly depend on oils. Monoolein adsorbs strongly to sugar in dodecane and forms more than one monolayer. It causes decrease in the sedi-

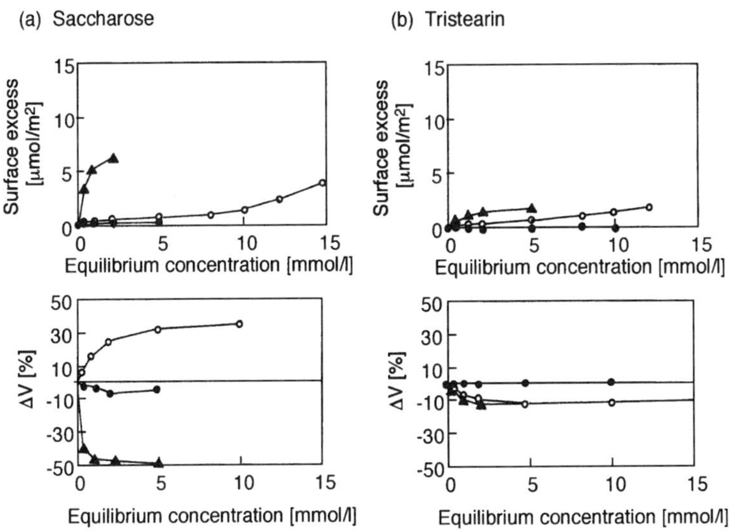

Fig. 19 A comparison of adsorption on, and sedimentation of, saccharose (a) and tristearin crystals (b) in different solvents at room temperature (~22°C). The adsorbed amount (Γ) and the relative change in sediment volume of the crystals (ΔV) are given as a function of equilibrium concentration of emulsifiers in the solvents. Monoolein in dodecane (▲), soybean oil (●), and decanol (○). In (a), the crystal volume is 1.25 mL, while the total sample volume is 6.1 mL for sedimentation. The sediment volume in the zero-samples is 5.1 mL in dodecane, 3.6 mL in soybean oil, and 3.2 mol in decanol, respectively. In (b), the crystal volume is 0.2 mL, while the total sample volume is 5.0 mL for sedimentation. The sediment volumes in the zero samples are 4.3 mL in dodecane, 4.0 mL in soybean oil, and 3.8 mol in decanol, respectively.

ment volume of the dispersion. Adsorption in soybean oil is much weaker and a loosely packed monolayer is formed. A similar pattern is observed for fat crystals. However, the adsorption to fat is weaker than that to sugar, and the adsorbed amounts are smaller, suggesting loosely packed layers. The adsorption and sedimentation data for sugar and fat crystals is presented in Tables 6 and 7. The monolayer adsorption, Γ_m (μmol/m^2), and the sedimentation data of the various emulsifiers for sugar and fat crystals are summarized in Table 8. On the basis of these data, estimates of the equilibrium constant k_a and free energies of adsorption ΔG_a (kJ/mole) were obtained as follows:

$$k_a = 10^6 \times k_c + \frac{\rho_{oil}}{M_{oil}}$$

$$\Delta G_a = -RT \ln k_a$$

where k_c is the equilibrium constant of the adsorption (L/mmol), ρ_{oil} is the density of the oil (g/mL), and M_{oil} is the molar mass of the oil (g/mol). These values are listed in Table 9.

The main conclusions drawn from the study are as follows:

1. Most of the emulsifiers adsorb weakly to fat crystals at high concentrations. At low concentrations, loosely packed layers are formed and decreased adhesion is observed. Unsaturated monoglycerides and phospholipids cause a decrease in adhesion for all concentrations examined.
2. The emulsifiers adsorb more strongly to sugar crystals than fat crystals and form tightly packed monolayers with hydrocarbon chains directed to the oil. The crystals are then stabilized sterically and the adhesion between them is weaker and sediments are more compact. Phospholipids reduce the adhesion between sugar crystals, resulting in much denser sediments.
3. Saturated monoglycerides in amounts over the solubilization limit tend to precipitate as a network between fat or sugar crystals, which causes bulky sediments and results in better stability against oiling-out.

B. Adhesion of Fat Crystals to Oil-Water Interface

Together with emulsifiers, fat particles play an important role in the stabilization mechanism of emulsions. As mentioned, a semisolid fat phase consisting of colloidal fat crystals exists in most low-fat spreads, such as margarine and butter (W/O microemulsions), and attaches to emulsion droplets. It is expected that submicron particles (less than 0.1 mm), with narrow size distribution and as spherical as possible, would need to be obtained in the oil phase in order to achieve good anchoring, with good wetting, at the water interface. Therefore, it

Table 6 Adsorption and Sedimentation Data for Sugar Crystals[a]

	Slope of adsorption isotherms (μmole/m^2) (mmol/L)	Slope of sedimentation curves % (mmol/L)	Maximum adsorption (μmole/m^2)	Surface per molecule at maximum adsorption (Å^2)	Number of monolayers at maximum adsorption	Change in sedimentation volume at maximum adsorption (%)
PC from SBO[b]	8.0	40	15.0	11	≥6	−30
Mixture of phospholipids	3.0	52	10.0	16	≥4	−40
Monoolein	1.2	10	5.0	33	~1	+30
Unsaturated MG	0.6	8	1.2	140	≪1	+25
Saturated MG	2.4	24		(precipitation)	≫1	+200
Polyglycerol EFA	2.0	20	4.5	37	~1	−70
Lactic acid EMG	0.8	6	2.5	66	<1	+10
Sorbitan EFA	1.4	40		(precipitation)	≫1	−30
Diacetyl tartaric acid EMG	0.6	38	2.0	83	<1	−45
Propylene glycol EFA	0.0	0	0	—	—	0

[a] Surface area per hydrocarbon chain is assumed to be around 35 Å^2 (70 Å^2 for phospholipids with two chains per emulsifier molecule).
[b] PC = phosphatidylcholine; SBO = soybean oil; MG = monoglycerides; EMS = esters of monoglycerides; EFA = esters of fatty acids.

Table 7 Adsorption and Sedimentation Data for Fat Crystals[a]

	Slope of adsorption isotherms (μmole/m²) (mmol/L)	Slope of sedimentation curves % (mmol/L)	Maximum adsorption (μmole/m²)	Surface per molecule at maximum adsorption (Å²)	Number of monolayers at maximum adsorption	Change in sedimentation volume at maximum adsorption (%)
PC from SBO[b]	1.0	24	4.0	41	~2	−15
Mixture of phospholipids	1.2	34	5.0	33	~2	−25
Monoolein	0.2	8	3.0	55	≤1	−10
Unsaturated MG	0.05	0	1.0	166	<1	−5
Saturated MG	0.6	30		(precipitation)	≫1	+350
Polyglycerol EFA	0.4	14	2.0	83	~1	+15
Lactic acid EMG	0.6	30	2.0	83	~1	+15
Sorbitan EFA	1.0	20		(precipitation)	≫1	+25
Diacetyl tartaric acid EMG	0.2	20	2.0	83	~1	+20
Propylene glycol EFA	0.0	0	0	—	—	0

[a] Surface area per hydrocarbon chain is assumed to be around 35 Å² (70 Å² for phospholipids with two chains per emulsifier molecule).
[b] PC = phosphatidylcholine; SBO = soybean oil; MG = monoglycerides; EMS = esters of monoglycerides; EFA = esters of fatty acids.

Table 8 Monolayer Adsorption Obtained from Adsorption Isotherms

	Monolayer adsorption Γ_m (μmol/m^2)		Surface area per adsorbed molecule a_m(A^2/molecule)[a]	
	Saccharose	Tristearin	Saccharose	Tristearin
Phosphatidylcholine from SBO[b]	c	c	—	—
Mixture of phospholipids	c	c	—	—
Monoolein	1.5	1.00	111	166
Unsaturated MG	1.0	0.75	166	221
Saturated MG	c	c	—	—
Polyglycerol EFA	4.5	2.00	37	83
Lactic acid EMG	2.5	2.00	66	83
Sorbitan EFA	c	c	—	—
Diacetyl tartaric acid EMG	2.0	2.00	83	83
Propylene glycol EFA	0.0	0.0	—	—

[a] Calculated from the monolayer adsorption.
[b] SBO = soybean oil; MG = monoglycerides; EMS = esters of monoglycerides; EFA = esters of fatty acids.
[c] Data are not precise enough to determine monolayer adsorption.

Table 9 List of Equilibrium Constants for Adsorption, k_a (estimated from Henry's model), and Free Energies for Adsorption, $-\Delta G_a$

	Equilibrium constant, k_a		Adsorption free energy $-\Delta G_a$ (kJ/mol)	
	Saccharose	Tristearin	Saccharose	Tristearin
Phosphatidylcholine from SBO[a]	—	—	—	—
Mixture of phospholipids	—	—	—	—
Monoolein	850	210	16.6	13.1
Unsaturated MG	640	70	15.9	10.4
Saturated MG	—	—	—	—
Polyglycerol EFA	470	210	15.1	13.1
Lactic acid EMG	340	320	14.3	14.2
Sorbitan EFA	—	—	—	—
Diacetyl tartaric acid EMG	320	110	14.2	11.5
Propylene glycol EFA	1	1	0.0	0.0

[a] SBO = soybean oil; MG = monoglycerides; EMS = esters of monoglycerides; EFA = esters of fatty acids.

is important to understand the interfacial behavior of fat crystals at the interface together with oil, water, and emulsifiers.

Wetting properties of fat crystals by oil and water is one probe to understand the microscopic structure of oil-water interface, and can be macroscopically characterized by contact angle measurements in a three-phase system as shown in Fig. 20 [55]. This property depends on the polymorphic form of fat crystals and is also influenced by addition of food emulsifier to the oil and/or to the water. When the contact angle measured through the oil phase is close to 0°, the crystals are nonpolar and located in the oil phase. When the angle is close to 180°, the crystals are polar and located in the water phase. Between 0 and 90°, they are attached to the oil-water interface from the oil side, between 90–180° from the water side. In the former case, W/O emulsion can be stabilized. On the other hand, O/W emulsion is expected to be stabilized by the latter case.

Table 10 shows the equilibrium contact angles at the water–fat crystal–oil three phases for different systems without additives. Advancing contact angles (Θ_a) correspond to fat crystals approaching the oil-water interface from the water side in the emulsion. Receding contact angles (Θ_r) correspond to fat crystals approaching the oil-water interface from the oil side. At room temperature, β polymorphs are completely wetted by oil ($\Theta \approx 0°$) without migration to the water phase. On the other hand, α and β' polymorphs are introduced to the oil-water interface from the oil side ($\Theta \approx 30°$). This suggests that the coverage energy per unit area of α and β' crystals is slightly higher than β. Tightly packed hydrocarbon surface of β' crystals produces completely nonpolar surfaces, while a looser packing in α and β crystals gives slightly polar surfaces. With increasing temperature, the contact angle for palmstearin α slightly decreases, probably due to increased mobility of hydrocarbon chain on the surface.

Table 11 shows the effect of emulsifiers (lecithins, monoglycerides and their esters, and ethoxylated emulsifiers) on the contact angle at fat crystal (palmstearin β')–oil–water interface. The emulsifiers made the crystals more polar (higher Θ). At an emulsifier level of 1–2%, the crystals became so polar that

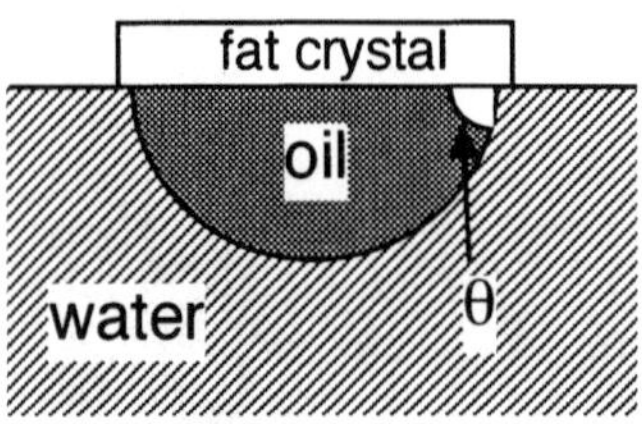

Fig. 20 Contact angle apparatus for measurement at the three phases boundary: fat crystal/oil/water [55].

Table 10 Contact Angle at Fat Crystal/Oil/Distilled Water Interface (no additives)

Crystal type	Polymorphic form	Temp. (°C)	Oil type	Advancing contact angle Θ_a(°)	Receding contact angle Θ_r(°)
Tristearin	α	22	SBO[a]	19–31	8–15
	β		SBO	0	0
	α		CPLSBO[a]	25–33	16
	β		CPLSBO	0	0
Palmstearin	α	22	SBO[a]	27–35	9–16
	β′		SBO	25–39	6–19
	α		CPLSBO[a]	20–25	5–12
	β′		CPLSBO	36–38	17–18
Palmstearin	α(+β′)[b]	40	SBO	16–20	10
	α(+β′)		butter oil	8	2
	β′		butter oil	15	5

[a] SBO = soybean oil; CPLSBO = chromatographically purified soybean oil (contains only triglycerides).

[b] At 40°C, fat crystal surface is under transformation from α to β′ polymorphic form, and a mixture of both forms most likely occurs.

Table 11 Contact Angles for Systems Containing Emulsifiers (1–2%) in Oil

Emulsifier	Receding contact angle Θ_r(°)	Advancing contact angle Θ_a(°)
None	12	32
Soya PC	143	160
Topcithin	45	120
Sterncithin	30	75
Metarin P	35	65
Monoolein	17	87
Saturated monoglycerides	100	125
Lactic acid esters of monoglycerides	85	132
Ethoxylated alkyl ether	69	135
Ethoxylated sorbitan monostearate	175	175
Ethoxylated (12) castor oil	90	90
Ethoxylated (20) castor oil	90	90

they could migrate into water. Only hydrophobic lecithins and unsaturated monoglycerides (monoolein) produced crystals which were preferably wetted by oil. Low concentrations of emulsifiers (0.1–0.2%) affected the crystal polarity to a minor degree, excepting polar emulsifiers such as hydrophilic lecithin (soya PC) and ethoxylated sorbitan monostearate (Tween 60). For a partially soluble emulsifier, such as saturated monoglycerides, crystal polarity increased with temperature due to improved solubility and surface activity. In this case, high temperature promoted migration of fat crystals into water.

C. Stability of Food Emulsions

As described, α-form fat crystals are significantly more hydrophilic than other β′ and β crystals and, therefore, will better wet the water interface and tend to anchor better at the W/O interface. Garti et al. crystallized the α form of tristearin microcrystals in the soybean oil phase by flash-cooling prior to making the emulsions, and studied their effect on the stability of emulsions in the presence of monomeric emulsifiers such as glycerol monooleate/lecithin and polyglycerol polyricinoleate (PGPR) [56,57]. The result stressed the need for both hydrophobic emulsifiers and submicronal fat particles to stabilize W/O emulsions. It is essential to have sufficient amounts of emulsifier in the oil phase to guarantee surface adsorption of the emulsifier onto the fat particles in order to reduce flocculation and growth processes and to stabilize the dispersions. Among the examined emulsifiers, PGPR had a contribution for the formation of submicron-sized tristearin crystals (α + β′) having narrower size distribution. As a result, these W/O emulsions have longer shelf-life and better stability. The role of PGPR seemed to prevent aggregation of submicron fat crystals in the oil phase rather than to control the formation of an adequate crystalline form. It seems that more α-form crystals can be obtained in the absence of emulsifier, but it does not guarantee the stabilization of the fat crystals in the dispersion and, therefore, the combination of the two is better.

For mixed emulsions and aerated emulsions, interfacial behavior is more complicated, especially when proteins are present at the interface. In aerated emulsions such as cake butters and bread mixes, proteins are thought to adsorb to the surfaces of any fat crystals adsorbed to the air-water interface [58]. Also, proteins at an oil-water interface would adsorb to fat crystals to effect on coalescence stability. Ogden and Rosenthal studied interactions between proteins and fat crystals in the interfacial region by measuring interfacial shear viscosity [59–61]. They showed that when proteins are present in the aqueous phase, the presence of tristearin crystals in an oil phase causes a synergistic increase in the interfacial shear viscosity. They also studied the effect of polymorphic form and amount of fat crystals. The β′ crystals caused bigger increase in the interfacial shear viscosity compared to β crystals. It suggests that β′ crystals may produce stronger crystal-crystal networks in the oil phase than β crystals. This result supports the fact that β′ crystals are superior to β crystals in stabilizing food emulsions and foams [62,63]. In aerated emulsions, the incorporated air bubbles tend to be much smaller when the fat is in the β′ polymorphs than when the crystals are in β, due to the smaller size of β′ crystals. As for the synergistic effect of proteins and fat crystals on the interfacial shear viscosity, they explained that a protein film at the interface anchors the adsorbed layers of fat crystals and the protein film moves together with a shear stress. The extent of the viscosity increase is related to the strength of the crystal-crystal interactions and the mass of crystals.

D. Sintering of Fat Crystals and Emulsifiers

In semisolid food products, fat crystals determine properties such as consistency, stability against oiling-out, and emulsion stability. Therefore, the formation of proper polymorphic form, size, and morphology of fat crystals during crystallization processes of production is necessary for producing desirable texture and melting sensation. After production, different crystallization processes proceed in the products. These postcrystallization processes include nucleation of new crystals and crystal growth, Ostwald ripening (dissolution of small crystals and growth of big ones), polymorphic transformation, migration of oil, or migration of small crystals. Crystal growth during postcrystallization may sometimes lead to formation of solid bridges (sintering) in narrow gaps of fat crystal networks [64,65].

Recent studies by Johansson and Bergenstahl indicate that sintering may be created by crystallization of fat phase with a melting point between that of the oil and the crystal [66]. They conducted a detailed study on the effect of type of sintering fat crystals and added emulsifiers. Two experimental techniques were utilized to study sintering effects:

1. Sedimentation experiments to measure qualitative comparison of particle adhesion in dispersions. As the adhesion between particles increases, they stick to each other and form large flocs and bulky sediments. An increased repulsion has the opposite effect; the particles do not stick to each other and pass each other more easily when they settle and form dense compact sediments.
2. Rheology measurements in which the Bingham yield stress was determined (the minimum stress or force which has to be applied on a sample to obtaining form flow through the whole sample, which corresponds to breaking of all bonds in the sample). The Bingham yield stress is expected to be most sensitive for the formation of solid bridges since these bridges are expected to be strong.

They used two types of high-melting fat crystals: pure tristearin (in its different polymorphic forms) and totally hydrogenated palm oil (palmstearin). Those were dispersed in soybean oil. Two fats with a melting range between the oil and the high-melting fat crystals (totally hydrogenated palm kernel oil and partially hydrogenated rapeseed oil) were used to form solid bridges in existing networks during the postcrystallization processes.

As for types of sintering fat crystals, their results indicate that β′ crystals will be sintered by β′ fat bridges, favored by rapid cooling, and β crystals will be sintered by β fat bridges, favored by slow cooling (Table 12). The necessity of the same polymorphic form of the crystal and bridge indicated that solid bridges, rather than bridges formed by small crystal nuclei, are formed. Figure 21 is the schematic representation of the different phenomena taking place during postcrystallization. The following three steps may occur simultaneously: nucleation of new crystals, crystal growth (a or b in Fig. 22), and formation of bridges between crystals. Formation of bridges may occur in two ways: true solid bridge (c) or bridges of small flocculated crystal nuclei (d).

Some emulsifiers influence the sintering process. The addition of technical lecithin and monoolein produced systematic changes in sediment volume and Bingham yield stress as shown in Figs. 22a, b. Addition of sintering fats (palm kernel oil) to the fat crystal (palmstearin) dispersions in soybean oil increased sediment volumes of fat crystals by more than 10%. With addition of lecithin, however, sediment volume was decreased by 20–25% in relation to the reference sample. On the other hand, the addition of monoolein has the opposite effect: it produced further increases in sedimentation volume. Thus, monoolein and palm kernel fat seem to have synergistic effects during sintering. Johansson and Bergenstahl explained that emulsifiers adsorb to fat crystals in oil and may enhance or present adsorption of sintering fats. Monoolein adsorbs weakly and forms a loosely packed layer on fat crystals in oil, causing a small decrease in crystal adhesion. The adsorbed layer is compatible with sintering triglycerides and may

Table 12 Conditions for Controlled Sintering

	β′ stable sintering fat	β stable sintering fat
β′ network	**Sintering** facilitated by rapid cooling	Sintering only if undercooled β′ bridge is formed (extremely rapid cooling)
β network	Sintering only if a β bridge is formed (extremely rapid cooling)	**Sintering** facilitated by slow cooling

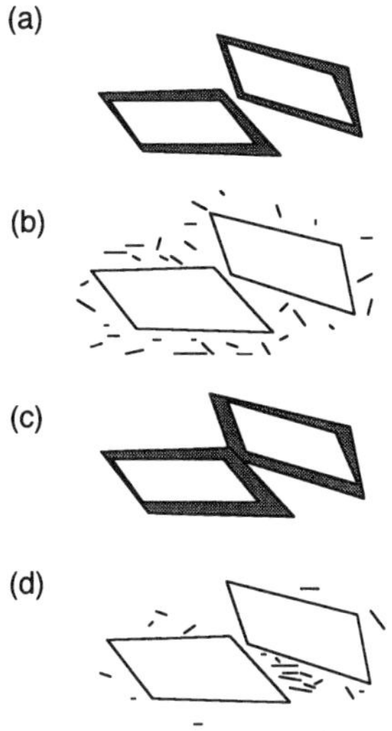

Fig. 21 A schematic representation of different phenomena taking place during postcrystallization. (a) nucleation of new crystals, (b) crystal growth, (c) solid bridge formation, and (d) formation of bridge by flocculation of small nuclei between two fat crystals [66].

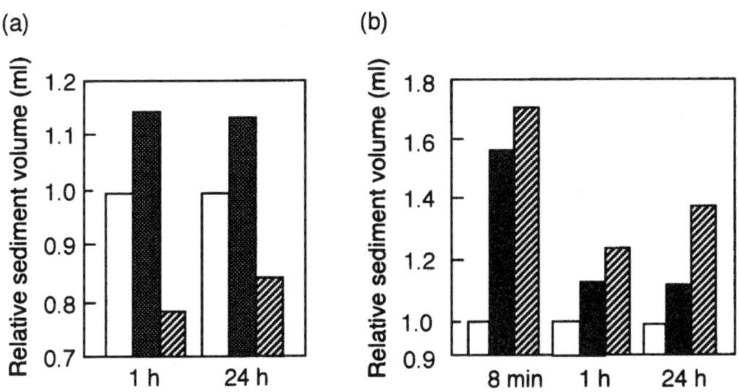

Fig. 22 Relative sediment volume of 5% palmstearin β′ crystals in soybean oil without additives (□), with 0.5% of palm kernel fat (▨), and with 0.5% of palm kernel fat + (a) 0.5% lecithin Metarin P or + (b) 1% monoolein (■). The water-bath temperature was lowered from 20 to 10°C during 1 hr and 24 hr in (a) and during 8 min, 1 hr, and 24 hr in (b) [66].

enhance their adsorption to fat crystals. The effect of monoolein is greater for slow cooling rates, when the degree of sintering tends to be smaller otherwise.

VII. CONCLUSIONS

This chapter described how emulsifiers affect crystallization processes of fats, including nucleation, growth, and phase transitions. In addition, surfactants affect rheological properties of fats by changing the network formation of fat crystals and, as a result, affect phenomena such as bloom, gloss, and cracking of fat coatings in chocolate or other fat-rich solid products. In dispersed systems based on dispersed fat, it was demonstrated that the emulsifiers affect the fat crystallization and the wetting and adhesion properties of the fats. It seems that one can at present better predict the behavior of fats in the presence of amphiphiles, which allows better control of the physical properties. However, detailed understanding of molecular level interactions between fats and emulsifiers has not been solved yet, and is a future task in this field.

REFERENCES

1. L. Hernqvist, Crystal structures of fats and fatty acids, in *Crystallization and Polymorphism of Fats and Fatty Acids* (N. Garti and K. Sato, eds.), Marcel Dekker, New York, 1988, pp. 97.
2. J. Riiner, *Lebensm. Wiss. Thechnol.*, *3*: 101 (1970).
3. R. L. Wille and E. S. Lutton, *J. Am. Oil Chem. Soc.*, *43*: 491 (1966).
4. L. deMan, J. M. deMan, and B. Blackman, *J. Am. Oil Chem. Soc.*, *66*: 1777 (1989).
5. A. E. Bailey, *Melting and Solidification of Fats and Fatty Acids*, Interscience, New York, 1950, pp. 117.
6. V. Boekel and P. Walstra, *Colloids and Surfaces A*, *3*: 109 (1981).
7. K. Boode, C. Bisperink, and P. Walstra, *Colloids and Surfaces A*, *61*: 55 (1991).
8. P. Walstra, Lipids, in *Advanced Dairy Chemistry*, Vol. 2, Ch. 4, 2nd ed. (P. F. Fox, ed.), Chapman and Hall, London, 1995.
9. M. van den Tempel, in SCI Monograph 32, SCI, London, 1968, pp. 22.
10. N. Garti, in Ref. 1, 1988, pp. 267.
11. A. G. Marangoni and D. Rousseau, *J. Am. Oil Chem. Soc.*, *73*: 991 (1996).
12. N. Garti, D. Lichtenberg, and T. Silberstein, *Colloids and Surfaces A*, *128*: 17 (1997).
13. H. Hauser, I. Pascher, and S. Sundel, *Biochem. Biophys. Acta*, *650*: 21 (1981).
14. I. Pascher, S. Sundell, and H. Hauser, *J. Mol. Biol.*, *153*: 807 (1981).
15. K. Larsson, *Ark. Kemi. 23*: 23 (1964).
16. I. Pascher, S. Sundel, and H. Hauser, *J. Mol. Biol.*, *153*: 791 (1981).
17. A. Hybl and D. L. Dorset, *Acta Crystallogr.*, *B27*: 977 (1977).
18. Palsfaard, Technical Memorandum, Unilever, 1952.

19. H. F. Bamford, K. J. Gardner, G. R. Howat, and A. F. Thomson, *The Manufacturing Confectioner for July*: 36 (1970).
20. D. Johansson and B. Bergenstahl, *J. Am. Oil Chem. Soc.*, *69*: 705 (1992).
21. T. Katsuragi, Interactions between surfactants and fats, in *Physical Properties of Fats, Oils, and Emulsifiers* (N. Widlak, ed.), AOCS Press, Champaign, IL, 2000, pp. 209.
22. D. C. Clark, P. J. Wilde, D. R. Wolson, and R. Wustneck, *Food Hydrocolloids*, *6*: 173 (1992).
23. D. Abran, F. Boucher, T. Hamanaka, K. Hiraki, Y. Kito, K. Koyama, R. M. Leblanc, H. Machida, G. Munger, M. Seidou, and M. Tessier, *J. Colloid Interface Sci.*, *128*: 230 (1989).
24. S. C. Sethi, S. D. Adyanthaya, S. D. Deshpande, R. G. Kelker, N. Natarajan, and S. S. Katti, *J. Surf. Sci. Technol.*, *2*: 103 (1986).
25. T. Kawaguchi, *Nagoya Kogyo Daigaku Gakuho*, *35*: 269 (1983).
26. S. Solans and M. J. Garcia-Cerma, *Curr. Opin. Colloid Interface Sci.*, *2*: 464 (1997).
27. T. Katsuragi, N. Kaneko, and K. Sato, *Colloids and Surfaces B*, *20*: 229 (2001).
28. M. A. Pes, K. Aramaki, N. Nakamura, and H. Kunieda, *J. Colloid Interface Sci.*, *178*: 666 (1996).
29. H. Kunieda, N. Kanei, I. Tobita, K. Kihara, and A. Tuki, *Colloid Polymer Sci.*, *273*: 584 (1995).
30. H. Kunieda, E. Ogawa, K. Kihara, and T. Tagawa, *Colloid Polymer Sci.*, *105*: 239 (1997).
31. T. Katsuragi, N. Kaneko, and K. Sato, *J. Jpn. Oil Chem. Soc. (Yukagaku)*, *49*: 265 (2000).
32. I. Niiya, E. Moire, M. Imamura, M. Okada, and T. Matsumoto, *Jpn. J. Food Sci. Technol.*, *18*: 583 (1969).
33. I. Niiya, T. Maruyama, M. Imamura, M. Okada, and T. Matsumoto, *Jpn. J. Food Sci. Technol.*, *20*: 182 (1971).
34. E. Sambuic, Z. Dirik, and M. Naudet, *Rev. Fr. Corps Gras*, *28*: 59 (1981).
35. J. S. Aronhime, S. Sarig, and N. Garti, *J. Am. Oil Chem. Soc.*, *64*: 529 (1987).
36. J. Schlichter-Aronhime, S. Sarig, and N. Garti, *J. Am. Oil Chem. Soc.*, *65*: 1144 (1988).
37. P. Elisabettini, A. Desmedt, and F. Durant, *J. Am. Oil Chem. Soc.*, *73*: 187 (1996).
38. P. R. Smith, D. J. Cebula, and M. J. W. Povey, *J. Am. Oil Chem. Soc.*, *71*: 1367 (1994).
39. P. R. Smith and M. J. W. Povey, *J. Am. Oil Chem. Soc.*, *74*: 169 (1997).
40. P. R. Smith, *Eur. J. Lipid Sci. Technol.*, 122 (2000).
41. T. L. Harris, in SCI Monograph 32, SCI. London, 1968, pp. 108.
42. S. Wahnelt, D. Teusel, and M. Tulsner, *Fat Sci. Technol.*, *93*: 117 (1991).
43. S. Wahnelt, D. Teusel, and M. Tulsner, *Fat Sci. Technol.*, *93*: 174 (1991).
44. N. R. Easton, D. J. Kelly, L. R. Barton, S. T. Cross, and W. C. Griffin, *Food Technol. Champaign*, *6*: 21 (1952).
45. J. W. DuRoss and W. H. Knightly, Proc. 19th Proc. Conf. Pennsylvania Manuf. Conf. Assoc., sect. 15, 1965.
46. J. V. Ziemba, *Food Eng. 38*: 76 (1966).
47. N. Garti, J. Schlichter, and S. Sarig, *J. Am. Oil Chem. Soc.*, *63*: 230 (1986).

48. D. P. J. Moran, *Rev. Int. Choc. (RIC)*, *24*: 478 (1969).
49. M. L. Herrera and F. J. Marques Rocha, *J. Am. Oil Chem. Soc.*, *73*: 321 (1996).
50. R. J. Baer, M. D. Wolkow, and K. M. Kasperson, *J. Dairy Sci.*, *80*: 3123 (1997).
51. H. D. Goff, *Int. Dairy J.*, *7*: 363 (1997).
52. E. H. Lucassen-Reynders, Ph.D. thesis, Agricultural University, Wageningen, The Netherlands, 1962.
53. D. Johansson and B. Bergenstahl, *J. Am. Oil Chem. Soc.*, *69*: 718 (1992).
54. D. Johansson and B. Bergenstahl, *J. Am. Oil Chem. Soc.*, *69*: 728 (1992).
55. D. Johansson, B. Bergenstahl, and E. Lundgren, *J. Am. Oil Chem. Soc.*, *72*: 921 (1995).
56. N. Garti, H. Binyamin, and A. Aserin, *J. Am. Oil Chem. Soc.*, *75*: 1825 (1998).
57. N. Garti, A. Aserin, I. Tiunova, and H. Binyamin, *J. Am. Oil Chem. Soc.*, *76*: 383 (1999).
58. A. Martinez-Mendoza and P. Sherman, *J. Disp. Sci. Technol.*, *11*: 347 (1990).
59. L. G. Ogden and A. J. Rosenthal, *J. Colloid Interface Sci.*, *190*: 38 (1997).
60. L. G. Ogden and A. J. Rosenthal, *J. Colloid Interface Sci.*, *168*: 539 (1994).
61. L. G. Ogden and A. J. Rosenthal, *J. Am. Oil Chem. Soc.*, *75*: 1841 (1998).
62. B. E. Brooker, *Food Structure*, *12*: 115 (1993).
63. H. Birmbaum, *The Bakers Digest*, *52*: 28 (1978).
64. P. Walstra, Fat crystallization, in *Food Structure and Behaviour* (J. M. V. Blanchard and P. Lillford, eds.), Academic Press, London, 1987, pp. 67.
65. I. Heertje, *Food Structure*, *12*: 77 (1993).
66. D. Johansson, and B. Bergenstahl, *J. Am. Oil Chem. Soc.*, *72*: 911 (1995).

7
Crystallization of Oil-in-Water Emulsions

Malcolm J. W. Povey
University of Leeds, Leeds, England

I. INTRODUCTION

Crystallization in emulsions is an increasingly important area both technically and scientifically. Agrochemicals, pharmaceuticals [1], ceramic manufacture, food, cosmetics [2], speciality chemicals, photographic emulsions are a few examples of industries in which emulsion crystallization is widely employed or in development. Ceramics manufacture may seem an unlikely candidate for the use of emulsions, but emulsion crystallization offers more uniform stochiochemistry, smaller ceramic particle size, and a superior fired ceramic [3,4]. Crystallization of ceramics is but one example of a new approach to materials processing based on emulsions [1,5–11]. In foods, emulsion crystallization was discovered accidentally, first as part of the butter-churning process [12] and then in margarine manufacture. Margarine and fatty spreads are oil-in-water emulsions in which a crystal network stabilizes and structures what is basically a liquid, imparting upon it solid properties. Margarine manufacture is a little more complicated than simple emulsion crystallization [13]. While nucleation is initiated in the dispersed oil phase of an oil-in-water emulsion, the emulsion is inverted under shear during the crystallization process so that crystallization completes when the oil forms the continuous phase. The result is a kinetically stable water-in-oil emulsion, which would otherwise be a thermodynamically stable oil-in-water emulsion. This has the interesting property of inverting back to a water continuous emulsion when the crystal network melts in the mouth. In ice cream, the role of crystallization in stabilizing and structuring the product is even more complex [14,15]. In pharmaceuticals [1] and agrochemicals, emulsions permit the delivery of expensive and possibly hazardous chemicals in a water continuous system. This makes it much easier to control dosage; it also reduces hazards through concentration reduction and the presence of a water barrier. Normally the drug or chemical is dissolved in the dispersed oil phase. Crystallization may be used to structure the emulsion, imparting upon it the properties of a soft solid so that it adheres to surfaces. Fat crystals play an important role in structuring emulsions [14–18].

As will be evident from the fact that this is but one chapter in a book on crystallization processes in fats and lipid systems, crystallization is a complex process. In bulk liquids guest molecules and foreign surface may catalyze crystal nucleation (see Chapter 1). This process is called *heterogeneous* nucleation (Fig. 1a). Once such a liquid is subdivided into very many particles (Fig. 1b) new phenomena become apparent. First, subdivision of the bulk liquid into a large number of smaller ones also partitions those nuclei that may catalyze heterogeneous nucleation among the droplets (Fig. 1b). As a result the number of catalytic impurities per droplet may vary, depending on particle size, between many per droplet to approximately zero. This is generally observed as a greatly increased supercooling (Fig. 2 shows this in an oil-in-water emulsion with an average volume-surface particle diameter of 0.8 μm). In this figure the bulk melting point of *n*-hexadecane is around 18.2°C. However, the emulsion droplets do not freeze until the temperature reaches 3.5°C. When the emulsion is heated it does not melt at its freezing temperature; instead it melts close to the melting point of the bulk liquid. When the number of catalytic impurities is less than one per droplet, the kinetics will initially be proportional to the volume of each drop and hence to the cube of droplet diameter.

Second, as a result of the creation of an enormous interfacial surface area, nucleation at the interface between the crystallizable, dispersed material and the noncrystallizable continuous phase becomes more probable. For example surfactant molecules such as Tween 20 contain a hydrophobic tail of lauric acid. If the droplets comprise trilaurin, then the surfactant may catalyze crystal growth from the surface [19,20]. Thus the probability of nucleation will be related to the surface interfacial area and hence to the square of droplet diameter. On the other hand, nucleation may occur due to a guest molecule such as monolaurin within the trilaurin, in which case the probability of crystallization relates to the droplet volume.

Third, small droplets with sizes smaller than about 5 μm in water undergo Brownian diffusion. As a result, collisions between droplets that have crystallized with ones that have not may catalyze crystallization (Figs. 3, 4). A refinement to this droplet collision process that is dealt with by Professor Walstra and colleagues (this volume, Chapter 8) involves the growth of needle crystals out of a droplet, the collision of this crystalline material with other droplets, and subsequent nucleation. This is called a *secondary nucleation* process since it cannot occur until *primary* nucleation, either homogeneous or heterogeneous, has occurred somewhere in the system.

Fourth, the presence of surfactant causes a level of solubilization of the dispersed liquid phase in the continuous phase (Fig. 3), enabling transport of the otherwise insoluble liquid between droplets. In the presence of a wide range of

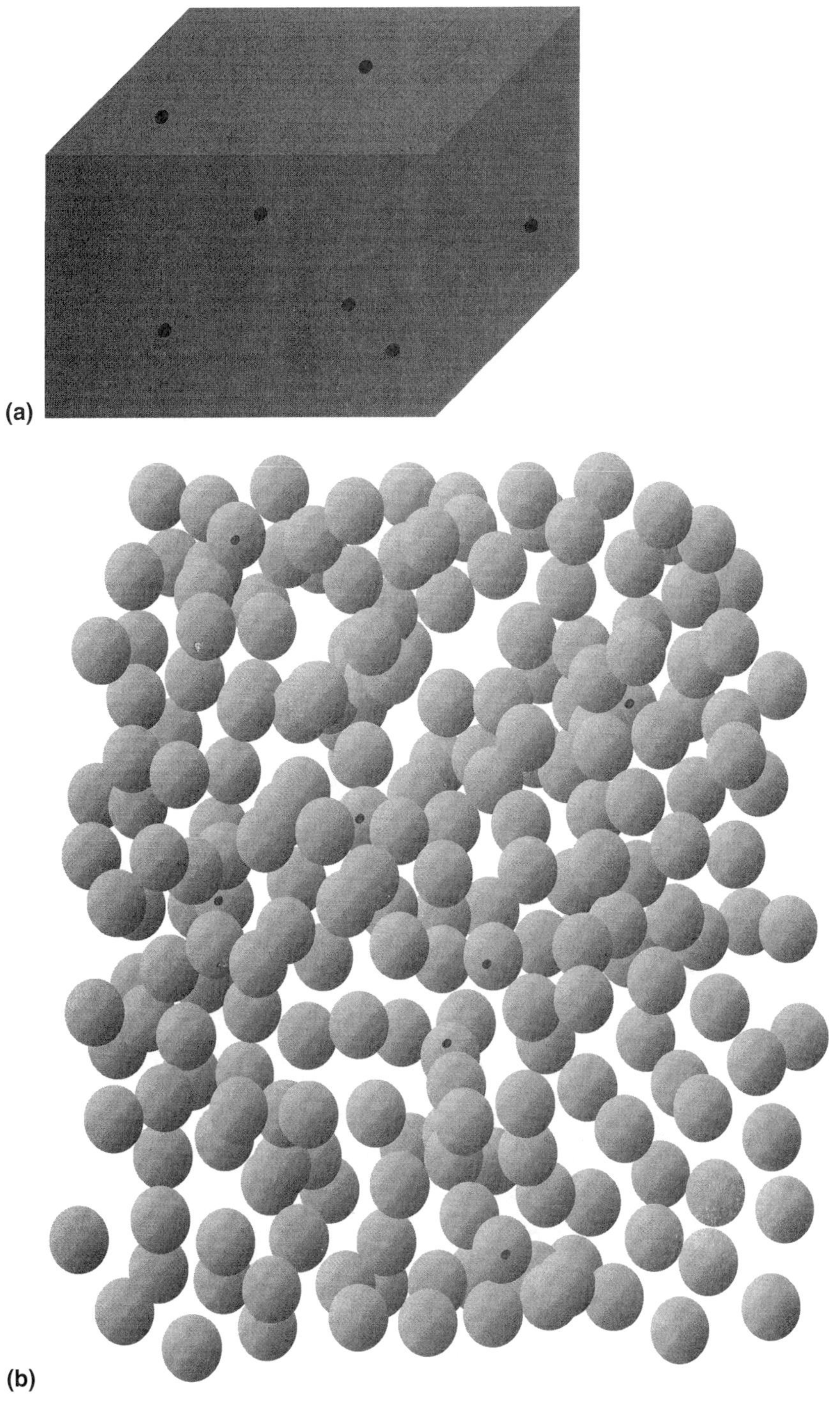

Fig. 1 (a) A volume of liquid contains seven catalytic impurities. The black dots represent catalytic impurities. (b) The same volume divided up into 112 parts.

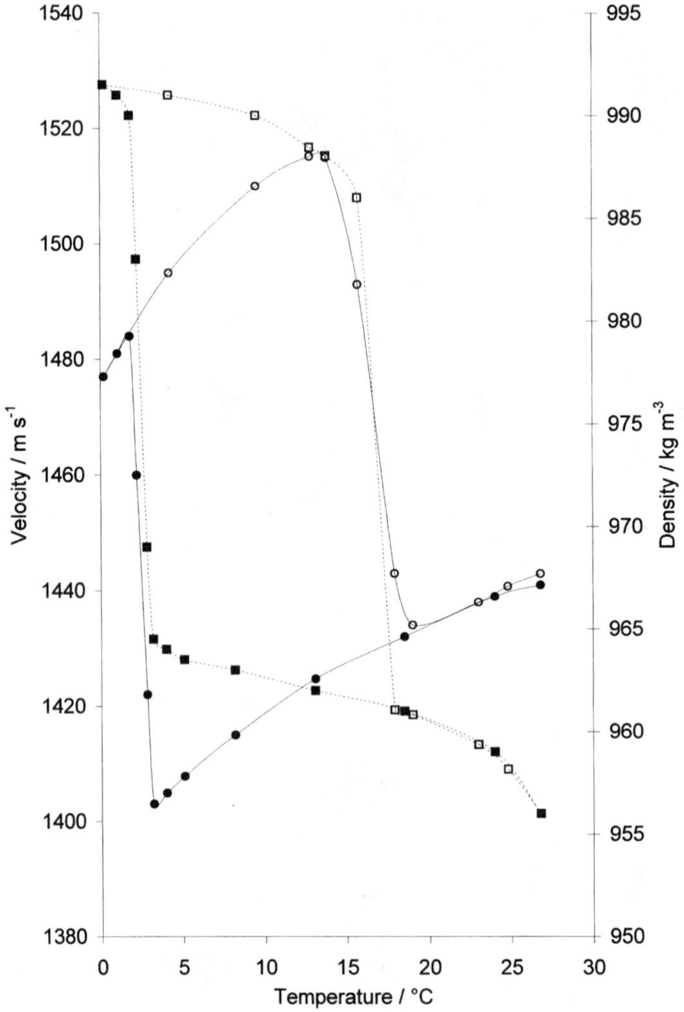

Fig. 2 Velocity of sound and density plotted against temperature for a 20 vol% *n*-hexadecane oil-in-water emulsions containing 2 wt % Tween 20. ●, velocity on cooling; ○, velocity on heating; ■, density on cooling; □, density on heating. (From Ref. 88)

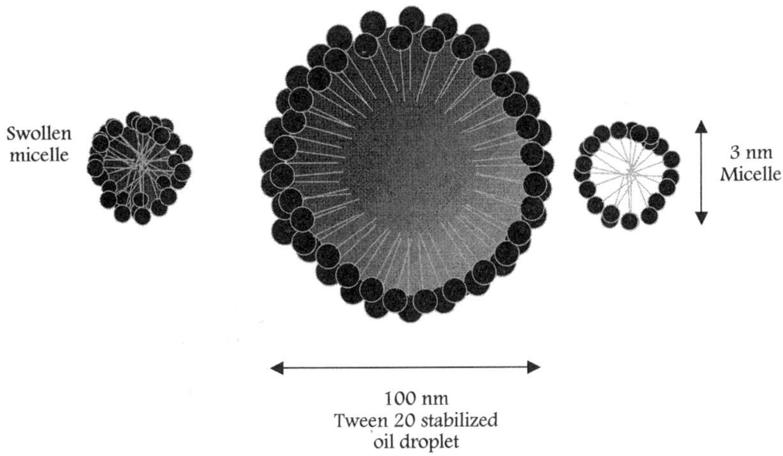

Fig. 3 An oil droplet stabilized with Tween 20 at its surface, together with a Tween 20 micelle and a swollen micelle containing oil.

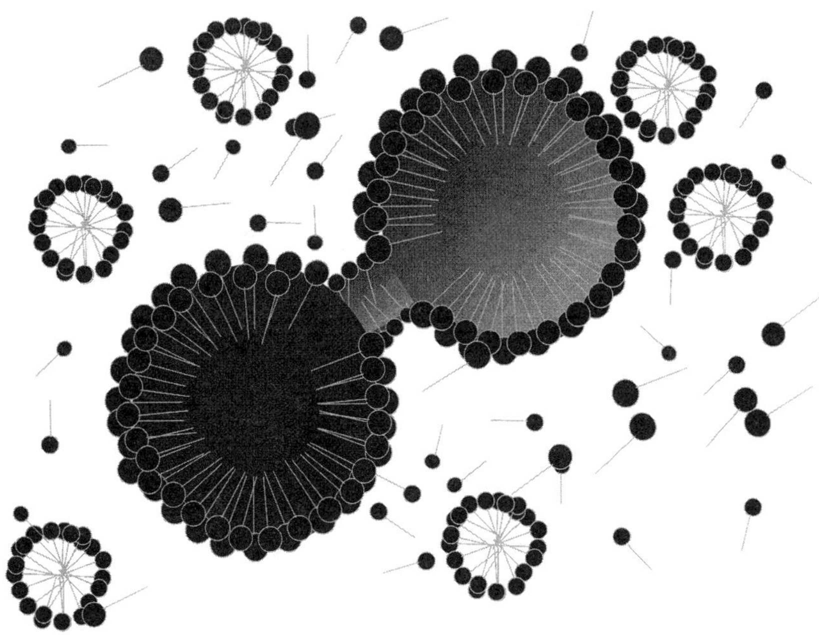

Fig. 4 Schematic of droplets undergoing collision-mediated nucleation.

particle sizes, a driving force exists which causes the large particles to grow at the expense of the smaller ones. This is called Ostwald ripening [21–26] and may have an impact on crystallization. For example, it can increase the likelihood of coalescence between two droplets, one crystallized and the other not, if they have very different sizes. This same transport mechanism may permit mixing of different oils between droplets of the same size but different composition [24,25]. Again this can have an impact on crystallization by altering the composition of each droplet. It has also been suggested that guest molecules and other catalytic impurities may be transported between droplets [27].

Fifth, the adsorbed layer at the oil-water interface may itself undergo crystallization [28,29], and, finally, heat transfer processes between the crystallizing dispersed phase and the continuous phase may greatly alter the polymorphic form and crystal habit [30–33].

As a result of the above, the following new factors are imparted by the subdivision of a bulk liquid into particles, having a very important bearing on crystallization, in addition to all the factors which must be accounted for in crystallization in a bulk liquid:

1. Particle size distribution, particle shape
2. Heat transfer between particles and the continuous phase
3. Nature of the boundary layer at the oil-water interface;
4. Viscosity of the continuous phase, contact angles (Fig. 5), and surface energies between all contacting surfaces
5. Kinetics of particle collision and mass transport, including mass transport of the dispersed phase and its minority components through the continuous phase
6. Surface nucleation

There are two practical aspects to crystallization in emulsions. First there is the technical application as outlined in the introduction. Second, emulsions form a very convenient framework within which to study crystallization itself. The earliest example of this is the work of Vonnegut [34]. Turnbull and coworkers formed dispersions of molten metal particles in order to study crystallization in bulk metals [35–39] and employed normal alkane oils as relatively simple model systems [40]. As indicated above, by subdividing a bulk fluid into smaller and smaller particles, it is possible to control the probability that a nucleus will be found in a given particle of fluid. A careful study of nucleation kinetics in emulsions can tell us a great deal about the nature of the crystal nucleation. For example, it is often possible to distinguish between homogeneous and heterogeneous nucleation. In the case of heterogeneous nucleation it is possible to determine the nature of the nucleus, its Gibbs free energy for nucleation, the concentration of nuclei, and even to get some idea of the size of nuclei.

Techniques used to study crystallization in emulsions include differential scanning calorimetry (DSC), differential thermal analysis (DTA), X-ray diffrac-

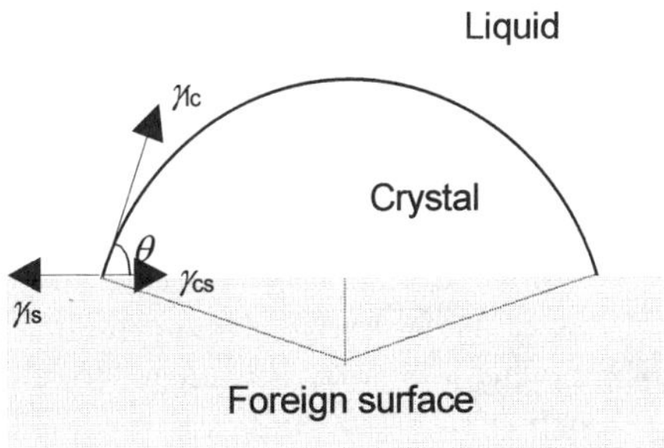

Fig. 5 Geometrical principle of heterogeneous nucleation. Vectors indicate the surface free energies; l = liquid, s = solid, and c = crystal.

tion, neutron diffraction, ultrasonic velocity measurement, and density measurement. These techniques will be discussed in the light of their application to the study of crystallization in emulsions.

In this chapter we are only concerned with crystallization processes within droplets and with nucleation. Elsewhere, Professor Walstra and his co-authors (this volume, Chapter 8) deal with the development of fat-crystal networks and the structuring of emulsions thereby. We will not concern ourselves with crystal growth and will assume that once nucleation occurs within an emulsion droplet its growth embracing the whole droplet is effectively instantaneous. This is a convenient simplification that glosses over the details of the crystallization process, particularly within larger droplets [41]. While the theory of crystal nucleation and growth is dealt with in detail by Professor Aquilano (this volume, Chapter 1), we sketch those aspects relevant to emulsion crystallization that have come to be called ''classical theory'' [27].

II. THEORY

A. Introduction

In the following discussion we assume that crystallization is initiated in the dispersed liquid phase and that the continuous phase of the emulsion cannot crystallize. Since the reverse process is very similar to crystallization in bulk materials it will not be considered here.

We adopt the nucleation model described in Chapter 1 and elsewhere [42], whereby the initial stage of crystallization involves nucleation. Although this theory is said not to be altogether accurate [43], it is sufficient for our purposes.

Homogeneous nucleation is a relatively rare process in emulsions. The reason for this is that subdivision of a pure material in the bulk into emulsion particles a few micrometers in size reduces the probability that liquid molecules will spontaneously assemble into an ordered domain of greater than the critical nucleus size. This probability is proportional to the amount of material and hence to the volume. In any case, it is possible to reduce the size of the droplets to below that of the critical nucleus size at any given undercooling. It is therefore normal to observe very considerable supercooling in emulsions. This supercooling will be a function of droplet size. For *n*-hexadecane dispersed as 0.8-μm droplets in water, a supercooling of 14.7 K (Fig. 2) is observed before crystallization begins. A supercooling of 26 K was measured in dispersed tristearin and tripalmitin [44,45]. In emulsified milk fat, supercooling of up to 20 K was observed [12]. This may be contrasted with a supercooling of only a few kelvins that is needed to induce crystallization in bulk fats. This small supercooling is explained by the presence of catalytic impurities in the bulk fat as indicated above.

On the other hand, guest molecules and surfaces may reduce the surface energy and/or volume entropy of the domain, reducing the critical size of the nucleus and greatly increasing the probability of a nucleus forming which is capable of growth. This is *heterogeneous* nucleation.

It is now possible to follow these nucleation processes in emulsions in considerable detail using recently developed techniques with pulsed nuclear magnetic resonance (pNMR) and ultrasound velocity measurement. As a result the kinetics of nucleation have been studied in detail, and quantities such as the chemical nature of nucleating material, its surface Gibbs energy, the concentration of nuclei, and even an upper limit on their size have been determined. Before attending to experimental issues, the theoretical foundation for the analysis of the experiments is sketched out. We adopt the classical theory [42,46] in what follows.

B. Crystal Nucleation in Bulk Liquids

The reader is referred to Chapter 1 for a detailed discussion of crystal nucleation in bulk liquids. The important result for emulsion crystallization is the nucleation rate for crystallization, J, given by the following semiphenomenological equation for the case of homogeneous nucleation. It is based on the product of the collision frequency in a system of N crystallizable molecules together with a kinetic barrier factor that delays crystallization and the entropy loss associated with the formation of a nucleus:

$$J = N\frac{k_B T}{h}\exp\left(\frac{-\alpha\Delta S_i}{R}\right)\exp\left(\frac{-\Delta G^*_{\text{nucleus}}}{k_B T}\right) \tag{1}$$

where $\exp(-\alpha\Delta S_i/R)$ is the probability that a fraction α of the molecule is in the right conformation to crystallize, and the loss of entropy ΔS_i on incorporation of material in a nucleus is given by

$$\Delta S_i = \frac{\Delta H_i}{T_{m,i}} \tag{2}$$

Here k_B is Boltzmann's constant, R is the gas constant, T is the crystallization temperature in kelvins; $T_{m,i}$ is the melting temperature of polymorph i, $\Delta G^*_{\text{nucleus}}$, is the activation Gibbs energy for the formation of a spherical nucleus, and ΔH_i is the enthalpy of fusion of polymorph i.

1. *Secondary Nucleation*

Once the first nucleus has formed and a crystal has begun to grow, other types of nucleation may come into play. Of course, primary nucleation may occur independently in many different places throughout the volume of liquid. However, the growing crystal may itself nucleate new centers of crystal growth by a variety of processes [12,18,27,41,47]. In the case of emulsions, for example, solid droplets can nucleate crystallization in liquid droplets through collision [48,21,24,25, 26,49–55,87, and see Secs. II.C and IV.A].

C. Nucleation in Emulsified Fats

In order to determine the effect of impurities on crystallization kinetics in emulsified fats it is necessary that the dispersed phase be sufficiently subdivided that at most one catalytic impurity is present per droplet (Fig. 1b). If subdivision is taken so far that the number of catalytic impurities per droplet approaches zero, then homogeneous nucleation will dominate.

Turnbull and co-workers [35–39] showed that nucleation rates can be determined by carrying out isothermal crystallization experiments at a series of undercooling and measuring the amount of solid material (ϕ) as a function of time (t). It is assumed that the average time needed for a nucleation event is much longer than the time needed for the droplet to achieve complete crystallization. Under these circumstances, the nucleation rate will be proportional to the number of droplets that do not contain crystals:

$$\frac{d\phi}{dt} = k_{\text{nucleation}}(1 - \phi) \tag{3}$$

The reaction rate constant $k_{\text{nucleation}}$ may be related to the nucleation rate J [Eq. (1)], according to the type of nucleation that occurs. For example, the reaction constant for homogeneous nucleation is simply related to the volume of material in the droplet (V_p) and the nucleation rate:

$$k_p = JV_p \tag{4}$$

On the other hand, if nucleation is catalyzed by impurities at the droplet surface, the rate constant k_s is proportional to the droplet surface area A_p:

Table 1 Different Types of Primary Nucleation in Emulsions

Type of nucleation	Example	Variables and equations[a]	Fit parameters
Homogeneous nucleation	Nucleation in pure mineral oils stabilized with sodium caseinate	$k_p = JV_p$ $J = N\frac{k_B T}{h}\exp\left(\frac{-\alpha\Delta S_i}{R}\right)\exp\left(\frac{-\Delta G^*_{\text{nucleus}}}{k_B T}\right)$ $\phi = 1 - \int_0^\infty \phi_d \exp(-k_p t)\, \mathrm{d}d$	J
Heterogeneous volume nucleation	Seed crystal nucleation in cocoa butter	$k_p = JV_p$, J must be determined experimentally from $\phi = \phi_m \frac{J_0 V_p t}{1 + J_0 V_p t}$, $\phi_m = 1 - \exp(-V_p N_c)$, and $\phi = 1 - \int_0^\infty \phi_d \exp(-k_p t)\, \mathrm{d}d$	J_0, N_c

Heterogeneous surface nucleation	(a) Tween 20 nucleation in triacylglycerol mixtures. (b) Interdroplet heterogeneous nucleation	(a) $k_s = JA_p$, J must be determined experimentally from $\phi = \phi_m \frac{J_0 V_p t}{1 + J_0 V_p t}$, $\phi_m = 1 - \exp(-V_p N_c)$, and $\phi = 1 - \int_0^{\infty} \phi_d \exp(-k_s t) \mathrm{d}d$ (b) Determine k_s from the slope of $\ln\left[\frac{1-\phi}{\phi}\right] = \ln\left[\frac{1-\phi_0}{\phi_0}\right] - k_s t$ Potential barrier to collision nucleation $= k_B T \ln(w)$ where the collision rate $w = \frac{k_s 3\eta}{8 k_B T n_0}$	J_0, N_c

[a] See text for definition of variables, explanation of equations, and assumptions.

$$k_s = JA_p \tag{5}$$

In this case Eq. (1) does not give J.

Nucleation may be catalyzed by impurities in the bulk in which case Eq. (4) will apply but J will once again *not* be given by Eq. (3). Solving Eq. (3) gives the volume fraction of droplets that have solidified as a function of time:

$$\phi = 1 - \exp(-kt) \tag{6}$$

The above equation applies in the case where all the droplets have the same size. This is rarely, if ever, true, and the equation must be rewritten to account for droplet size distribution:

$$\phi = 1 - \int_0^\infty \exp(-kt)\, \mathrm{d}d \tag{7}$$

where ϕ_d is the differential volume fraction of droplets with sizes between d and $d + \mathrm{d}d$.

In the case of heterogeneous nucleation, the number of catalytic impurities will vary according to droplet size (Table 1). It is therefore not possible to model the nucleation process with a single nucleation rate for all droplet sizes. Moreover, the droplets containing the highest number of catalytic impurities will crystallize first. As crystallization proceeds, smaller droplets will crystallize and the rate will fall with time. Finally, crystallization will reach a plateau ϕ_m which is less than the total amount of crystallizable material since many of the smaller droplets will contain no catalytic impurities at all. The maximum achievable volume fraction of solid droplets may be related to the number of catalytic impurities per volume by assuming a Poisson distribution of catalytic impurities throughout the volume of oil:

$$\phi_m = 1 - \exp(-V_p N_c) \tag{8}$$

where N_c is the number density of catalytic impurities. Equation (8) is plotted in Fig. 6 for two values of the number density of catalytic impurities found in cocoa butter at 15.8°C and 14.2°C. N_c may be strongly dependent on temperature if the catalytic impurities are mono- or diacylglycerols. This is because they are present in small quantities that must first crystallize from the mixture of liquid triacylglycerols from which the oil is constituted. This appears to be the situation in cocoa butter seed crystals for example (see below). However, this may not be the case if other catalytic impurities are present. For example, we would not expect this to be the case if the lauric acid moieties in Tween 20 were responsible for catalytic nucleation at the surface of the oil droplet (Fig. 7). Equation (8) assumes that the catalytic impurities are distributed at random between the droplets. This may not be the case if the impurities themselves are of a similar or greater size than many of the droplets. The emulsification process may actually exclude large catalytic impurities from smaller droplets.

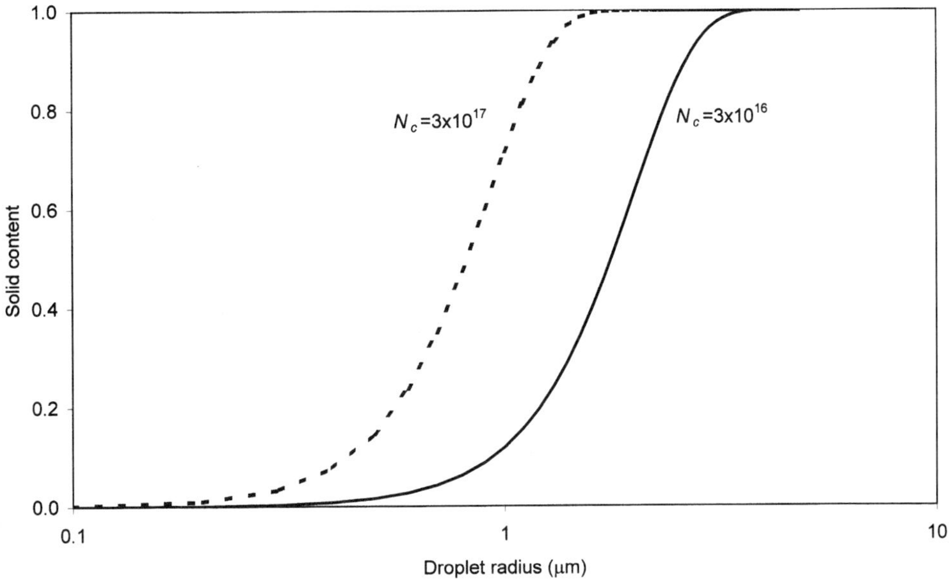

Fig. 6 Solid content of a cocoa butter–in–water emulsion modelled as a function of time for ——— 14.2°C (N_c = 3.08 × 10^{17}m^{-3}) and ——— 15.8°C (N_c = 3.08 × 10^{16}m^{-3}).

Walstra and van Beresteyn [12] showed that the nucleation rate is approximately described by

$$J = J_0\left(1 - \frac{\phi}{\phi_m}\right) \tag{9}$$

From Eqs. (3)–(8) we obtain the volume fraction of droplets containing crystals in the case of heterogeneous nucleation as a function of time:

$$\phi = \frac{1 - \exp[J_0 V_p t(\phi_m - 1)/\phi_m]}{1 - (1/\phi_m)\exp[-J_0 V_p t(\phi_m - 1)/\phi_m]} \tag{10}$$

where J_0 is the maximum nucleation rate. The previous equation may be written approximately as

$$\phi = \phi_m \frac{J_0 V_p t}{1 + J_0 V_p t} \tag{11}$$

The differences between the above three types of *primary* nucleation are summarized in Table 1.

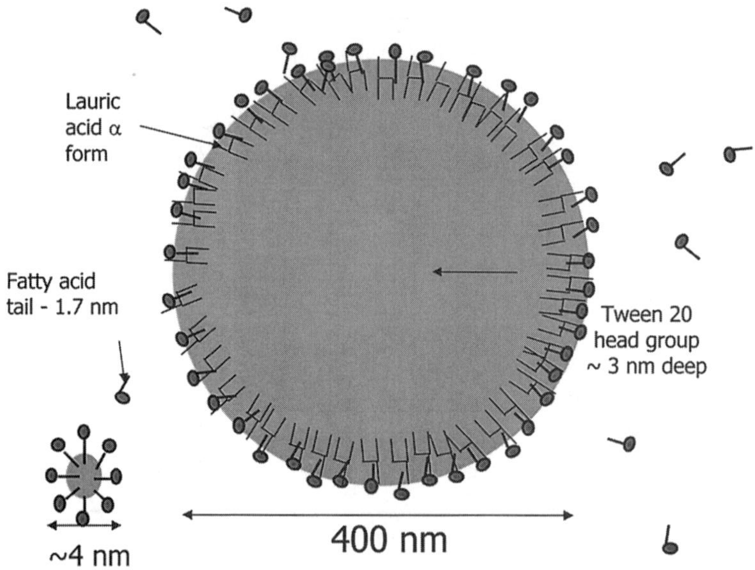

Fig. 7 Artist impression of Tween 20 adsorbed at the surface of a trilaurin oil droplet and ordering the trilaurin molecules.

D. Surface Energy and Contact Angle

As indicated, surface energy and the resultant contact angle are fundamental considerations in nucleation and crystal growth. In Fig. 5 (see also this volume, Chapter 1), the geometric principle relating the various contact angles together is shown. At an interface between two colliding droplets or between an adsorbed surface layer, a drop, and the continuous phase, Fig. 5 requires modification. Many workers have discussed the complex issues that arise with regard to crystal wetting in emulsions [56–64]. Here we confine ourselves to remarking that the surface energy concept itself may fail when the entities involved (nuclei, surfactant molecules and micelles, and oil molecules) are all of comparable sizes in a system which may be kinetically stable but certainly is not in thermodynamic equilibrium.

III. EXPERIMENTAL METHODS

A. Density Measurement

Turnbull used density measurement successfully in his work on normal alkane liquid crystallization in emulsions [40]. The main contemporary method for density measurement is the vibrating tube densitometer. In the version produced by

Paar (Gratz, Austria) a glass U tube is vibrated between two small electromagnets and the vibration frequency detected electronically. This is an elegant, accurate, and fast method of measuring density. However, care needs to be taken if there is a big density difference between the continuous and dispersed phases. In this case the dispersed phase may vibrate out of phase with the continuous phase, and the frequency is no longer a reliable indicator of density. For oil-in-water emulsions, this is not a problem. The vibrating tube is contained within a small glass thermostatic bath whose temperature can be controlled by pumping water through it. Thus this is a convenient method for following crystallization in emulsions since the phase change is generally accompanied by a density change. An example of data obtained using the Paar densitometer is given in Fig. 2.

Older methods of determining density include the density bottle and dilatometry [45,65,66].

B. X-ray Techniques

X-ray diffraction detects the order present in crystalline material and is very sensitive to crystallization. Since phase changes generally occur suddenly, it is sometimes necessary to use sophisticated X-ray diffraction instrumentation with a temperature-controlled stage and arrays of detectors. In this case information can be obtained about crystal polymorphs [67–76]. In addition, accurate data about the long spacing necessitates measurements at small angles, which requires specialist equipment and X-ray sources such as synchrotron radiation. X-ray diffraction has also been successfully used to determine the proportion of droplets that have solidified [27].

C. Pulsed Nuclear Magnetic Resonance

pNMR is the most widely used technique to determine the amount of solid fat in emulsions, replacing wide-line NMR [77,78] as the preferred method for solid content determination in fats and fat-containing materials. A detailed account of its application to food emulsions and fat crystallization may be found in [79]. A comparison of pNMR, calorimetry, and DSC appears in [76]. Protons possess a small magnetic moment that is caused to precess by a radio frequency (RF) pulse. In pNMR, the pulse is applied at right angles to a large static magnetic field that orients the magnetic field of the protons, which then precess like tops. The precession generates a detected RF field, the decay of which is related to the physical environment within which the protons find themselves. This decay in the signal gives the relaxation time for proton precession. The large difference in relaxation time between protons in a solid and a liquid environment is the means by which pNMR is used to measure solid content in fats. The signal, which is proportional to the total amount of protons, is measured directly after applying the pulse. It is measured again after all the protons in the solid material have relaxed and the ratio gives the amount of solid fat [80]. In an emulsion the large amount of water complicates the measurement. It is possible to differentiate between the signal from the protons in the water and those in the liquid oil, and in this case all measurements are made once the proton precession in the solid phase has relaxed. The advantage of measuring only the signal from the liquid is that polymorphic modification of the solid phase does not affect the results.

The solid content of a crystallizing emulsion can be determined by measuring the liquid signal of the emulsion (S_{em}), the liquid signal of a nonsolidified emulsion (S_{liq}), and the liquid signal of a completely solidified emulsion (S_{sol}):

$$\phi = \frac{S_{em} - S_{liq}}{S_{sol} - S_{liq}} \tag{12}$$

Kloek [81] has shown that the accuracy of pNMR in determining a 20% v/v hardened palm oil–in–water sample was 1.5% and in 10% v/v hardened palm oil–in–water it was 15%. A comparison of pNMR with ultrasound velocity measurement is given in Fig. 8. pNMR has been used to study the high-pressure behavior of fat crystallization in emulsions [82].

D. Thermal and Calorimetric Techniques

Thermal and calorimetric techniques involve the detection of a temperature difference between the sample and a standard as the temperature is changed. Endothermic and exothermic phase transitions cause an increased temperature differential from which the heat flow and therefore the enthalpy of the phase change may be inferred. The temperature difference is related to the rate at which heat is evolved or absorbed in the sample. These techniques consequently suffer at

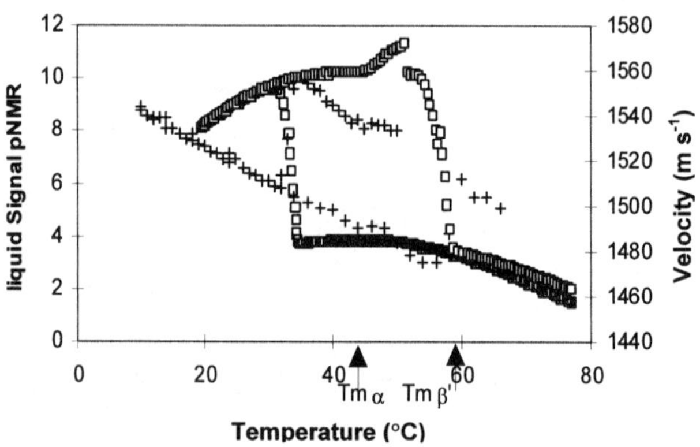

Fig. 8 Cooling-heating cycle of 20% v/v hardened palm oil–in–water measured by the liquid pNMR signal (+) and the ultrasound velocity (□). $d_{32} = 0.54$ μm, emulsifier was 2% w/w sodium caseinate. (Adapted from Ref. 81)

low crystallization rates since the heat flux may be too low to allow accurate determinations. Worse still, it can be very difficult to measure the baseline from which phase transitions are measured because the time needed to achieve thermal equilibrium in fats is too long [81]. However, DSC has been used successfully to study emulsion crystallization [31,32,75,76,81,83,84].

E. Neutron Diffraction Techniques

Neutron diffraction relies on the scattering of neutrons by atomic nuclei. It is similar in many respects to X-ray diffraction but unlike X-ray diffraction is not so sensitive to atomic number. It is therefore well suited to diffraction studies of low-atomic-weight carbon-hydrogen compounds such as fats. It has the further advantage that the phase of the signal diffracted from hydrogen is opposite to that of all other atoms. Deuteration, in effect the addition of a neutron to the proton in hydrogen, reverses this phase effect. Consequently, there is a range of elegant techniques that make neutron diffraction uniquely suited to the study of the hydrogen atoms and their distribution in fats [85]. Unfortunately, there are only a very few places where neutron diffraction can be carried out, due to its very high cost.

F. Ultrasound Techniques

A detailed account of ultrasound techniques applied to emulsion crystallization may be found in Ref. 86. Here only the briefest of details are given, sufficient to enable understanding of experimental details. Ultrasound is a new technique for determining solid fat content and is especially well suited to the measurement of emulsions. Recently, a number of groups have adopted the technique for the study of crystallization in bulk and emulsified fats [22,23,26,50,51,81,87–93]. The technique is perhaps the most accurate of all the experimental techniques available for the study of crystal nucleation in emulsified fats with sharp melting/freezing profiles and is capable of detecting very small amounts of solid material. In the case of hardened palm oil, Kloek [81] demonstrated that ultrasound velocity measured solid content to within an accuracy of 0.3% in a 20% emulsion. This may be compared with the 1.5% accuracy achieved by pNMR in the same emulsion. If the concentration of oil in the emulsion was reduced to 10%, the corresponding figure for pNMR rose to 15% and for ultrasound velocity it became 1%. A comparison of pNMR and ultrasound velocity measurement is shown in Fig. 8. Note that the melting transition causes an anomaly in the velocity of sound that arises due to absorption of sound by the melting process [86,93–96]. This anomaly may also appear in the freezing transition if the supercooling is just a few degrees K. It will occur if the coupling between the acoustic field and the phase transition is of the same order in energetic terms as the difference in Gibbs energy between the two coexistent phases. This energy is typically a few k_BT. In this case energy flows from the acoustic field into the phase transition process as the acoustic field flips the system backward and forward between the two energetically matched phases. This process may be used to study the phase transition, its energetics, and relaxation frequency through ultrasound attenuation measurements [93–97].

Ultrasound measurement of the solid content of fats invariably involves a measurement of the velocity of sound at a specified frequency. This sound velocity measurement is normally carried out using a technique in which a pulse of sound is generated at one face of an emulsion and then flies through the liquid to a detector. Alternatively, the generator may also act as the detector and the pulse is reflected at the far face of the liquid (Fig. 9). Care must be taken to degass samples and to ensure accurate thermostating and temperature measurement, if the full accuracy of the method is to be achieved. An example of data obtained from the apparatus of Fig. 9 is shown in Fig. 8.

The relationship between ultrasound velocity and solid content is

$$\phi = \left[\frac{v^{-2} - v_L^{-2}}{v_S^{-2} - v_L^{-2}} \right] \phi_f \tag{13}$$

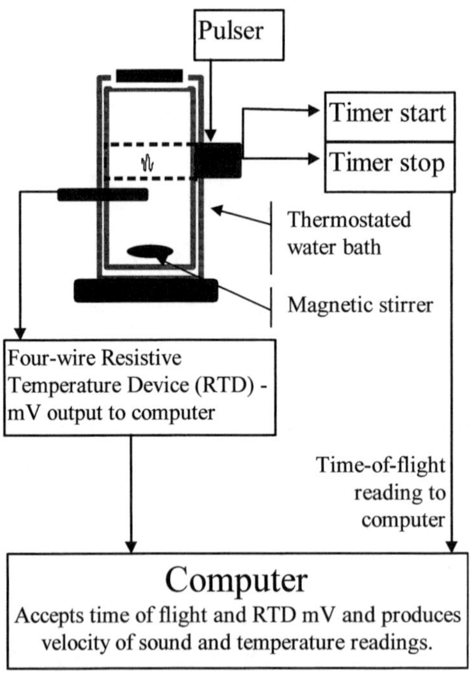

Fig. 9 Schematic of an ultrasonic velocity meter. (Adapted from Ref. 86)

where v is the velocity in the emulsion, v_L is the velocity measured when there is no crystalline material present, v_S is the velocity measured when all crystallizable material is solid, and ϕ_f is the total amount of fat in the emulsion. An example of the determination of solid content using Eq. (13) is shown in Fig. 10.

In the case of fats with a wide range of melting behavior and complicated

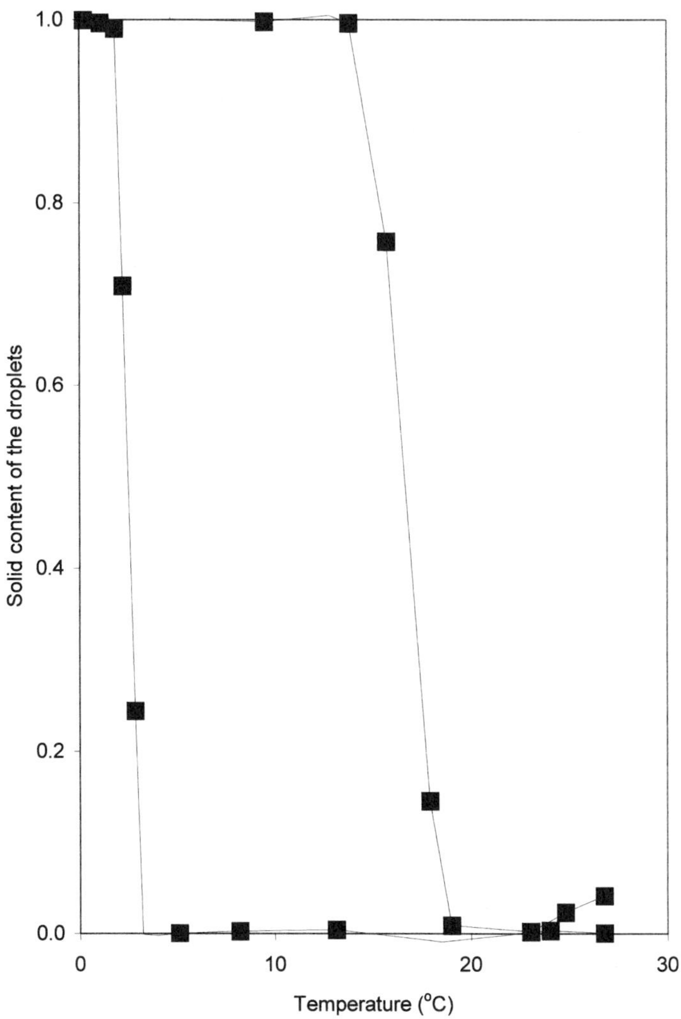

Fig. 10 Solid content plotted against time for the 20% v/v *n*-hexadecane oil-in-water emulsion of Fig. 2.

polymorphic behavior, ultrasound velocity measurement can still be used. However, it is generally necessary in this case to identify the polymorphs and triacylglycerols involved in each part of the melting/freezing curve and characterize the ultrasonic properties of the pure material. Of course, this is not always possible.

G. Other Techniques

Other techniques include photomicrography [44], wide-line NMR [78], and electron spin resonance [98].

IV. MEASUREMENTS OF CRYSTALLIZATION IN AN OIL-IN-WATER EMULSION

A. Studies of Nucleation by Isothermal Crystallization

As indicated, it is often useful to study bulk crystallization by examining the kinetics of nucleation in an emulsion. Vonnegut [34] and then Turnbull and coworkers [35–40] first demonstrated the basic outline of this technique. In addition to the ability to characterize the concentration of crystal nuclei, emulsion crystallization is well suited to ultrasound velocity measurement.

Cocoa butter oil-in-water emulsions exemplify the technique here. Cocoa butter has been chosen rather than a pure mineral oil because it is typical of fat that is widely used in foods and because it demonstrates a number of interesting phenomena. The emulsion is first prepared by melting the fat and emulsifying it using 0.8 wt% Tween 20 as the emulsifier. This is done at 70°C since the fat must be liquid. Details are given in Ref. 99. An emulsion with a concentration of 20 wt% cocoa butter was produced with a volume-surface mean particle size (d_{32}) of 0.26 μm.

Cocoa butter has six polymorphic forms instead of the three typical of much fat [73]. Partial glycerides dominate nucleation in most fats; however, in cocoa butter there is very strong evidence that it is ''seed crystals'' playing this role [100–106].

The cocoa butter emulsion is cooled at the rate of 5.0°C hr^{-1} from 45°C down to just below 0°C. The solid content is plotted in Fig. 11 as a function of temperature. For the moment consider only the sodium caseinate–stabilized emulsion. Sodium caseinate is a protein that adsorbs to the surface of the oil droplet and forms a relatively thick film. It is an effective steric stabilizer and is not expected to catalyze acylglycerol nucleation through any form of molecular similarity or incorporation into a compound crystal. The data clearly indicates that there are at least three crystallization processes at work, since there are three different rates of crystallization. In the case of the Tween 20–stabilized emulsion, careful inspection (below) indicates that there are two crystallization processes,

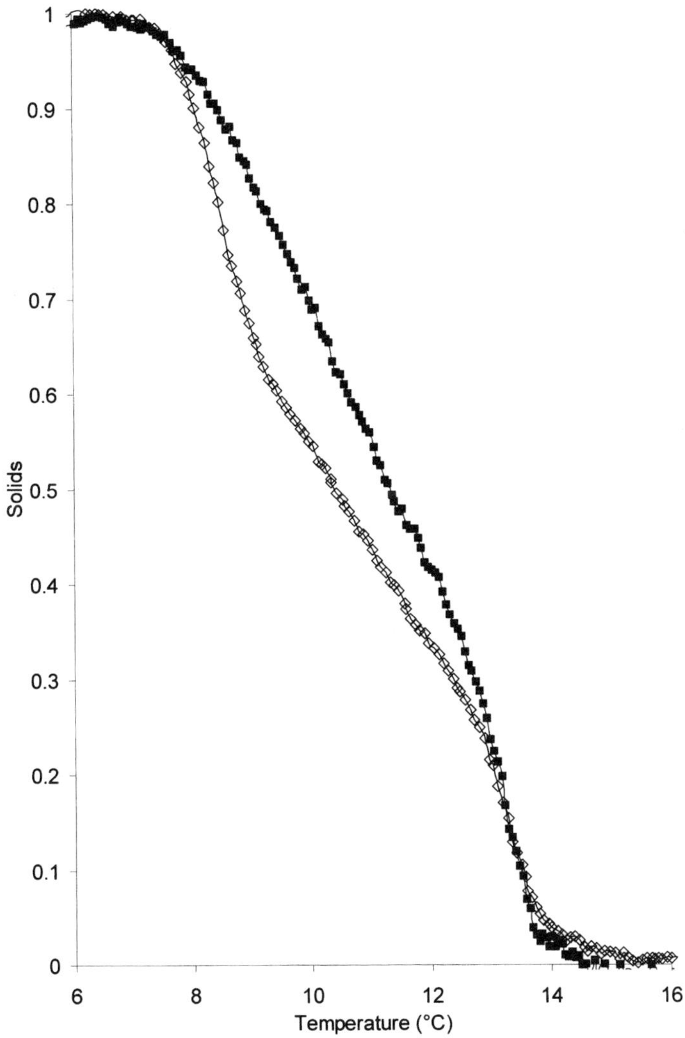

Fig. 11 Plot of solids against temperature for 20.75% (v/v) WACB-in-water emulsions cooled at 5.0°C hr^{-1} (■, 0.8% Tween 20 and ◇, 1.0% sodium caseinate). (From Ref. 99)

nucleation due to seed crystals, followed by collision-mediated secondary nucleation.

The emulsions were crashed-cooled at the rate of 5°C min^{-1} to various undercoolings at which temperature they were thermostated. The emulsion reached thermal equilibrium within 2 min, and the data in Fig. 12 is the resulting evolution of solid content at the various undercoolings for the Tween 20–stabilized emulsion. A heterogeneous volume nucleation model accounting for the actual particle size distribution [Eq. (7)] is shown in Fig. 12. While this model fits the later part of the solid content curve, it clearly does not explain the initial part of the nucleation process. In Fig. 13 the first 10 min of crystallization are plotted, and very precisely the heterogeneous volume nucleation model now fits the data. The kinetic data from this fit are given in Table 2. This is a strong indication that cocoa butter seed crystals catalyze crystallization in the emulsion; the number of nuclei per droplet is significantly less than one, a necessary requirement if we are to use the simplified nucleation model. We can assume that the solid content achieved at the end of the initial phase of nucleation corresponds to the number of droplets containing ‘‘active’’ seed crystals. At 14.2°C the heterogeneous model fitted yields a value of 0.072 catalytic impurities per droplet (or approximately 1 catalytic impurity per 14 droplets). If the assumption of seed crystals catalyzing the nucleation process holds, then the vast majority of droplets

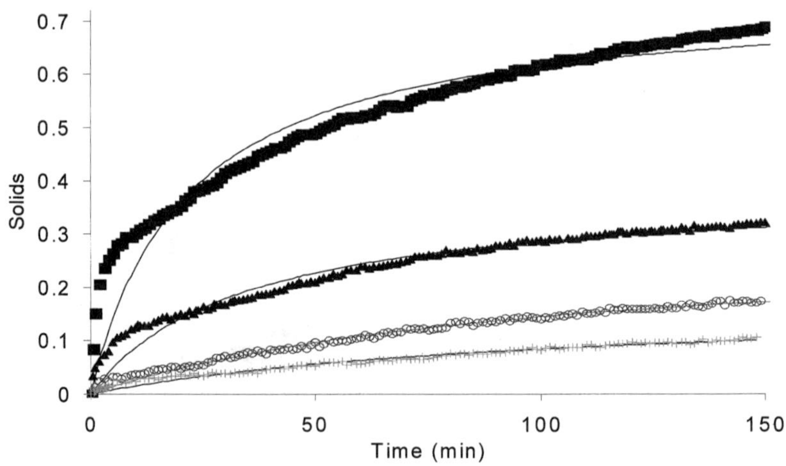

Fig. 12 Plot of solids against time for 20.75% (v/v) WACB-in-water emulsions (0.8% v/v Tween 20) crystallized isothermally at ■, 14.2 °C; ▲, 15.0 °C; ● 15.5 °C and +, 15.8°C. Heterogeneous volume particle-size-distribution models (continuous lines) are fitted over the time period 0–150 min. (From Ref. 99)

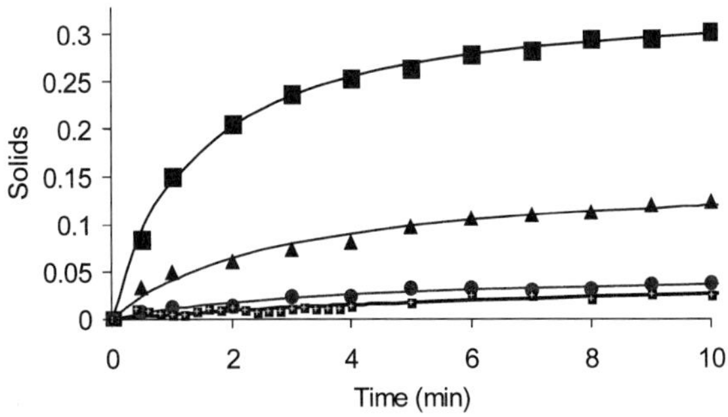

Fig. 13 Plot of solids against time for 20.75% (v/v) WACB-in-water emulsions (0.8% Tween 20) crystallized isothermally at ■, 14.2 °C; ▲, 15.0 °C; ● 15.5 °C and +, 15.8°C. Heterogeneous volume particle-size-distribution models are fitted over the time period 0–10 min. (From Ref. 99)

Table 2 Quantitative Data Obtained from Fitting Volume Particle-Size-Distribution Nucleation Models to Data in Fig. 12

Temp (°C)	Quantity	0–10 min Volume nucleation
14.2	$J(\mathrm{m^{-3}s^{-1}})$[a]	4.58×10^{16}
	N ($\mathrm{m^{-3}}$)	3.08×10^{17}
	n ($\mathrm{droplet^{-1}}$)	0.403
15.0	J ($\mathrm{m^{-3}s^{-1}}$)	2.26×10^{16}
	N ($\mathrm{m^{-3}}$)	1.08×10^{17}
	n ($\mathrm{droplet^{-1}}$)	0.142
15.5	J ($\mathrm{m^{-3}s^{-1}}$)	1.72×10^{16}
	$N(\mathrm{m^{-3}})$	3.16×10^{16}
	n ($\mathrm{droplet^{-1}}$)	0.041
15.8	J ($\mathrm{m^{-3}s^{-1}}$)	7.22×10^{15}
	N ($\mathrm{m^{-3}}$)	3.08×10^{16}
	n ($\mathrm{droplet^{-1}}$)	0.040

[a] J is nucleation rate ($\mathrm{m^{-3}s^{-1}}$), N is the number of crystal nuclei per unit volume of oil ($\mathrm{m^{-3}}$), and n is the number of nuclei per droplet.

Source: Ref. 99.

will not crystallize by this mechanism. Since the solid content rises to 100%, some other mechanism must be at work and this will be considered later.

There is a huge difference in volume between the largest and smallest droplets (2.87×10^{-23} m^3 and 9.98×10^{-18} m^3, respectively), of the order of 1×10^5 m^3. Hence, it is possible that seed crystals are either fairly large in relation to our emulsion droplets and will not fit into a large proportion of them or the actual number of "active" seed crystals is much smaller than the number of droplets in the emulsion. Whichever is the case, there is a higher probability that they will be located in the larger emulsion droplets.

It is possible to estimate an upper limit on the size of the seed crystals from our data. Seed crystals larger than the largest droplet will not be incorporated into a droplet. Instead, they will form an isolated emulsion particle contributing little to the initial crystallization rate. Any active seed crystals must therefore be smaller than the maximum oil droplet size. Increasing droplet size in small increments tested this hypothesis. The concentration of seed crystals, measured as the number of nuclei per unit volume of oil, estimated from the model changed little as the mean size of the emulsion was changed from 0.26 to 2 μm. (Larger size droplets could not be used because the emulsion destabilized due to creaming.) This strongly suggests that the seed crystals are smaller than 0.26 μm, the mean size of the emulsions containing 0.8% v/v Tween 20. To estimate the mean size of the seed crystals ($d_{\mathrm{seedcrystal}}$) from the number of nuclei per oil droplet and the rough value of the concentration we use the formula

$$d_{\mathrm{seedcrystal}} = \sqrt[3]{\frac{\phi y}{{}^{4}\!/_{3}\pi N_{\mathrm{seedcrystal}}}} \tag{14}$$

where $N_{\mathrm{seedcrystal}}$ is the number of seed crystals per unit volume (Table 2), ϕ is the volume fraction of oil, and y is the volume fraction of seed crystal material in the cocoa butter. This gives a maximal value of 0.14 μm for the average diameter of cocoa butter seed crystals. Here we used an estimate of the upper limit for the concentration of seed crystal material of 1.0% v/v [100].

In a further series of experiments, seed crystals were removed from, and added to, cocoa butter, so that we could vary the concentration of seed crystal material. Plots of solids against time are shown in Fig. 14 following isothermal crystallization at 15.0°C for samples designated as standard, seeds(+), and seeds(−). Seed material is added to those named seeds(+) and removed from those named seeds(−). Heterogeneous volume particle-size-distribution models are again fitted against experimental data in Fig. 14 and the quantitative values determined are shown in Table 3. As expected, the seeds(+) emulsions give the greatest numerical values of catalytic impurities per droplet, 0.267, as compared to 0.165 for the standard and 0.127 for the seeds(−) sample.

There is considerable evidence that droplet collision is a significant factor in crystallization in emulsions. This is likely to occur through breakdown in the

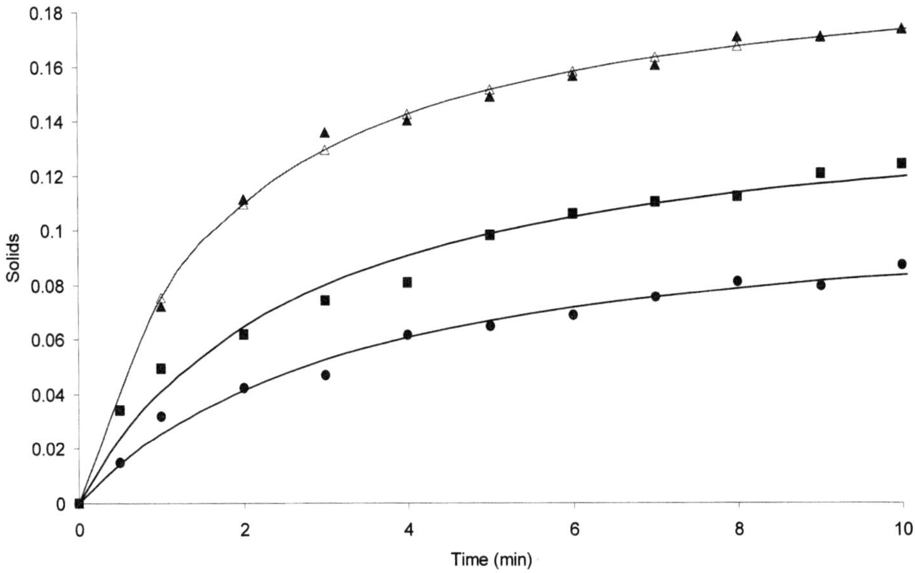

Fig. 14 Plot of solids against time for 20.75% (v/v) WACB-in water emulsions (0.8% Tween 20) crystallized isothermally at 15.0°C. Heterogeneous volume particle-size-distribution models are fitted against samples ■, standard; ▲, seeds (+) and ●, seeds (−) over the first 10 min solidification. (From Ref. 99)

Table 3 Quantitative Data Obtained from Fitting Volume and Surface Particle-Size-Distribution Nucleation Models to the Data in Figs. 14 and 15

Sample	Quantity	0–10 min Volume nucleation	10–150 min Surface nucleation
Seeds (+)	J ($m^{-3}s^{-1}$)[a]	3.43×10^{16}	6.14×10^{7}
	N (m^{-3})	1.55×10^{17}	2.86×10^{11}
	n ($droplet^{-1}$)	0.203	2.019
Standard	J ($m^{-3}s^{-1}$)	2.26×10^{16}	8.95×10^{7}
	N (m^{-3})	1.08×10^{17}	1.24×10^{11}
	n ($droplet^{-1}$)	0.142	0.875
Seeds (−)	J ($m^{-3}s^{-1}$)	1.81×10^{16}	9.11×10^{7}
	N (m^{-3})	7.65×10^{16}	9.35×10^{10}
	n ($droplet^{-1}$)	0.100	0.660

[a] J is nucleation rate ($m^{-3}s^{-1}$), N is the number of crystal nuclei per unit volume of oil (m^{-3}), and n is the number of nuclei per droplet.
Source: Ref. 99.

stabilizing surfactant film at the oil-water interface, followed by ''bridging'' phenomena. McClements and Dungan [49] have shown that an attractive potential of a few kT exists between closely approaching droplets in the case of Tween 20–stabilized emulsions. Evidently, some of these nucleating collisions also induce coalescence. This suggests that, in some cases at least, the postulated surfactant ''bridge'' [50] (Fig. 4) becomes a conduit for liquid material supplying material to the crystallized drop. Neither the fully crystallized nor the liquid oil emulsions displayed any significant change in particle size throughout the initial phase of the experiment. It seems reasonable to assume therefore that only those collisions that result in a nucleation event ($\sim$1 in 10^7 collisions [50]) may then give rise to coalescence. Slow heterogeneous nucleation has been observed in emulsions [22,26,48–51,55,87], and the dependence of this process on surfactant (Fig. 11) has also been demonstrated [51,53,87]. For example, the much increased nucleation rate in the presence of Tween 20 is consistent with the weaker stabilizing effect of this surfactant seen previously [51]. The nucleation rates that are measured in our experiment are consistent with those observed for similar emulsions and surfactants [51,87]. To test the droplet collision hypothesis, a surface fit, post 10 min crystallization (Fig. 15, Table 3), gave very similar nucleation rates for all

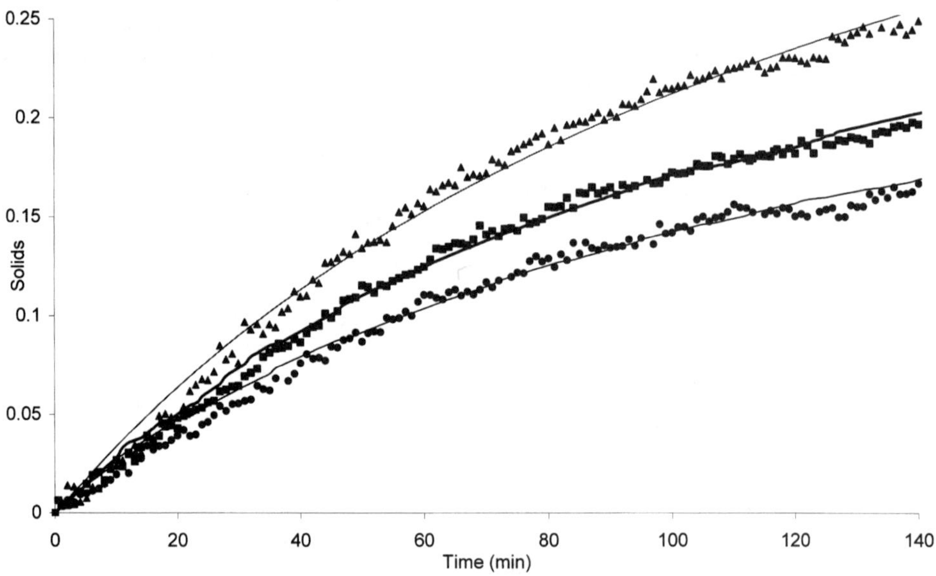

Fig. 15 Plot of solids against time for 20.75% (v/v) WACB-in-water emulsions (0.8% Tween 20) crystallized isothermally at 15.0°C. Heterogeneous surface particle-size-distribution models are fitted after 10 minutes (time and solids content considered to be zero at this point) for samples ■, standard; ▲, seeds (+) and ●, seeds (−). (From Ref. 99)

three samples, confirming that droplet collision is the likely *secondary* nucleation mechanism in the later stages of crystallization.

In summary then, three types of nucleation dominate crystallization in the cocoa butter emulsions:

a. A primary nucleation process due to seed crystals. This dominates the first few minutes of crystallization.
b. A secondary nucleation process arising from droplet collision. This dominates the rest of the crystallization process. However, the kinetics of the process differs according to the surfactant used. This will be considered in more detail later.
c. A tertiary nucleation process whose origin is unknown is clearly present in the caseinate-stabilized emulsion of Fig. 11. The crystallization rate suddenly accelerates as the solids content rises above 60%. This might, for instance, be related to the destabilization of the protein at the interface by the solidification of the droplets, reducing the energy barrier to collision-mediated nucleation.

B. Surfactant Effects

Surfactants may modify the crystallization process in a number of ways:

1. Modification of the Interfacial Tension

The interfacial tension of the droplet surface may be modified, altering the energy balance [Eq. (1)]. The surface of the droplet may itself act as a catalytic impurity. Depending on the wetting of the solid surface by the nucleus, a smaller liquid-crystal interface needs to be created. Figure 5 [6] shows a geometrical consideration of the heterogeneous nucleation process. The volume and surface area are dependent upon the liquid-solid contact angle θ. For $\theta = 180°$, no wetting of the foreign surface by the crystal occurs and the type of nucleation is homogeneous. Here the activation Gibbs energy is at its greatest and much greater undercoolings result. If $\theta = 0°$ no energy barrier for nucleation exists and instantaneous phase separation occurs. For values between 0° and 180° the free activation energy for heterogeneous nucleation is given by Zettlemoyer [107]:

$$\Delta G_{3D} = f(\theta)\Delta G_{3D,hom} \quad \text{with} \quad f(\theta) = \tfrac{1}{4}(2 + \cos\theta)(1 - \cos\theta)^2 \tag{15}$$

This corresponds to the following conditions on the quantity ϵ given by the surface free energy terms γ_{ls}, γ_{lc}, γ_{cs}:

$$\varepsilon = \gamma_{cs} - (\gamma_{cs} + \gamma_{ls}) \tag{16}$$

$$\gamma_{ls} = \gamma_{cs} + \gamma_{ls}\cos\theta \tag{17}$$

(a)

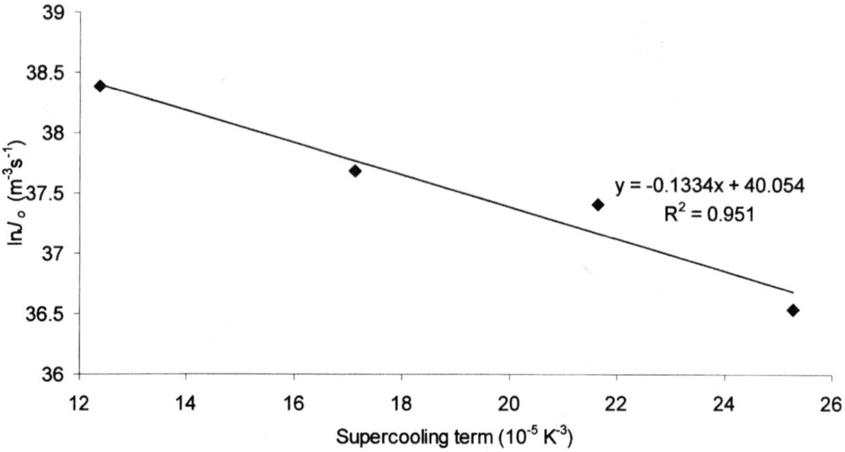

(b)

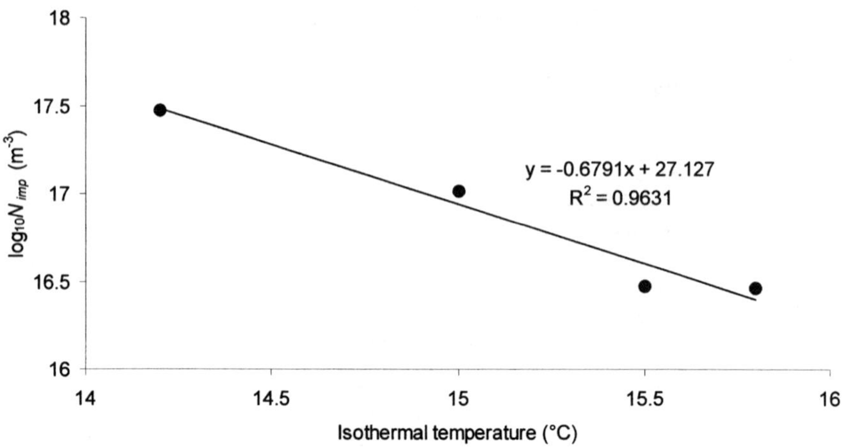

Fig. 16 (a) Plot of ln J_0 against the supercooling term $1/(T\Delta^2 T)$ used to derive the Gibbs surface free energy for the cocoa butter emulsion of Fig. 13. (From Ref. 99) (b) Plot of $\mathrm{Log}_{10}N_{\mathrm{imp}}$ against isothermal temperature for the cocoa butter emulsion of Fig. 13. (From Ref. 99)

For ε > 0, surface nucleation cannot occur, but ε < 0 favors surface nucleation. A polar oil emulsion system will be more favorable to nucleation at the droplet surface. This is because polar oils have a greater affinity for water than nonpolar oils, and hence ε is more likely to be negative.

The interfacial tension can be determined from isothermal crystallization measurements of the sort shown in Figs. 11 to 15. Figure 16 illustrates the determination of the Gibbs surface free energy and concentration of catalytic impurities from data such as is contained in Fig. 13. From Fig. 16, we obtain a value of 0.13 mJ m^{-2} for the Gibbs free energy of the nucleating layer in the cocoa butter seed crystals [99]. This is a very small value, much smaller even than that for octadecane (9.64 mJ m^{-2} [40]), but this may not be surprising given that the seed crystal is supposed to grow by successive nucleation of lower-melting acylglycerols. Fig. 17 gives a slope of 0.68 $m^{-3}\,K^{-1}$, corresponding to the temperature sensitivity of the seed crystals/nuclei within the emulsion droplets.

2. Steric Barrier Effects

The surfactant acts as a barrier (Table 4 and the discussion in Sec. IV. C), preventing coalescence and transient bridging of droplets. Thus the surfactant will alter the probability that a droplet collision leads to a secondary nucleation event. This barrier may be destabilized or altered by the phase change in the freezing

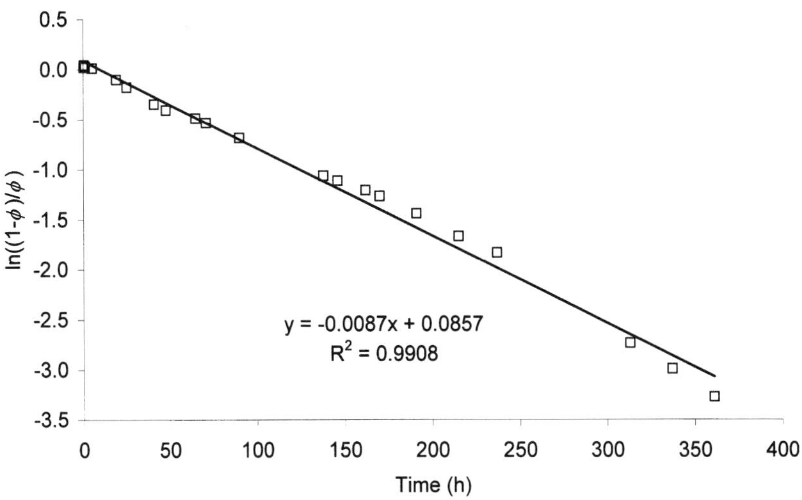

Fig. 17 A reduced log plot from Fig. 18 used to determine the rate constant k_{1s} for collision-mediated nucleation. In this case it is 0.0087/3600 = 2.42 × 10^{-6} sec^{-1}. (From Ref. 50)

Table 4 Kinetic Data for Collision-Mediated Nucleation in a Range of Emulsions Containing Different Surfactants

	k_s (10^{-6} sec^{-1})	T (K)	ϕ	η (Pa s)	d_{32} (μm)	n_0 (10^{20} m^{-3})	n_c (10^{20} sec^{-1} m^{-3})	n_c/n_0 (sec^{-1})	$1/w$ (μsec)	E/k_BT
0.5 wt% beta casein[51]	0.18	279.2	0.2	0.007	0.4	0.0597	0.523	9	0.0205	17.70
0.9 wt% beta lactoglobulin[51]	0.66	279.2	0.2	0.007	0.36	0.0819	0.984	12	0.0549	16.72
0.65 wt% SDS[51]	1.30	279.2	0.2	0.007	0.36	0.0819	0.984	12	0.108	16.04
0.8 wt% Tween 20[99]	100	288.2	0.2075	0.007	0.26	7.21	7880	1093	0.0915	16.21
0.94 wt% Tween 20[87]	1.7	277.2	0.125	0.007	0.37	0.0471	0.324	7	0.247	15.21
1.55 wt% Tween 20[87]	3.10	277.2	0.125	0.007	0.37	0.0471	0.324	7	0.451	14.61
2 wt% Tween 20[50]	2.42	279.2	0.2	0.007	0.36	0.0819	0.984	12	0.185	15.50
2.16 wt% Tween 20[87]	5.20	277.2	0.125	0.007	0.37	0.0471	0.324	7	0.757	14.09
3.07 wt% Tween 20[87]	7.40	277.2	0.125	0.007	0.37	0.0471	0.324	7	1.08	13.74
5.23 wt% Tween 20[87]	8.10	277.2	0.125	0.007	0.37	0.0471	0.324	7	1.18	13.65
7.07 wt% Tween 20[87]	9.20	277.2	0.125	0.007	0.37	0.0471	0.324	7	1.34	13.52
2 wt% Tween 20 +0.3 wt% xanthan[22]	1.80	279.2	0.2	0.1	0.37	0.0754	0.0584	1	2.32	12.97

droplets ([27] and Sec. IV.B.5). It is worth noting in Table 4 that the two situations that certainly produce depletion flocculation, 7 wt% Tween 20 and 2 wt% Tween 20 + 0.3 wt% xanthan, give the lowest energy barrier to secondary nucleation.

3. Surfactant as Catalytic Impurity

The hydrocarbon tails of adsorbed surfactant molecules penetrate into the oil droplet and catalyze crystallization [19,20,90,108]. This will depend on stereochemical and chemical affinity between the surfactant and the oil. Such a possibility has already been discussed in the introduction. A schematic of this in the case of Tween 20 and trilaurin is shown in Fig. 7. If we assume that the solid solution extends to a depth equivalent to two lauric acid moieties then it will have a thickness (dr) of 3.4 nm. The fraction of the total volume of oil organized in such a way will be approximately $3dr/r$ of a droplet of radius r. This is approximately 5% of the volume in the case of a droplet 400 nm in diameter. There is some evidence that the adsorbed layer forms a solid solution with a depressed melting point [94].

4. Crystallization of the Surfactant

The surfactant monolayer at the oil-water interface may itself "crystallize," causing a sharp fall in the interfacial tension [109,110], favoring surface nucleation. This is the case with monoglycerides used in ice cream manufacture where cooling below the transition temperature of the monoglyceride destabilizes the adsorbed protein at the oil-water interface [15]. This increases the likelihood of crystallization and the partial coalescence of droplets, especially at the air-water interface. Small crystals may also act as a surface stabilizer [16,17].

5. Destabilization of the Surface Layer

The droplets may expand on crystallization due to the development of an extensive defect structure [111,112]. The resulting depletion of surfactant at the surface may lead to increased coalescence, partial coalescence, and collision-mediated nucleation. The instability of crystallized emulsions has been widely observed [45,84,113]. This may explain the tertiary crystallization process referred to in Fig. 11 and Sec. IV.A.

6. Ostwald Ripening and Droplet Coalescence

Emulsions with a relatively narrow distribution of particle size (e.g., 1 μm ± 0.8 μm) possess a small thermodynamic driving force for Ostwald ripening [14,21]. Ostwald ripening occurs because the surfactant solubilizes the otherwise insoluble liquid to a degree, within the continuous phase. Solubilization will only occur to a significant extent if the surfactant concentration in the continuous phase exceeds the critical micelle concentration (CMC). Solid droplets cannot undergo size reduction due to solubilization [11] and Ostwald ripening, but solid droplets in contact with slightly smaller liquid droplets may well undergo significant Ostwald ripening during a collision event. This may explain why only a small minority of droplets exhibits coalescence. As the particle size distribution widens, this coalescence effect will rapidly become more important.

7. *Mixing of Oil*

Even when there is no net driving force for Ostwald ripening, exchange of fluid will take place between droplets, altering their composition in a system containing a heterogeneous mixture of droplets of different oils [24,25,51–54,114].

C. Droplet-Collision-Mediated Secondary Nucleation

In the case of droplet collisions we have suggested the following kinetic model [22,26,51,55,87]. Unambiguous evidence exists for this mechanism (Fig. 18). Assume that a ''reactive'' pair encounter occurs between a single liquid droplet and a single solid droplet with the degree of reactivity quantified by a second-order rate constant k_{ls}. Further assume that the newly formed solid droplets have the same reactivity as the original solid droplets. The change in the fraction of liquid oil (1-ϕ) remaining after time t is

$$\frac{d(1-\phi)}{dt} = -k_{ls}\phi(1-\phi) \tag{18}$$

Integration then gives

$$\ln\left[\frac{1-\phi}{\phi}\right] = \ln\left[\frac{1-\phi_0}{\phi_0}\right] - k_{ls}t \tag{19}$$

k_{ls} can be determined from the slope of $\ln[(1-\phi)/\phi]$ versus time [88]. This rate constant will be related to the number of collisions per second (n_c), the fraction of collisions leading to nucleation ($1/w$), and the number of droplets per unit volume of emulsion (n_0).

Smoluchowski [115] showed that the frequency of collisions in a dilute quiescent emulsion is

$$n_c = 8\pi D d n_0^2 \tag{20}$$

where D is the diffusion coefficient for a single particle. For spherical particles this becomes

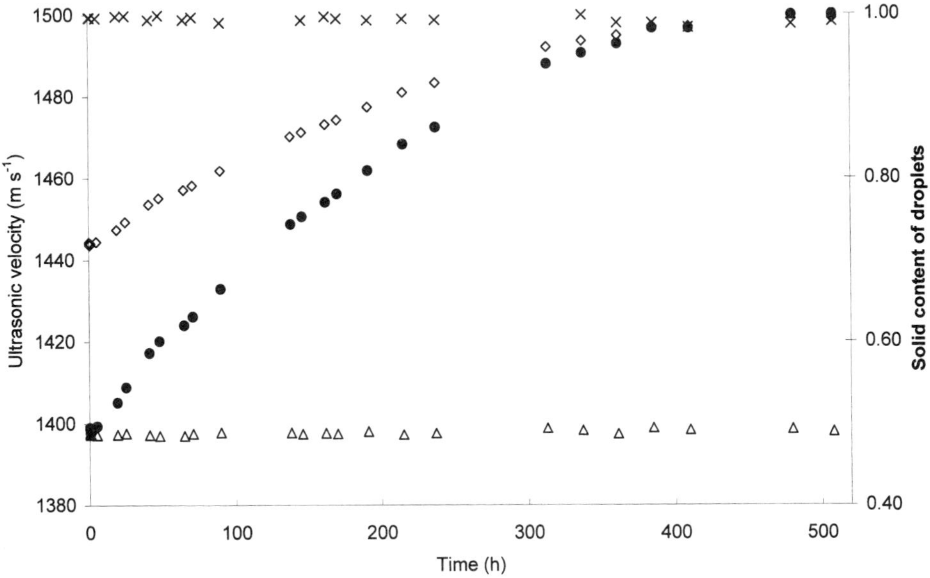

Fig. 18 Plot of ultrasound velocity and solid content -20 wt % n-hexadecane oil-in-water emulsion stabilized with 2.5 wt% Tween 20, $d_{32} = 0.35$ μm at 6°C. △, original emulsion, droplets unfrozen; x, droplets fully frozen by cooling to 2°C for 2 hr before returning to 6°C; ◇, a 50/50 mix of the previous two emulsions, ●, the solid content of the 50/50 mixture of liquid and solid droplets. (From Ref. 50)

$$n_c = \frac{8k_B T n_0^2}{3\eta} \tag{21}$$

where η is the viscosity of the continuous phase. For the emulsion of Fig. 16, the rate constant is $k_{ls} = 2.4 \times 10^{-6}$ s^{-1} and n_c/n_0 is 12 sec^{-1}. Thus $1/w$ is 2×10^{-7}; i.e., only one in 5×10^6 collisions is apparently effective in causing nucleation. The magnitude of the potential barrier to collision-mediated nucleation is $k_B T \ln w$ which is approximately $16k_B T$. A detailed discussion of collision-mediated nucleation appears in [79]. However, it is interesting to compare the crystallization rate constant for various surfactants (Table 4 [22,50,51,87,99]). Note in particular the relatively low value for the potential barrier ($E/k_B T$) in the presence of xanthan and the highest concentration of Tween 20. There are examples in Table 4 of emulsions which have undergone depletion flocculation, which suggests that this type of flocculation greatly increases the probability of collision-mediated nucleation, per collision. However, since the xanthan also raises the viscosity of the continuous phase by two to three orders of magnitude, the number of collisions per second falls drastically; the result is an overall nucleation rate (k_{ls}) which is not very different from the lower concentration Tween 20–stabilized emulsions.

The sterically stabilizing, large molecule surfactants, such as casein and beta lactoglobulin, have the highest values of E/k_BT, consistent with their acting to prevent direct contact between the oil in colliding droplets. On the other hand, the highest concentrations of the small amphiphilic surfactant Tween 20 show the lowest value of E/k_BT, together with the xanthan-containing emulsion. In these two cases, depletion flocculation will occur, bringing the droplets into close proximity with each other, and hence it is unsurprising that the barrier to collision-mediated crystal nucleation should be lowered under these circumstances.

D. Particle Size and Viscosity Effects

The particle size and size distribution have a very big effect on the nucleation rates for all types of nucleation [91]. In the case of collision-mediated nucleation the particle size also combines with the continuous phase viscosity to control the collision rate. This raises the difficult question of the measurement of particle size distribution. This is not the place to discuss the merits of different particle size and viscosity measurement techniques. Suffice it to say that serious errors are easily made [116].

V. CONCLUSION—THE FUTURE

With the right techniques and accurate models a great deal of progress may be made in understanding crystal nucleation and growth in emulsions, despite its complexity. Outstanding issues include the detailed understanding of collision-mediated nucleation, and the interaction among the droplet boundary layer, the crystallizing/melting droplet, and the continuous water phase. We can expect to see emulsion nucleation studies play a central role in elucidating crystal nucleation mechanisms in bulk liquids. Technically, the detailed understanding of fat and lipid crystal nucleation and growth is likely to lead to new possibilities for structuring oil-in-water emulsions and to new products.

ACKNOWLEDGMENTS

Scott Hindle (Leeds) provided the cocoa butter data, which is presented with permission from Unilever. William Kloek (Wageningen) is thanked for the use of data from his Ph.D. thesis. I am grateful to Kevin Smith (Unilever) for helpful discussions and for collaboration in the cocoa butter measurements.

REFERENCES

1. F. Espitalier, B. Biscans, J. R. Authelin, and C. Laguerie, *Chem. Eng. Res. Design*, *75*: 257 (1997).
2. J. Wang and G. H. J. Lee, *Soc. Cosmetic Chem.*, *48*: 41 (1997).
3. T. Hirai, S. Hariguchi, I. Komasawa, and R. J. Davey, *Langmuir*, *13*: 66503 (1997).
4. T. Hirai, N. Okomoto, and I. Komasawa, *Langmuir*, *14*: 66483 (1998).
5. R. J. Davey, J. Garside, A. M. Hilton, D. McEwan, and J. W. J. Morrison, *Nature*, *375*: 664 (1995).
6. R. J. Davey, J. Garside, A. M. Hilton, D. McEwan, and J. W. J. Morrison, *Crystal Growth*, *166*: 971 (1996).
7. R. J. Davey, A. M. Hilton, and J. Garside, *Chem. Eng. Res. Design*, *75*: 245 (1997).
8. R. J. Davey, A. M. Hilton, J. Garside, M. de la Fuente, M. Edmondson, and P. J. Rainsford, *Chem. Soc. Faraday Trans. 1*, *92*: 1927 (1996).
9. M. P. Pileni, *Langmuir*, *13*: 32666 (1997).
10. J. J. Bibette, *Coll. Int. Sci.*, *147*: 474 (1991).
11. A. D. Dinsmore, J. C. Crocker, and A. G. Yodh, *Current Opinion Coll. Int. Sci.*, *3*: 5 (1998).
12. P. Walstra and E. C. H. van Beresteyn, *Neth. Milk Dairy J.*, *29*: 35 (1975).
13. A. J. Haighton, *J. Am. Oil Chem. Soc.*, *42*: 397 (1976).
14. H. D. Goff, *J. Dairy Sci.*, *80*: 26200 (1997).
15. H. D. Goff, *Int. Dairy J.*, *7*: 363 (1997).
16. N. Garti, J. Aserin, and I. Tiunova, *J. Am. Oil Chem. Soc.*, *76*: 383 (1999).
17. N. Garti, H. Binyamin, and J. Aserin, *J. Am. Oil Chem. Soc.*, *75*: 18251 (1998).
18. P. Walstra, T. van Vliet and W. Kloek, in *Advanced Dairy Chemistry*, *Vol 2*; *Lipids*, 2nd ed. (N. Fox, ed.) Chapman and Hall, London, 1995, p. 179.
19. P. R. Smith, *The Molecular Basis for Crystal Habit Modification in Triglycerides*, Ph.D. thesis, Leeds University, 1995.
20. P. R. Smith and M. J. W. Povey, *J. Am. Oil Chem. Soc.*, *74*: 169 (1997).
21. W. Ostwald, *Z. Phys. Chem.*, *22*: 289 (1897).
22. E. Dickinson, J. G. Ma, and M. J. W. Povey, *J. Chem. Soc. Faraday Trans 1*, *92*: 12135 (1996).
23. E. Dickinson, M. I. Goller, D. J. McClements, and M. J. W. Povey, in *Food Polymers, Gels and Colloids*, Royal Society of Chemistry, London, 1991, p. 171.
24. D. J. McClements, S. R. Dungan, J. B. German, and J. E. Kinsella, *J. Phys. Chem.*, *97*: 73048 (1993).
25. D. J. McClements, S. R. Dungan, J. B. German, and J. E. Kinsella, *Coll. Surf. A*, *81*: 203 (1993).
26. D. J. McClements, E. Dickinson, and M. J. W. Povey, *Chem. Phys. Lett.*, *172*: 449 (1990).

27. A. B. Herhold, D. Ertas, A. J. Levine, and H. E. King Jr, *Phys. Rev. E*, *59*: 69465 (1999).
28. T. Hianik, S. Kupcu, S. Sletyr, U. B. Rybar, P. Krivanek, and U. Kaatze, *Coll. Surf. A*, *147*: 331 (1999).
29. A. J. Fillery-Travis, L. H. Foster, and M. M. Robins, *Biophysical Chem.*, *54*: 253 (1995).
30. J. P. Dumas, M. Krichi, M. Strub, and Y. Zeraouli, *Int. J. Heat Mass Trans.*, *37*: 737 (1994).
31. J. P. Dumas, Y. Zeraouli, and M. Strub, *Thermochim. Acta*, *236*: 227 (1994).
32. J. P. Dumas, Y. Zeraouli, and M. Strub, *Thermochim. Acta*, *236*: 239 (1994).
33. J. P. Dumas, Y. Zeraouli, M. Strub, and M. Krichi, *Int. J. Heat Mass Trans.*, *373*: 747 (1994).
34. B. Vonnegut, *J. Coll. Interface Sci.*, *3*: 563 (1948).
35. D. Turnbull, *J. Appl. Phys.*, *21*: 10228 (1950).
36. D. Turnbull, *J. Chem. Phys.*, *20*: 411 (1952).
37. D. Turnbull and R. E. Cech, *J. Appl. Phys.*, *21*: 804 (1950).
38. D. Turnbull and J. C. Fisher, *J. Chem. Phys.*, *17*: 71 (1949).
39. D. Turnbull, *J. Appl. Phys.*, *21*: 10228 (1950).
40. D. Turnbull and R. L. Cormia, *J. Chem. Phys.*, *34*: 820 (1961).
41. P. Walstra, *Prog. Colloid Polym. Sci.*, *108*: 4 (1998).
42. M. Volmer, *Kinetik der Phasenbildung*, Steinkopf, Leipzig, 1939.
43. J. Lyklema, *Fundamentals of Interface and Colloid Science*, Academic Press, London, 1991.
44. L. W. Phipps, *Trans. Faraday Soc.*, *60*: 1873 (1964).
45. W. Skoda and M. J. van den Tempel, *Coll. Sci.*, *18*: 585 (1963).
46. S. Özilgen, C. Simoneau, J. B. German, M. J. McCarthy, and D. S. Reid, *J. Sci. Food Agric.*, *61*: 101 (1993).
47. P. Walstra, in *Food Structure and Behaviour* (J. M. V. Blanshard and P. J. Lillford, eds.), Academic Press, San Diego, 1987, p. 67.
48. E. Dickinson and D. J. McClements, *Advances in Food Colloids*, Blackie, Glasgow, 1995, p. 211.
49. D. J. McClements and S. R. Dungan, *J. Coll. Interface Sci.*, *186*: 17 (1997).
50. D. J. McClements, E. Dickinson, and M. J. W. Povey, *Chem. Phys. Letts.*, *172*: 449 (1990).
51. D. J. McClements, E. Dickinson, S. R. Dungan, J. E. Kinsella, J. G. Ma, and M. J. W. Povey, *J. Coll. Interface Sci.*, *160*: 293 (1993).
52. D. J. McClements, S. R. Dungan, J. B. German, and J. E. Kinsella, *Food Hydrocoll.*, *6*: 415 (1992).
53. D. J. McClements, S. R. Dungan, J. B. German, C. Simoneau, and J. E. Kinsella, *J. Food Sci.*, *58*: 1148 (1993).
54. D. J. McClements, S. R. Dungan, J. B. German, and J. E. Kinsella, *J. Coll. Interface Sci.*, *156*: 425 (1993).
55. D. J. McClements, S-W. Han, S. R. Dungan, and M. J. McCarthy, *J. Am. Oil Chem. Soc.*, *71*: 13859 (1994).
56. K. Boode, C. Bisperink, and P. Walstra, *Coll. Surf. A*, *61*: 55 (1991).
57. K. Boode, P. Walstra, and A. E. A. Degrootmostert, *Coll. Surf. A*, *81*: 139 (1993).

58. P. Walstra and K. Boode, *Coll. Surf. A*, *81*: 121 (1993).
59. D. Johannsson and B. Bergenståhl, *J. Am. Oil Chem. Soc.*, *72*: 205 (1995).
60. D. Johannsson and B. Bergenståhl, *J. Am. Oil Chem. Soc.*, *72*: 933 (1995).
61. D. Johannsson, B. Bergenståhl, and E. Lundgren, *J. Am. Oil Chem. Soc.*, *72*: 921 (1995).
62. B. Bergenståhl and J. Alander, *Current Opinion Coll. Interface Sc.*, *2*: 590 (1997).
63. L. G. Ogden and A. J. Rosenthal, *J. Coll. Interface Sci.*, *191*: 38 (1997).
64. L. G. Ogden and A. J. Rosenthal, *J. Am. Oil Chem. Soc.*, *75*: 18417 (1998).
65. American Oil Chemists Society, *Official Tentative Methods of the American Oil Chemists Society*, *Section A*, 1973.
66. K. van Putte, L. F. Vermaas, J. C. van den Enden, and C. den Holler, *J. Am. Oil Chem. Soc.*, *52*: 179 (1975).
67. E. S. Lutton, *J. Am. Oil Chem. Soc.*, *67*: 524 (1945).
68. T. Malkin, *Prog. Chem. Fat Lip.*, *2*: 1 (1954).
69. D. Chapman, *Chem. Rev.*, *62*: 433 (1962).
70. K. L. Larsson, in *The Lipid Handbook* (F. D. Gunstone, J. L. Harwood, and F. B. Padley, eds.), Chapman and Hall, London, 1986, p. 321.
71. L. Hernqvist, *Polymorphism of Fats*, Ph.D. thesis, University of Lund, 1984.
72. S. de Jong, *Triacylglycerol Crystal Structures and Fatty Acid Conformation*, Ph.D. thesis, University of Utrecht, 1980.
73. K. Sato, *Adv. Appl. Lipid Res.*, *2*: 213 (1996).
74. K. Sato, *Prog. Colloid Polym. Sci.*, *108*: 58 (1998).
75. H. Bunjes, K. Westesen, and M. H. J. Koch, *Int. J. Pharm.*, *129*: 159 (1996).
76. I. T. Norton, C. D. Lee-Tuffnell, S. Ablett, and S. M. Bociek, *J. Am. Oil Chem. Soc.*, *62*: 12374 (1985).
77. P. B. Mansfied, *J. Am. Oil Chem. Soc.*, *48*: 4 (1971).
78. S. Shanbhag, M. P. Steinberg, and A. I. Nelson, *J. Am. Oil Chem. Soc.*, *48*: 11 (1971).
79. E. Dickinson and D. J. McClements, *Advances in Food Colloids*, Blackie, Glasgow, 1995, p. 145.
80. K. van Putte and J. C. van den Enden, *J. Am. Oil Chem. Soc.*, *51*: 316 (1964).
81. W. Kloek, *Mechanical Properties of Fats in Relation to Their Crystallization*, Ph.D. thesis, Wageningen University, 1998.
82. E. G. Schutt, E. Frede, and W. Buchheim, *Kieler Milchwirtschaftliche Forschungsberichte*, *47*: 209 (1995).
83. J. Zhao and D. S. Reid, *Thermochim. Acta*, *246*: 405 (1994).
84. C. Simoneau, M. J. McCarthy, D. S. Reid, and J. B. German, *J. Food Eng.*, *19*: 365 (1993).
85. M. J. W. Povey, in *Developments in the Analysis of Lipids* (J. H. P. Tyman and M. H. Gordon, eds.), Royal Society of Chemistry, Cambridge, 1993, p. 161.
86. M. J. W. Povey, *Ultrasonic Techniques for Fluids Characterization*, Academic Press, San Diego, 1997.
87. E. Dickinson, M. I. Goller, D. J. McClements, S. Peasgood, and M. J. W. Povey, *J. Chem. Soc. Faraday Trans.*, *86*: 11478 (1990).
88. E. Dickinson, F-J. Kruizenga, M. J. W. Povey, and M. van der Molen, *Coll. Surf. A*, *81*: 273 (1993).

89. Y. Hodate, S. Ueno, J. Yano, T. Katsuragi, Y. Tezuka, T. Tagawa, N. Yoshimoto, and K. Sato, *Coll. Surf. A*, *128*: 217 (1997).
90. N. Kaneko, S. Ueno, T. Katsuragi, and K. Sato, *J. Crystal Growth*, *197*: 263 (1999).
91. D. Kashchiev, N. Kaneko, and K. Sato, *J. Coll. Interface Sci.*, *208*: 167 (1998).
92. M. J. W. Povey, in *New Physico-Chemical Techniques for the Characterization of Complex Food Systems* (E. Dickinson, ed.), Chapman and Hall, London, 1995, p. 196.
93. E. Dickinson, D. J. McClements, and M. J. W. Povey, *J. Coll. Interface Sci.*, *142*: 103 (1991).
94. D. J. McClements, M. J. W. Povey, and E. Dickinson, *Ultrasonics*, *31*: 433 (1993).
95. V. A. Akulichev and V. N. Bulanov, *Sov. Phys. Acoust.*, *27*: 377 (1982).
96. V. A. Bulanov, *Sov. Phys. Acoust.*, *25*: 202 (1979).
97. D. J. McClements, M. J. W. Povey, and E. Dickinson, *Ultrasonics International 91 Conference Proceedings*, Pergamon, Oxford, 1991, p. 107.
98. O. L. Mikhalev, I. N. Karpov, E. B. Kazarova, and M. V. Alfimov, *Chem. Phys. Letts.*, *164*: 96 (1989).
99. S. Hindle, M. J. W. Povey, and A. H. Smith, *J. Coll. Interface Sci.*, *232*: 370 (2001).
100. P. S. Dimick, in *L. Alliance 7*, CEDUS, 1994.
101. I. Hachiya, T. Koyano, K. Sato, and T. H. Gouw, *J. Am. Oil Chem. Soc.*, *66*: 1757 (1989).
102. I. Hachiya, T. Koyano, and K. Sato, *J. Am. Oil Chem. Soc.*, *66*: 1763 (1989).
103. T. R. Davis and P. S. Dimick, *J. Am. Oil Chem. Soc.*, *66*: 14948 (1989).
104. T. R. Davis and P. S. Dimick, *J. Am. Oil Chem. Soc.*, *66*: 14883 (1989).
105. D. H. Arruda and P. S. Dimick, *J. Am. Oil Chem. Soc.*, *68*: 385 (1991).
106. P. S. Dimick and D. M. Manning, *J. Am. Oil Chem. Soc.*, *64*: 16639 (1987).
107. A. C. Zettlemoyer, *Nucleation*, Marcel Dekker, New York, 1969.
108. X. Z. Wu, E. B. Sirota, S. K. Sinha, B. M. Ocko, and M. Deutsch, *Phys. Rev. Lett.*, *70*: 958 (1993).
109. N. Krog and K. L. Larsson, *Fat Sci. Tech.*, *94*: 55 (1992).
110. T. Hianik, S. Kupcu, S. Sletyr, U. B. Rybar, P. Krivanek, and U. Kaatze, *Coll. Surf. A*, *147*: 331 (1999).
111. D. J. Cebula, M. J. W. Povey, and D. J. McClements, *J. Am. Oil Chem. Soc.*, *67*: 76 (1990).
112. A. Hvolby, *J. Am. Oil Chem. Soc.*, *51*: 50 (1974).
113. J. N. Coupland, E. Dickinson, D. J. McClements, M. J. W. Povey, and C. de Rancourt de Mimmerand, in *Food Colloids, Polymers: Stability and Mechanical Properties* (E. Dickinson and P. Walstra, eds.), Royal Society of Chemistry, Cambridge, 1993, p. 243.
114. D. J. McClements, S. R. Dungan, J. B. German, and J. E. Kinsella, *J. Coll. Interface Sci.*, *156*: 425 (1993).
115. V. M. Smoluchowski, *Z. Phys. Chem.*, *92*: 129 (1917).
116. C. Javanaud, N. R. Gladwell, S. J. Gouldby, D. J. Hibberd, A. Thomas, and M. M. Robins, *Ultrasonics*, *29*: 331 (1991).

8
Fat Crystal Networks

Pieter Walstra
Wageningen University, Wageningen, The Netherlands

William Kloek* and Ton van Vliet
Wageningen University, and Wageningen Centre of Food Sciences, Wageningen, The Netherlands

I. INTRODUCTION

Natural oils consist of several different triacylglycerols (TAGs) and thus have a melting range which may extend from, say, −40°C to temperatures above 0°C. If part of the oil is crystalline at room temperature, the word *fat* rather than *oil* is commonly used. A fat consists of TAG crystals and liquid oil. The temperature at which the last of the crystals melt, called the final melting point or clear point, is rarely above 40°C for natural fats, but may be higher for fats that have been modified, e.g., by hydrogenation or by fractionation. Some of the individual TAGs in a fat often have a melting temperature far above the clear point, e.g., 72°C for β-tristearoylglycerol (SSS). The higher melting TAGs then are partly dissolved in the oil.

A fat is a plastic material, meaning that it looks like a solid but can be readily deformed or, more precisely, can be made to flow by exerting an external force. The crystals in the fat form a continuous network, thereby giving firmness to the system. The oil presents a continuous liquid phase. The properties of the network are of considerable practical importance.

---

* Current affiliation: DMV International, Veghel, The Netherlands.

1. The *consistency* or, more precisely, the mechanical properties of the fat depend on the geometry of the network and on the interaction forces between the crystals. It concerns shape retention of the fat and its plastic deformability during storage and handling and in the mouth. This is of special importance for such products as butter, margarine, shortening, and cocoa butter.
2. *Oiling off* can occur in plastic fats, which means that oil separates under gravity. This primarily depends on the size of the meshes in the crystal network.
3. The crystal network can *immobilize* other structural elements in the system. A case in point is the aqueous droplets in margarine, which would otherwise encounter each other and then coalesce.
4. A crystal network can be the main cause of *partial coalescence* in oil-in-water emulsions if the oil droplets also contain fat crystals. A solid network in such droplets often leads to fat crystals slightly protruding from the globules, thereby triggering coalescence of two closely encountering globules. Full coalescence is not possible because of the crystal network and irregular clumps are formed. Very stable oil-in-water emulsions can become extremely unstable to partial coalescence as soon as continuous crystal networks are formed inside the droplets. Partial coalescence is what happens during such processes as churning of cream.

Many factors affect the formation and the properties of crystal networks in multicomponent fats. The phenomena are so intricate that very little quantitative understanding has been gained. Considerable work has been done earlier [1], but this was all descriptive. Only recently has more fundamental work been done [2,3]. Most of the quantitative results in this chapter are taken from our own work [3].

II. NETWORK FORMATION

A. Events Occurring

When an oil becomes supersaturated with respect to some of the constituting TAGs, the following events, in principle, occur:

- Crystal nuclei are formed.
- The nuclei grow to form crystals.
- The crystals, when large enough, form aggregates.
- Aggregation leads to formation of a continuous network of crystals.
- The network may alter in several ways, often involving sintering (“growing together”) of adjacent crystals.

Some of these events are illustrated in Fig. 1.

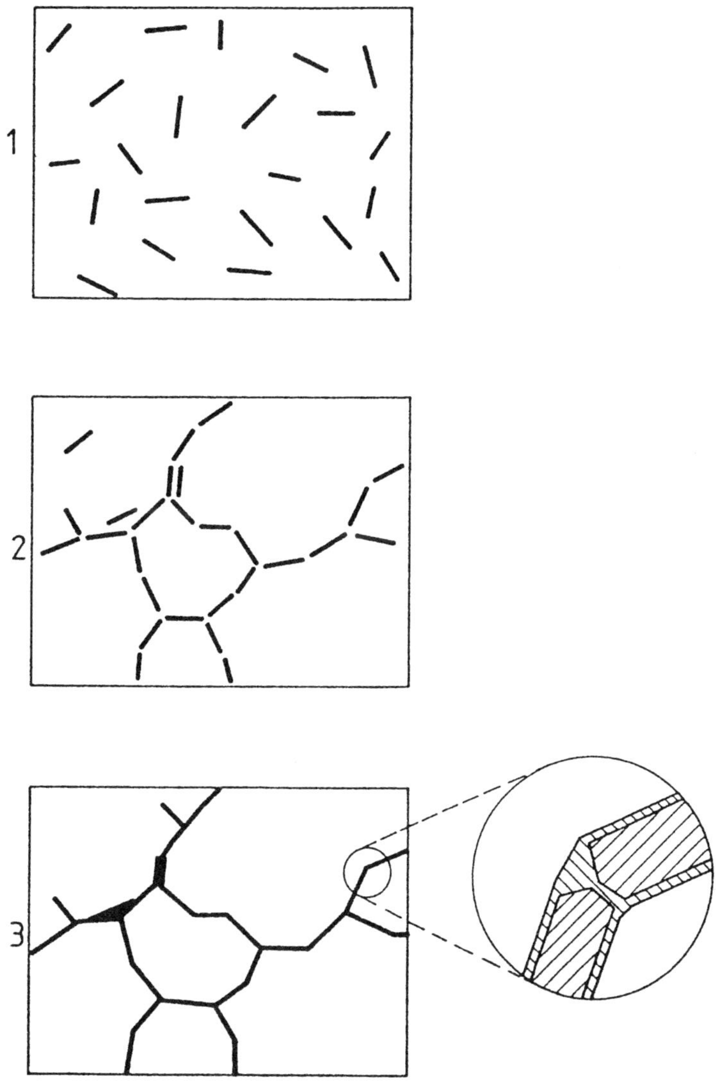

Fig. 1 Various stages during crystallization showing aggregation, network formation, and sintering of crystals [1].

1. *Solubility*

To obtain a supersaturated system, the oil is generally cooled. For a single TAG dissolved in TAG oil, the solubility x_S is given by the Hildebrand equation

$$\ln x_S = \frac{\Delta H_m}{R}\left(\frac{1}{T_m} - \frac{1}{T}\right) \tag{1}$$

where x is expressed as mole fraction, ΔH_m is the molar enthalpy of melting of the solute, R is the gas constant, T is absolute temperature, and T_m is the melting temperature of the pure solute [1,4]. For a component i in a mixture of TAGs that has a mole fraction x_i, incorporation of this value in the left-hand side of Eq. (1) yields for T the clear point T_c of that component.

Equation (1) assumes ideal behavior; it generally holds well for a single TAG dissolved in a liquid TAG oil. It should be taken into account that ΔH_m and T_m (and thereby T_c) depend on the polymorphic form of the crystal; generally, both parameters increase in magnitude in the order α, β′, β. Another complication is that Eq. (1) is not valid for multicomponent fats exhibiting compound crystallization. For binary compound crystals, theory has been developed [5]. This has also been done for more complicated mixtures [4], although application would need precise knowledge of fat composition and of characteristic parameters of the components. A mixture of a few closely related TAGs may in practice behave like a single component and satisfy Eq. (1) fairly well, provided that ΔH_m and T_m are experimentally determined, and making sure that the same polymorph prevails in all cases [3]. It should, however, be realized that multicomponent fats are virtually never in thermodynamic equilibrium [1].

2. *Supersaturation*

The driving force for crystallization is given by the difference in chemical potential between a supersaturated solution (mole fraction x) and a saturated one (x_S):

$$\Delta\mu = RT \ln\left(\frac{x}{x_S}\right) = RT \ln \beta \tag{2}$$

where $\ln \beta$ is called the supersaturation. Assuming that the solubility can be derived from Eq. (1), $\ln \beta$ can be obtained. Figure 2 gives some examples of the supersaturation of HP (a mixture of saturated TAGs obtained by complete hydrogenation of palm oil) in oil as a function of temperature and of fraction HP in two polymorphic forms. It is seen that (1) supersaturation increases with decreasing temperature and with increasing x; (2) it is, for a given temperature and composition, always greater for the more stable polymorph; and (3) that

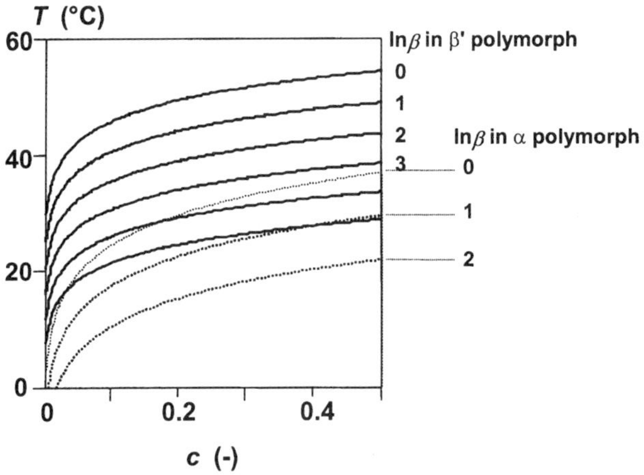

Fig. 2 Calculated iso–ln β curves of HP/SF systems in the α- and β′-polymorph as a function of the mass fraction HP (c) and of temperature (T). Parameters used for α: $\Delta H_m = 98\ \text{kJ} \cdot \text{mol}^{-1}$, $T_m = 315.0$ K; for β′: $\Delta H_m = 161\ \text{kJ} \cdot \text{mol}^{-1}$, $T_m = 330.5$ K.

considerable supersaturation in the β′ form is possible without the α form being supersaturated, the more so at smaller x.

Nucleation is discussed, e.g., in Chapter 1, and later in this chapter. Suffice it to say that in most practical conditions, heterogeneous nucleation will occur. Homogeneous nucleation generally needs a supercooling of 20–25 K below the α clear point, or about 35 K below the β clear point of the fat crystals in the mixture [1]. In practice, supercooling is often less. Heterogeneous nucleation is caused by catalytic impurities; their number concentration can be determined by studying crystallization rate in emulsions of the fat in an aqueous solution, where the droplet size distribution is varied [3,6; Chaps. 1, 7]. By and large, the greater the supersaturation (the deeper the supercooling), the greater are the number of nuclei, hence of crystals, formed and the smaller the resulting crystals.

Crystal growth is discussed in Chapters 1 and 3. The growth rate depends on supersaturation and on several other factors. Relations generally differ between crystal faces and so does growth rate. Hence, growth conditions determine crystal shape. They may also affect crystal size, since rapid growth leads to a fast decrease in supersaturation, hence to a rapidly decreasing rate of nucleation.

3. *Aggregation*

As soon as the crystals attain a certain minimum size they readily aggregate due to van der Waals attraction [2,3]. There is virtually no repulsion between the crystals, except hard core repulsion at very small mutual distance (0.5 nm or less). The aggregation leads to formation of fractal aggregates, which then form a continuous network (Fig. 1), as explained in Sec. II.B. The network gives the fat elastic properties, whereas a dispersion of free crystals (or of nontouching aggregates of crystals) will merely have a higher viscosity than that of the oil.

Aging of bonds between aggregated particles often occurs. In fat crystal aggregates, sintering may occur, because additional crystalline material may be deposited onto the aggregated solids (Fig. 1). This is further discussed in Sec. III.

As discussed in Sec. II.C, the various process steps described *overlap* in time.

B. Fractal Aggregation

Random aggregation of spherical particles that encounter each other due to Brownian motion and then stick results in aggregates of a fractal nature [7,8]. It has been shown that for such diffusion-limited aggregation the relation between the number of particles in an aggregate N_p and the radius of the aggregate R (defined as the radius of the smallest sphere that can contain the aggregate) is given by

$$N_p = \left(\frac{R}{a}\right)^D \tag{3}$$

where a is the radius of the primary particles and D is called the fractal dimensionality; invariably $D < 3$. Since the number of sites in an aggregate (N_s) is proportional to R^3, the volume fraction of particles in an aggregate is given by

$$\varphi_{ag} \equiv \frac{N_p}{N_s} = \left(\frac{R}{a}\right)^{D-3} \tag{4}$$

Assuming that D is constant, which is often true, this relation implies that φ_{ag} decreases with increasing R; i.e., large aggregates are more tenuous (rarefied, open-structured) than small ones. Another property of fractal aggregates is that their structure is self-similar, implying that they have on average the same structure when seen at various length scales (if $>a$). The qualification ''on average'' is needed because the aggregates are stochastic fractals and thus exhibit random variation in structure.

Simulation of unhindered aggregation of equal-sized spheres leads to $D \approx 1.75$, but various factors may cause D to be higher, up to about 2.4; these include rearrangements occurring in the aggregates, the presence of an activation free energy for aggregation (which is then said to be reaction-limited), or any disaggregation occurring [7].

For the fractal dimensionality of aggregating crystals of SSS in oil, volume fraction of crystals $\varphi = 0.005$, a D value of 1.7 has been obtained, as measured by light scattering [9]. Similarly, for dispersions of HP crystals, $D = 1.7–1.8$ was obtained, also by light scattering [3]. In the latter experiment, the crystals were allowed to form at very high shear rates, preventing their aggregation, after which shear was stopped. For fat crystals, aggregation probably is diffusion-limited, but some rearrangement may be possible, which would cause an increase of D. In practice, $D < 1.8$ is rarely observed for particle aggregates and the low D values obtained may thus be somewhat suspicious. However, rotational, besides translational, diffusion occurs, which would lead to somewhat smaller dimensionalities then those calculated by most simulation methods (J. van Opheusden, personal communication). If aggregation occurs during (moderate) agitation, the resulting aggregates tend to have higher D values [9,10].

If aggregation goes on unhindered, the total volume occupied by the aggregates will keep increasing, according to Eq. (4). When φ_{ag} becomes about equal to the original volume fraction of particles in the dispersion φ, the aggregates, in principle, touch each other and a gel is formed [8]. At this point of gelation, the average radius of the aggregates is given by

$$\langle R_g \rangle = a\ \varphi^{1/(D-3)} \tag{5}$$

The gel now has a fractal character at length scales between a and R_g; at shorter and longer scales, the dimensionality generally is Euclidian, i.e., equal to 3.

The fractal dimensionality can be obtained from parameters that depend on φ, and a good example is the permeability coefficient B of the gel. This parameter follows from the Darcy equation for the superficial flow rate v of liquid through a porous material due to a pressure gradient ∇p:

$$v \equiv \frac{Q}{A} = \frac{B \nabla p}{\eta} \tag{6}$$

where Q is the volume flow rate, A is the cross-sectional area involved, and η is the viscosity of the percolating liquid. For a fractal gel, B is assumed to be given by [8,11]

$$B = \left(\frac{a}{K}\right)^2 \varphi^{2/(D-3)} \tag{7}$$

where K is a constant for any system; it was observed to be of order 10 for some gels built of uniform small spheres [11]. By determination of permeability of gels of various φ, D can be derived if log B is linear with log φ (Sec. II.D).

An important consequence of the mechanism of gel formation mentioned (unhindered random aggregation) is that there is in principle no minimum φ value below which no gel is formed (often called a percolation limit). However, for small φ, the gel tends to be very weak. We observed an elastic shear modulus G' of about 1 Pa or less for a volume fraction of fat crystals of 0.005. Since G = stress over strain, and fat crystal networks are observed to yield at strain values of, say, 0.3 (Sec. III.B), the stress at which the network breaks down is then of the order of 0.1 Pa (corresponding to the pressure exerted by a water "column" 10 μm high under gravity). This implies that the presence of such a network will not be observed under most conditions.

There is one additional condition for a continuous network to be formed throughout the whole volume. This is that the smallest dimension d of the vessel containing the dispersion should be larger than 2 R_g; otherwise, there would not be enough particles to make aggregates of diameter $\geq d$. This becomes important for particles in emulsion droplets. From Eq. (5), the condition becomes [11]

$$d > 2a\varphi^{1/(D-3)} \tag{8}$$

Assuming emulsion drops of diameter 3 μm, crystals of an equivalent radius of 0.1 μm and $D = 2$, the minimum φ would be 0.07. Formula (8) can only be approximate, because the basic equations are valid for spherical particles. Nevertheless, the trend predicted is observed. It has been observed that partial coalescence of fat globules only occurs if the crystal network extends throughout the globule and that, indeed, solid fat fractions of the order of 0.1 are generally needed to that end [12]. Partial coalescence depends on a variety of other factors, but that is outside the realm of this chapter.

C. Kinetics

The properties of the crystal network formed depend on the time scales needed for nucleation, crystal growth, and gel formation due to aggregation of crystals.

1. Nucleation

Classical nucleation theory is largely due to Volmer [13]. It states that the rate J (in $m^{-3} \cdot s^{-1}$) at which nuclei are formed, in a solution or melt containing N molecules per cubic meter that may crystallize, is given by [3,14]

$$J = N\frac{kT}{h}\exp\left(-\frac{\alpha\Delta S^{\ddagger}}{R}\right)\exp\left(-\frac{\Delta G^{\ddagger}}{kT}\right) \tag{9a}$$

$$\Delta G^{\ddagger} = \frac{f\Phi\gamma^3V^2}{(\Delta\mu)^2} \tag{9b}$$

$$\Delta\mu = \frac{\Delta H_m(T_c - T)}{T_c} = RT\ln\beta \tag{9c}$$

where k, T, h, and R have their normal meanings; α is the fraction of a molecule that has to be incorporated in a nucleus before it "sticks"; $\Delta S^{\ddagger}$ is the molar entropy decrease of the molecules going from the melt or solution to the crystal; $\Delta G^{\ddagger}$ is the activation Gibbs (free) energy for the formation of a nucleus; the numerical factor f depends on the geometry of the crystal nucleus; γ is the interfacial Gibbs energy between crystal and liquid; V is the molar volume of the crystallizing material; T_c is the clear point of the crystallizing component; see Eqs. (1) and (2). The relative volume of a nucleus Φ equals unity for homogeneous nucleation; for heterogeneous nucleation a smaller number of molecules is needed to form a nucleus, due to the presence of a catalytic impurity of suitable surface properties and shape, and Φ then is proportionally smaller.

Although the validity of Eq. (9) is questionable (see e.g., Ref. 14), it is to some extent useful for the interpretation of crystal nucleation in fats and it has been applied by several workers [3,15–17]. It is generally assumed that nucleation occurs in the α polymorph, unless the temperature is above the α clear point; cf. Fig. 2. It virtually always concerns a mixture of TAGs, and one should take the values of J, T_c, and ΔH_m for the component(s) that will crystallize first; in particular, T_c should be the clear point of the component(s) in the mixture studied, not the melting point of the pure component (T_{m}). In many cases nucleation gives rise to compound crystals (hence the plural "components"). Before nucleation is finished, appreciable crystal growth may occur, and the parameters J and ln β then decrease during the process.

Nucleation in fats has generally been studied in fat emulsions (see also Chapter 7). If the emulsion droplets are small enough, nucleation often is homogeneous (i.e., $\Phi = 1$) and its rate becomes appreciable at $T_{\mathrm{c}} - T \approx 20$ K. Deeper supercooling then increases the nucleation rate tremendously. At higher temperatures, heterogeneous nucleation often occurs. In that case the factor Φ in $\Delta G^{\ddagger}$ will be smaller, thereby leading to greatly enhanced J values, provided that sufficient catalytic impurities are present. Often, their number concentration N_{cat} determines nucleation rate, and it varies widely. For instance, the presence of monoglycerides (which can result from lipolysis) may greatly enhance N_{cat} [15,17]. Its dependence on temperature is often given by the empirical relation

$$\log N_{\mathrm{cat}} \approx A - BT \tag{10}$$

where A and B are constants for any sample. This means that N_{cat} would be roughly proportional to the supersaturation ln β. In most cases, heterogeneous nucleation prevails in bulk fats at moderate supersaturation, and its rate increases with increasing supersaturation (or decreasing temperature).

The system studied by Kloek [3] consisted of fully hydrogenated palm oil, dissolved at various concentrations in sunflower oil; it will be denoted as x%HP/SF. For nucleation in the α polymorph it appeared that Eq. (9) was well obeyed, assuming homogeneous nucleation; the fit parameters resulting were $\alpha \approx 1.0$

and $\gamma \approx 4.1$ mJ·m^{-2}, both reasonable values. However, the nucleating material was not one of the main components [e.g., tripalmitoylglycerol (PPP)] but small quantities of higher melting TAGs (e.g., SSS). The crystals of SSS then appear to act as very active catalytic impurities for the other crystals to be formed. This phenomenon is very common in mixed TAGs: as soon as some crystals have been formed, the supercooling for other crystals to form appears to be quite small [6]. For nucleation in the β′ polymorph, nucleation appeared to be heterogeneous, with $\alpha \approx 0.8$ and $\Phi \approx 0.8$.

Some results on nucleation rate are given in Fig. 3. It is seen that most systems show a steep dependence of J on T for instance, d ln $J/\mathrm{d}T = -2.7$ K^{-1}. Notice also the large effect of T_c. The deeper supercooling needed for SSS in paraffin oil as compared to the HP/SF systems is presumably due to a larger

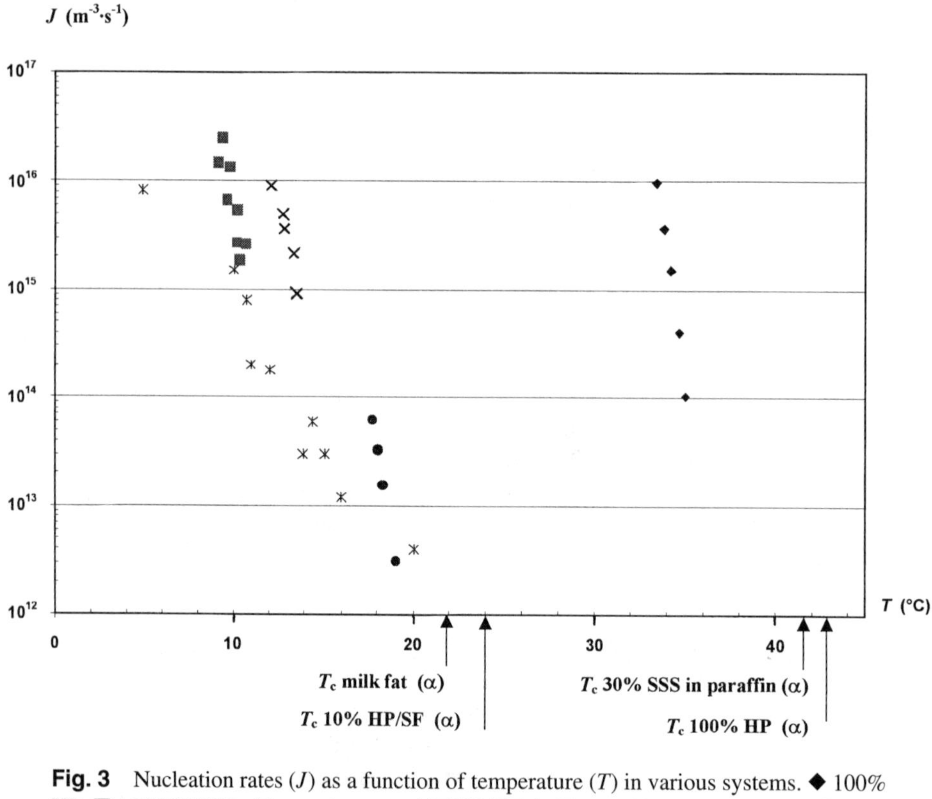

Fig. 3 Nucleation rates (J) as a function of temperature (T) in various systems. ◆ 100% HP; ■ 10%HP/SF in Na-caseinate; × 10%HP/SF, in Tween 20; * SSS/milk fat; ● 30% SSS in paraffin oil. After various sources.

value of γ between crystal and oil. This indicates that the composition of the oil phase can markedly affect nucleation rate. The composition of the emulsifier used can also affect nucleation rate, since some emulsifiers, especially those containing long-chain saturated fatty acid chains, may cause the presence of catalytic impurities at the internal surface of the emulsion droplets [15]. Presumably, this causes the difference between the 10%HP/SF emulsions made with caseinate and Tween 20. It is seen that milk fat, with its very wide TAG composition, shows a less steep dependence of nucleation rate on temperature ($\mathrm{d}\ln J/\mathrm{d}T \approx -0.8\ \mathrm{K}^{-1}$). Application of Eq. (9) to such a system is doubtful, partly because of the extensive compound crystallization occurring. Nevertheless, it is clear that primary nucleation was predominantly heterogeneous [6].

A far more serious objection against the use of Eq. (9) is the occurrence of true *secondary nucleation*. It has been concluded [17] that extensive secondary nucleation can occur in some fats if the supersaturation is appreciable, even in the absence of any agitation. This means that as soon as a crystal is formed, other crystals are nucleated nearby. Since the emulsion studies effectively determine the onset of crystallization in the droplets, such secondary nucleation then is not noticed. Its effect on crystal number, and thereby on size, can, however, be very large. It was concluded to be especially important in multicomponent TAG mixtures. Copious secondary nucleation can occur in milk fat (e.g., increasing the number of crystals by the factor 1000), less so in a classical margarine fat, and far less in the HP/SF mixtures mentioned; it was absent in mixtures of paraffins [17].

2. *Crystal Growth*

Growth rate of TAG crystals is very slow [1]. This may have several causes.

1. The incorporation of a TAG molecule into a crystal lattice implies a very large loss in conformational entropy, and a great number of conformational changes have to occur before a molecule fits into the lattice. Before it is for the most part incorporated, it may readily become detached from the crystal surface, and it may take a relatively long time before a molecule finally fits in a vacant site. For pure SSS in trioleylglycerol (OOO), linear growth rates v_c of the order of 10^{-8} to $10^{-7}\ \mathrm{m{\cdot}s^{-1}}$ have been observed for ln β ranging from 0.5 to 1.5. The growth rates were orders of magnitude smaller than the dissolution rates at roughly the same $|\Delta\mu|$ [18]; cf. Fig. 5. Although we are unaware of conclusive experimental results, growth rate at the same ln β should be slower for a more stable polymorph.

2. In a multicomponent fat, the competition between similar molecules for a vacant site in a crystal lattice is fierce. Presumably, several nonfitting molecules may have to become detached from the crystal surface before a fitting one becomes incorporated; even di- and monoacylglycerols may compete. Moreover,

for a group of TAGs that can all crystallize and that are of comparable T_m and ΔH_m, the supersaturation will be smaller for a given temperature and total mole fraction if the group consists of a greater number of different molecules. Altogether, multicomponent fats crystallize far slower than a solution of just one TAG at the same (average) supersaturation.

The slowness is to some extent compensated by the formation of compound crystals. For those, the ''joint'' supersaturation of the components to be incorporated is greater and fewer nonfitting molecules can now compete for lattice sites. Consequently, in multicomponent fats copious compound crystallization often occurs.

3. The point just made implies that the crystals formed are not in equilibrium with the liquid phase. This is because compound crystallization occurs far more readily and more extensively in the α form than in the β′ form; it can hardly occur in the β polymorph. Consequently, several rearrangements among crystals may occur, leading to more stable polymorphs. Because it concerns compound crystals, all rearrangements go along with compositional changes and have to occur via the liquid phase. Particularly at fairly high mass fractions of crystals (say, >0.5), such rearrangements can be very slow. In milk fat, for instance, it takes many hours after cooling to about 15°C before the amount crystallized appears to be constant, and α crystals can persist for years.

4. As in all crystallization from solution, the driving force $\Delta\mu$ decreases during the process because of depletion. The rate of decrease depends much on the nucleation rate. Faster nucleation gives more crystals, hence a larger specific crystal surface area, hence a faster mass rate of crystallization for the same v_c and a faster decrease of $\Delta\mu$.

5. Crystallization causes release of a large quantity of heat of fusion. Assuming ΔH_m to be about 200 $J \cdot g^{-1}$ and the specific heat of the fat to be about 2 $J \cdot g^{-1} \cdot K^{-1}$, crystallization of half of the fat in an isolated system causes a 50 K temperature increase. Consequently, supersaturation would rapidly decrease until it became zero. In an emulsified fat or in bulk fat in very small quantities (laboratory scale), heat can readily be removed. For large quantities special measures have to be taken, since a fairly solid crystal network is rapidly formed. Scraped-surface heat exchangers are generally used. This implies intensive agitation, upsetting the relations assumed for isothermal quiescent nucleation and crystallization.

6. Crystal growth is under some conditions diffusion-limited. However, the growth rates observed are quite small (see, e.g., Fig. 5) and—since the diffusion coefficient is presumably about 10^{-11} $m^2 \cdot s^{-1}$—it can readily be calculated that diffusion of the TAG molecules to the crystal surface is not limiting. This might be different if a very small fraction of liquid fat is left (and the temperature is low as well).

Nearly all of these factors may affect the growth rate of different crystal faces to a different extent, thereby influencing crystal shape.

Because of these intricacies, it makes little sense to apply basic theory on crystallization rate to mixed fats. We will just consider some results on the mentioned HP/SF systems crystallizing in the β′ polymorph [3], where complications 2, 3, 5, and 6 are relatively small. It concerned isothermal crystallization at various fractions of HP and at various initial supersaturation (ln β_o).

Often, the Avrami equation is used for describing crystal growth. Assuming nucleation rate J and linear growth rate v_c to be constant, and the crystals to be and remain spherical and well separated, the equation is [19]

$$\varphi(t) = \tfrac{1}{3}\pi v_c^3 J t^4 \tag{11}$$

where φ is the volume fraction of crystals. It can only give reasonable fits with the results at the very beginning of the crystallization, say $\varphi < 0.01$.

Good fits were obtained with a modified Gompertz equation, which is often used to model bacterial growth [20]. The equation can be written as

$$\varphi(t) = \varphi_{\max} \exp\left\{-\exp\left[e\frac{S_{\max}}{\varphi_{\max}}(t_{\text{ind}} - t) + 1\right]\right\} \tag{12}$$

where $S_{\max}$ is the maximum slope of the curve. The nominal induction time t_{ind} follows from extrapolation of that slope to the X-axis; actually $\varphi(t_{\text{ind}}) = 0.066\varphi_{\max}$. The slope and the induction time are the adjustable parameters of the equation. Examples are given in Fig. 4, and it is seen that t_{ind} varied from about

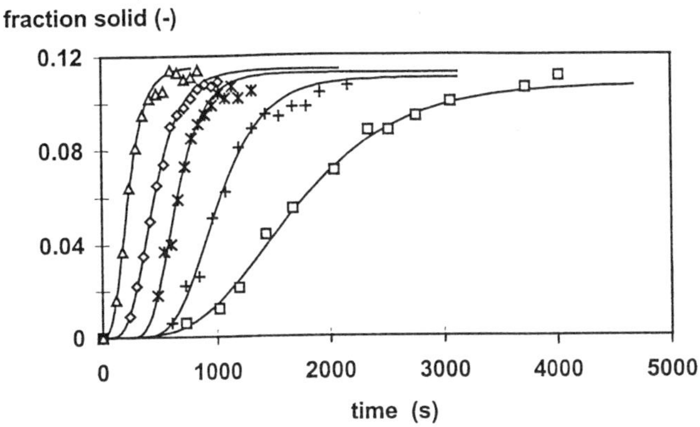

Fig. 4 Isothermal crystallization curves of 12%HP/SF systems at various initial supersaturation ln β_0: □ 2.25; + 2.50; * 2.75; ◆ 3.00; △ 3.25. Curves were fitted to Eq. (12).

1 min at ln $\beta_0 = 3.25$ to about 14 min at ln $\beta_0 = 2.25$. This is because nucleation rate strongly depends on supersaturation.

By taking the nucleation rates as fitted to Eq. (9) (with $\alpha = 0.8$ and $\gamma = 3.8\ \text{mJ}\cdot\text{m}^{-2}$) and the fitted results of $\varphi(t)$, a numerical simulation of the nucleation and growth process was made, which permitted determination of the linear growth rate v_c as a function of supersaturation for various ln β_0. Results are shown in Fig. 5. It is seen that the curves agree well with one another and that log growth rate is about linear with ln β for most of the supersaturation range. It is also observed that the linear growth rate is smaller by nearly three orders of magnitude than that observed for pure SSS [18], mentioned earlier in this section.

3. Crystal Size

From the Avrami equation (11) it can be derived [3] that the apparent radius of a spherical crystal—more precisely the volume-surface average radius—would be given by $a \approx \frac{1}{4} v_c t$. Taking for t the induction time t_{ind} and remembering that Eq. (12) gives $\varphi = 0.066\varphi_{max}$ at t_{ind}, we can calculate a^*, being the apparent radius at t_{ind}. Assuming that the number of crystals does not materially alter after t_{ind}, we can calculate the number concentration of crystals N from $4\pi a^3 N/3 = \varphi$. The final radius a_∞ is then derived as $(1/0.066)^{1/3} a^*$. From the results on HP/SF given in Fig. 4, we obtain that as

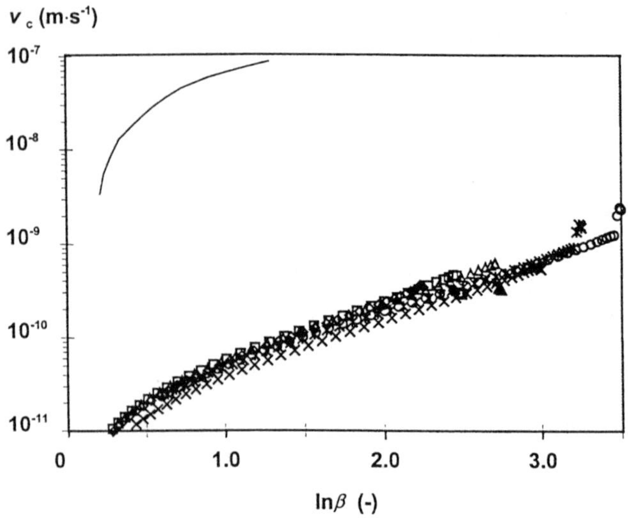

Fig. 5 Growth rate (v_c) of TAG crystals as a function of supersaturation (ln β). Points: results on 12%HP/SF at various initial supersaturation, ln β_0 2.25–3.50 [3]. Curve: results for pure SSS in the β-polymorph [18].

ln β_0 ranges from	2.25 to 3.25
t_{ind} varies from	825 to 60 s
a^* from	0.2 to 0.1 μm
N from	0.3 to 3 μm^{-3}
and a_∞ from	0.5 to 0.2 μm

For a higher supersaturation, the crystals formed are smaller. This is because d ln J/d ln β > d ln v_c/d ln β. A more sophisticated mathematical treatment, taking nucleation rate as a function of the changing supersaturation into account, leads to virtually the same results for a_∞ [3]; these are given in Fig. 6. The concentration of HP had, for the same ln β_0, very little effect on a_∞.

Most fat crystals as observed in practice are thin and fairly long platelets. Electron microscopy shows, for instance, a ratio of length : width : thickness of 50:10:1 [21]. If we call thickness δ, it follows that $\delta/a = 0.20$, considering a to be the radius of a sphere of equal volume. For $a = 0.1$ μm the crystal dimensions then are about $1 \times 0.2 \times 0.02$ μm; for $a = 0.5$ we obtain $2.5 \times 1 \times 0.1$. These are reasonable values; they may be somewhat better than order-of-magnitude estimates.

Several complications may disturb the relations given. The most important one may be the occurrence of secondary nucleation. As mentioned, the fairly simple system of HP dissolved in SF will probably exhibit very little secondary nucleation, but in fats of a wider compositional range it may be considerable [17]. The number of crystals actually formed may well be higher by two orders of magnitude than the number predicted via Eq. (9), provided that ln β_0 is large. Consequently, the size of the crystals formed would be far smaller than predicted. We are not aware of a quantitative treatment involving secondary nucleation.

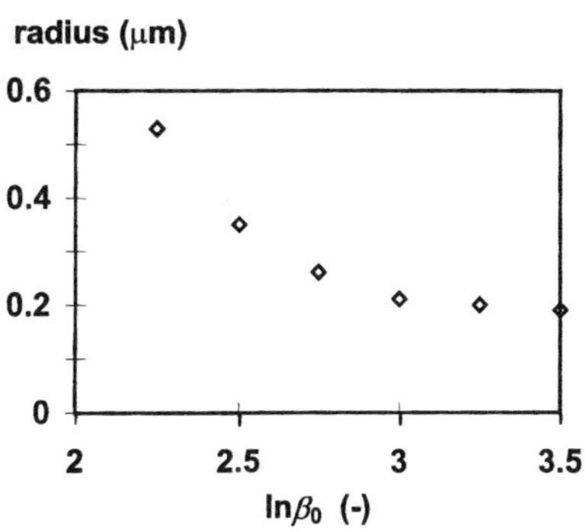

Fig. 6 Calculated number average crystal radius, assuming the crystals remain spherical, after crystallization of 12%HP/SF at various initial supersaturation (ln β_0).

There is some circumstantial evidence that a wide compositional range also leads to relatively thinner crystals. This may partly be due to the persistence of unstable polymorphs; for instance, β′ crystals tend to be more anisometric than those of β.

Another point is that we have only considered fairly high supersaturation ($\ln \beta_0 > 2$). At much smaller $\ln \beta_0$ very large crystals can form. It often concerns spherulites, composed of branched, radially oriented needle-like crystals, interspersed with oil. These generally do not form continuous crystal networks.

Finally, slow recrystallization may occur, and in most cases this also leads to coarser crystals, such as large spherulites. After all, recrystallization especially occurs at temperatures not very far below T_c, implying that the driving force for recrystallization ($\Delta\mu$) is quite small.

4. Aggregation Time

As mentioned in Sec. II.A, aggregation of fat crystals will occur due to van der Waals attraction as soon as they have obtained a given size. For spherical particles of equal radius a the colloidal interaction free energy between them is given by [14]

$$\Delta G_{\text{ci}} = -\frac{A_{\text{H}}a}{12h} \tag{13}$$

provided the interparticle distance $h \ll a$. Presumably, the Hamaker constant $A_{\text{H}} \approx 6.10^{-21}$ J, or 1.5 kT for fat crystals in TAG oil [3]. Assuming $h \approx 0.4$ nm at contact, spheres of radius 12 nm then yield $\Delta G_{\text{ci}} = -3\ kT$, sufficient to cause almost permanent contact as soon as two particles meet. This implies that fat crystals will start to aggregate very soon (often <1 min) after their nucleation.

The aggregation rate would then follow Smoluchowski theory for fast perikinetic aggregation [22], where particles meet due to Brownian motion and stick as soon as they make contact. For spheres of equal size the decrease in number concentration N due to aggregation (counting aggregates as single particles) is given by

$$-\frac{\mathrm{d}N}{\mathrm{d}t} = 8\pi D a N^2 = \frac{4\ kT}{3\eta} N^2 \tag{14}$$

where η is the viscosity of TAG oil (~0.08 Pa·s) and D is the diffusion coefficient of the crystals in oil. It is often convenient to use the time needed to halve the initial number of particles N_0, given by

$$t_{0.5} = \frac{3\eta}{4kTN_0} = \frac{\pi\eta a^3}{kT\varphi} \tag{15}$$

For fat crystals in oil, Eq. (15) yields, for $N_0 = 1\ \mu m^{-3}$, $t_{0.5} \approx 15$ s. However, some conditions underlying the Smoluchowski equations are not fulfilled for fat crystals. It is assumed that the hydrodynamic radius of a particle (which determines the diffusion coefficient D) equals its collision radius, which is only true for a sphere. Fat crystals are very anisometric, which causes a large deviation from this assumption. The overall effect is probably that the aggregation is faster than predicted. On the other hand, the strong anisometry may prevent that all of the encounters between crystals result in sticking, because the van der Waals attraction—even if sufficient for spheres to stick on contact—is insufficiently strong for certain encounter configurations of crystals (e.g., between pointed ends) to ensure sticking. This tends to decrease aggregation rate.

Experimental results were obtained by Kloek on HP crystals in oil [3]. Size of crystals or aggregates was roughly derived from results of forward light scattering under shear. During nucleation and crystal growth, the suspension of crystals was sheared at such a rate that any crystal aggregates were immediately disrupted, and then the shear was suddenly stopped. The time needed to form aggregates of a few particles could so be determined and it ranged between 10 and 100 s. The crystals had an equivalent radius of about 0.1 μm, and, from Eq. (15), $t_{0.5} \approx 12$ s was calculated. This implies that Eqs. (14) and (15) predict at least correct orders of magnitude.

In most practical situations, the relations are more intricate. Crystals nucleate and grow while aggregation goes on. By inserting correct values for nucleation and growth rates as a function of time, a full solution may be obtained in several cases by numerical integration. The problem is, of course, that all the rates involved vary with ongoing crystallization.

Another problem is that the aggregation of fat crystals soon leads to the formation of a continuous network, i.e., a gel in which the particles are immobilized. To obtain an expression for the gel point, we have to integrate Eq. (14) from $t = 0$ to t_g, where the latter corresponds to the time needed to reach an aggregate of size R_g. This yields

$$t_g = t_{0.5}(q^D - 1) \qquad q = \frac{R_g}{a} \tag{16}$$

where R_g is given by Eq. (5).

Experiments on 12% HP/SF were done at various ln β, where the shear modulus was measured as a function of time [3]. This allowed fairly precise determination of t_g. For ln β ranging from 2.75 to 3.50, t_g varied from 270 to 60 sec. These values are of the order of 10 $t_{0.5}$. By comparing t_g with the determined values of the volume fraction of crystals φ as a function of t, gel formation was seen to occur at $\varphi \approx 0.01$. This is slightly above the value of φ at the induction time (about 0.007) and is far below its final value of 0.11.

5. Further Considerations

It is clear that in most situations the various processes involved in network formation overlap. Figure 7 gives two examples of results on the system studied by Kloek [3]. Nucleation time would be shorter than growth time but of the same order of magnitude, and nucleation will thus go on while crystals grow. Gelation time is of the same order of magnitude, implying that a continuous network is formed long before crystallization is ''complete.'' Sintering of aggregated crystals and of junctions between crystals in a network may take longer times [23]; see Section III.

All these times depend, however, on a number of variables, and often in a different manner. Some important variables are temperature, concentration, and agitation.

a. Temperature. A lower temperature means a deeper supercooling, hence a larger ln β and (much) faster nucleation and crystal growth rates; nucleation rate tends to increase more (Fig. 7). The consequence will be smaller crystals in the early stages, which would mean somewhat faster aggregation (smaller $t_{0.5}$). Since φ will soon be larger, also the gel time would clearly be shorter and the crystal size at the moment of gel formation will be (somewhat) smaller.

However, these conclusions are not generally valid because of some disturbing factors. Temperature may affect the polymorph formed, which has several consequences. Secondary nucleation may occur, upsetting the relations governing crystal size.

b. Concentration. If more crystallizable material is present, the supersaturation will be greater, other conditions being equal. However, if ln β_0 is kept constant (implying a higher temperature), relative nucleation and growth rates would not materially change. Because of a higher concentration of crystallizable material, a greater number of nuclei, and thereby of crystals, will be formed. This implies that a gel would be formed at an earlier stage in the crystallization process. Moreover, relative aggregation rate would be increased, because of the smaller viscosity of the oil at a higher temperature. Altogether, the initial network formed may be different; this is discussed in Sec. II.D.

c. Agitation. Stirring as such would hardly affect nucleation and growth of crystals, since these phenomena are not diffusion limited. Indirectly, agitation may have enormous effects on crystallization time and crystal size, since it can greatly affect the cooling regime. Mild agitation, e.g., shearing at a rate of 5 s^{-1}, may speed up aggregation of crystals, since orthokinetic (velocity gradient determined) aggregation then would be faster than perikinetic aggregation. At high strain rates, e.g., $>100\ s^{-1}$ [3,10], the resulting viscous forces cause disaggregation, as long as the fat crystals are bonded by van der Waals attraction only. In

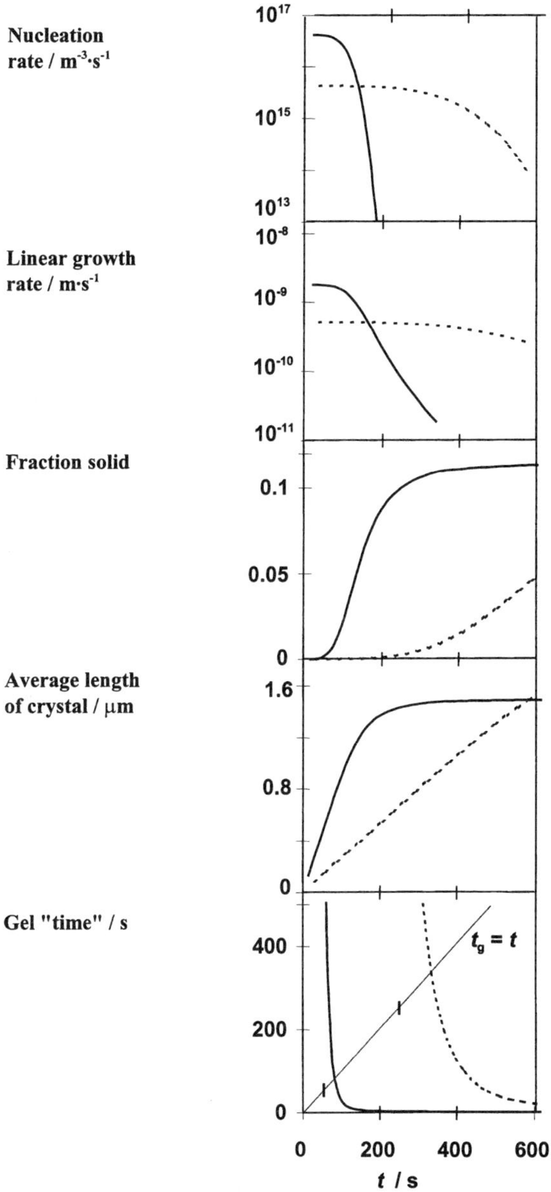

Fig. 7 Values of various crystallization parameters as a function of time (t) after cooling of 12%HP/SF systems at two values of the initial supersaturation (ln β_0 = 3.5, ——; ln β_0 = 2.75, – – –). The results are calculated from experimental data on nucleation and growth. The results on the gel ''time'' are calculated from Eq. (16), with D = 1.7, assuming size and concentration to be fixed at the values as they would be at the times on the axis. The observed gelation times are indicated by vertical dashes.

practice, prediction of these effects is far from easy, especially because the prevailing strain rates tend to vary enormously with place and time.

D. Network Structure

It is very difficult to obtain quantitative data on the structure of a fat crystal network. Various types of microscopy have been applied, but it is virtually impossible to avoid making artefacts. It is clear from micrographs that a continuous network is formed [21], and in some cases the shape and size of the building blocks, i.e., the fat crystals, can be observed (see Sec. II.C.3). However, a detailed and reliable picture of the arrangement of the solid fat throughout space cannot be obtained in many cases. Often separate crystals cannot be discerned and the fat network looks like a fairly regular porous structure.

For fractal gels, relations between macroscopic parameters and volume fraction may offer a means of study. Permeability would then be the method of choice, since it is almost purely dependent on geometrical characteristics. The low values (1.7–1.8) of the fractal dimensionality of fat crystal aggregates point to no or very little mutual rearrangement of the crystals once aggregated [24], which would simplify the analysis. Some results on log permeability versus log volume fraction solid [3] are in Fig. 8. In these studies, the initial supersaturation was kept constant, implying that the temperature had to be varied. It is seen that the negative slopes are quite high, 6.0 and 6.7. Application of Eq. (7) then yields $D \approx 2.7$, very far removed from the value of 1.75 obtained for the initial aggregates.

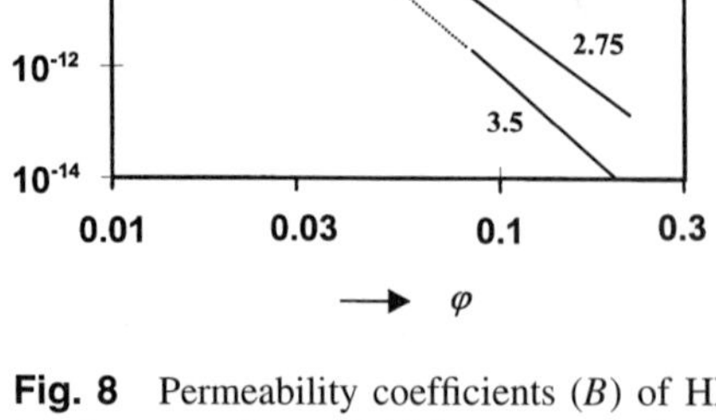

Fig. 8 Permeability coefficients (B) of HP/SF systems as a function of fraction solid (φ) for two values of initial supersaturation ($\ln \beta_0$), given near the lines. The range of the results as roughly expected for fractal networks is also given.

Moreover, it was concluded in Sec. II.C that, for the systems studied, a network is already formed when the volume fraction of crystals is of order 0.01. One would expect that a further deposition of solid material onto an existing network would not materially alter the permeability. This was observed, for instance, for heat-set gels of the protein β-lactoglobulin, where a gel is formed in an early stage of formation of aggregatable particles, and where the permeability weakly decreased when more material was deposited [25], by a factor of about 0.8 for a threefold increase in φ, very different from the results on fats.

Extrapolating the lines in Fig. 8 to $\varphi = 0.01$ results in $B \approx 10^{-6}$ m^2, an impossibly large value: applying Eq. (7) it would follow that $a/K = 30$ μm. The fat crystals are certainly not spheres. Instead of a, the length of a crystal may be taken as the size parameter, and it would be about 10 times that of the equivalent sphere radius (about 0.2 μm). Assuming, moreover, that $K = 1$ (rather than 10, as observed for gels of caseinate particles), Eq. (7) leads, for $D = 1.7$, to a permeability of $5 \cdot 10^{-9}$ m^2. In practice, φ_{gel} may be somewhat <0.01 and D slightly >1.7; then a value of $B = 10^{-8}$ may be obtained, but that is to be seen as an absolute maximum. It is far more likely that B would be smaller. Taking these results into account, high and lower estimates of B versus φ are given in Fig. 8, and they are very far removed from the observed values. Since the values of B at the initial very small φ cannot be far wrong, the curves must be strongly curved at low φ values, but experimental data are not available.

Other measurements on a variety of fats [12] show similar results, i.e., negative slopes of 6.2 to 8.6. Extrapolation to $\varphi = 0.01$ gave $B = 10^{-2}$–10^{-6} m^2. In these studies temperature rather than initial supersaturation was kept constant.

We are thus left with a large discrepancy between ''theory'' and results. The main problem may be that we do not know how the deposition of crystallizing material on the existing network occurs. One may envisage three possible mechanisms:

1. The aggregated crystals merely grow, as depicted in Fig. 1. It would only lead to a weak decrease in B.
2. New crystals are formed and these aggregate with the existing network. It appears likely that they would preferentially orient themselves parallel to the strands in the network. Also this would not strongly decrease the initial value of B.
3. New crystals are formed and these form aggregates that subsequently ''block'' the pores in the existing network. This would greatly decrease B values.

Mechanism 1 surely occurs, as follows from the strong increase in consistency after network formation under nearly all conditions (see Sec. III). Formation of new crystals depends on nucleation occurring. If the network is formed at a stage where only a very small part of the material is crystallized, ln β has decreased only a little below its initial value and nucleation would still go on at a reasonable rate. Aggregation is also a fast process, and it may well be that Mechanism 3 then is the main cause for decrease of B during ongoing crystallization.

A more detailed look at the data on which Fig. 8 is based gives the following picture. The percentage of crystallizable material (HP) ranged from 8% to 14%. The same systems were also subjected to rheological measurements and it was observed that the gel time—taken as the time where a modulus of 1 Pa was reached—ranged from 106 to 18 s. That corresponds with φ_{gel} values from 0.0094 to 0.0027, and the fraction of the crystallizable material crystallized at the gel time was 0.12 to 0.015. At that point, the main structure of the network is presumably determined. According to Eq. (5), two variables affect the value of R_{gel}, and thereby the value of B (assuming D to be constant). At a smaller value of φ_{gel}, which would occur at a higher proportion HP, R_{gel} would be larger; this would be counteracted by the crystal size (a) being smaller (see below). The first effect is undoubtedly the overriding one, which then implies a larger value of B at the gel time for a higher fraction HP. This is contrary to the dependence of the final B on fraction HP. However, for a large fraction HP, the supersaturation will only be slightly decreased at the gel time, implying that crystals can still nucleate and grow. Moreover, the concentration of aggregatable crystals will then be fairly high and aggregation fairly fast. Both trends will promote the occurrence of Mechanism 3. For a smaller fraction HP, these events would occur to a lesser extent, leading to a smaller decrease in B compared to its initial value. Recall that nucleation rate depends very strongly on ln β, even more than does growth rate (hence, the smaller crystal size just mentioned).

It is also seen in Fig. 8 that a larger initial supersaturation gives a smaller permeability at the same φ. This may be explained in a similar manner. Figure 7 shows that at higher ln β_0 the fraction solid increases much faster after the gel point, because the value of ln β decreases more slowly with ongoing crystallization. This implies, again, a greater possibility for the formation of aggregates that can ''fill'' the initial pores.

Altogether, Mechanism 3 is the only one that we can think of that may explain the results obtained, albeit in a semiquantitative manner. It implies that the linear relation between log B and log φ cannot be ascribed to the final network being fractal. On the contrary, the deposition of aggregates of crystals in the larger pores of the (initially fractal) network would make it far more homogeneous. The latter point is not inconsistent with electron micrographs of fats [21]. Far more study is needed to establish the mechanism(s) of network formation and the quantitative relations involved.

III. MECHANICAL PROPERTIES

Fats generally show pseudoplastic behavior during deformation. At small stresses elastic and viscoelastic deformations occur, which are in the realm of rheology. At larger stress, yielding and flow occur, which are more related to fracture mechanics. For fats with a very high solid fraction and/or at very fast deformation, true fracture may occur, but this has hardly been studied. At fairly small supersaturation, large and roughly isometric fat crystals (or crystal aggregates) can be formed, especially spherulites [26]; the interaction forces between these structural elements are relatively weak, and the resulting dispersion tends to have a very weak consistency or even behave like a true liquid (negligible elastic response).

We first discuss small deformation behavior. The modulus, be it elastic or viscoelastic, then is the most important parameter. The results are rarely of practical significance, but the theory is rather well developed, through which the results can yield information about bond properties. Second, large deformation behavior is discussed, and such results are of great practical importance; the interpretation is, however, very difficult. Finally, some results on fats that have been obtained by fairly-large-scale processing are given.

This is not the place for a comprehensive treatment of the mechanical properties of pseudoplastic systems. Basic theory on various rheological methods, together with the interpretation of results, is given in virtually all texts on rheology. Methods to determine large deformation and fracture properties are treated, e.g., by van Vliet and Luyten [27]. Some fundamental aspects of the response of plastic fats to (large) stresses are discussed by de Bruijne and Bot [28].

A. Small Deformation

1. Dynamic Shear Moduli

Mostly dynamic, i.e., oscillatory, measurements in shear are performed at one or a series of frequencies ω (in radians per second). The sample is sheared, often between concentric cylinders, and the relation between stress σ and shear strain γ (relative deformation) is recorded. The modulus G equals stress over strain; it is a measure of the stiffness (not of the strength) of the system. If the sample behaves in a purely elastic manner, stress and strain are in phase; if it is viscoelastic, a phase lag δ is observed. The larger δ is, the more prominent is the viscous part of the rheological behavior. One now speaks of a complex modulus

$$\tilde{G} = G' - iG'' = G' - i\eta\omega \tag{17}$$

where G' is called the storage modulus, G'' is the loss modulus, and η is the viscosity. The absolute value of the modulus $|\tilde{G}|$ equals $(G'^2 + G''^2)^{0.5}$. The loss tangent is given by

$$\tan \delta = \frac{G''}{G'} \tag{17a}$$

The smaller the value of tan δ, the smaller is the proportion of mechanical energy applied that is lost, i.e., dissipated into heat.

One can only speak of a true modulus as long as the strain is proportional to the stress. In that case the behavior is said to be linear. For plastic fats, the linear region, i.e., the maximum strain for which linearity is observed, is of order 10^{-4}; this is a very small value. Outside the linear region irreversible changes in the crystal network are induced.

Examples of the evolution of the storage modulus with time after applying a given supersaturation of HP/SF systems are in Fig. 9a. Comparison with Fig. 7 indicates that the modulus rises fast after the gel point has been reached. The loss tangent is very small; for $\omega = 1\ \text{rad·s}^{-1}$, tan $\delta \approx 0.01$. The loss tangent is larger for a smaller ω, but even at $\omega = 10^{-3}\ \text{rad·s}^{-1}$, tan δ equals only about 0.2. A constant value is reached very soon after the gel point [3], implying that the system already behaves almost purely elastically from a very early stage.

Figure 9a also shows that the final values of the modulus are higher for a larger initial supersaturation, despite the final fraction solid being the same. Figure 9b gives examples of the dependence of G' on the volume fraction of solid fat φ, where the initial supersaturation ln β_0 was kept equal for each case. It is

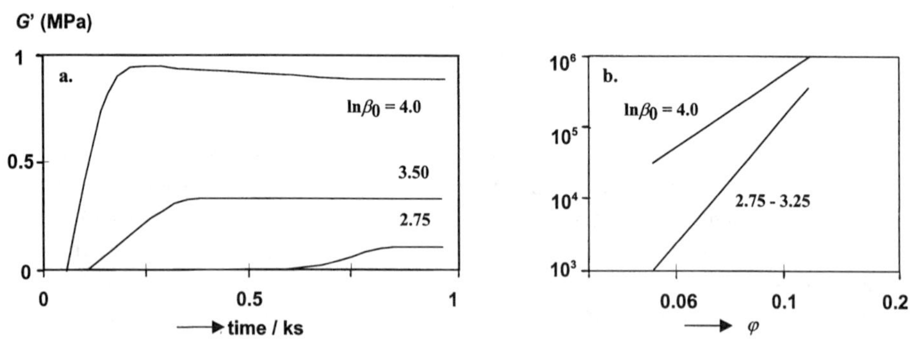

Fig. 9 Storage shear modulus (G'), determined at 0.1 Hz, of HP/SF systems, at various initial supersaturation (ln β_0), indicated near the curves. (a) 12%HP/SF at various times after cooling. (b) Values after 2 hr for various fractions solid (φ). (After results by Kloek [3].)

Table 1 Values of the Scaling Exponent μ and the Preexponential Factor A of the Storage Modulus $G' = A\varphi^{\mu}$, for Various Fats and Conditions during Nucleation (main polymorph, supersaturation level $\ln \beta$)

Type of fat	Polymorph	$\ln \beta$	μ	log (A/Pa)	Ref.
HP/SF	β'	moderate	5.5–7.0	10.6–11.8	3
HP/SF	β'	high	4.1	9.8	3
HP/SF	α	high	4.3–4.5	8.0–8.6	3
SSS/olive oil	α	high	4.1	8.0	9
Margarine fat	α	fairly high	4.3	9.1	12
Milk fat	α, β'	variable	6.4	10.1	12

seen that for a higher β_0 higher moduli are found at the same φ (although the differences were not significant for $\ln \beta_0 = 2.75$ to 3.25). It is seen that the modulus very steeply depends on φ, the slope of the log-log plots ranging between 4.1 and 6.5. Other fats also exhibit such high slopes (Table 1).

The results discussed so far relate to situations where all nucleation and growth of crystals was in the β' modification. In HP/SF systems with nucleation occurring in the α modification the moduli tended to be somewhat higher. The initial supersaturation was, however, very high, which in itself tends to give high moduli. Moreover, polymorphic transitions ($\alpha \rightarrow \beta'$ and $\alpha \rightarrow \beta$) will take place, which may significantly affect the moduli. For nucleation in the β polymorph, the crystal networks formed are generally weak, presumably because initial supersaturation tends to be small, and large, fairly isometric, crystals are formed.

2. Network Models

Various models have been developed to explain the elastic modulus of particle networks. Most equations for the modulus can be written as the product of three factors: (1) a numerical factor characteristic for the geometry of the network; (2) a function due to the interaction forces involved, generally in Pa; and (3) a function of the volume fraction of building blocks φ. The latter function generally has the simple form φ^{μ}, where μ is a positive number.

The first model appears to be due to van den Tempel [29]. He reasoned that only van der Waals attraction and hard core repulsion act between fat crystals in oil. He further assumed, for simplicity, the crystals to be fairly isometric particles in a regular cubic network. The result is

$$G = \frac{5}{24\pi}\frac{A_{\mathrm{H}}d^{0.5}}{h_0^{3.5}}\varphi \tag{18}$$

Reasonable values are as follows: for the Hamaker constant of van der Waals attraction between fat crystals in oil, $A_H = 6 \cdot 10^{-21}$ J (calculated from refractive indices [3,29]); for the hard core distance between particles, $h_0 = 0.5$ nm; and for the equivalent sphere diameter of the crystals, $d = 100$ nm. For a volume fraction $\varphi = 0.1$, this gives $G = 4 \cdot 10^6$ Pa, which seems to be a reasonable value. Several objections can, however, be raised against the equation, the main one being that it predicts a very wrong dependence of G on φ: the exponent μ would equal 1 rather than the observed values >4. Consequently, far too high G values are calculated for low φ. This must, at least partly, be due to many of the particles not carrying a stress when the network is deformed. This is because a particle network resulting from random aggregation cannot be regular (cf. Sec. II.B). Moreover, other interactions may and will be involved.

The crystals are not spheres, but very anisometric particles that can be characterized, with some oversimplification, by a length x, a width y, and a thickness z. On deformation, such crystals can bend. Kloek has worked out a theory [3] where it is assumed that the only interaction forces acting between particles are, again, van der Waals attraction and hard core repulsion. The moduli obtained for a regular network are not greatly different from those from Eq. (18), but the deformability of the network due to bending of the crystals is considerable. The result for the maximum macroscopic shear strain γ, defined as the bending deflection of a single crystal divided by its length x, is

$$\gamma = 0.08 \frac{A_H}{h_0^2} \frac{1}{E_b} \frac{x^2}{y^{1/2} z^{5/2}} \tag{19}$$

Taking A_H and h_0 as above, the bending modulus of the crystal $E_b = 10^9$ Pa, and x, y, and z equal to 1, 0.2, and 0.05 μm, respectively, results in $\gamma = 0.1$, i.e., about 10^3 times the value observed for the linear region; see also Fig. 10. This is another argument against a model in which separate rigid crystals are merely held together by van der Waals forces.

Theories for the elastic modulus of fractal particle gels have been developed [11,30], reviewed [3,11,24], and applied to fat crystal networks [3]. The theories are based on Eq. (5) and on the self-similarity of fractal aggregates. The latter point implies the number n of contact points between aggregates after gel formation is constant (with some statistical variation); n is often taken to be 6. This implies that only a proportion of the particle-particle bonds in the gel carry a stress on its deformation, the proportion being smaller for a smaller φ.

For fat crystals aggregated due to van der Waals attraction, the equation would become

$$G = F \frac{nA_H}{h_0^3} \varphi^{k/(3-D)} \tag{20}$$

The numerical factor F is generally unknown, which reduces Eq. (20) to a scaling law. The value of k in the exponent is a matter of debate; it is related to the flexibility of the stress-carrying strands of particles in the network. In theory, it ought to be somewhat larger than 4, assuming the strands to be fully flexible and fractal. In practice, values of $k > 3$ are rarely observed; $k = 3$ corresponds to strands that have one hinge point within the fractal aggregates that make up the gel, but are otherwise stiff. Completely stiff strands lead to $k = 2$.

It is clear that Eq. (20) cannot explain the results observed. To begin with, the fractal dimensionality D of the aggregates is observed to equal about 1.75 (Sec. II.D), and this leads to values for the exponent $\mu = k/(3 - D)$ of 2.4 for $k = 3$ and of 3.2 for the unlikely value of $k = 4$. In practice, $\mu = 4.1$ to 7 is observed (Table 1). Moreover, the values observed for G are too small. The highest values will be calculated for $k = 2$ (stiff strands), for which case also the factor F is known; it then equals 0.0064. Taking the same values for A_H and h_0 as before, this gives for $\varphi = 0.12$, $G = 6.10^4$ Pa, whereas values of G' close to 10^6 Pa are observed for 12% HP/SF (Fig. 9b).

In Sec. II.D it was concluded that the crystal networks obtained are not fractal, and this is thus confirmed by the observations on the moduli. It was also stated that in an early stage of the process, the network may be fractal. We consider the results on 12% HP/SF at the stage where $\varphi \approx 0.01$, where $G' \approx 10^3$ Pa [3]. Using Eq. (20) with $k = 2$, we then arrive at $G = 1150$ Pa, which fits the results. However, for $\varphi = 0.01$, the ratio $R_g/a = 40$, according to Eq. (5). This means that the stress-carrying strands are many particles long, and it is difficult to envisage such strands of particles to be very stiff. On the other hand, any calculations involving $k \geq 3$ result in far too small values of G. Moreover, even at the time at which $\varphi = 0.01$ is reached, $\tan \delta$ is already as small as 0.3 [3]. Altogether, we have to conclude that already at a very early stage in its formation, the network is quite stiff. Unfortunately, results on the magnitude of the linear region at very small φ are not available; we suppose that it will be larger than in the final networks obtained.

3. Sintering

Several of the mentioned observations point to sintering of the crystals held in the network. This is not a new conclusion, as it has been assumed long ago; see Fig. 1, which goes back to much older observations. A specific study has been done by Johansson and Bergenståhl [23], who made dispersions of crystals of pure TAGs in oil. These do aggregate to form a gel, but do not show significant sintering. They studied the effect of adding small amounts of crystallizable TAG mixtures. These could then cause considerable sintering, as measured, for instance, by an increase in modulus. Sintering occurred if the newly added TAGs crystallized in the same polymorph and if the melting point in solution was below that of the original crystals. For crystals in the β′ polymorph, very rapid cooling, i.e., a very fast increase in the supersaturation ln β, enhanced sintering; presumably, at a slower increase in ln β new crystals were formed rather than deposition of the crystallizable material on the crystals present. For β crystals, a slow increase in ln β promoted sintering, presumably because a high ln β will induce crystallization in the β′ polymorph.

Translating these results into the observations on the HP dispersions, it may be concluded that sintering can certainly occur. All crystallization was in the β′ polymorph, and HP consists of a number of TAGs of different melting points. Apparently, sintering can already occur in a very early stage.

Kloek [3] further studied sintering by comparing crystallization of HP with that of a pure TAG (PSP). Results are summarized in Table 2. It is seen that the modulus for PSP is much smaller than that for HP. Calculating G from Eq. (20), with $F = 0.04$, $k = 3$, and $D = 1.75$, leads to a value of 46 kPa, roughly as observed for PSP. This suggests that no sintering had occurred in the PSP fats, which also fits the higher loss tangent compared to that of HP. Working and keeping the sample at rest after working hardly affected the results on PSP, but in the HP sample the loss tangent was increased and the modulus was greatly reduced by working; it increased again during keeping. This strongly suggests that many of the sintered bonds are broken by working and that sintering of HP occurs again afterward, albeit far more slowly than it did during initial crystallization.

4. Concluding Remarks

It was concluded in Sec. II.D that the network of fat crystals formed will initially be fractal, but that its structure greatly changes when more material crystallizes. This also fits the observations on the modulus, at least in a qualitative sense. Three hypothetical mechanisms for network change were distinguished. The first

Table 2 Storage Moduli G' and Loss Tangents (tan δ) at 1 Hz of Fats Containing 10% Solids of HP or Pure PSP, in Sunflower Oil[a]

Solid fat	Treatment	G'/kPa	tan δ
PSP	at rest	35	0.11
PSP	worked, after 1 hr	30	0.09
PSP	worked, after 20 hr	32	0.08
HP	at rest	703	0.02
HP	worked, after 1 hr	39	0.09
HP	worked, after 20 hr	110	0.08

[a] Values were determined at 10°C at rest, 1 hr after working the sample, and 20 hr after working.

is that the existing crystals would merely grow. For crystals that are already bonded due to van der Waals attraction, this would lead to sintering, if there is some spread in composition of the crystallizing TAGs. The second mechanism involves formation of new crystals that would subsequently become bonded to the existing network, for the most part in an orientation parallel to the existing strands. These two mechanisms would explain the small values of tan δ and the high values of the moduli observed, but not their very strong dependence on φ nor the very small extent of the linear region.

The third mechanism envisaged is the formation and aggregation of new crystals, where the aggregates would become incorporated in the network, especially ''filling'' the larger pores. It was concluded that this will especially happen if the supersaturation stays high for a while after formation of the network. It can explain the very strong negative dependence of the permeability B of the fat on φ and the smaller values of B for a higher ln β. Figure 9 shows that the modulus increases very strongly with increasing φ and increases with increasing ln β. This clearly fits with Mechanism 3. It would also explain the linear region being very limited, because the filling of the larger pores in the network with crystal aggregates would greatly decrease the freedom of bending the strands of the original network.

It will depend on several conditions to what extent the assumed three mechanisms will occur, and this will, of course, affect the quantitative relations observed. Moreover, polymorphic transitions occurring after network formation may significantly alter the mechanical properties of the network. Consequently, it will be very difficult to predict more than trends.

B. Large Deformation

Large-strain experiments can be applied in various deformation modes, which tend to give different results. Moreover, several disturbing factors may arise. Special care should be taken to avoid significant deformation of the sample when bringing it in the measuring body, since such prior deformation may materially alter the mechanical properties, the change being at least partly irreversible. To avoid this, uniaxial compression of cylindrical samples is often applied. However, friction of the sample with the plates tends to greatly increase the observed stress and extrapolation to infinite sample height is necessary to obtain meaningful data. The results may also depend on the strain rate. The most fundamental problem may be that the deformation of the sample is generally markedly inhomogeneous, making it all but impossible to apply unequivocal theory. Most results should thus be considered with care, giving trends rather than absolute data.

Figure 10 gives examples of the response of three different materials on fairly large strains. It is seen that the apparent moduli decrease and the apparent loss tangents increase as the strain increases and as the sample is deformed repeat-

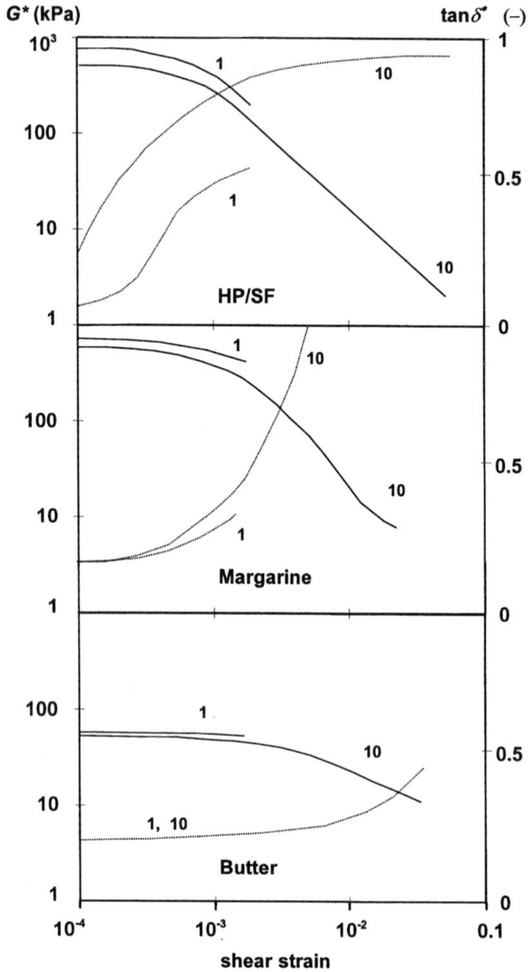

Fig. 10 Apparent shear modulus (G^*, decreasing curves) and apparent loss tangent ($\tan \delta^*$, increasing curves), as a function of shear strain of three different materials (indicated). The samples where sheared 10 times at ever-increasing maximum shear; the first and the tenth curves are given.

edly. (We speak of apparent modulus, etc., because stress and strain are no longer proportional.) Most of the change is reversible, which implies that (chains of) crystals are bent; some bending and stretching of van der Waals bonds may also be involved. As the strain applied increases, a significant part of the changes becomes irreversible and this signifies that bonds are broken that are not re-formed upon release of the stress. Thereby, the structure of the network is altered. However, considerable differences are observed among materials. The region of linear deformation and the reversibility increase in the order HP-margarine-butter, while the extent of the changes decreases in that order. Butter shows an almost truly elastic response up to a strain of about 1%, while the HP system begins to change close to 0.01% deformation. This is presumably due to the butterfat crystals being more slender, hence more bendable, than those of HP; see Eq. (19).

Figure 11 gives an idealized stress-strain curve for a margarine-type fat of fairly small solid content, say $\varphi = 0.1$. Four regimes can be usefully distinguished.

1. Here the behavior is *linearly elastic*, which means that the strain is proportional to the stress. The loss tangent is small, ≤ 0.1. The deformation is fully reversible, which implies that no bonds between crystals are broken.

2. For larger strain the behavior becomes *nonlinear*, and it is increasingly viscoelastic. The loss tangent increases to about 1; see also Fig. 10. Although most of the deformation is still reversible, irreversibility increases with increasing strain. This would mean that an increasing number of bonds is broken. Presumably, van der Waals bonds are broken and re-form, whereas fairly weak sintered

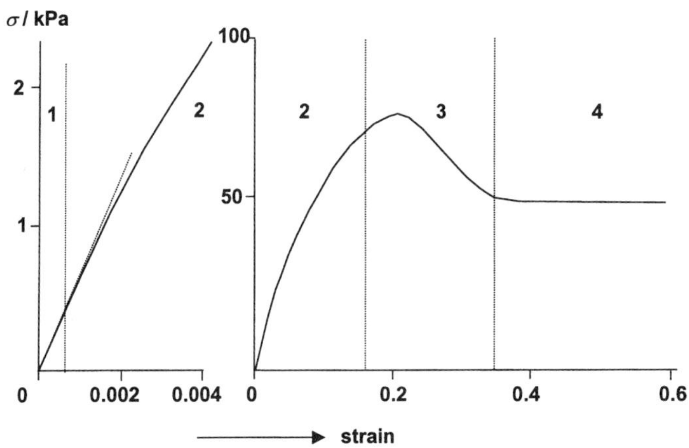

Fig. 11 Idealized relation between stress (σ) and strain of pseudoplastic fat. See text.

bonds are also broken but do not re-form, at least not within a short time span (several minutes). It should be realized that the inhomogeneity of the system leads to a wide variation in bond strength.

3. This is the regime of *stress overshoot*, a phenomenon typical for yielding of the material. *Yielding*, i.e., the transition from viscoelastic to viscous behavior (flow), only occurs in materials composed of a ''solid'' network interspersed with a continuous liquid phase. In local planes, all of the bonds in the network are broken, but the ''cracks'' are immediately filled with liquid, and the system remains continuous. Often, fairly extended slipping planes are formed in the fat. This implies that the velocity gradient varies greatly throughout the sample. In most systems exhibiting yielding, including plastic fats, removal of the stress leads to fast recovery of a part of the elastic properties. This means that significant interaction forces must remain between the solid structural elements, be it due to colloidal attraction or to strong friction caused by ''entanglements.'' Anyway, the structure of the system is greatly changed by yielding.

4. Finally, a regime of *plastic flow* is reached. The material flows, but it is very viscous (the Bingham elongational viscosity is often between 0.1 and 1 MPa·s), strain-rate thinning, and exhibits some elasticity. The high viscosity is primarily due to the presence of fairly large remnants of the once continuous crystal network, which take up a large effective volume fraction. Moreover, the irregular shape and spiky surface of these structural elements leads to their ''entanglement'' during flow and, thereby, to elastic properties. Possibly, van der Waals attraction between crystals or crystal aggregates also contributes to these properties.

In practice, the stress needed to cause yielding of the fat is especially important, since cutting, spreading, and shaping all involve yielding. Ideally, one may determine the deformation rate (flow rate) as a function of the stress applied; an example is in Fig. 12a. Above a certain stress σ_y the material starts to flow, but the value of σ_y tends to be smaller if the time scale of the experiment is longer. Often, the flow curve becomes virtually linear, and by extrapolation the Bingham yield stress σ_B can be obtained. It turns out that σ_B correlates well with the firmness (often called hardness) of the fats as determined by penetrometer, extrusion, or wire-cutting tests [31].

Figure 12b shows how firmness tends to depend on solid fat content. Both the minimum solid fat content needed and the shape and steepness of the curves vary. Important causative variables presumably are extent of sintering, homogeneity of the network, and crystal size and shape.

In practice, firmness tends to increase considerably when lowering the temperature. In the first place it causes an increase in the fraction solid, although this greatly depends on the steepness of the melting curve of the fat. In the second place, it tends to increase sintering. In several fats, a 10 K decrease may cause

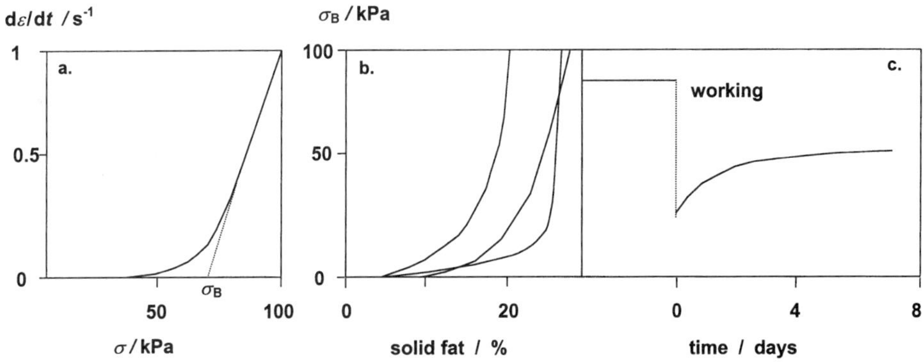

Fig. 12 Large deformation properties of pseudoplastic fats; approximate relations, meant to illustrate trends. (a) Deformation rate ($d\epsilon/dt$) as a function of stress (σ), and definition of the Bingham yield stress σ_B. (b) Examples of firmness (σ_B) as a function of solid fat content. (c) Example of the effect of working and subsequent setting on firmness (σ_B). (Adapted from Ref. 1.)

an increase of σ_B by about an order of magnitude. Also the temperature history may markedly affect firmness, but simple rules cannot be given [31].

Many fats increase in firmness during keeping and the rate of increase tends to be faster at a higher storage temperature. The cause appears to be further sintering. This especially occurs in a fat of a wide compositional rate, such as milk fat [31]. Such a fat generally contains a considerable amount of compound crystals, especially in the α polymorph. During keeping at a not-too-low temperature, these are partly transformed, via melting and recrystallization, into crystals of other composition; this tends to go along with (partial) transition to the β′ polymorph. Such recrystallization appears to promote sintering [1,31]. Its rate can be greatly enhanced by even small fluctuations in storage temperature, since these cause melting and recrystallization.

Recrystallization involving transition to the β polymorph tends to lead to a decrease in firmness, often to a considerable extent. This is because β crystals tend to be rather isometric and not very prone to sintering. The latter may be because hardly any compound crystal formation can occur in the β polymorph. The presence of diacylglycerols greatly hinders the transition to β, and addition of these components to a fat is an effective way to prevent its softening on storage [32].

Figure 12c shows how working can decrease the firmness of a plastic fat; this is called work softening. The extent of work softening can vary widely: a relative decrease in firmness between 10 and 75% can be observed, according to type of fat and intensity of working. Table 2 also gives some results on the effect of working on the modulus as determined at very small strain. It was concluded above that the difference between the two fats was due to differences in extent of sintering. It is also seen in Fig. 12c that considerable setting, i.e., increase in firmness, can occur after working. Initially, setting will be caused by aggregation due to van der Waals attraction of the crystals and crystal network fragments into a continuous network. At longer time scale (say, after 1 h) sintering due to recrystallization presumably is responsible. Its cause must be the same as in the previous paragraph; moreover, the breaking of sintered bonds during working may lead to sharp crystal edges and thereby to some Ostwald ripening. Again, considerable variation is observed in the rate and maximum extent of setting; compare, e.g., Fig. 12c with Table 2 results on HP.

It is clear from the above considerations that predicting the yield properties of a plastic fat is very difficult. Nevertheless, some trends can be explained.

C. Fats Crystallized under Agitation

From the previous discussion, it may be clear that the formation of a firm crystal network needs crystallization at a high supersaturation, hence, at high supercooling; hence, cooling must be very fast. However, the latter is not easily realized on an industrial scale. Consequently, surface-scraped heat exchangers (''votators'') are commonly used. The oil is pumped through a cylinder that is deeply cooled from the outside, while the contents are strongly agitated by means of a stirrer with blades that scrape the cylinder wall. Naturally, fat crystals are formed at the wall and are scraped off and mixed through the liquid. Average residence times in the heat exchangers are generally between 0.5 and 1.5 min. All or most of the crystallizable fat is solidified at the outlet of the heat exchanger. This means that any crystal networks that tend to form would be strongly disturbed.

To establish the effect of this process on the properties of the resulting fat, Kloek [3] did some studies with HP/SF systems. The temperature in the heat exchanger was adjusted to ensure that at the exit ln $\beta = 4$ for HP in the β′ polymorph in all cases; consequently, crystallization occurred at very high supersaturation. The flow rate was either 20 or 50 kg/h, corresponding to average residence times of 1.2 and 0.5 min, respectively. In the latter case, about 80% of the crystallizable fat was crystallized at the exit of the heat exchanger; at the slower flow rate all HP had crystallized. This means that crystallization proceeds quite fast, compared to isothermal crystallization at rest (cf. Fig. 7).

Some typical results are given in Fig. 13. It is seen that both the modulus and the yield stress one day after processing are higher for the faster flow rate, and that they do not increase during further storage; the product obtained at the

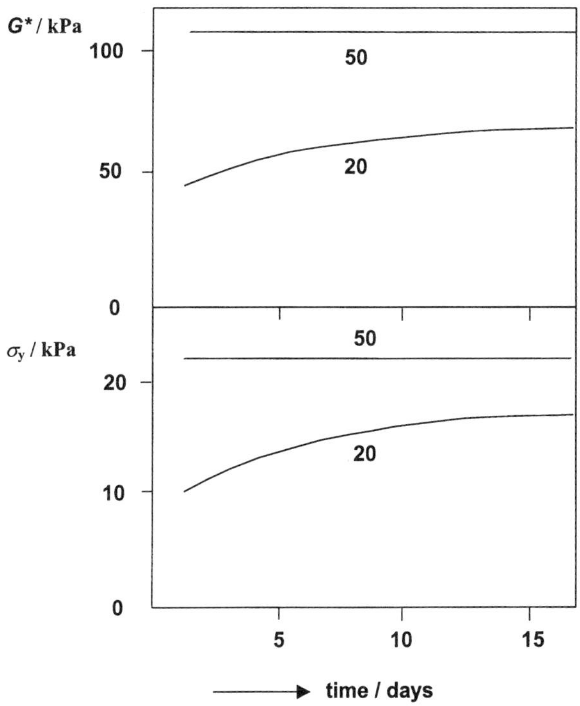

Fig. 13 Results on the apparent shear modulus (G^*) and the yield stress (σ_y) of 15%HP/SF systems, cooled and scraped in a heat exchanger, as a function of time after treatment. The flow rate through the heat exchanger (in kg/h) is indicated. See text.

slower flow rate does increase in consistency with time. The explanation is presumably that the large velocity gradients in the heat exchanger not only speed up formation and aggregation of the crystals, but also cause the crystal aggregates to deform in such a way that they become more compact and isometric. After leaving the apparatus, these aggregates will further aggregate to form a continuous network. If the crystallization then is already complete (as appears to be the case at the slower flow rate), sintering of the bonds between aggregates will not readily occur, and in any case proceeds slowly. If not all HP is crystallized at the exit, the bonds between the aggregates can sinter, thus producing a firmer fat; such sintering is complete within a day (and probably much earlier).

The percentage of HP in the system was also varied (φ = 0.08 to 0.15). Within this range, log-log plots of modulus or yield stress were about linear. The slopes were for yield stress about 1.9 at 50 kg/h and 1.1 at 20 kg/h. Note that these slopes are very much smaller than those for the moduli of fats crystallized at rest (Table 1). The slopes (μ values) for the apparent moduli where somewhat higher, about 2.45 for the fats obtained at 50 kg/h and 1.25 at 20 kg/h.

For the latter case, fairly isometric and compact aggregates appear to be the building blocks of the final network, which would not greatly alter upon storage. The bonds between the aggregates would be far weaker than those in the aggregates, which typically fits the theory for a fractal ''weak-link regime'' as defined by Shih et al. [33]. The relation between modulus and volume fraction would then be as in Eq. (20) with $k = 1$, assuming the size of the aggregates forming the gel to be governed by Eq. (5). However, the latter assumption cannot be correct in the present case, since the ''aggregates'' are in fact a kind of compacted network remnants. Consequently, it appears unwarranted to apply fractal theory in this case, as well as in the other cases discussed.

IV. SUMMING UP

Most natural or partially hardened fats can be processed so as to become a pseudoplastic material, consisting of a solid crystal network interspersed with oil. Prediction of the network structure and of the resulting macroscopic properties is very difficult because of the great number of aspects involved and of variables affecting the result. This is mainly because

- Crystallization of multicomponent fats is very intricate.
- Several events (nucleation, crystal growth, aggregation, gel formation, sintering) occur simultaneously, and the rate of each can vary.
- The network can be disturbed by agitation.
- The system is generally not in thermodynamic equilibrium and several slow changes can occur.

Figure 14 summarizes the most important events, their time scales, and the resulting structures. The network formed is initially of a fractal nature, but it is subsequently changed considerably by ongoing crystallization. It becomes more regular, although it certainly does not become truly homogeneous, and its firmness greatly increases.

Not all of the resulting macroscopic properties mentioned in Sec. I were discussed. *Oiling off* seems not to have been studied systematically; the main variable is presumably the permeability, discussed in Sec. II.D [see Eq. (6)]. It especially occurs if the crystal network is relatively coarse, which may be the case when part of the solid fat has melted, or if external pressure is high, as when flexible boxes of fat (margarine) are piled up. *Immobilization* of structural elements like aqueous droplets is fairly simple and hardly needs discussion. It is often observed that some fat crystals are oriented in the oil-water interface [21].

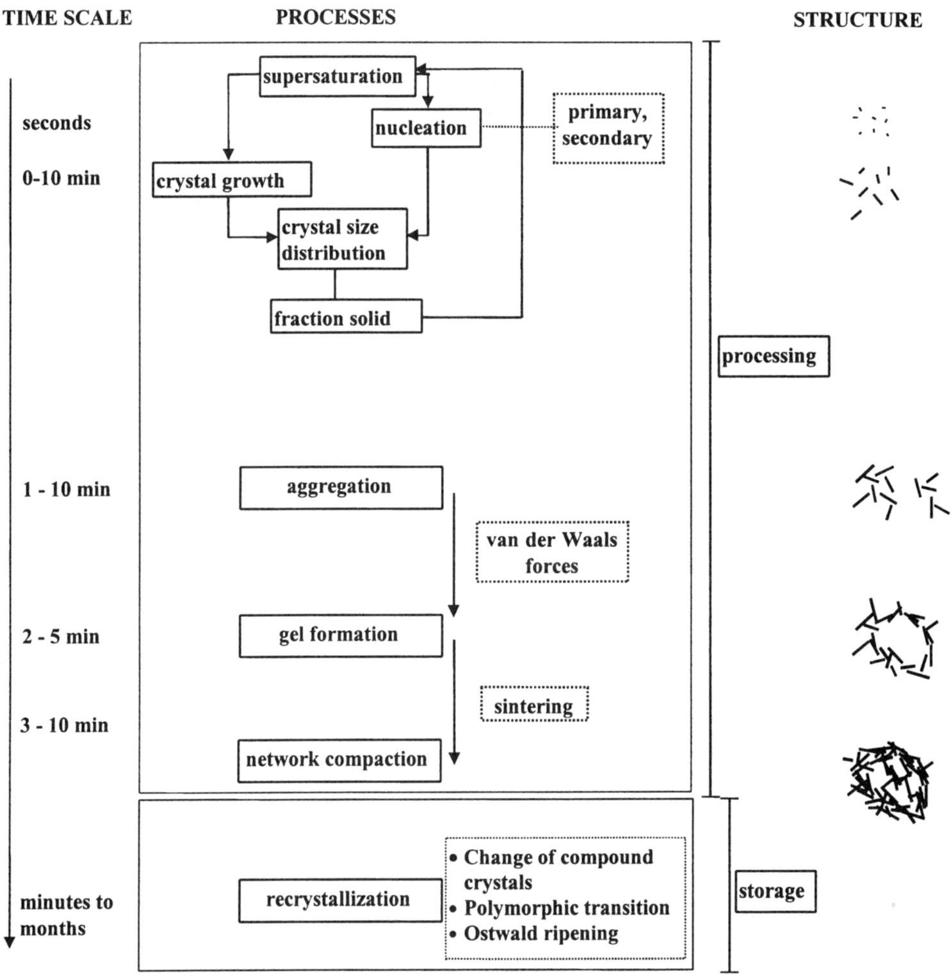

Fig. 14 Schematic presentation of the various processes occurring during crystallization and keeping of a fat. The time scales are for 10%HP/SF, crystallizing at rest at moderate supersaturation.

Partial coalescence of partly solid emulsion droplets is outside the scope of this book. This leaves the *mechanical properties* of bulk fats.

The modulus of various fats is far better understood than the large deformation behavior, such as the magnitude of the yield stress. It is often implicitly assumed that these parameters are closely correlated. Fortunately, the correlation can be reasonably good for plastic fats, but it is not nearly perfect and one may not assume that a higher modulus simply means a proportionally greater yield stress. Moreover, a modulus tells nothing about other parameters, such as the (extent of) elastic deformability, which may vary by almost two orders of magnitude among fats.

Often, a high yield stress is desired for a product containing only a little crystalline fat. To achieve this, the supersaturation should be high during most of the formation and aggregation of crystals, especially when aggregation occurs under quiescent conditions. These conditions lead to fairly small crystals that form a continuous fractal network long before crystallization is complete, implying that further crystals and crystal aggregates are formed that ''fill'' the pores in the network and make it much stiffer. Sintering of bonds between adjacent crystals, as caused by ongoing crystallization or recrystallization, then provides a very stiff and strong network, which is, however, prone to extensive work softening. A high supersaturation needs very rapid cooling, and this can only be achieved on an industrial scale by vigorous agitation during cooling, which does disturb the network formed. In such a case, sintering after the agitation has stopped is essential to obtain a high yield stress.

Crystals of one pure TAG in oil do aggregate and form a network, but the van der Waals bonds involved are relatively weak. Multicomponent fats, which are virtually always used in industrial practice, behave quite differently. The compositional range can vary widely—from, say, 10 to thousands of different TAGs, and this has considerable influence on the macroscopic properties. In a qualitative sense a wide compositional range has the following effects [1,31]:

1. The occurrence of copious secondary nucleation, which means that a great number of nuclei are formed even if nucleation is heterogeneous and the number of catalytic impurities is small.
2. The linear growth rate of the crystals is relatively slow. This, in combination with point 1, nearly always leads to formation of small crystals.
3. Probably, the crystals formed are relatively thin and can readily be bent upon deformation. This means that the elastic deformation of the fat can be relatively large. The evidence for the crystal shape is largely circumstantial, and more research would be needed.
4. The melting range is very wide, which means that the amount of solid fat is greatly dependent on temperature and so is the firmness. Moreover, a slight temperature increase causes some melting, and upon subsequent slight cooling newly formed crystalline material will be deposited on existing crystals, thereby causing sintering. Consequently, the fat may markedly increase in firmness due to small temperature fluctuations.

5. Extensive compound crystallization occurs. This has several consequences:

- The composition and amount of crystalline fat depend on the temperature history, and this may significantly affect fat firmness.
- Compound crystals are especially formed in the α polymorph and, though less, in the β' polymorph. These crystals are far more prone to sintering than are β crystals, and the latter are often not readily formed if the compositional range is very wide.
- The crystals are never in thermodynamic equilibrium, which means that slow recrystallization can occur at constant temperature. Since this process will for the most part proceed via local melting, the newly formed crystalline material will be deposited onto existing crystals. As mentioned, this means sintering. Consequently, the firmness of the fat will increase during storage, especially after working of the fat.

Another important compositional variable is the concentration of mono- and diacylglycerols, and possibly other substances. Diacylglycerols counteract the transition $\beta' \rightarrow \beta$ [32]. Monoacylglycerols and some other surfactants can adsorb onto TAG crystals [2], and have been shown to diminish the colloidal interaction forces between (pure) SSS crystals [29]. It is, however, questionable whether this is of importance in multicomponent fats, where strong sintering occurs. Moreover, the presence of monoglycerides may have other effects, notably the promotion of heterogeneous nucleation [15,17]. All these factors may affect network strength and stability.

In conclusion, the events occurring during formation of a fat crystal network, the variables involved, and their effects on the macroscopic properties of plastic fats, are fairly well understood in a qualitative sense. Trends can be explained. For the relatively simple system HP/SF, which has been studied most extensively, some quantitative relations fit into fairly simple theoretical models. However, a complete quantitative understanding of the structure and properties of a fat, as a function of material and process variables, needs considerable further study.

REFERENCES

1. P. Walstra, in *Food Structure and Behaviour* (J. M. V. Blanshard and P. Lillford, eds.), Academic Press, London (1987), pp. 67–85.
2. D. Johansson, *Colloids in Fats*, Ph.D. thesis, Lund University, Sweden (1994).
3. W. Kloek, *Mechanical Properties of Fats in Relation to Their Crystallization*, Ph.D. thesis, Wageningen University, The Netherlands (1998).

4. L. H. Wesdorp, *Liquid-Multiple Solid Phase Equilibria in Fats*, Ph.D. thesis, Technical University Delft, The Netherlands (1990).
5. M. Knoester, P. de Bruyne, and M. van den Tempel, *J. Crystal Growth*, *4*: 309–319 (1972).
6. P. Walstra and E. C. H. van Beresteijn, *Neth. Milk Dairy J.*, *29*: 35–65 (1975).
7. P. Meakin, *Adv. Colloid Interface Sci.*, *28*: 249–331 (1988).
8. L. G. B. Bremer, T. van Vliet, and P. Walstra, *J. Chem. Soc. Faraday Trans. 1*, *85*: 3359–3372 (1989).
9. R. Vreeker, L. L. Hoekstra, D. C. den Boer, and W. G. M. Agterof, *Colloids Surf.*, *65*: 185–189 (1992).
10. W. Kloek, T. van Vliet, and P. Walstra, in *Food Colloids*: *Proteins*, *Lipids and Polysaccharides* (E. Dickinson and B. Bergenståhl, eds.), Royal Society of Chemistry, London (1997), pp. 168–18.
11. L. G. B. Bremer, *Fractal Aggregation in Relation to Formation and Properties of Particle Gels*, Ph.D. thesis, Wageningen University, The Netherlands (1992).
12. K. Boode, P. Walstra, and A. E. A. de Groot-Mostert, *Colloids Surf. A*, *81*: 139–151 (1993).
13. M. Volmer, *Kinetik der Phasenbildung*, Steinkopf, Leipzig (1939).
14. J. Lyklema, *Fundamentals of Interface and Colloid Science*, Vol. I, Academic Press, London, 1991.
15. W. Skoda and M. van den Tempel, *J. Colloid Sci.*, *18*: 568–584 (1963).
16. L. W. Phipps, *Trans. Faraday Soc.*, *60*: 1873–1883 (1964).
17. P. Walstra, *Progr. Colloid Polym. Sci.*, *108*: 4–8, 1998.
18. W. Skoda and M. van den Tempel, *J. Crystal Growth*, *1*: 207–217 (1967).
19. M. Avrami, *J. Chem. Phys.*, *8*: 212–224 (1939).
20. M. H. Zwietering, L. Jongenburger, F. M. Rombouts, and K. van't Riet, *Appl. Environm. Microbiol.*, *56*: 1875–1881 (1990).
21. I. Heertje and M. Pâques, in *New Physico-chemical Techniques for the Characterization of Complex Food Systems* (E. Dickinson, ed.), Blackie, London, 1995, pp. 1–52.
22. L. G. B. Bremer, P. Walstra, and T. van Vliet, *Colloids Surf. A*, *99*: 121–127 (1995).
23. D. Johansson and B. Bergenståhl, *J. Am. Oil Chem. Soc.* *72*: 1091–1099 (1995).
24. P. Walstra, Proc. Royal Society-Unilever INDO-UK Forum, 5th Conference, Mysore, India (2000).
25. M. Verheul and S. P. F. M. Roefs, *Food Hydrocolloids*, *12*: 17–24 (1998).
26. K. P. A. M. van Putte and B. H. Bakker, *J. Am. Oil Chem, Soc.* *64*: 1138–1143 (1987).
27. T. van Vliet and H. Luyten, in Ref. 21, pp. 157–176.
28. D. W. de Bruijne and A. Bot, in *Food Texture*: *Measurement and Perception* (A. J. Rosenthal, ed.), Aspen, Gaithersburg, MD (1999), pp. 185–227.
29. M. van den Tempel, *J. Colloid Sci.*, *16*: 284–296 (1961).
30. L. G. B. Bremer and T. van Vliet, *Rheol. Acta*, *30*: 98–101 (1991).
31. P. Walstra, T. van Vliet, and W. Kloek, in Advanced Dairy *Chemistry*. Vol. 2: *Lipids* (P. F. Fox, ed.), Chapman and Hall, London (1994), pp. 179–211.
32. L. Hernqvist and K. Anjou, *Fette Seifen Anstrichmittel* *85*: 64–66 (1983).
33. W-H. Shih, W. Y. Shih, S-I. Kim, J. Liu, and I. A. Aksay, *Phys. Rev. A*, *42*: 4772–4779 (1990).

9

Lipid Crystallization and Its Effect on the Physical Structure of Ice Cream

R. Adleman
Pillsbury Co., Minneapolis, Minnesota

R. W. Hartel
University of Wisconsin, Madison, Wisconsin

Ice cream is a complex, partially frozen matrix made up of air, lipids, ice crystals and unfrozen water, proteins, stabilizers, emulsifiers, and sugars [1]. A typical composition of ice cream mix is given in Table 1. Three phases and multiple physical states are present in ice cream. The continuous phase is a highly concentrated aqueous solution comprised of soluble and colloidal milk proteins, lactose, added sugars, milk salts, and stabilizers. Fat is dispersed throughout the continuous phase as an oil-in-water emulsion, stabilized by proteins and added emulsifiers. Ice crystals are also dispersed throughout the matrix, constituting about 60% of the mass of the ice cream at hardening room temperature (−20°C). Incorporated air cells typically make up between 50% to 100% of the volume of ice cream.

Many complex interactions between ingredients occur during processing. These interactions are responsible for the overall quality, texture, and flavor of ice cream. One of the most important phenomena contributing to quality and texture of ice cream is crystallization. It is important to control crystallization of three components in ice cream; ice, lactose, and lipid [2]. As discussed in this chapter, lipid crystallization plays an important role in the development of structure during processing of ice cream.

The process for making ice cream is outlined in Table 2. Also described in this table are the relevant crystallization phenomena related to the lipids in

Table 1 Composition of a Typical Ice Cream Mix

Component	% By weight
Water	60–64
Sugar	13–18
Fat	10–18
Milk solids nonfat	7.5–11.5
Stabilizers/emulsifiers	0.3–0.5

ice cream that occur in each step of the process. Making ice cream involves first blending the ingredients to form ice cream mix. Next, the mix is pasteurized to kill pathogenic bacteria and reduce the number of spoilage microbes.

Immediately following pasteurization, the mix is homogenized to reduce the fat globule size, forming an emulsion. The mix is immediately cooled to refrigeration temperature and aged for 3 to 24 hr. After aging, the mix is frozen in a scraped surface heat exchanger that simultaneously allows a fixed amount of air to enter the mix as it freezes, whipping the mix to create a foam. The partially frozen ice cream is drawn from the freezer at the desired temperature and air incorporation level. The ice cream is then packaged and immediately placed in a hardening freezer, where the final structure is established. Each step in the process of ice cream making has an effect on the final product. Important molecular and structural changes occur in many of the above steps, particularly

Table 2 State of Lipid during Processing of Ice Cream

Process step	State/interaction of lipid
Mix ingredients	Liquid fat
Pasteurize/homogenize	Liquid fat
	Decreased fat globule size
	Rearrangement of surface-active components
Rapid cooling and aging	Generation of supercooled liquid fat
	Crystallization of lipid to semicrystalline state
Freezing	Destabilization of emulsion
	Coalescence and partial degradation
	Further crystallization of lipid
Hardening	Further crystallization of lipid
Storage and distribution	Recrystallization due to temperature fluctuations
	Melting, growth, ripening, etc.

in relation to the fat phase, which play a significant role in the structure and quality of the end product.

Lipid in ice cream can come from several different sources [1]. In the United States, all frozen products labeled ''ice cream'' use milk fat as the only source of fat. The most common source of milk fat is from cream or milk, but anhydrous milk fat or butter can be used as well. In some other countries, the use of alternative sources of fat, such as palm oil or palm kernel oil, is permitted in ice cream. Use of these fats in the United States is only permitted in frozen dairy desserts labeled ''mellorine'' product.

Fat is an essential constituent that contributes to the overall structure, texture, and mouthfeel of ice cream. It generally makes up 10–18% by weight of ice cream. Milk fat, the main source of fat in ice cream products, is also one of the most complex fats in terms of its composition and behavior. About 98% of milk fat is in the form of triacylglycerols (TAG), with the remainder as mono-(MAG) and diacylglycerols (DAG), phospholipids, glycolipids, and sterols. Over 400 fatty acids have been found in milk fat [3]. In particular, milk fat has high levels of palmitic, oleic, and stearic acids [4]. Composition of milk fat can vary slightly depending on season, feed, breed, and location, although distinct trends in composition have been observed. The most common fatty acids in milk fat are saturated C4–C18 chains, oleic, and linoleic acids, which can arrange in varying combinations on the glycerol backbone (refer to Chapter 11 in this volume).

Since there is substantial variability in fatty acid arrangement of the TAG, melting of milk fat occurs over a very wide range of temperatures, often quoted as being −40°C up to 40°C [5]. Thus, there is liquid fat virtually throughout all stages of ice cream making. During pasteurization and homogenization, all fat is liquid because the temperature can reach up to 80°C. However, when the mix has been frozen and hardened at −30°C, milk fat is primarily crystalline with only a small percentage of liquid fat. At intermediate temperatures, the ratio of solid to liquid fat varies between these extremes. The extent of fat in the liquid form is extremely critical during certain stages of ice cream manufacture [6,7].

Fat in homogenized milk naturally exists in globules with a mean diameter of 3–4 μm. The fat is enclosed within the milk fat globule membrane (MFGM), which is produced by the secretory cells in the lumen of the cow. The MFGM is approximately 10 nm thick [8] and is made up of proteins, lipoproteins, enzymes, and cholesterol. The MFGM serves to emulsify the fat in the watery matrix of the milk and protects it from enzymatic degradation. The natural membrane in milk consists of 41% proteins (casein and whey proteins), 30% phospho- and glycolipids, 14% neutral glycerides, 13% water, and 2% cholesterol [9]. In ice cream, the fat globule membrane rearranges after homogenization during the aging step to attain the lowest possible energy state [10,11]. This fat globule membrane is thought to consist of casein micelles and subunits, denatured proteins, bound water, inorganic ions, emulsifiers, and high-melting TAG just below the membrane surface [12]. The arrangement of the fat globule membrane in ice cream mix greatly affects the behavior of the ice cream during freezing.

One way to optimize usefulness of milk fat as an ingredient in food is by fractionating to obtain milk fat with narrower ranges of melting temperatures [13]. Anhydrous milk fat can be separated, using melt fractionation, solvent fractionation, or supercritical fractionation, into various fractions with different melting profiles. Low-melting fractions, with melting points below 20°C, have a higher percent of short-chain and unsaturated fatty acids than anhydrous milk fat [14]. Middle- and high-melting fractions have higher levels of long-chain, saturated fatty acids. Although milk fat fractions have been studied in great depth in butter spreads and chocolate, very few studies have reported the effect of milk fat fractions in ice cream [7,15,16].

Other sources of fat in ice cream potentially include vegetable oils, such as palm kernel oil, palm oil, safflower oil, or coconut oil. Vegetable fats function to enhance structural and textural behavior and to decrease costs incurred from using milk fat. However, these fats do not provide the same dairy flavor as does fat derived from milk sources. This may be explained by the abundance of short-chain fatty acids in milk fat that contribute to its dairy taste [17].

The lipid phase plays a critical role in the organoleptic properties of ice cream, as it imparts a creamy texture and smooth mouthfeel. Additionally, the fatty acid composition in milk fat contributes to the overall dairy flavor of ice cream.

El-Rhaman et al. [15] studied the effect of milk fat fractions on sensory characteristics of ice cream. Ice cream made with regular cream scored an overall higher hedonic acceptance than ice cream made with fractionated milk fat. Ice cream made with low-melting milk fat fractions had the lowest overall acceptance. The correlation of acceptability to peroxide value was negative, indicating that oxidation was a problem, particularly with the lower-melting fractions and anhydrous milk fat. Texturally, the ice creams made with cream or very-high-melting fractions had better acceptability than ice creams made with low-melting fractions or anhydrous milk fat.

Berger and White [8] reported that ice cream made with cottonseed oil had ''cold'' eating properties, meaning that the ice cream did not have the typical rich, smooth, creamy mouthfeel obtained when fat coats the mouth. Sensory and analytical measurements showed more lipid oxidation, off-flavors, and aftertaste in ice creams made with vegetable oils compared to those made with milk fat. This was due to the increased level of unsaturated fatty acids compared to milk fat, which makes the lipid more susceptible to oxidation from the oxygen incorporated during whipping.

I. LIPID CRYSTALLIZATION DURING PROCESSING OF ICE CREAM

Important lipid crystallization phenomena occur during nearly all steps of ice cream manufacture. During mixing, pasteurization, and homogenization, the fat is entirely liquid. Homogenization decreases the average fat globule size, increasing the globule surface area and changing the surface-active components. During cooling and aging of ice cream mix, the liquid fat becomes supercooled and then crystallizes. When mix is frozen, destabilization of the emulsion occurs, with partial degradation and coalescence of the fat globules. As the temperature decreases during freezing and hardening, further crystallization of the lipid occurs. The rates of crystallization and destabilization of the emulsion during freezing are particularly important for controlling the quality of the ice cream.

A. Pasteurization and Homogenization

Pasteurization is carried out at high temperatures to kill pathogenic bacteria and reduce the number of spoilage organisms. Immediately following, or sometimes even during, pasteurization, the mix is homogenized to decrease the fat globule size. It is important that all fat be liquid in this step so that adequate shearing and size reduction of the fat globules take place. Homogenization reduces the fat globule size to less than 2 to 3 μm and forms a more highly dispersed and stable emulsion. If the mix is not homogenized, the air cells and ice crystals will be very large [18] and there is a greater chance that churning of the fat will occur during freezing due to the large fat globule size. Once homogenized, the fat exists as a fine emulsion with a globule size ranging from 0.04 to 3.0 μm with a mean size of 0.45 μm [19].

Since homogenization reduces the average fat globule size, there is an increase in surface area of the fat globule by about four to six times [20]. Therefore, there is not enough of the natural milk fat globule membrane to surround the fat. However, milk proteins (caseins) adsorb onto the surface of the globule, making up a new membrane [19] which is adequate to stabilize the emulsion. When emulsifiers are added prior to homogenization, interactions at the lipid-aqueous interface change and interfacial tension decreases.

When a bulk lipid source, such as anhydrous milk fat or vegetable fat, is used in ice cream mix, homogenization is very important because fat globules must be stabilized by the dairy proteins contained in the added serum solids in the mix. For instance, skim milk, milk powders, or condensed skim milk are often used in mixes, in addition to emulsifiers, to obtain properly stabilized fat globules. Additional emulsifiers are often required in these cases to make good ice cream, even though an adequate emulsion can be formed from just the proteins that are present [20].

B. Cooling and Aging

Immediately following homogenization, the mix is cooled rapidly to about 4°C for aging. The kinetics of fat crystallization during this rapid cooling period are not well known. However, a certain amount of supercooling takes place that leads to crystallization of the lipid phase during aging.

It is well known that aging for 3 hr or more is essential to the overall physical structure of ice cream. Ice cream made from unaged mix has been shown to be thin, cold, and lacking in body [21]. During aging, emulsifiers and proteins adsorb on the surface of the fat globule, stabilizers become hydrated, and lipid crystallization takes place. Figure 1 shows a schematic of the structure of the fat globule/serum interface in ice cream mix before and after aging. The difference in surface coverage is apparent. The fat globule before aging has a rough surface, due to the predominant coverage by proteins. After aging, the surface of the globule is smooth [6], due to the displacement of proteins by the emulsifiers. Electron micrographs showing these differences can be seen in Buchheim [22] or Goff [6].

During aging, emulsifiers and proteins compete for adsorption sites on the surface of the fat globule. Immediately after homogenization, the surface of the fat globules is covered with milk proteins. Over time, emulsifiers in the mix replace proteins at the surface. Goff et al. [23] showed that fat globules in aged ice cream mixes had very little protein at the surface compared to those in aged ice cream mixes made without added emulsifiers. Barfod et al. [10] showed that adsorbed protein decreased during aging to a greater extent when emulsifiers were present than when they were absent. They also showed that interfacial tension of model mix systems with emulsifiers decreased as temperature decreased, whereas mixes without emulsifiers had little to no change in interfacial tension, regardless of changes in temperature. This indicates that at low temperatures, emulsifiers adsorb preferentially onto the surface of the fat globule, displacing the proteins and, thus, reducing the interfacial tension. Numerous studies have shown that added emulsifiers reduce interfacial tension between lipid and aqueous phases at elevated temperatures where the lipid is melted [11,16,24–27]. However, the lipid in a fat globule in aged ice cream mix is partially crystalline and the surface characteristics of this globule are not completely understood.

The amount of liquid fat present when the mix enters the freezer has significant impact on final structure of the ice cream [6,28]. The behavior of the fat in the mix during freezing, including coalescence, aggregation,and dispersion, is affected by the amount of liquid fat within the globules [18].

Since the fat in ice cream mix is in emulsion form, its crystallization behavior is similar to that of other emulsified systems. The state of the emulsion greatly affects the crystallization behavior of the fat. The temperature required for crystallization of an emulsified system is lower than for a nonemulsified system of

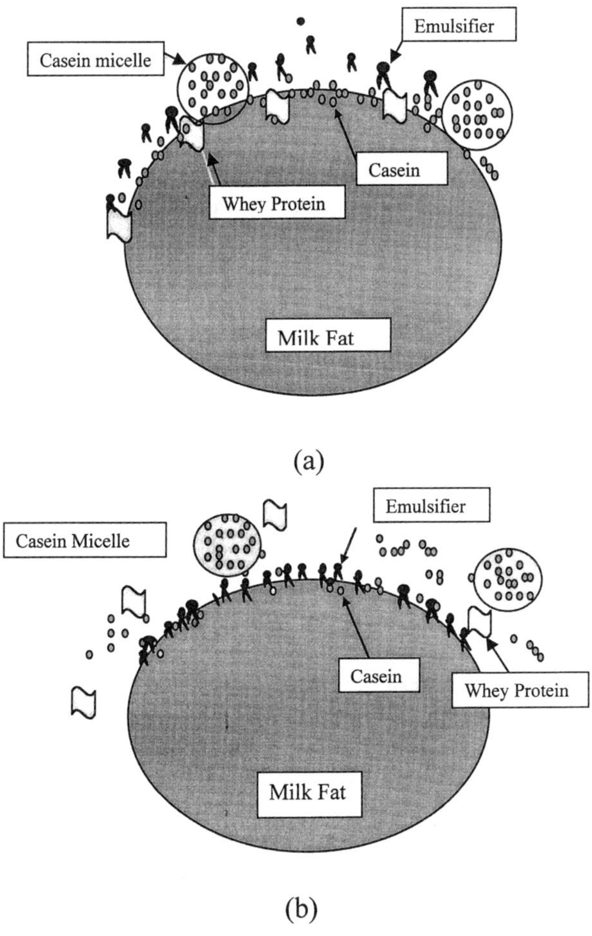

Fig. 1 Schematic of ice cream mix (a) before aging and (b) after aging [16].

the same composition and depends on the type of emulsifier [29]. In bulk fat, there are many impurities that can act as nucleation sites. Skoda and van den Tempel [29] believed that the non-TAG components (MAG/DAG, or products of lipolysis) act as catalytic impurities and, thus, are nucleating sites for crystallization. Walstra and van Beresteyn [30] studied crystallization in milk fat emulsions. In samples where non-TAGs were removed, the proportion of solid to liquid fat was less than in samples that contained MAG and DAG or where rancidity (hydrolysis of the TAG to form fatty acids) was allowed to take place. Thus, they concluded that heterogeneous nucleation takes place at catalytic sites based on the non-TAG components in bulk milk fat.

In homogenized cream, there are approximately 10^8 fat globules per milligram of fat in the dispersed phase [31]. In frozen ice cream, there are about 1.5 $\times 10^{12}$ globules/g, giving a total surface area of 12 m^2 [19]. In order for heterogeneous nucleation to occur, a nucleating site must be present or form in each globule. Due to the high number of globules in an emulsion, there are relatively few nucleating sites present in any single fat globule. Thus, an emulsified fat must undergo substantial supercooling to crystallize. Compared to bulk fat, more time and a lower temperature are needed for the fat to crystallize when in the emulsified state. Barfod et al. [10] observed that the presence of saturated MAG enhanced the rate of crystallization more than unsaturated MAG. McClements et al. [32] showed similar effects of emulsifier type on crystallization in emulsions.

Recently, Groh [27] studied the influence of emulsifier type on crystallization temperature of fat (a blend of coconut and palm fat) at an aqueous interface. The addition of a MAG/DAG blend caused an increase in crystallization temperature as the sample was cooled. In the absence of MAG, the pure fat system supercooled by 13°C prior to crystallization. Upon addition of 1% MAG/DAG, substantially less supercooling was attained prior to crystallization, which indicates that the MAG/DAG promoted lipid crystallization in this case. The nature of the alkyl chain of the MAG/DAG blend and the ratio of MAG to DAG were also found to influence crystallization. Longer-chain fatty acids (stearate versus palmitate) promoted crystallization at lower supercooling values (at higher temperature). Both a higher ratio of saturated to unsaturated fatty acids in the MAG/DAG blend and a higher ratio of MAG to DAG were found to enhance fat crystallization.

Walstra and van Beresteyn [30] also found that the diameter of the fat globule affected the level of supercooling required for crystallization. Finer emulsions resulted in greater supercooling, and the time needed to obtain initial nuclei was proportional to the globule volume. As an emulsion is dispersed into smaller and smaller droplets, the chance of a catalytic impurity being present in each globule decreases. Since one nucleus must be formed for every emulsion droplet, a finer emulsion results in a lower temperature needed to form a nucleus in every droplet. Thus, a larger fat globule size results in more crystalline fat at a given temperature because there is a better chance of an impurity or catalytic site to be present in each fat globule [29].

In milk fat, since there is a wide range of melting points, not all of the fat will crystallize during aging. Figure 2 shows the development of solid fat content (SFC) over time during the first few hours of aging for several ice cream mixes made with anhydrous milk fat (AMF), or milk fat fractions [16]. In this study, mixes were cooled from 71 to 5°C within 10 min, and the development of SFC in the fat phase of the mix was measured with nuclear magnetic resonance (NMR)

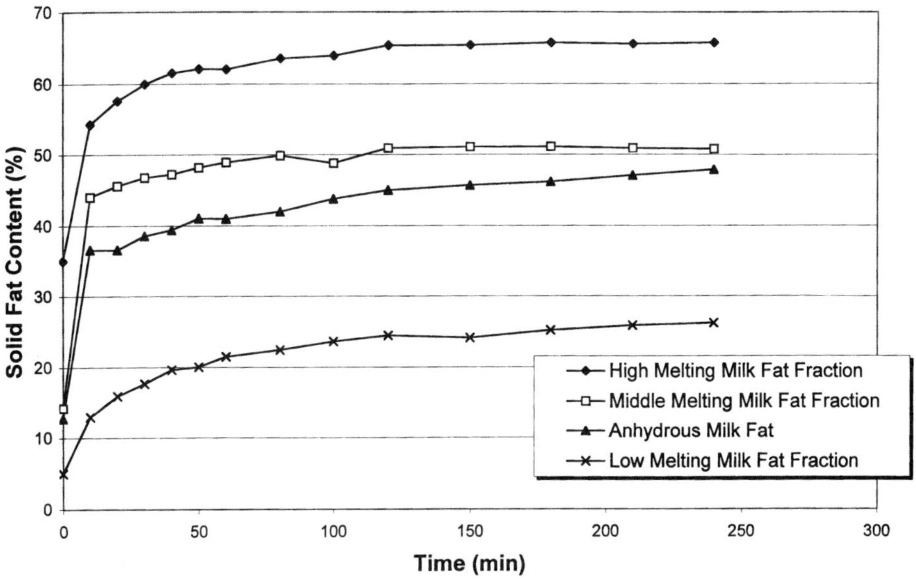

Fig. 2 Effect of fat phase on development of solid fat content during aging at 5°C of ice cream emulsified with mono- and diglycerides added at 0.2% [16].

spectroscopy. Figure 3 shows the development of SFC over time during aging of ice cream mixes made with vegetable oil, using a similar technique [10]. Similar results for ice cream mix made with vegetable oil were shown by Akehurst [33] using differential scanning calorimetry (DSC). Upon cooling, there is an initial rapid increase in SFC over the first several minutes of aging. After approximately 2 hr, the SFC values of mixes made with high-melting (HMF) and middle-melting (MMF) milk fat fractions began to level off. The SFC of mixes made with AMF or low-melting (LMF) milk fat fraction were increasing slowly even after 4 hr of aging. In Figure 4, the SFC profile of ice cream mix after 4 hr of aging is compared to the SFC profile of the same bulk fat crystallizing at 5°C [10]. The fat in the emulsified ice cream system is less solidified than that of the bulk fat, in agreement with the above theories on emulsion crystallization.

The rate of crystallization in the aging process is dependent on many factors, including the type of fat used, dispersion of the emulsion, cooling rate, and aging temperature. Figure 5 shows how the initial crystallization rate for ice cream mixes in a simulated cooling and aging process varied with the type of fat used in the mix [16]. Initial rate was measured as the change in SFC (in %) during the first 10 min of cooling and aging. The initial crystallization rate increased with increased melting point and SFC of the bulk fat at 5°C of the fat

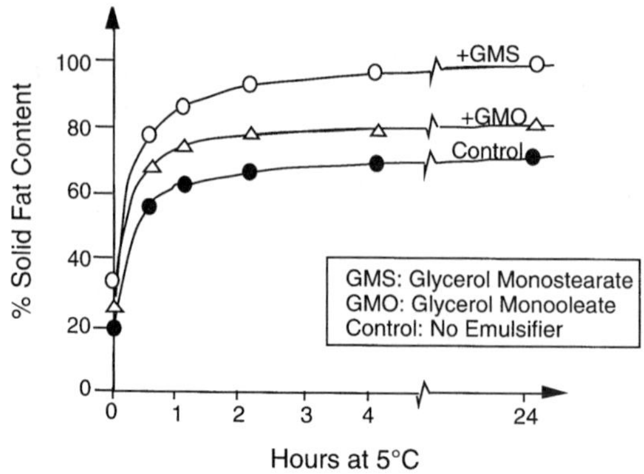

Fig. 3 Development of solid fat during aging at 5°C of ice cream mix made with vegetable oil [10].

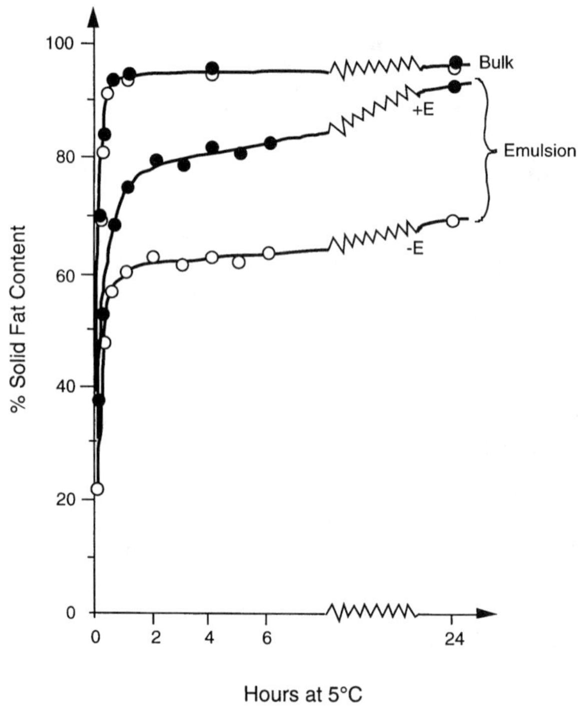

Fig. 4 Development of solid fat during aging at 5°C of ice cream mix made with (+E), and without (−E) emulsifiers, compared to fat in bulk phase [35] (with permission from Elsevier Science).

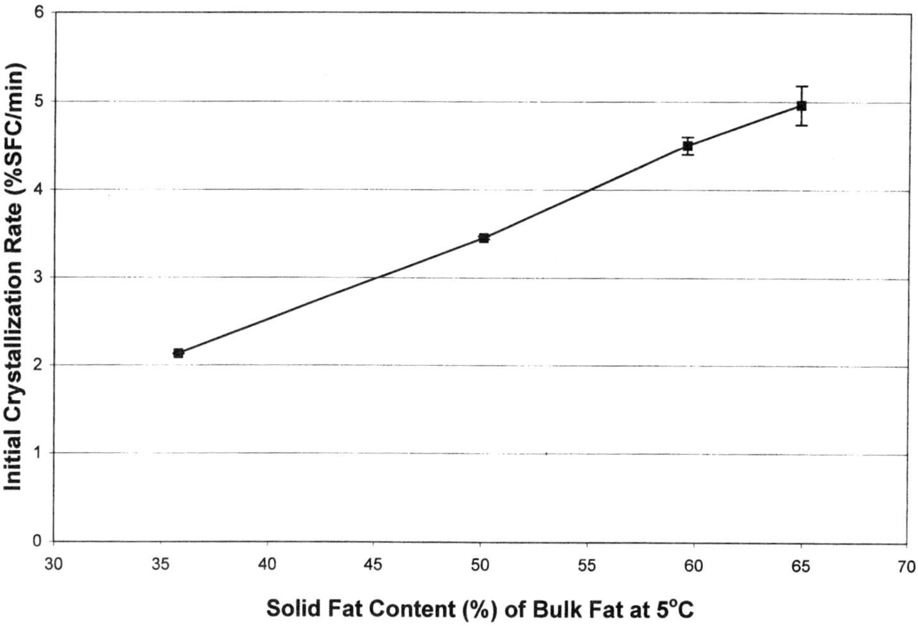

Fig. 5 Initial crystallization rate at 5°C of ice cream mixes made with varying milk fat sources, based on solid fat content of milk fat component in bulk phase at 5°C [16].

source. The four points represent ice cream mixes made with LMF, AMF, MMF, and HMF (in increasing order of SFC at 5°C). These differences in crystallization rate emphasize the effect of the type of fat on the crystallization kinetics in the ice cream mix. Fats that have a higher melting point generally crystallize faster than those with a lower melting point, in accordance with the higher content of long-chain, saturated fatty acids [13].

Emulsifier type and level of addition have also been shown to influence the rate of fat crystallization during aging and the total amount of crystallized fat. Figure 6 shows SFC after 4 hr for mixes made with varying milk fat sources and at two emulsifier (a commercial blend of MAG/DAG and polysorbate 80) levels [16]. Not surprisingly, SFC in the ice cream mix increased with increasing SFC of the fat source. That is, bulk fats with higher SFC resulted in ice cream mixes with higher SFC. The SFC also differed according to emulsifier addition level. Higher emulsifier levels generally resulted in a slightly higher SFC ($p <$ 0.05). This was most likely due to increased emulsifier coverage on the fat globule membrane providing additional nucleating sites for crystallization to take place. Groh [27] found a similar result for MAG. However, Adleman [16] found no

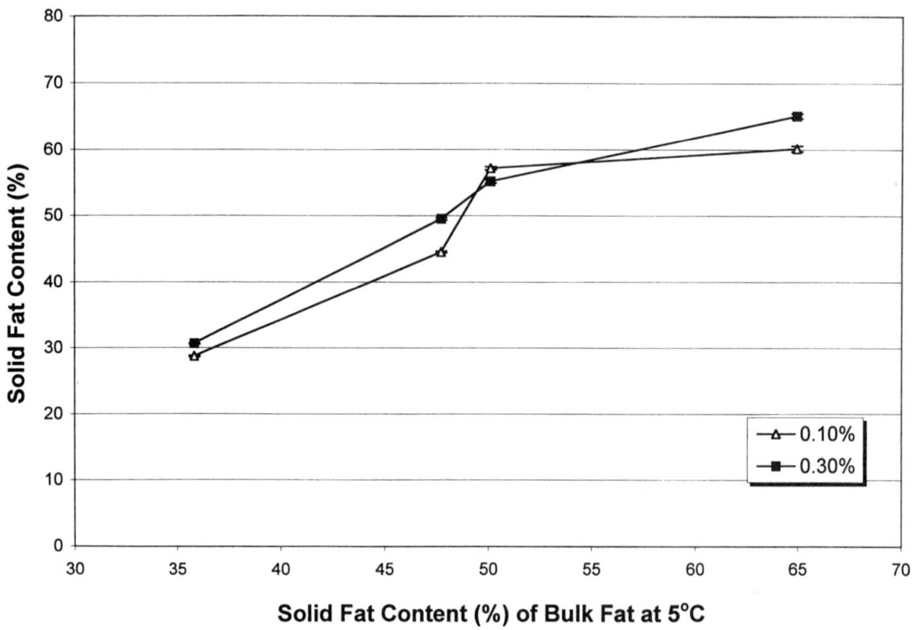

Fig. 6 Solid fat content of ice cream mixes made with different addition levels of a commercial emulsifier blend (80% mono- and diglycerides plus 20% polysorbate 80) after aging at 5°C for 4 hr compared to solid fat content of bulk fat at 5°C [16].

significant differences in SFC after 4 hr of aging between mixes made with three different emulsifiers: a MAG/DAG blend, polysorbate 80, or the commercial blend of MAG/DAG (80%) and polysorbate 80 (20%).

Numerous researchers have found that emulsifiers in ice cream can act as nucleation sites for fat crystallization [10,15,20,27,34]. Addition of emulsifiers was found to cause an increase in SFC of the mix, as seen in Figs. 3 and 4. According to Barfod et al. [10], the effect of emulsifiers on fat crystallization and other phenomena is due to interfacial interactions. The lipophilic chains of the emulsifier protrude into the fat globule membrane, potentially creating a nucleation site for localized crystallization. Berger [21] and Barfod [35] showed that emulsifiers with saturated hydrocarbon chains were better initiators of fat crystallization than those with unsaturated hydrocarbon chains. This can be seen in Fig. 3, where the MAG with stearic acid (GMS) had a faster rate of aging than the MAG with oleic acid (GMO).

Another potentially important phenomenon in the study of lipid crystallization during the aging process is the polymorphism of the fat. Milk fat is stable in the β′ form under most conditions and this is the form found in ice cream made with all milk fat. However, addition of other fats to ice cream may influence the polymorphic form. Berger and White [36] found that ice cream made with butterfat and palm kernel oil initially crystallized in the α form. An unpublished study by Wright and Bell (cited in Ref. 36) found that after 90–100 min of aging at 5°C, a mixture of α and β′ polymorphs was found. In ice cream mix made with all palm kernel oil, the fat was entirely in the β′ polymorph within 40 min. X-ray diffraction studies showed that the α crystals of palm oil were the most stable, butter oil was less stable, and palm kernel oil was least stable and had the shortest life-span [36]. Barfod et al. [10], however, found that no polymorphic changes took place during aging of ice cream made with vegetable fat in the presence or absence of emulsifiers. Apparently, the polymorphic changes of lipids during the manufacture of ice cream are not entirely understood, and further research is needed, particularly in relation to butter fat, to understand the impact of polymorphism of the fat phase on the physical properties of the finished ice cream.

C. Freezing

During the freezing process, incorporation of air cells and crystallization of ice occur. Also, the majority of fat destabilization (also called fat agglomeration) takes place due to the combination of shear forces, freezing of ice, and concentration of the unfrozen phase. During aging, the emulsifiers adsorb to the fat globule, displacing the proteins. This causes a decrease in interfacial tension, which, along with solidified fat near the surface, makes the globule membrane weaker and, thus, more susceptible to breakage. During air incorporation, an interface forms between the air and the serum phase. During freezing and destabilization, fat globules agglomerate and adhere to the air cells as they form.

1. Fat Destabilization

Fat destabilization occurs when the membrane of the fat globule is broken due to the combination of shear force from the mechanical actions of the blades and decreasing temperatures in the freezer. Several attributes of the emulsion are often used to characterize the extent of fat destabilization. One of the earliest methods used to quantify extent of fat destabilization in ice cream is based on the change in turbidity of melted mix compared to the original mix. This method requires that both the mix and the melted ice cream be diluted to such an extent that turbidity can be measured. For the same level of dilution, the turbidity of the melted ice cream is greater than that of the diluted mix, which indicates that some of the fat globules have been disrupted. Another method used to characterize the extent of fat destabilization is the amount of extractable, or "free," fat in the melted ice cream. Based on the extent of disruption, the amount of fat that is easily extracted by solvent increases and this can be used to characterize destabilization. The third method is based on measuring the fat globule size distribution by using a light scattering detector. The fat globules in melted ice cream have a larger mean size than the fat globules in the original mix and the distribution of sizes is bimodal, which indicates that some of the original droplets have agglomerated into larger units. Although each of the methods provides a measure of destabilization, there is often little agreement between them. Each method measures a slightly different aspect of destabilization and there is no agreement as to which is the best, or most accurate, method.

Previous researchers have speculated that liquid fat leaks out of the globule and disperses around the air cell interface [8,18,37], although the mechanism for this is not clear. The amount of easily extractable fat present in frozen ice cream can be significantly higher than that present in ice cream mix. This suggests that whatever happens during destabilization of the fat in the ice cream freezer makes the fat more readily available for solvent to remove. The presence of ''free'' fat at the air cell interface has been suggested by Berger et al. [19] based on the plate-like appearance of fat evident there. Berger [21] speculated that the free (liquid) fat that leaks out of the globules acts as a cementing agent for fat globules and air cells during whipping, thus increasing the whippability of the ice cream. According to Berger et al. [19], free fat and fat globules coat the air cells, providing the structure that is important to whippability and melt-down behavior. However, recent observations with a newly developed scanning electron microscopy technique by Goff et al. [38] strongly suggest that there is no layer of ''free'' fat at the air cell–serum interface. They suggest that it is primarily agglomerated fat globules and a protein layer that serve to stabilize the air cells in ice cream. By inference, this suggests that no liquid fat escapes from the disrupted globules, although the presence of liquid fat within the globule is certainly important in the mechanism of destabilization. The lack of liquid fat in the globule has been shown to lead to limited destabilization [6].

Although the exact nature of the air cell–serum interface remains somewhat in question, it is clear that destabilization of the emulsion during freezing is critical to proper development of the air cell structure in ice cream. If there is too much whipping, the air cells are destroyed, too much fat is destabilized and the fat is churned, forming undesirable butter granules. Higher levels of liquid fat in the globules present at draw, as well as lower freezer temperatures, result in a greater amount of fat destabilization [8]. More destabilized fat results in an ice cream that is stiffer and drier in appearance than an ice cream made with less destabilized fat [39,40]. Additionally, an increase in destabilization reduces the melt-down rate of ice cream [39]. Tharp et al. [40] state that the agglomerated fat provides rigidity to the air cells and allows ice cream to maintain its shape during melt-down.

Since agglomerated fat provides strength to the air cell [40], it is possible that this strength supports the formation of smaller air bubbles and prevents coalescence into larger air bubbles during freezing and whipping. Smaller air bubbles would decrease the space between air bubbles, and the matrix would have greater ability to hold the melted mix in place. The air bubbles would also provide a degree of insulation that would slow the melting of ice crystals.

The nature of fat crystallization within the globule also may influence destabilization. King [41] theorized that a layer of high-melting TAG formed on the inside of the fat globule, next to the membrane, and that layers of crystalline fat extended consecutively into the core of the globule. Using freeze-etching and electron microscopy, Berger and White [36] found a shell of crystalline fat at the globule surface. They also found that liquid fat had leaked out of the fat globules, causing aggregation of several fat globules together. It was speculated that the crystalline shell found in fat globules was not continuous, but contained many overlapping ''stacking faults'' that trapped liquid fat between the crystals. Thus, they believed that the fat crystals actually formed a suspension within the globule. They also suggested that this crystalline shell actually weakened the fat globule membrane, leaving it more susceptible to breakage and, thus, to destabilization. More recently, van Boekel and Walstra [42] have shown that fat crystals at the surface of an emulsion are responsible for coalescence of droplets of partially crystalline fat. Fat crystals emanating from the surface of an emulsified droplet disrupt the globule membrane of a colliding droplet and allow coalescence of the two droplets. Darling and Birkett [43] also consider the mechanical properties of the crystals as an important factor in determining extent of destabilization. More malleable crystals, such as α polymorphs and mixed or compound crystals, will mold with the surface and have less penetration out of the emulsion droplet than harder crystals. They suggest that ''fats with a broad triglyceride composition will tend to produce more stable emulsions, for a given solids content, than fats containing a narrow range of co-crystallizing triglycerides.''

The extent of crystallinity within the fat globule affects the plasticity of the globule and its sensitivity to shear [44]. Figure 7 shows the effect of milk fat fractions on extent of destabilization, as measured by spectrophotometry, during batch freezing [16]. The ice cream made with low-melting milk fat fraction had a higher extent of fat destabilization throughout freezing than the ice cream made with high-melting fraction. Figure 8 shows the extent of destabilization at the end of batch freezing versus the SFC at 5°C of the milk fat fraction in its bulk form. The extent of destabilization decreased with the SFC of the milk fat component used in the ice cream. Apparently, increased SFC in the globule at freezing reduces shear sensitivity of the globule and results in less fat destabilization.

El-Rhaman et al. [15] found that the highest level of destabilization in ice creams made with added emulsifier occurred in samples made with cream.

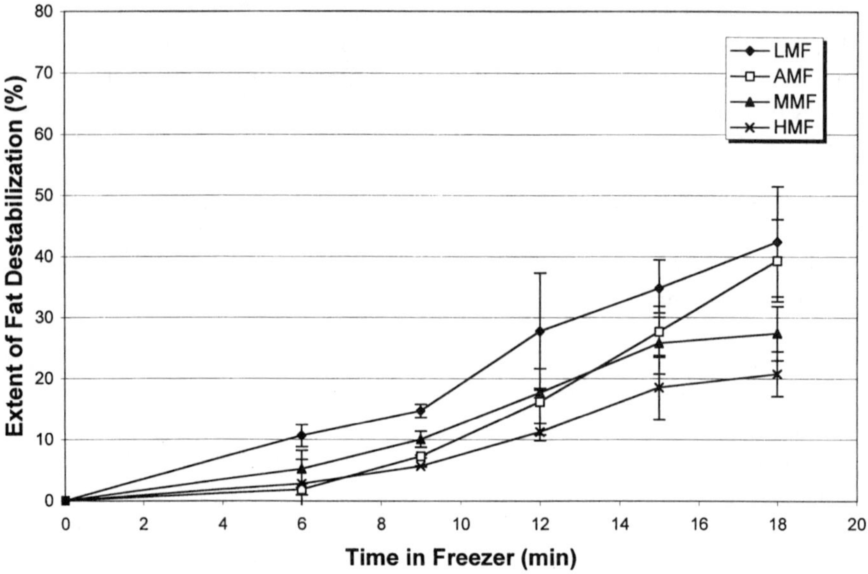

Fig. 7 Extent of fat destabilization during batch freezing of ice cream mixes made with different fats and emulsified (0.2%) with a commercial emulsifier blend (80% mono- and diglycerides plus 20% polysorbate 80). LMF = low-melting milk fat fraction; AMF = anhydrous milk fat; MMF = middle-melting milk fat fraction; HMF = high-melting milk fat fraction [16].

Ice creams made with anhydrous milk fat and low-melting milk fat fraction (with added emulsifier) were not significantly different from each other, but had slightly lower fat destabilization than ice cream made with cream. Ice cream made with very-high-melting milk fat fractions had the lowest extent of destabilization.

Although emulsifiers provide some emulsification properties, the primary role of emulsifiers in ice cream is to destabilize the globule surface [39]. Emulsifiers aggregated on the surface of fat globules have been shown to decrease the interfacial tension of the droplet, which leaves the fat globule more susceptible to rupturing [6]. Therefore, addition of emulsifiers has typically been found to increase the amount of destabilization that occurs during freezing, as clearly shown by Goff et al. [23]. El-Rhaman et al. [15] found that ice creams made without added emulsifier had significantly less destabilization than ice creams made with added emulsifier. In addition, Adleman [16] showed that an increase in emulsifier levels from 0.1 to 0.3% caused an overall increase in destabilized fat, although the extent of this effect was dependent on the type of emulsifier

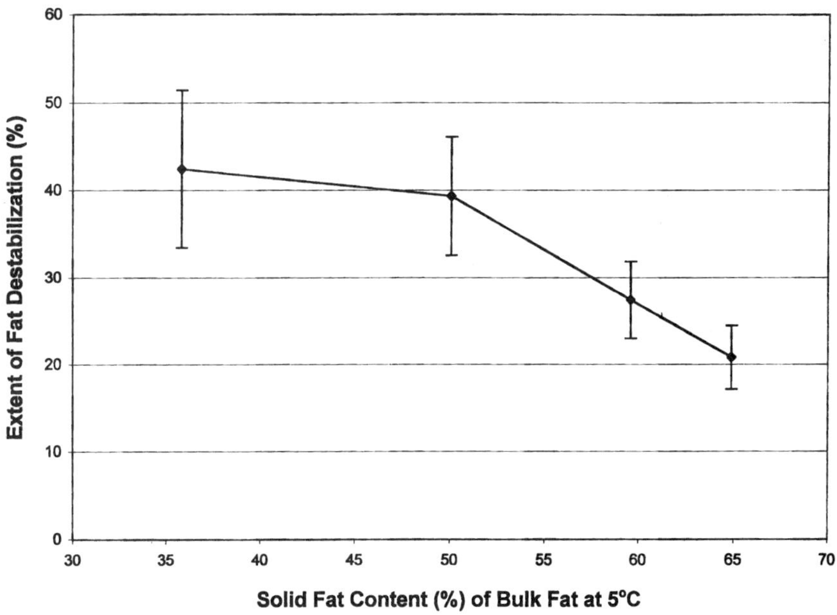

Fig. 8 Extent of fat destabilization in ice cream made with different fats and emulsified (0.2%) with a commercial emulsifier blend (80% mono- and diglycerides plus 20% Polysorbate 80) after 18 min of batch freezing compared to the solid fat content of the bulk fat at 5°C [16].

used and the nature of the fat phase. The effect of polysorbate 80 on fat destabilization in ice cream made with a low-melting milk fat fraction was quite large. At low levels (0.02%) of polysorbate 80, the extent of fat destabilization increased to about 20% at the end of batch freezing (18 min). However, in the presence of 0.06% polysorbate 80, the extent of fat destabilization had increased to over 70% at the end of freezing under the same conditions. The MAG/DAG emulsifier showed the same effect, although the magnitude was much smaller. In addition, the effect of emulsifier level on extent of destabilization was more pronounced for ice creams made with lower-melting fats. The ice cream made with the high-melting milk fat fraction had only low levels of destabilization and these were hardly affected by emulsifier level [16]. However, Groh [27] found that an ice cream mix with high emulsifier level actually had less fat destabilization (measured by laser diffraction to obtain particle size distribution).

At low levels of addition, emulsifiers promote destabilization and provide significant quality enhancement of the frozen ice cream. However, emulsifiers added at too high a level can promote creaming of the emulsion as butter granules form during shearing [6]. Thus, during freezing, a delicate balance is required between the nature of the fat globules (solid fat content, emulsifier type, and addition level) and the shear forces applied in the freezer to yield the desired level of fat destabilization.

2. *Air Incorporation*

Air incorporation, which generally occurs during freezing, is an essential process for high-quality ice cream. Air cells give ice cream its characteristic light texture. Mechanical action produced by the blades inside the freezer provides whipping action to produce air bubbles in the form of a foam. Air incorporation is measured as percent overrun. It is estimated that for an ice cream at 100% overrun, there are 8.3×10^6 air cells per gram of ice cream [39]. A mean diameter of 60 μm has been reported [12,39,45], although more recent data suggest a slightly smaller mean size [27,46]. Reported size distributions range from a smallest diameter of 5 μm [12,45] to largest diameter of 300 μm [12].

The nature of the fat emulsion plays an important role in the incorporation of air during manufacture of ice cream. Ice cream without any fat whips to a much higher overrun than ice cream made with fat [20]. According to Sommer [47], as air is incorporated into a mix during freezing, the distance between air cells decreases. At a point, this distance becomes so small that the air cells break and coalesce. The strength of the air cell walls is weakened by fat because the thinness of the lamellae between the air cells is limited by the presence of fat globules and coalesced fat. The lamellae cannot form as thin of a layer as it could if fat were not present and, thus, the fat suppresses foam formation [12]. Additionally, the presence of fat at the air-mix interface due to destabilization displaces proteins that were at the air interface, and this also suppresses foam formation [12].

The nature of the fat used in ice cream mix may also impact the level and rate of air incorporation. Figure 9 shows the development of overrun during batch processing of ice creams made with varying milk fat sources [16]. The ice cream made with low-melting milk fat fraction (LMF) had an initial lag in overrun development for the first 7 min of freezing. Over the initial 13 min of freezing, the ice creams made with HMF or MMF had higher overrun than those made with LMF or AMF. Overall, the ice cream made with LMF maintained the lowest overrun development throughout the first 13 min of freezing, after which it increased sharply and had a slightly higher overrun than any other ice cream. After 16 min, differences in overrun between samples were small, although there were significant differences in the amount of overrun for ice cream made with LMF versus ice cream made with HMF ($p < 0.05$). Apparently, the availability of liquid fat is important for air cell stabilization, just as it is for fat destabilization.

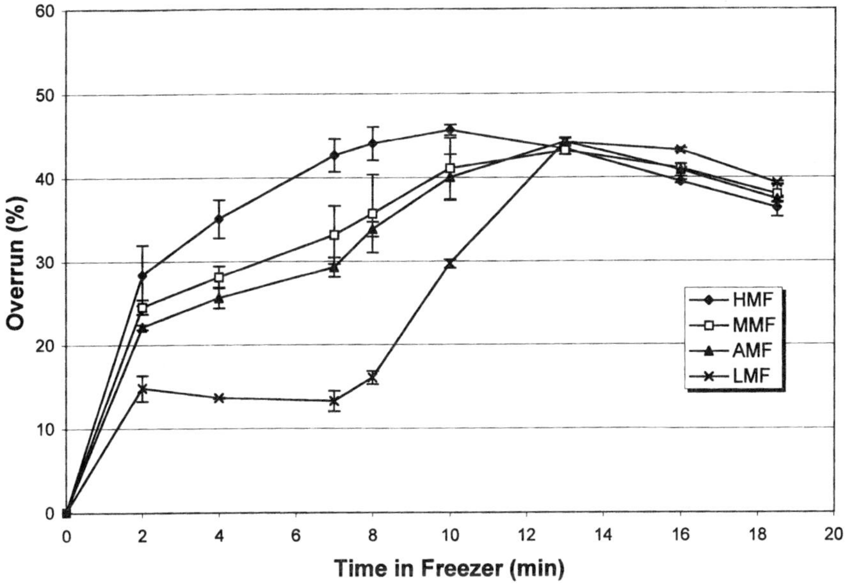

Fig. 9 Development of overrun during batch freezing of ice cream mixes made with different fats and emulsified (0.2%) with a commercial emulsifier blend (80% mono- and diglycerides plus 20% Polysorbate 80). LMF = low-melting milk fat fraction; AMF = anhydrous milk fat; MMF = middle-melting milk fat fraction; HMF = high-melting milk fat fraction [16].

Pilhofer et al. [48] also found that the presence of solid fat (high-melting milk fat fraction) during foam formation in a model system caused increased collapse in air cells, yet provided more overall stability to the foam than a lower-melting milk fat fraction. This was attributed to the crystals in the higher-melting milk fat fraction piercing the air bubbles as they were forming and limiting the extent of air incorporation. The larger, less uniform air bubbles found in the foams made with high-melting fractions were thought to support this theory.

The emulsifier level also influences the rate of air incorporation, although this effect depends on the nature of the fat used. Adleman [16] showed that increased level of emulsifier generally led to slower incorporation of air, perhaps due to the availability of more liquid fat to suppress the foam. However, the maximum amount of air incorporated during freezing was not affected by emulsifier level, although the time at which maximum overrun was reached increased as emulsifier level increased. Interestingly, this effect was greater for the MAG/DAG emulsifier than for polysorbate 80, and greater for ice cream samples made with higher-melting milk fat components.

As air is incorporated into the ice cream during whipping in the freezer, the size of the air bubbles also changes. Figures 10 and 11 show the average air bubble sizes during freezing for ice creams made with low-melting and high-melting milk fat fractions, respectively, as estimated with an optical microscopy technique [16]. Average air bubble size decreased over the first few minutes of freezing, with greater numbers of large air bubbles at the beginning of freezing. After a few minutes in the batch freezer, the mean size of the air bubbles was significantly smaller as large air bubbles were whipped into smaller bubbles. During the later stages of freezing, this study showed that mean size generally began to increase again, potentially due to coalescence of bubbles at longer times. At the beginning of freezing, the air bubble size of ice cream made with HMF was smaller than air bubbles in ice cream made with LMF, although the statistical significance of this result needs further verification. In a subsequent study performed on ice cream made with cream, and not with reconstituted milk fats, bubble size continued to decrease throughout the duration of freezing (unpublished work in the author's lab). Thus, further work is necessary to quantify the change in air bubble size with extent of fat destabilization and air incorporation.

Interestingly, the mean air bubble size generally correlated well with the overrun development in Adleman's [16] study. As overrun increased, mean air

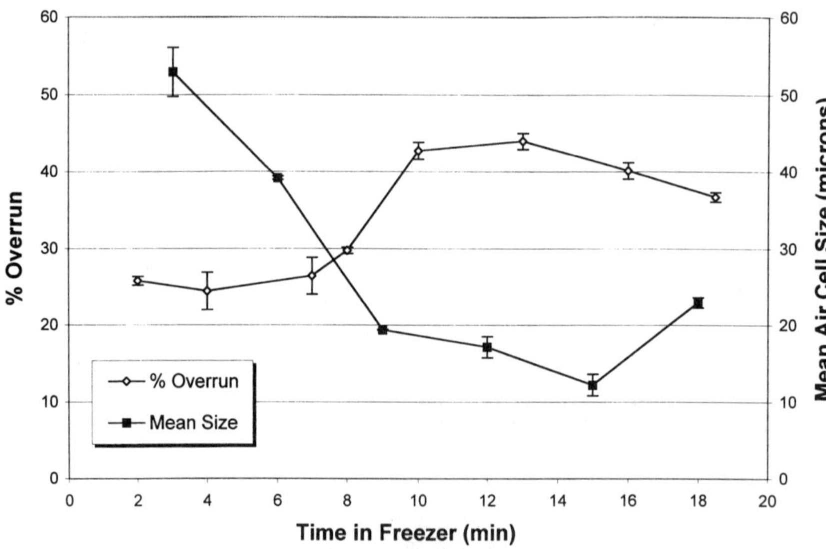

Fig. 10 Development of overrun and air cell size during batch freezing of ice cream mix made with low-melting milk fat fraction and emulsified (0.2%) with a commercial emulsifier blend (80% mono- and diglycerides plus 20% polysorbate 80) [16].

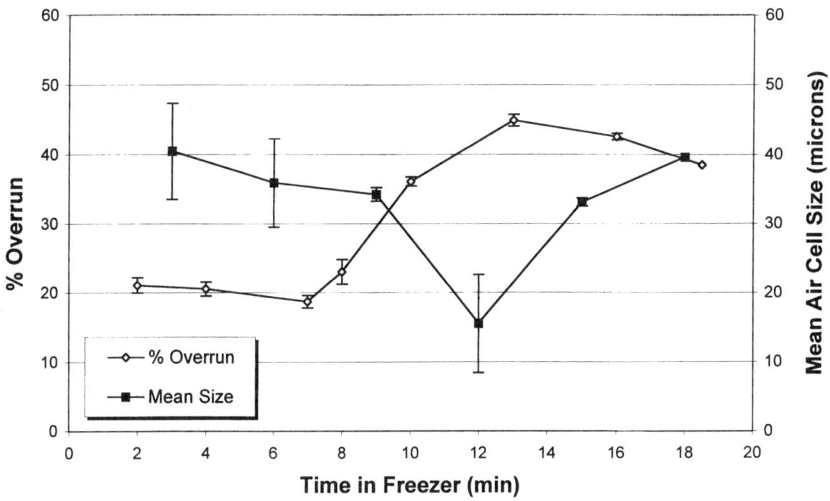

Fig. 11 Development of overrun and air cell size during batch freezing of ice cream mix made with high-melting milk fat fraction and emulsified (0.2%) with a commercial emulsifier blend (80% mono- and diglycerides plus 20% polysorbate 80) [16].

bubble size decreased and the maximum in overrun during freezing generally correlated well with the smallest air bubble size. These results suggest the following process. As air is initially incorporated into ice cream during freezing, the air bubbles continuously get smaller due to shearing in the freezer. However, once the number of air bubbles becomes sufficiently high, coalescence begins to occur due to the continued agitation. This leads to larger air bubble sizes and reduced overrun. Again, more recent work on ice cream made with cream suggests that this trend is not always followed. Further work to verify the effects of the globule interface on air incorporation is necessary.

The nature of the fat phase also influences air incorporation, dependent on the extent of destabilization of the emulsion [16]. The air cells in ice cream made with HMF were unstable and exhibited a rapid increase in size during the later stages of freezing (Fig. 11), perhaps due to the high crystallinity of the fat. Destabilized fat surrounding the air cells, when in the crystalline state, may penetrate the air bubbles and cause them to collapse or coalesce with other air cells. Thus, larger air cells were found in ice cream made with HMF at the end of freezing. Mulder and Walstra [18] found that an increased solid fat content in cream caused a decrease in foam stability during whipping. Additionally, Pilhofer et al. [48] found that hard fractions of milk fat affected formation of foams and their stability, with hard fractions contributing to a faster collapse of the foam than soft fractions.

Vegetable fats also have an effect on fat destabilization during freezing of ice cream, since they have such a low melting point and do not crystallize in the same way as milk fat. Tong et al. [37] studied the effects on overrun and fat destabilization of increasing safflower oil concentration and decreasing milk fat level in ice cream mixes. With increasing safflower oil concentration, they found a depressed overrun development, but little effect on extent of fat destabilization. They postulated that the increasing ratio of safflower oil, a polyunsaturated fat with a very low melting point, caused the fat globule membrane to be more mobile and, thus, less susceptible to breakage at lower temperatures. In the same study, they also found that fat destabilization increased as freezing temperature decreased. This may also be caused by an increase in crystalline fat at lower temperatures, which causes the fat globule to be more susceptible to shearing and destabilization.

D. Draw and Hardening

There are several parameters used to describe ice cream as it is drawn from the freezer [1,40]. Stiffness describes how well the ice cream stands up and how flowable it is. Dryness is related to how glossy (or "wet") the ice cream appears at draw. Different stiffness and dryness levels are necessary depending on the application of the ice cream, such as whether it is being used for a novelty product or to fill bulk containers.

One factor that affects stiffness and dryness is the nature of the fat structure. According to Tharp et al. [40], wetness is directly related to fat destabilization because destabilized fat provides a somewhat rigid structure around the air cell, which contributes to thickness. It also allows for smaller air cells that maintain their integrity throughout the freezing process.

Destabilization also slows the rate of melt-down. The nature of the fat phase that is constructed around and between the air cells during manufacture of ice cream influences the rate at which ice cream melts down when exposed to ambient temperatures. This melt-down characteristic is an important organoleptic property of ice cream [40].

Typically, melt-down rate is measured by placing a standard measure of ice cream on a wire mesh at ambient temperature. A receiving vessel below the sample collects the liquid that melts from the ice cream sample during melt-down. The time for appearance of the first drop as well as the amount of liquid collected over time are measured, and provide a characteristic melt-down profile for ice cream (Fig. 12). Tharp et al. [40] suggests a measure of shape, based on the ratio of height to width, as another indicator of melt-down of ice cream.

Although many compositional factors (water, protein, sugars) influence the rate of melt-down of ice cream, the lipids and emulsifiers play a critical role [40]. The amount and type of fat used, as well as the processing conditions that lead

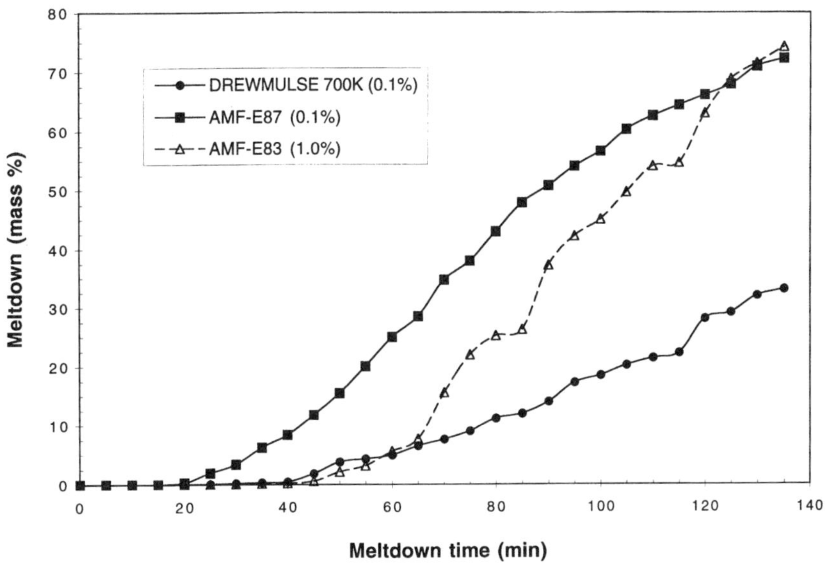

Fig. 12 Melt-down curves for ice creams made with different emulsifiers. Drewmulse 700K is a commercial blend of 80% mono- and diglycerides with 20% polysorbate 80; AMF-E87 and AMF-E83 are mono- and diglyceride blends.

to lipid structure, can all impact ice cream melt-down rate; however, very little quantitative work is available to relate the nature of the lipid phase with the rate of melt-down. Recently, Sakurai et al. [49] studied the effects of different fat types on melt-down of ice cream. Melt resistance of ice creams produced from mix made with different fats under identical processing conditions decreased in the order of cream, vegetable fat, and butter.

El-Rhaman et al. [15] found no significant differences in hardness between ice creams made with cream, anhydrous milk fat (AMF), low-melting milk fat (LMF), or very-high-melting milk fat (VHMF). This was attributed to the fact that the majority of the fat in all of these samples was in crystalline state at hardening temperatures. In those ice creams made with added emulsifier, ice cream made with VHMF had the slowest rate of melt-down. However, there were no significant differences found between melt-down rates of ice creams made with AMF, LMF, or cream. This is likely due to the very high melting point of the VHMF (reported as 44.9°C). El-Rhaman et al. [15] suggested that the VHMF provided additional standup properties at temperatures above the melting point of ice cream than did lower-melting fractions or unfractionated milk fat, since much of the fat in VHMF was still in crystalline form at room temperature. These results agree with other studies that looked at hardened fats in ice cream, such as palm or coconut oil [50].

Tharp et al. [40] showed that melt-down rate is inversely proportional to the extent of fat destabilization. That is, more coalescence of the emulsion during freezing (higher fat agglomeration index) leads to slower melt-down of ice cream. The agglomerated lipid emulsion surrounding the air cells provides a structure that retains water even at temperatures above freezing. These agglomerated fat particles provide strength to the air cell lamellae [51], which may help to stabilize smaller air cells [40]. These smaller air cells may enhance the capillary effect as ice cream melts and retain the melted water within the structure. However, there may be an upper limit of fat agglomeration where the large size of agglomerated fat particles destroys the integrity of the air cells and limits the overrun attainable [24].

El-Rhaman et al. [15] also found that addition of emulsifiers caused slower rates of melt-down than ice cream made without emulsifier, and suggested that this effect was related to the extent of fat destabilization. They speculated that the destabilized fat that coats the air cell also provides structural integrity. The air cell is more likely to stay intact as the ice cream melts due to the destabilized fat, thus giving a slower melt-down rate.

The type and level of added emulsifier added to the ice cream mix also affects the melt-down rate of ice cream. In Fig. 12, melt-down curves are compared for two emulsifiers. The commercial emulsifier (Drewmulse 700K) contains 80% MAG/DAG with 20% polysorbate 80, whereas the AMF-E87 emulsifier contains only an MAG/DAG blend. The polysorbate 80 clearly provides substantial protection against melt-down compared to the MAG/DAG blend. This may be due to differences in the nature of the coalesced fat globules at the air cell interface. Based on SEM and optical microscope images of ice cream structure, the coalesced fat globules are found to coat the air cell surface and provide stability of the air cell structure. However, the exact nature of the fat globule structure at the air interface and into the serum phase of the ice cream is not clear. Leo [52] suggested that the fat globule structure was dependent on the type of emulsifier used in the ice cream mix. He claimed that, with monoglycerides, the coalesced fat and emulsifiers form a ''plate''-type structure around the air cells, whereas with polysorbate the fat and emulsifier formed ''grape''-like clusters between the air cells. Perhaps the intertwined structure formed from the grape-like clusters leads to the enhanced melt-down characteristics of ice cream made with polysorbate 80. Some researchers [10,25,53] have shown that ice creams made with unsaturated MAG had less melt resistance than ice creams made with saturated MAG. Ice creams made with polysorbates had the greatest melt resistance. In Adleman's [16] study, addition of 1% emulsifier (MAG/DAG) to the ice cream mix delayed the onset of melt-down (Fig. 12), after which the rate of melting was increased substantially compared to the commercial emulsifier with polysorbate 80. These results verify that greater levels of fat destabilization generally lead to greater resistance to melt-down of ice cream [54].

II. SUMMARY

Despite recent advances in our understanding of the complex structure of ice cream, there is much to learn about the exact nature of the various structures (air, ice, fat, etc.) and how this structure impacts the organoleptic properties of ice cream. In particular, the lipid phase plays an important role in ice cream structure. The rate and extent of lipid crystallization during processing of ice cream is dependent on the nature of the fat phase. Processing conditions, lipid source, and emulsifier type and level, can all influence the crystallization behavior of ice cream and lead to differences in organoleptic properties, which contribute to its final structure, quality, and eating properties.

REFERENCES

1. W. S. Arbuckle, *Ice Cream*, 3rd ed., Van Nostrand, New York, 1986.
2. R. W. Hartel, in *Phase/State Transitions in Foods* (M. A. Rao and R. W. Hartel, eds.), Marcel Dekker, New York, 1997, pp. 327–368.
3. M. A. Amer, D. B. Kupranycz, and B. E. Baker, *J. Amer. Oil Chem. Soc.*, *62(11)*: 1551–1557 (1985).
4. W. W. Nawar, in *Food Chemistry*, (O. R. Fennema, ed.), Marcel Dekker, New York, 1985, pp. 139–244.
5. N. V. Lovegren, M. S. Gray, and R. O. Feuge, *J. Amer. Oil Chem. Soc.*, *50*:129–131 (1973).
6. H. D. Goff, *Int. Dairy J.*, *77*:363–373 (1997).
7. H. D. Goff, J. W. Sherbon, and W. K. Jordan, *J. Dairy Sci.*, *71* (Suppl. 1):80 (1988).
8. K. G. Berger and G. W. White, *J. Food Technol.*, *6*:285–294 (1971).
9. H. D. Belitz and W. Grosch, *Food Chemistry*, Springer-Verlag, Berlin, 1987, pp. 377–402.
10. N. M. Barfod, N. Krog, G. Larsen, and W. Buchheim, *Fat Sci. Technol.*, *93(1)*:24–29 (1991).
11. J. L. Gelin, L. Poyen, J. L., Courthadon, M. Le Meste, and D. Lorient, *Food Hydrocolloids*, *8*:299–308 (1994).
12. K. G. Berger, G. W. Bullimore, G. W. White, and W. B. Wright, *Dairy Industries*, *37*:419–425 (1972).
13. K. E. Kaylegian and R. C. Lindsay, in *Handbook of Milk Fat Fractionation Technology and Applications*, Amer. Oil Chem. Soc. Press, 1995, p. 525.
14. J. M. DeMan and M. Finoro, *Can. Inst. Food Sci. Technol. J.*, *43(4)*:167–173 (1980).
15. A. M. A. El-Rhaman, S. A. Madkor, F. S. Ibrahim, and A. Kilara, *J. Dairy Sci.*, *80*: 1926–1935 (1997).

16. R. A. Adleman, M. S. thesis, University of Wisconsin-Madison, 1998.
17. J. S. Im, R. T. Marshall, and H. Heymann, *J. Food Sci.*, *69(6)*:1222–1226 (1994).
18. H. Mulder and P. Walstra, *The Milk Fat Globule*, The Universities Press, Belfast, Ireland, 1974.
19. K. G. Berger, G. W. Bullimore, G. W. White, and W. B. Wright, *Dairy Ind.*, *38*: 493–497 (1972).
20. E. L. Thomas, *Food Technol.*, *35(1)*:41–48 (1981).
21. K. G. Berger, in *Food Emulsions* (K. Larsson, ed), Marcel Dekker, New York, 1990, pp. 367–429.
22. W. Buchheim, in *Proceedings of the Penn. State Ice Cream Centennial Conference* (M. Kroger, ed.), Penn. State University, 1992, pp. 281–297.
23. H. D. Goff, M. Liboff, W. K. Jordan, and J. Kinsella, *Food Microstructure*, *6*:193 (1987).
24. R. Govin and J. G. Leeder, *J. Food Sci.*, *36*:718–722 (1971).
25. H. D. Goff and W. K. Jordan, *J. Dairy Sci.*, *72*:18–29 (1989).
26. J. Chen, E. Dickinson, and G. Iverson, *Food Structure*, *12*:135–146 (1993).
27. B. F. Groh, Ph.D. dissertation, Technical University of Munich, Germany, 1998.
28. H. D. Goff, Ph.D. dissertation, Cornell University, Ithaca, NY, 1988.
29. W. Skoda and M. Van den Tempel, *J. Colloid Sci.*, *18*:568–584 (1963).
30. P. Walstra and E. C. H. van Beresteyn, *Neth. Milk Dairy J.*, *29*:35–65 (1975).
31. D. Precht. In *Crystallization and Polymorphism of Fats and Fatty Acids* (N. Garti and K. Sato, eds). Marcel Dekker, New York, 1988, pp. 305–361.
32. D. J. McClements, S. R. Dungan, J. B. German, C. Simoneau, and J. E. Kinsella, *J. Food Sci.*, *58(5)*:1148–1151, 1178 (1993).
33. E. E. Akehurst, *Proc. Soc. Analy. Chem.*, *9*:91 (1972).
34. S. Olsen, *Scand. Dairy Inform.*, *6(45)*:63 (1992).
35. N. M. Barfod, in *Characterization of Foods:Emerging Methods* (A. G. Gaonkar, ed.). Elsevier Science, London, 1995, pp. 59–91.
36. K. G. Berger and G. H. White, *Dairy Ind. Int.*, *41*:236–241, 243 (1976).
37. P. Tong, W. K. Jordan, and G. Houghton, *J. Dairy Sci.*, *67*:779–793 (1984).
38. H. D. Goff, E. Verespej, and A. K. Smith, *Int. Dairy J.*, *9(11)*: 817–829 (1999).
39. R. T. Marshall and W. S. Arbuckle, *Ice Cream*, 5th ed., Chapman and Hall, New York, 1996.
40. B. W. Tharp, B. Forrest, C. Swan, L. Dunning, and M. Hilmoe, Internat. Ice Cream Symp. Athens, Greece, 1997.
41. N. King, *Dairy Ind.*, *10*:1052–1055 (1950).
42. M. A. J. S. Van Boekel and P. Walstra, *Colloids and Surfaces*, *3*:109–118 (1991).
43. D. F. Darling and R. J. Birkett, in *Food Emulsions and Foams* (E. Dickinson, ed.), Royal Society of Chemistry, London, 1987, pp. 1–29.
44. D. Johansson, B. Bergenstahl, and E. Lundgren, *J. Amer. Oil Chem. Soc.*, *72(8)*: 939–950 (1995).
45. K. G. Berger and G. H. White, in *Food Microscopy*, (J. G. Vaughn, ed)., Academic Press, London, 1979, pp. 499–528.
46. K. B. Caldwell and H. D. Goff, *Food Structure*, *11*:1–9 (1992).
47. H. Sommer, *The Theory and Practice of Ice Cream Making*, H. Sommer. Madison, WI, 1951.

48. G. Pilhofer, H. C. Lee, M. J. McCarthy, P. S. Tong, and J. B. German, *J. Dairy Sci. 77*:55–63 (1994).
49. K. Sakurai, S. Kokubo, K. Hakamata, M. Tomita, and S. Yoshida, *Milchwissenschaft. 51(8)*:451 (1996).
50. E. A. Flack, *Ice Cream Frozen Confectionary*, *39(4)*:232 (1988).
51. H. D. Goff, *Modern Dairy*, *673*:15–16 (1988).
52. A. Leo, in *Proceedings of the Penn. State Ice Cream Centennial Conference* (M. Kroger, ed.), Penn. State University, 1992, pp. 309–315.
53. B. M. C. Pelan, K. M. Watts, I. J. Campbell, and A. Lips, in *Food Colloids* (E. Dickinson and B. Bergenstahl, eds.), Royal Society of Chemistry, London, 1997, pp. 55–66.
54. N. Krog, Internat. Ice Cream Symp., Athens, Greece, 1997.

10
Crystallization of Palm Oil and Its Fractions

Kevin W. Smith
Unilever Research Colworth, Sharnbrook, Bedfordshire, England

I. INTRODUCTION

Palm oil is an important raw material obtained from the mesocarp (fruit flesh) of *Elaeis guineensis*. It currently represents about 23% of the world output of the major vegetable oils and fats [1]. The last four decades have seen a dramatic increase in the world production of palm oil, especially from Malaysia, with significant growth occurring in Indonesia and, to a lesser degree, in Ivory Coast. For example, palm oil production in Malaysia has increased more than 100-fold from 92,000 tonnes in 1960 to 10.55 million tonnes in 1999 and now accounts for more than 50% of the world production (Source: PORLA). Figure 1 shows the steady growth in palm oil production over the last 10 years. The growth of palm oil in the world oils and fats market has led to many studies, particularly into its crystallization behavior and, more recently, into its nutritional value.

Palm oil has chemical and physical properties that render it particularly suitable for modification by hydrogenation, interesterification, or fractionation. This chapter will focus on the fractionation of palm oil. There are three principal processes used: with detergent, with solvent, and with neither. Dry fractionation, carried out without solvent, is relatively cheap and widely used to tailor the properties of the oil. Detergent fractionation is a modification of the dry method. Wet fractionation, carried out with solvent, is less widespread than dry, being a more expensive process. However, its cleaner separation of the components of palm oil means that it is necessary to use it where a more precise composition of the fractions is required.

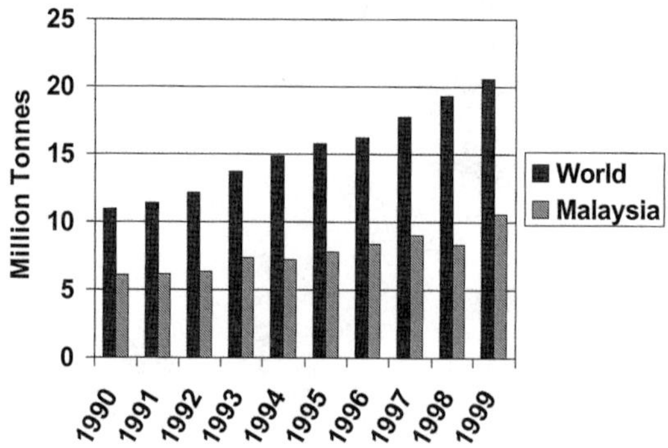

Fig. 1 Palm oil product over the last decade.

Palm oil has a complex physical behavior, as do its fractions, which are used as frying (and salad) oils, margarine hardstocks, and confectionery fats.

II. PHYSICOCHEMICAL PROPERTIES OF PALM OIL

A. Chemical Composition

In common with most other vegetable oils, palm oil is composed of acylglycerols, principally triacylglycerols (TAGs). The specific arrangement of acyl groups within the molecules and the specific proportions of each acylglycerol give rise to the physical properties of palm oil. Other, minor, components modify this behavior.

1. Fatty Acids

In fatty acid terms, palm oil is relatively simple, containing about 44% palmitic (P, hexadecanoic) acid and around 39% of oleic (O, *cis*-octadec-9-enoic) acid [2]. The remaining 16% is composed principally of stearic (St, octadecanoic) acid (5%) and linoleic (L, *cis-cis*-octadec-9,11-enoic) acid (10%). Table 1 shows fatty acid ranges. Palm oil thus has almost equal amounts of saturated and unsaturated acids present. Unlike most other vegetable oils, palm oil has a relatively high proportion of saturated acids in the 2-position of the glycerol (12%).

Free fatty acids present in palm oil are due to enzymatic hydrolysis in the fruit and, later, in the oil. They lie between 3% and 5%. Free fatty acids form eutectics with the TAGs leading to a softer fat [3] and will also influence the crystallization. Fortunately, they are readily removed during refining, yielding oils with around 0.1% of remaining FFA.

Table 1 Typical Fatty Acid Composition Ranges for Palm Oil

Fatty acid	Total oil range (wt %)	2 Position range (wt %)
12:0	nd–0.2[a]	tr–0.1
14:0	0.8–1.3	0.4–0.7
16:0	42.3–46.3	12.2–15.1
16:1	tr–0.3	tr
18:0	4.0–6.3	1.0–1.8
18:1	36.7–40.8	62.1–62.7
18:2	9.3–11.9	19.6–22.8
18:3	0.1–0.4	tr–0.4
20:0	0.1–0.4	tr–0.2

[a] nd = not detected; tr = trace (<0.05%).
Source: Refs. 2 and 4.

2. *Triacylglycerols*

Each TAG molecule incorporates three fatty acids. Hence, even with the relatively few fatty acids types present, there are an enormous number of possible TAGs. Nevertheless, only a few TAGs make up the majority of palm oil [2,4]. Table 2 summarizes the TAG composition. The main TAGs present are 1,3-dipalmitoyl-2-oleoylglycerol (POP, 22%) and *rac*-1-palmitoyl-2,3-dioleoylglycerol (POO, 22%). Other important TAGs are *rac*-1,2-dipalmitoyl-3-oleoylglycerol (PPO, 5%), tripalmitoylglycerol (PPP, 5%), *rac*-1-palmitoyl-2-oleoyl-3-stearoylglycerol (POSt, 5%), 1,3-dipalmitoyl-2-linoleoylglycerol (PLP, 7%), trioleoylglycerol (OOO, 5%), *rac*-1-palmitoyl-2-linoleoyl-3-oleoylglycerol (PLO, 7%), and *rac*-1-palmitoyl-2-oleoyl-3-linoleoylglycerol (POL, 3%). Other triacyglycerols are present at less than 3%.

Table 2 Triacylglycerol Composition of Palm Oil

TAG	(wt%)	TAG	(wt%)	TAG	(wt%)	TAG	(wt%)	TAG	(wt%)
All 0 db	7.4	All 1 db	36.8	All 2db	34.0	All 3 db	16.1	All >3 db	5.6
PPP	5.1	MOP	0.9	POO	20.3	OOO	4.4	LOO	1.8
PPSt	1.2	POP	23.7	StOO	2.4	POL	4.1	OLO	1.2
PStP	0.3	POSt	5.7	OPO	1.0	PLO	5.6		
		PPO	4.4	PLP	6.5				
		PStO	0.2	StLP	1.6				

[a] M = myristic; P = palmitic; St = stearic; O = oleic; L = linoleic; db = double bonds.

Table 3 Typical Triacylglycerol Composition of Palm Oil by Argentation HPLC

TAG class[a]	Weight %
SSS	8.8
SOS	30.9
SSO	7.1
SLS	8.8
SSL	2.2
SOO	22.2
OSO	2.1
SLO	10.0
OOO	3.5
>3 db	4.5

[a] S = saturated; O = oleic; L = linoleic; db = double bond.

The TAG composition can be viewed also in terms of the unsaturation of the molecules. Argentation HPLC can analyze fats in this way [5]. Table 3 shows the composition of palm oil analyzed by this method. Harder palm oils have higher levels of trisaturated (SSS), 1,3-disaturated-2-monooleoyl (SOS), and *rac*-1,2-disaturated-3-monooleoyl (SSO) TAGs. Conversely, softer oils will be richer in 1,3-disaturated-2-monolinoleoyl (SLS) and *rac*-1-monosaturated-2,3-monooleoyl (SOO) TAGs. These five groups of TAGs play a very important role in fractionation. Due to the difference in melting point between these groups (greater degree of saturation leading to a higher melting point), it might be expected that palm oil can be readily separated according to these classes. In practice, however, fractionation is generally into two fractions. For confectionery application, a greater degree of separation is required and three fractions are taken. Of course, depending on the specific conditions, palm oil can be separated into many fractions but these will not necessarily show clear separations between the TAG classes. Understanding the composition of palm oil is important since the crystallization characteristics greatly depend upon its TAGs and their specific interactions.

3. Minor Components

In its raw, unrefined state palm oil contains significant amounts of partial acylglycerols as well as FFA. Since significant FFA levels are found, similar levels of diacylglycerols (DAGs) would be expected. Levels of DAG, however, are somewhat higher than would be expected and lie between 5% and 8%. They are not present solely due to deterioration of the fruit and oil postharvest (although many are due to this); some exist as intermediates to TAG synthesis in vivo. Monoacylglycerols (MAGs) are generally present at less than 1% in the crude oil and even lower in refined oil. Up to 8% DAG can remain in the refined oil, although typical values are about 4%. The main DAGs present are dipalmitoylglycerol (PP), palmitoyloleoylglycerol (PO), and dioleoylglcyerol (OO) [6]. Minor components such as DAG can influence the crystallization behavior of the oil, as will be seen, and knowledge of their influence is important in palm oil fractionation. Setting the levels of such minor components is an important part of the palm oil specification [7].

Other minor components exist in palm oil at much lower levels. Among these are the carotenoids, tocopherols, and tocotrienols, which are antioxidants. Palm oil also contains sterols and phospholipids. All of these minor components together add up to less than 1% in refined oil [2]. The main phospholipids present in palm oil (20–2000 ppm) are, in order of level, phosphatidyl ethanolamines, phosphatidyl inositols, and phosphatidyl glycerols [2]. Other phospholipids amount to less than 10% of total phospholipids. Most of these components are removed during refining.

B. Physical Characteristics

1. *Melting*

The mixture of TAGs in palm oil gives rise to a melting range rather than a single melting point. Some of the TAGs have high melting points (e.g., PPP at 69°C), others have low melting points (e.g., OOO at 4°C), and still others melt at intermediate temperatures (e.g., POP at 38°C). Together in the palm oil they give rise to a melting profile in which the amount of solid fat decreases as the temperature increases. Palm oil is semisolid at room temperature and has a final melting point between 35 and 45°C. The amount of solid at any given temperature can be measured using wide-line nuclear magnetic resonance (NMR) spectrometry. An example of the solid profile obtained is given in Table 4. Palm oil has a broad melting range compared to many other seed oils, and it is this property that makes it an ideal target for fractionation.

2. *Crystallization*

The mixture of TAGs means that palm oil does not crystallize at a single temperature. Instead, increasing levels of solid fat will crystallize as the temperature is reduced further below the final melting point of the palm oil. Differential scanning calorimetry (DSC) can be a useful tool for studying crystallization (as well as melting). Figure 2 shows a typical thermogram of palm oil crystallization and melting. Despite the compositional complexity, only two groups of peaks are

Table 4 Typical Melting Profile for Palm Oil

Temperature (°C)	Solid content[a] (%)
10	46
15	43
20	33
25	21
30	13
35	9
40	4
45	0

[a] Solid content measured by pulse NMR with Bruker P20i Minispec.

evident, illustrating the easy separation of two fractions, one solid and one liquid, from the initial oil [8]. By progressive removal of the high-melting solid, it has been demonstrated conclusively that these two peaks indeed represent the distinct solid and liquid fractions [9]. At its most fundamental, the aim of fractionation is to separate these two phases. Thus the fractionation process involves the crystallization of a solid phase under defined conditions and separation of this solid phase from the remaining liquid phase.

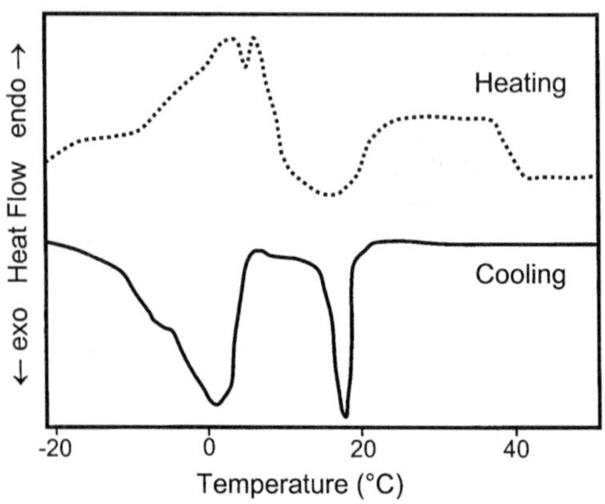

Fig. 2 Differential scanning calorimetry (DSC) thermograms of crystallization and melting of palm oil. Measured using Perkin Elmer DSC7. Oil melted for 2 min at 100°C prior to rapid cooling to 60°C then scanning at 5°C/min to −30°C and reheating at the same rate.

3. Polymorphism

Traditionally TAGs are considered to have three basic crystal forms or polymorphs. In order of increasing melting point and thermodynamic stability these are named α, β′, and β. In addition, a further form, lower in stability than α, is found in many TAGs and is called sub-α (or β'_2). Palm oil exhibits all four of these basic polymorphic forms [10,11], although for some time it was thought that the most stable form was β′. However, this is not too surprising since palm oil is relatively stable in the β′ form.

Classification of the crystal structures is on the basis of peaks in the powder x-ray diffraction pattern. The lateral spacing between the acyl chains (short spacing) gives rise to characteristic peaks. Figure 3 is a view along the acyl chains showing the differences in short spacings of the different polymorphs. In some fats, both β′ and β may exhibit subforms having similar chain packing but different thermodynamic stability and melting point. In such cases a subscript is used, which increases with decreasing stability; i.e., β'_1 is the most stable and β'_2 is less stable. There may be differences in the layer spacing (long spacing) also. TAGs can pack in double- or triple-chain-length arrangements as illustrated in Fig. 4. The long spacing is indicated by a suffix. Thus, for example, palm oil is relatively stable in the β'_1-2 polymorph.

Very rapid cooling of the liquid gives rise to the sub-α form (=β'_2), which transforms into the α form on heating. A slower cooling yields the α form, which may also transform into the β′ form on heating. Cooling to a temperature above the α melting point forms the β′ polymorph. The β form is obtained either by heating and transforming the β′ form or by slow cooling. Under the right condi-

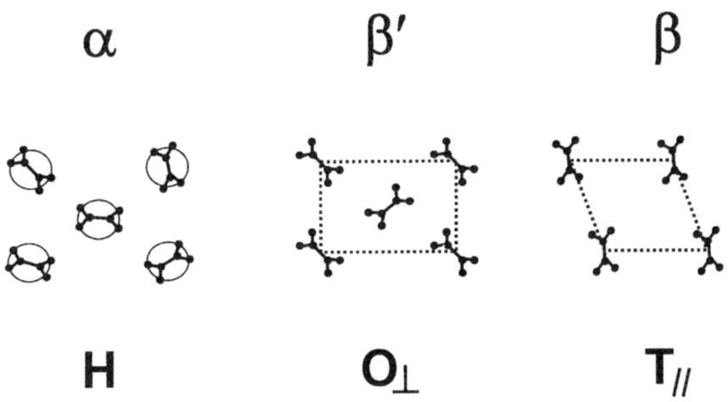

Fig. 3 Acyl chain packing (short spacings) in the principal triacylglycerol polymorphic forms. Viewed looking onto the ends of the chains. In the α form the zigzag of the acyl chains are randomly oriented and the chains are packed in a hexagonal arrangement. In the β′ polymorph the zigzags of the acyl chains are perpendicular and packed orthorhombically. In the β form the zigzags are parallel with each other and the chains are packed triclinically.

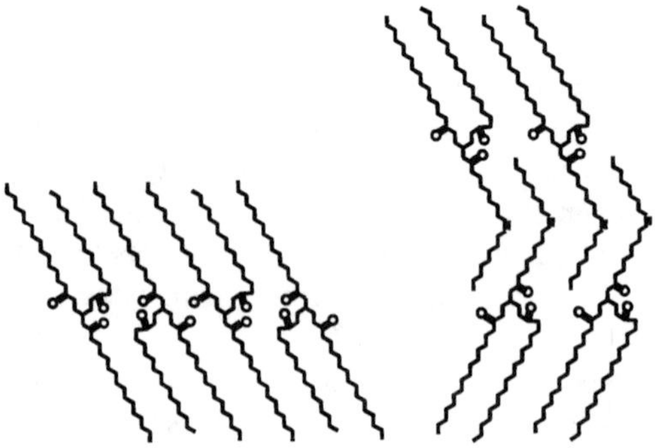

Fig. 4 Schematic representation of layer packing (long spacing) in triacylglycerols. (a) Example of double-chain-length packing. (b) Example of triple-chain-length packing.

tions, the polymorphs undergo irreversible transformation from the less stable polymorphs into the more stable, i.e., sub-$\alpha > \alpha > \beta' > \beta$.

4. Phase Behavior

The specific crystallization, melting, and polymorphic behavior of palm oil is due to the phase behavior of its constituent TAGs. Each TAG may have specific interactions with other TAGs. At the very least a TAG in the liquid state may act as a solvent for an otherwise solid TAG. The binary phase diagrams of some of the principal palm oil TAGs show eutectics and peritectics. POP with PPP shows monotectic behavior with partial solid solution. This complexity is reflected in the characteristics of palm oil itself.

As noted previously, palm oil is rich in POP, POO, and PPP. This leads to the existence of a eutectic mixture having two solid solution phases. The higher melting phase is mainly POP and PPP while the lower-melting phase consists principally of POO. Both PPP and POP are stable in the β polymorph, but, due to specific interactions, it is the β'_1-2 polymorph which is generally stable under the fractionation conditions when crystallized from the melt. However, from a solvent it is possible to obtain the β-3 crystal form, which is the most stable polymorph of POP. In the solvent-free system neither the stable polymorph of POP (β-3) nor the stable polymorph of PPP (β-2) is usually formed, although careful control of the process can lead to β crystallization.

PPO forms a compound with POP which has a double-chain-length structure [12,13] in both β' and β forms. This factor may inhibit the formation of the stable form of POP, which has a triple-chain-length structure. In addition, the

PPP does not crystallize alone but forms a solid solution with POP in which it does not appear to form its stable crystal.

III. FRACTIONATION PROCESSES

Fractionation, or fractional crystallization, is a thermomechanical process that separates an oil into two fractions, each with a specific melting behavior caused by modification of the TAG composition. Fractionation is thus a process producing at least two fractions, unlike other fat modification processes such as interesterification and hydrogenation. Fractionation is based on the melting points of the individual TAGs and on the particular solid phases that they form. The complex mixture of TAGs in palm oil, its polymorphism and the influence of the conditions under which crystallization occurs mean that many patents have been obtained for ever better separations or applications.

The principal goals of fractionation are to generate a liquid oil with improved properties (e.g., cloudpoint) or to produce a fraction with a narrower composition and melting behavior. The former may find use as a frying or a salad oil while the latter, depending on its composition, may be used as a margarine hardstock or in confectionery fats.

A number of authors have reviewed the area of fractionation in general and palm oil fractionation in particular and are worth reading for further information [14–24].

There are three principal techniques applied to crystallize fractions of oils and fats including palm oil: dry fractionation, detergent fractionation, and wet fractionation. In the first process, crystallization occurs from the melt while in the last it occurs from a solvent. Detergent fractionation is a modification of the dry method in which an aqueous detergent solution is utilized to separate the solid and liquid phases.

A. Dry Fractionation

Dry fractionation is the crystallization of fractions of the oil from the melt and is a technique commonly applied to a number of different fats. The fractionation is often a batch process but it may be carried out semicontinuously. The oil is generally fully refined although crude palm oil can be fractionated. First the oil is heated to a temperature that is sufficiently high, usually 70°C, to completely melt all solid fat and remove all nuclei. This ensures that the fractionation is controlled and reproducible. A number of slightly different techniques are applied, according to the specific equipment utilized. The largest difference between crystallization techniques is whether the cooling is fast or slow. The largest producers of equipment employing fast cooling are CMB and Extraction de Smet

[25]. Tirtiaux use a slow crystallization [14], which appears to offer better separation of the solid (stearin) and liquid (olein) phases. Typically the palm oil will be cooled, using water (although other media can be used), to the desired temperature, usually in the region of 15–20°C. The oil is stirred or agitated during this cooling phase and nucleation and crystal growth begin. Once the desired fractionation temperature is reached, stirring may be discontinued and the oil is held until sufficient crystallization has occurred.

The separation of the stearin from the olein may be effected in several ways. Popular filtration systems currently used are Florentine filters (stainless steel belt), rotary drum filters, and membrane filters. The separation stage is an important part of the process. For a number of reasons, a significant amount of liquid phase can be held up in the stearin. This reduces the yield of olein and influences the properties of the recovered stearin. Clearly, the greater the amount of olein that remains in the filter cake, the softer (i.e., lower solid fat at a given temperature) will be the resulting stearin.

In an ideal world it would be possible to completely separate the stearin from the olein. However, separation turns out to be one of the most difficult parts of the process. The structure of the crystals of solid fat is such that olein can be incorporated into gaps within them. In addition, in the filter cake, gaps between the crystals themselves can occlude olein [26]. Dry fractionation followed by rotary drum or Florentine filters will lead to a higher amount of olein in the filter cake compared to a membrane filter. The latter allows high pressures (up to 50 bar [27]) to be applied to the filter cake effectively squeezing out more of the olein (reducing olein levels in the filter cake from about 60% to 40–45% or even less) [15,28]. Removing more or less olein from the cake permits solid fractions of varying hardnesses to be obtained.

B. Detergent Fractionation

In a variation on the dry fractionation process, detergent can be added to aid in the separation of the stearin and olein. This is the Lanza or Alfa-Laval Lipofrac system. The initial oil is often crude rather than refined. As in dry fractionation the oil is fully melted before being cooled to the fractionation temperature (again about 20°C) in the presence of detergent and magnesium sulfate. Once crystallization has occurred, an aqueous detergent solution (0.5% sodium lauryl sulfate with magnesium or sodium sulfate) is added. This wets the fat crystals, allowing them to become suspended in the aqueous phase. Centrifugation separates the olein phase from the aqueous phase containing the stearin. The aqueous phase, together with the suspended solid fat, is heated almost to boiling point, which melts the solid fat and breaks the emulsion. This is then centrifuged to recover the melted stearin. Both the stearin and the olein phase are washed with hot water to remove the detergent and then dried under vacuum. This gives an improved separation compared to the purely dry process (leaving 35–50% olein in the stearin) except when dry fractionation is combined with a membrane filter press. This combination has rendered the detergent process almost obsolete since it reduces problems of effluent, does not involve additives, and is a lower-cost process.

C. Wet Fractionation

Wet fractionation, crystallization from a solvent, allows a cleaner separation of the stearin from the olein. This means that it possible to obtain a ‘‘purer’’ solid fraction and an increased yield of olein (see, for example, Ref. 29). Wet fractionation is much more expensive than dry or detergent fractionation. Solvent recovery systems must be utilized and fractionation temperatures are much lower (requiring deeper cooling). Due to the high costs, relative to dry and detergent processes, the wet fractionation technique is usually reserved for high-value applications, such as confectionery fats, like cocoa butter equivalents. Solvents commonly used are hexane and acetone but other solvents, such as isopropyl alcohol (IPA), have been proposed [30,31]. Hexane is used where a high-quality olein is desired since DAGs, which appear to cause clouding [32,33], will be removed in the stearin [34]. Acetone is preferred when the goal is to produce a POP-rich fraction, when a clear separation is required, and low DAGs are wanted in the stearin. The (refined) oil is dissolved in the solvent and the resulting miscella is cooled, usually in scraped-surface heat exchangers. Since temperatures are much lower than in dry fractionation, water alone is not a suitable cooling fluid and brine or ammonia may be used. Once the stearin has crystallized, the slurry is filtered under vacuum. Clean, chilled solvent may be used to wash the filter cake to remove entrained olein. The solvent is removed from both fractions by distillation. The separation of the solid and liquid phases is much better than in dry fractionation, in part because any entrained liquid phase in the filter cake is not pure liquid olein but a solution of olein in solvent. Stearin crystallized from solvent is in the most stable, β, form.

The fractionation process is somewhat different when IPA is used as the solvent. As the temperature is lowered, there is a demixing between the oil and IPA [30,31]. Lowering the temperature further leads to crystallization of the stearin, which becomes suspended in the IPA. Since the density of the crystals is reduced by the IPA, separation of the solid and liquid phases can be carried out by decantation. Similarly, a solvent system can be added to the partly crystallized oil such that it is intermediate in density between solid and liquid, effecting a separation between the two fractions [35]. This type of technique comes very close to the detergent separation in concept and operation.

D. Multiple Fractionation

In certain situations, particularly the preparation of confectionery fats, a middle-melting fraction is desired. For cocoa butter equivalents a palm mid fraction

(PMF) is desired, which is readily produced by wet fractionation. A two-stage fractionation can be carried out in two ways: (1) either a stearin enriched with respect to the trisaturated TAGs (principally PPP) is removed and the olein is refractionated to yield a second stearin, called palm mid fraction, rich in disaturated-monounsaturated TAGs (mainly POP); or (2) a stearin enriched with respect to both trisaturated TAGs (principally PPP) and disaturated-monounsaturated TAGs (mainly POP) is removed and refractionated to yield the palm mid fraction as the second stage olein. Note, in the first case the mid fraction is the second-stage stearin, while in the second case it is the second-stage olein. Figure 5 illustrates this. The required composition for palm mid fraction is relatively narrow and requires very careful control of both fractionation stages. The value of palm mid fraction lies in its sharp melting point (Fig. 6).

Clearly, there is no need to be restricted to a two-stage fractionation. Multiple fractionations can lead to purer fractions and are useful with dry fractionation to produce a high-quality palm mid fraction as well as high-IV (iodine value), low-cloudpoint oleins [16,36].

Naturally, when more than one fractionation are carried out it is not necessary that all stages are carried out with the same technique. Berger and Tan [37]

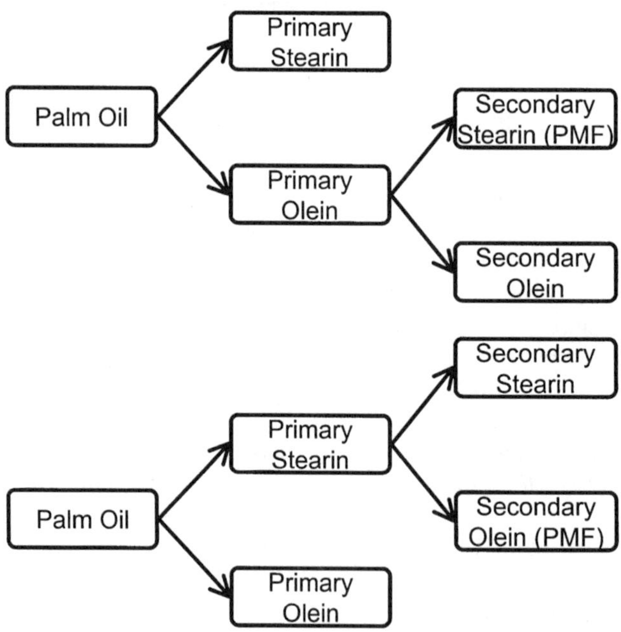

Fig. 5 Schematic of alternative two-stage fractionations to produce palm mid fraction.

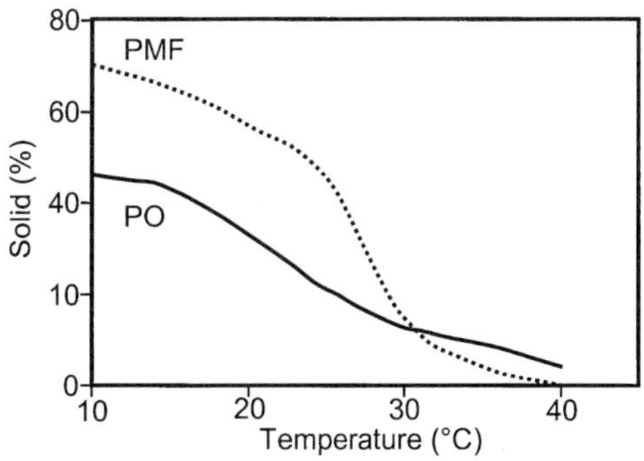

Fig. 6 Melting profile of a palm mid fraction. Solid content measured by pulse NMR with Bruker P20i Minispec.

have investigated a combined fractionation method in which a stearin and olein produced by detergent fractionation were each refractionated in *n*-hexane to yield four fractions.

E. Separation Efficiency

The separation of solid and liquid is dependent on a number of factors. The crystal size distribution, the polymorphic form, and the crystal shape significantly affect the efficiency of the separation. This separation efficiency impacts both the quality of the stearin and the yield of olein. The size, shape, and polymorphism of the solid arise from the specific conditions under which crystallization takes place, i.e., the particular fractionation conditions applied.

During the crystallization process, the stearin crystals generally grow as spherulites. Depending on the density of these spherulites there will be incorporation of more or less olein. In stirred systems, especially, the growing crystals can agglomerate, trapping olein within the particles as they do so. This trapped olein can be virtually impossible to remove by any separation method. However, olein is also trapped between the agglomerates during filtration. This olein is much easier to remove by using high pressure or by washing (e.g., with chilled solvent in wet fractionation).

Thus the separation efficiency of the solid and liquid depends on the amount of olein that is trapped between the solid particles *and* within the particles. The relative proportion of these two pools of olein will vary depending on the fractionation conditions (dry or wet, stirred or unstirred, etc.). Bemer and Smits [26] show that in a system with a scraped-surface crystallizer agglomerates will not form. This means that very little olein is trapped within the particles. The disadvantage is that the particles are small and efficient filtration is difficult. If coaxial cylinders are used to agitate the slurry via Taylor eddies, agglomerates will form in which up to half of the trapped olein is within the particles. However, the advantage of the agglomerates is that they are larger and lead to an easier removal of the interparticle olein. Thus, if the intraparticle olein can be kept to a minimum, the overall separation efficiency can be improved.

IV. FACTORS AFFECTING CRYSTALLIZATION

The kinetics of fat crystallization has been likened, in part, to the aggregation and flocculation of colloids. Berg and Brimberg [38] develop equations and rate constants that appear to relate to the equilibrium solid content at a given temperature.

A. Phase Behavior and Polymorphism

The crystallization of palm oil follows a sigmoidal curve of increasing solid at a temperature of 25°C [39]. Crystallization at lower temperatures is also sigmoidal but shows a second sigmoidal increase in solid, i.e., crystallization occurs in two stages. This is confirmed with isothermal differential scanning calorimetry.

Measurement of crystallization of palm oil by DSC shows two clear peak areas, despite the many TAGs present. The lower temperature peak (around 5°C) does not vary much with composition but the high-temperature peak (around 15°C) does alter. The changes in this peak have been attributed to PPP [40], higher levels of PPP leading to a larger sharper peak at a higher temperature. Addition of SSS TAGs to a cocoa butter equivalent leads to a similar observation [41].

Using DPT x-ray diffraction and DSC, Berger and Wright [42] produce data that demonstrates the complexity of palm oil polymorphism and crystallization. They detect what appears to be a low-melting β phase, which they postulate is due to a PLP-rich phase since it melts between 21–30°C (PLP has a β melting point of 28.5°C). The usual palm oil β phase melts at around 38°C.

Deffense [14] notes that *rac*-1,2-disaturated-3-monounsaturated (SSU) TAGs tend to be enriched in the stearin, with respect to the original palm oil, during dry fractionation. Conversely, the symmetrical isomer tends to go into the olein. The enrichment of SSU (mainly PPO) in the stearin phase is due to its intersolubility with the other crystallizing TAGs. The enrichment is dependent, of course, on the polymorphic form of the solid. In general, less stable TAG crystal forms, being less compact, are able to accommodate foreign TAG molecules more easily than stable crystal forms. Thus, in the α form, most TAGs exhibit monotectic phase behavior with continuous solid solutions. In the β form, the molecules are more tightly packed, leaving little leeway for accommodation of different molecules. The phase behavior of TAGs in the β phase exhibits eutectics, peritectics, and solid solution discontinuities.

TAGs that are similar in characteristics (such as acyl chain length and unsaturation) will form solid solutions with one another. A solid solution is a homogeneous crystal in which some of the molecules are replaced in part with foreign molecules—hence the need for similarity in molecular size and shape. The foreign molecules need not have the same polymorphism. Mixed crystals may also form. In this case, the two component molecules have the same polymorphism. This intersolubility behavior makes it difficult to obtain a clean separation between certain pairs of TAGs.

In certain circumstances (cooling rapidly to −40°C), the two fractions of palm oil appear to crystallize in different polymorphs. The higher-melting (PPP dominated) fraction crystallizes in the β form while the lower-melting fraction (dominated by POO) crystallizes in the α form, a phenomenon called "double crystallisation" by Deroanne [17]. This is similar to the observations of Gibon et al. [43–45] in pure TAGs. Using differential scanning calorimetry and powder x-ray diffraction they have shown a monotectic behavior of POP with PPP. They demonstrate the occurrence of a mixture of β′ (POP-PPP mixed crystal) and β (POP) phases. Thus, two solid phases exist, each with its own polymorphism.

The formation of mixed crystals leads to a poorer separation of solid from liquid. If the crystallization occurs rapidly, crystal structures may be less tightly packed than at equilibrium (or under slower crystallizing conditions) allowing the incorporation of TAGs that would normally remain in the liquid phase [15]. Clearly, these TAGs cannot be removed by filtration but contribute to a softening of the stearin.

It is generally considered that the solid is most easily separated if it is crystallized in the β′ form [10]. For optimum filtration, the crystals of stearin should be firm, uniform, and spherical [17]. β′ crystals form compact spherulites suitable for filtration. Crystals in the so-called intermediate form are more open and compressible, blocking the filter. β crystals are needle-like and clog the filter.

In order to ensure β′, rather than β, crystallization, it is necessary to completely melt the oil by heating 10–20°C above the melting point. Insufficient heating allows the "memory effect" to cause crystallization in the β form.

Seed crystals may be added to the cooled oil to initiate crystallization at low degrees of supercooling. These seeds may be in the β′ form (crystallized between 20 and 28°C) or in the β form (crystallized between 30 and 35°C) [46].

Harris et al. [47] describe a semicontinuous process in which the crystallized stearin represents less than 50% of the thermodynamic equilibrium value of solids. In the process a kinetically stable steady state is achieved in such a way that more than 60% of the stearin phase may be in the β polymorph while permitting easy filtration using a filter press. A small proportion of semisolid oil is periodically removed from a large batch for separation in a filter press. It is replaced with an equal amount of fully liquid oil, which is seeded by the crystals already present in the crystallizer. Since there is semicontinuous seeding, crystallization can be carried out at a low degree of supercooling, yielding crystals that are readily filterable.

While β′ is desired from the separability point of view, it has a disadvantage over β in that greater intersolubility is possible due to a less compact structure. Thus, if the solid is filterable, β crystallization would favor purer fractions.

The appearance of the crystals formed at different crystallization temperatures is characteristic. At 297 K spherulites form that do not appear dendritic like those that form at 299 K and above [48,49]. These crystals are thought to be due to the α form (with some β′) and β form, respectively [11]. This same demarcation in temperature at 299 K has also been observed by plotting induction time against temperature [9].

Kawamura analyzed the crystallization of the β form of filtered palm oil (heterogeneous nuclei removed) using the approach of Avrami and showed that, under isothermal conditions, nucleation is sporadic (i.e., occurring throughout the crystallization process) and that growth is polyhedral [48,49]. The primary nucleation rate of palm oil (whether β′ or β) is strongly temperature dependent and plots linearly against reciprocal squared logarithm of the supersaturation [50] according to an Arrhenius-type equation. The primary nucleation rate of the β′ form is much faster than that of the β form below about 35°C. Hence, low degrees of supercooling lead to β crystallization while greater degrees of supercooling yield the β′ form.

B. Minor Components

Most of the minor components in palm oil remain in the liquid phase. Exceptions are phospholipids and metals [14]. The presence of phospholipids in the stearin must be due, in part, to their higher melting point and to their co-crystallization with the TAGs. Monostearoylglycerol and monopalmitoylglycerol added to palm oil led to a reduction in the induction time. Addition of lecithin (i.e., phospholipids) to palm oil had no effect on crystallization on its own. Water, itself, also had no influence on the start of crystallization but in the presence of lecithin the crystallization of palm oil was retarded [51–53].

Failures in the fractionation with certain crystallizers have been attributed to inhibition of nucleation by impurities [54]. This later nucleation allows greater supercooling to be achieved before nucleation occurs, leading to a rapid nucleation when it finally occurs. In turn, this plethora of nuclei leads to many crystals of diverse sizes that cause poor filtration. The impurities are likely to be DAGs. Removal of the DAGs from palm oil followed by addition of specific DAGs shows that they inhibit the crystallization, both nucleation and growth [55]. The most effective inhibitor was PP (both 1,2 and 1,3). 1,2-PO had no influence while the 1,3-isomer did inhibit to some degree. The inhibition was very temperature dependent and was not large at high degrees of supercooling. Berger and Wright [42] showed that DAGs also retarded the transformation of the α polymorph into β'.

Repeated heating of palm oil (which occurs frequently during processing, transport, and use) has been shown to influence the crystallization behavior [56]. These changes are likely to be due to oxidation of minor components as well as of the TAGs themselves (especially unsaturated).

C. Additives

One of the main uses for palm olein is as a cooking oil. At low temperatures, however, it has a tendency to become cloudy which, given time, can sediment in the bottle. These crystals are in the β polymorph and are enriched with respect to DAGs (predominantly PP) [32,33]. Deliberate addition of PP (both 1,2 and 1,3) to palm olein decreased the time for crystallization (1,3 more than 1,2) while PO appeared to delay crystallization somewhat [57]. OO had little effect. In view of the fact that the crystallizing solid is PP it is perhaps not surprising that additional quantities decrease the time. The retardation of crystallization by PO may be due to a crystal poisoning effect. Sulaiman et al. [58] do not consider clouding to be due to DAGs, at least in the short term. Rather they identify PPP, dipalmitoyl-myristoyl-glycerol (MPP), and dipalmitoyl-stearoyl-glycerol (PPS) as being responsible for the clouding at relatively high temperatures (30°C) and short times (24 hr). However, according to Swe et al. [59] the responsible TAGs are POP and POSt.

The effect of various additives on the cloudpoint (or resistance to crystallization) of palm oleins has been investigated [60]. Additives include sorbitan tristearate, polyglycerol esters, and lecithin. At 20°C all additives delayed the onset of crystallization. However, at lower temperatures (5 and 15°C) one type of sorbitan tristearate at 0.1% induced crystallization. A solid polyglycerol ester was found to be best at delaying the onset of cloudiness. Sorbitan tristearate has also be found to inhibit polymorphic transformations in palm oil [11].

Additives can be used to improve the separation efficiency by influencing the crystal morphology. One class of such additives is membrane lipids [61]. Polysaccharide esters of fatty acids may also be used to influence crystallization so as to increase the separation efficiency by excluding olein from the growing crystals [62]. The polysaccharide is an inulin or phlein in which at least 50% of the hydroxyls are esterified with fatty acids. Instead of polysaccharide, a copolymer may be used [63], again esterified with fatty acids. Optimum results are obtained where the fatty acids match those of the crystallizing TAGs.

Comb polymers, commonly used to prevent waxing out in diesel fuel, are very effective crystallization modifiers in palm oil [64]. Significant effects can be seen at addition levels of 500 ppm (0.05%). They modify the appearance and structure of the crystals, altering them from spiky spherulites into smooth, dense balls. Separation of solid from liquid is much improved. Comb polymers appear to reduce the number of nuclei, leading to few large crystals, rather than many small ones. The additive would appear to inhibit nucleation. In addition, it seems that the polymer slows down the growth rate of the fastest growing face of the crystal (i.e., the one facing out from the center of the spherulite). Slowing down this face permits the other faces of the crystals, which make up the spherulite, to grow further within the spherulite itself [64] (Fig. 7).

Unfortunately, comb polymers are not food grade. The effect on palm oil crystallization of several natural molecules comprising an acyl chain attached to a bulky group has been investigated by Smith [64]. One such molecule was extracted from bovine brains. Figure 8 shows typical palm oil spherulites taken from a filter cake. Note the open ''fluffy'' appearance, occluding much olein within. Figure 9 shows crystals following fractionation in which brain lipid has been added at level of 0.2%. Spherulites still form but they are much more regular and much denser, occluding less olein and filtering more easily.

Phosphatidyl ethanolamines, phosphatidyl inositols, lysophosphatidyl ethanolamines, and phosphatidyl cholines (whether from soyabean or egg yolk) all give rise to large spherulites when added to palm oil prior to fractionation [64]. The separation efficiency was worse in the presence of phosphatidyl cholines, compared a control, despite the formation of large spherulites. This is possibly due to an inhibition of nucleation (hence few, large crystals) but growth of more

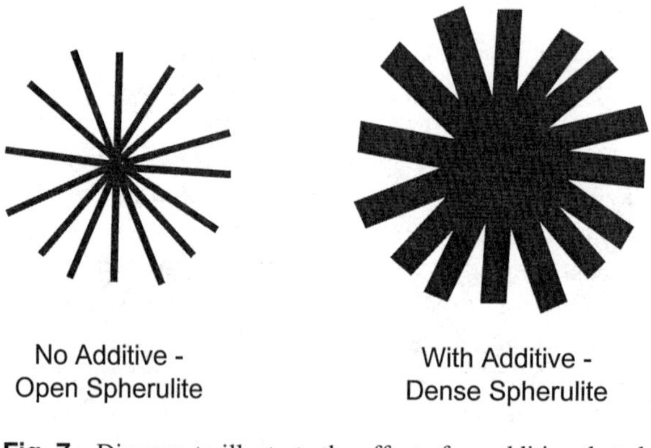

Fig. 7 Diagram to illustrate the effect of an additive that slows the growth of the fastest growing face, thus allowing the interior faces to grow to yield a denser spherulite.

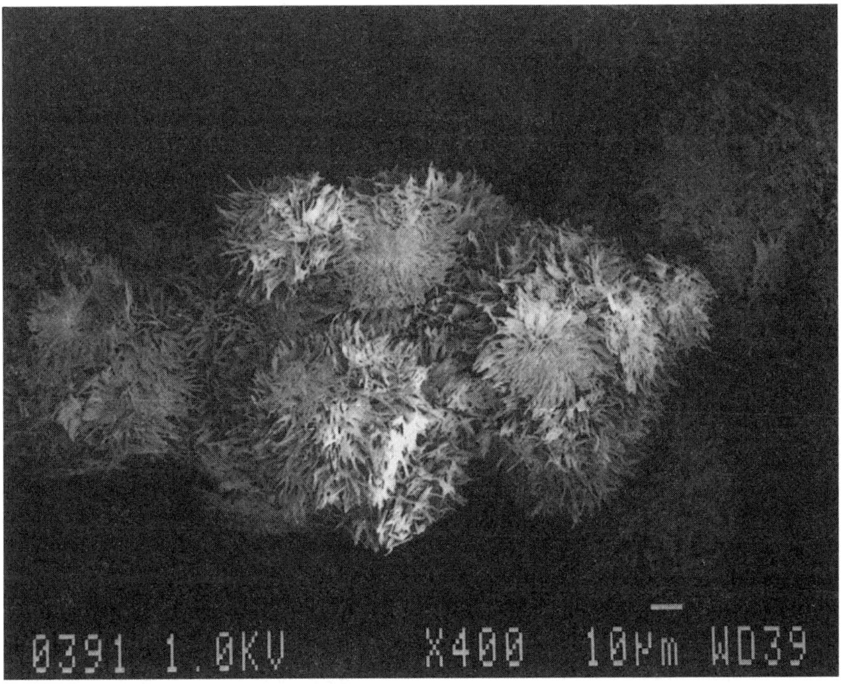

Fig. 8 Electron micrograph of spherulites taken from a filter cake.

open spherulites. Phosphatidyl cholines from different sources modify the crystallization differently, perhaps due to the variation in the acyl chains between sources.

D. Postcrystallization Processes

When palm oil fractions are used in products a slow crystallization or recrystallization can take place after manufacture. During production, cooling conditions may be applied that do not lead to thermodynamic equilibrium between the composition of the solid and the liquid phases. This has important ramifications for the product texture. Recrystallization can occur into a new solid phase as thermodynamic equilibrium is approached. The stable form of POP has a triple-chain-length structure and this property can lead to reduced solubility in other TAGs. An example of this is the graininess development in margarines, where a β_1 POP-rich phase crystallizes at the expense of the preexisting solid [40]. This is not helped by the fact that in any semisolid fat, if the crystal size distribution is broad,

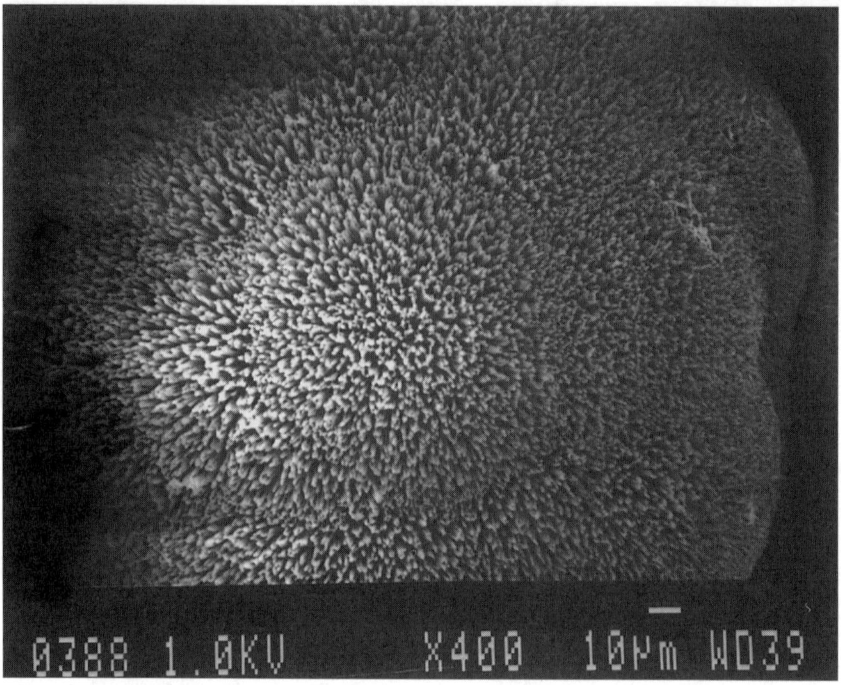

Fig. 9 Electron micrograph of spherulites taken from a filter cake after crystallization in the presence of 0.2% brain membrane lipids.

there will be a tendency for the large crystals to grow larger still at the expense of the smaller ones (a process called Ostwald ripening). In another example, fat crystals can sinter (i.e., link together) through bridges formed, apparently, by TAGs with melting points between those of the existing solid and liquid phases [65]. These fat bridges nucleate from the existing crystals. Thus β′ crystals are bridged by β′ solid and β crystals by β solid.

Palm mid fractions are produced by a two-stage (at least) fractionation. In order to obtain the sharpest melting product it is usually necessary to fractionate from solvent. Depending on the precise temperatures used for each stage, palm mid fractions can be produced with slip melting points from 29°C up to 42°C [66]. Palm mid fractions are produced to serve as components in cocoa butter equivalents or for use in their own right as confectionery fats. In terms of TAG composition, they are similar to cocoa butter in being mainly SOS. The principal TAG present in palm mid fraction is POP. Other significant TAGs are PPO, POO, and PPP. This latter can be reduced to around 1% by using a suitable fractionation temperature.

Like cocoa butter, the stable polymorphic form of palm mid fraction is β (corresponding to that of its major component). However, POP forms a 1:1 compound with PPO [12,13]. The complexity which this adds to the system means that the transformation rate from β′ to β is reduced. This is not desirable in a cocoa butter equivalent and thus low levels of PPO are sought.

The compound formation between POP and PPO, however, can be used to advantage. Increasing the level of PPO in a palm mid fraction can lead to a product that is strongly metastable in the β′ form, effectively rendering the fat nontemper. This has allowed the design of sharp-melting, β′ stable fats [67].

E. Modeling

The kinetics of fat crystallization has been likened, in part, to the aggregation and flocculation of colloids. Berg and Brimberg [38] develop equations and rate constants that appear to relate to the equilibrium solid content at a given temperature. In an alternative approach, several researchers have applied Avrami's analysis of polymer crystallization [68–70] to fat. The Avrami exponent is an integer between 1 and 4. The value of the integer depends on a combination of nucleation and growth factors. A value of 4 indicates that the nucleation is sporadic (i.e., occurring throughout the crystallization process) and growth is polyhedral (in all dimensions). This value of 4 is found when the equation is applied to palm oil (filtered to remove impurities) crystallization measured using DSC under isothermal conditions [48]. Other workers [71] have utilized laser light equipment to determine the induction time for the onset of crystallization, allowing a calculation of the nucleation rate and activation energy by application of the Fisher-Turnbull equation. Results agreed well with earlier work [9,39], demonstrating the validity of the approach.

Palm oil crystallizes in spherulites. These appear to start with a single tiny crystal. Daughter crystals are nucleated on the surface of this crystal and grow outward, often nucleating daughter crystals of their own. Los et al. [72], building on older models of crystal growth in palm oil [50], have developed a model for the spherulitic growth of palm oil that is partially empirical. From their data and model they conclude that the crystal growth is limited by diffusion, contradicting previous results, which they attribute to the quality of the past data. Thus, the addition of solid to the spherulite occurs in a small region near the apparent surface of the spherulite. Solid is not added to the "sides" of the daughter crystals in the interior of the spherulite because the supersaturation at this point is low. This leads to the open structure of the spherulites and to the occlusion of olein. If crystallization at the growth front (apparent surface of the spherulite) is retarded, perhaps due to an additive (see above), there is greater time for diffusion of the crystallizing molecular species to take place into the interior. Thus the interior supersaturation is increased and a greater growth from the sides of the daughter crystals may take place. In this way, denser spherulites can be formed having less occluded olein. The model is a good beginning to understanding a complex process.

V. SUMMARY

The world production of palm oil has increased greatly and will continue to increase, forming an even greater proportion of the world vegetable oil market. Demand will continue to increase as consumption rises, especially in developing countries. Although the market expansion in the past has prompted much research, there is still much to understand regarding this complex fat. The drive continues to be one of reducing costs yet achieving improved separation. Additives may well prove to be the way forward, either in the final product or as a fractionation aid.

REFERENCES

1. Economic Research Service, U.S. Department of Agriculture Market and Trade Economics Division, *Oil Crops Yearbook*, available electronically via www.econ.ag.gov October(OCS-1999): (1999).
2. FB Padley, *The Lipid Handbook*, 2nd ed. (F. D. Gunstone, J. L. Harwood, F. B. Padley, eds.), Chapman & Hall, London, 1994, pp. 82–84.
3. FVK Young, *Crit. Rep. Appl. Chem. 15*: 39–70 (1987).
4. B Jacobsberg, *Oléagineux 30(6)*: 271–276 (1975).
5. BSJ Jeffrey, *J. Am. Oil Chem. Soc. 68(5)*: 289–293 (1991).
6. WL Siew, WL Ng, *J. Am. Oil Chem. Soc. 72(5)*: 591–595 (1995).
7. MGA Willems, FB Padley, *J. Am. Oil Chem. Soc. 62(2)*: 454–460 (1985).
8. FCH. Oh, KG Berger, *Palm oil product technology in the eighties* (E. Pushparajah, M. Rajadurai, eds.), The Incorporated Society of Planters, Kuala Lumpur, 1983, pp. 419–431.
9. WL Ng, *J. Am. Oil Chem. Soc. 67(11)*: 879–882 (1990).
10. C Deroanne, JM Marcoen, JP Watelet, M Severin, *J. Thermal Anal. 11*: 109–119 (1977).
11. K Kawamura, *J. Am. Oil Chem. Soc. 57(1)*: 48–53 (1980).
12. A Minato, J Yano, S Ueno, K Smith, K Sato, *Chem. Phys. Lipids 88*: 63–71 (1997).
13. A Minato, S Ueno, K Smith, Y Amemiya, K Sato, *J. Phys. Chem. B 101(18)*: 3498–3505 (1997).
14. E Deffense, *J. Am. Oil Chem. Soc. 62(2)*: 376–385 (1985).
15. E Deffense, *Rev. Fra. Corps Gras 36(5)*: 205–212 (1989).
16. E Deffense, *Lipid Technology March*: 34–38 (1995).
17. C Deroanne, *Progress in Food Engineering (Conference Proceedings) (Lebensmittel Wissenschaft Technologie Series 7)* (C Cantarelli, ed.), 1983, pp. 499–506.

18. W Hamm, SCI Lecture papers series. Society of Chemical Industry, Belgrave Sq., London, 1994.
19. W Hamm, *Trends Food Sci. Technol. 6(0)*: 121–126 (1995).
20. W Hamm, *Trans IChemE 74(Part C)*: 61–72 (1996).
21. MJ Kokken, *Rev. Fra. Corps Gras 38(11/12)*: 367–376 (1991).
22. T Thiagarajan, *Trends Food Sci. Technol.* 449–458 (1988).
23. R Timms, *Chem. Ind. 20 May* 342–345 (1991).
24. RE Timms, *Society of Chemical Industry Lecture Paper Series No. 0039*, Society of Chemical Industry, London, UK, 1995.
25. M Bernardini, R Sassoli, *Palm oil product technology in the eighties* (E Pushparajah, M Rajadurai, eds.), The Incorporated Society of Planters, Kuala Lumpur, 1983, pp. 291–302.
26. GG Bemer, G Smits, *Ind. Crystall. 81*: 369–371 (1982).
27. T Willner, *Society of Chemical Industry Lecture Paper Series No. 0038*, Society of Chemical Industry, London, UK, 1994.
28. M Kellens, *Oléagineux Corp Gras Lipides 5(6)*: 421–426 (1998).
29. ARM Ali, MS Embong, BK Tan, *Proc. Malays. Biochem. Soc. Conf. 12*: 179–185 (1986).
30. L Koslowsky, *Oléagineux 27(11)*: 557–560 (1972).
31. L Koslowsky, *Oil Palm News 15*: 24–29 (1973).
32. Ng WL Siew WL, *J. Sci. Food Agric. 70*: 212–216 (1996).
33. PZ Swe, YB Che Man, HM Ghazali, *J. Am. Oil Chem. Soc. 72(3)*: 343–347 (1995).
34. RE Timms, *Palm oil product technology in the eighties* (E Pushparajah, M Rajadurai, eds.) The Incorporated Society of Planters, Kuala Lumpur, 1983, pp. 277–290.
35. ASH Ong, PL Boey, CM Ng, *J. Am. Oil Chem. Soc. 60(10)*: 1755–1760 (1983).
36. JS Baker, RM Weitzel, *EU Patent EP 0 189 669 A1* (1986).
37. KG Berger, BK Tan, RJ Hamilton, B Jacobsberg, *J. Am. Oil Chem. Soc. 57(2)*: 124A (1980).
38. TGO Berg, UI Brimberg, *Fette Seifen Anstrichmittel 85(4)*: 142–149 (1983).
39. WL Ng, CH Oh, *J. Am. Oil Chem. Soc. 71(10)*: 1135–1139 (1994).
40. A Watanabe, I Tashima, N Matsuzaki, J Kurashige, K Sato, *J. Am. Oil Chem. Soc. 69(11)*: 1077–1080 (1992).
41. DJ Cebula, KW Smith, *J. Am. Oil Chem. Soc. 69(10)*: 992–998 (1992).
42. KG Berger, WB Wright, *PORIM Occasional Paper (17)*: 1–10 (1986).
43. V Gibon, Docteur en Sciences Thesis, Universitaires Notre Dame de la Paix, Namur, 1984.
44. V Gibon, F Durrant, *J. Am. Oil Chem. Soc. 62(4)*: 656–657 (1985).
45. V Gibon, F Durrant, C Deroanne, *J. Am. Oil Chem. Soc. 63(8)*: 1047–1055 (1986).
46. A Dieffenbacher, *US patent 4, 594, 194* (1986).
47. JB Harris, CNM Keulemans, LA Milton, EJG Roest, *World patent WO 96/05279* (1996).
48. K Kawamura, *J. Am. Oil Chem. Soc. 56(8)*: 753–758 (1979).
49. K Kawamura, *J. Agri. Chem. Soc. Japan (Nippon Nogeikagaku Kaishi) 57(5)*: 475–485 (1983).
50. KPAM van Putte, BH Bakker, *J. Am. Oil Chem. Soc. 64(8)*: 1138–1143 (1987).
51. M Naudet, E Sambuc, *J. Am. Oil Chem. Soc. 57(2)*: 117A (1980).

52. E Sambuc, M Naudet, *Oléagineux 25(12)*: 559–563 (1980).
53. E Sambuc, Z Dirik, M Naudet, *Rev. Fra. Corp Gras 28(2)*: 59–65 (1981).
54. YB Che Man, PZ Swe, *J. Am. Oil Chem. Soc. 72(12)*: 1529–1532 (1995).
55. WL Siew, WL Ng, *J. Sci. Food Agric. 79*: 722–726 (1999).
56. T Haryati, YB Che Man, PZ Swe, *J. Am. Oil Chem. Soc. 74(4)*: 393–396 (1997).
57. WL Siew, WL Ng, *J. Sci. Food Agric. 71*: 496–500 (1996).
58. MZ Sulaiman, NM Sulaiman, S Kanagaratnam, *J. Am. Oil Chem. Soc. 74(12)*: 1553–1558 (1997).
59. PZ Swe, YB Che Man, HM Ghazali, LS Wei, *J. Am. Oil Chem. Soc. 71(10)*: 1141–1144 (1994).
60. NA Idris, Y Basiron, H Hassan, R Bassin, *Elaeis 5(1)*: 47–64 (1993).
61. PR Smith, M van der Kommer, *World patent WO 95/04123* (1995).
62. PHJ van Dam, C Winkel, A Visser, *World patent WO 96/20266* (1996).
63. M van den Kommer, PR Smith, A Visser, C Winkel, *World Patent WO 95/04122* (1995).
64. PR Smith, *Ph.D. dissertation, The University of Leeds*, 1995.
65. D Johansson, B Bergenstahl, *J. Am. Oil Chem. Soc. 72(8)*: 911–920 (1995).
66. S Khatoon, DK Bhattacharyya, *Oléagineux 41(11)*: 519–522 (1986).
67. F Duurland, K Smith, *Lipid Technol. 7(1)*: 6–9 (1995).
68. M Avrami, *J. Chem. Phys. 7(1)*: 1103–1111 (1939).
69. M Avrami, *J. Chem. Phys. 8(1)*: 212–224 (1940).
70. M Avrami, *J. Chem. Phys. 9(1)*: 177–184 (1941).
71. CC Let, K Sato, *23rd World Congress and Exhibition of the International Society for Fat Research 101(18)*: 3498–3505 (1999).
72. J Los, P Bennema, W van Enckevort, F Hollander, R Beltman, PHJ van Dam, CNM Keulemans, *23rd World Congress and Exhibition of the International Society for Fat Research: Abstracts 7(1)*: 15 (1999).

11

Advances in Milk Fat Fractionation

Technology and Applications

R. W. Hartel
University of Wisconsin, Madison, Wisconsin

K. E. Kaylegian
Wisconsin Center for Dairy Research, Madison, Wisconsin

Fractionation is a common processing technology used to enhance the value of natural fats. It involves separation of components of a natural fat based on some physical or chemical property. By separating a natural fat into components with specific chemical composition and physical properties, value-added ingredients can be produced from a commodity with lower value. The potential for milk fat fractionation has been explored for many years. Kaylegian and Lindsay [1] recently provided a detailed review of types of fractionation techniques and the nature of the fractions produced from milk fat. In this chapter, we will discuss recent advances in our understanding of the principles that underlie fractionation technologies and the applications of milk fat fractions as value-added ingredients in the food industry.

To help understand how molecular components of milk fat may be fractionated, it is necessary to review the chemical composition of milk fat. The primary component of milk fat, as for all natural fats, is the triacylglycerol (TAG). Milk fat contains 96–98% TAG with the remainder made up of diacylglycerols (DAG), monoacylglycerol, (MAG), free fatty acids (FFA), phospholipids, sterols, and other polar lipids [2,3]. Milk fat contains over 400 different species of TAG with fatty acids from C2:0 to C24:0. The main fatty acids present in milk fat and their approximate composition are shown in Table 1. Milk fat contains high levels of palmitic (C16:0) and oleic (C18:1) fatty acids, but what makes milk fat unique is the high content of short-chain fatty acids, particularly butyric acid (C4:0).

Table 1 Fatty Acid Composition (weight %) of Milk Fat

Fatty acid	Composition
C4:0	3.3
C6:0	1.6
C8:0	1.3
C10:0	3.0
C12:0	3.1
C14:0	9.5
C16:0	26.3
C16:1	2.3
C18:0	14.6
C18:1	29.8
C18:2	2.4
C18:3	0.8

Source: Ref. 3.

These short-chain fatty acids have an important role in dairy flavor, but also impact crystallization of milk fat by influencing the arrangement of milk fat TAG in the crystal lattice. The manner in which the TAG interact is determined not only by fatty acid content but also by how the fatty acids are arranged on the glycerol molecule of the TAG. Although positional analysis of TAG in milk fat is very complex, Christie [3] provides a breakdown of the likelihood of finding a particular fatty acid at a particular site in the TAG, as shown in Table 2. The distribution of fatty acids on the glycerol molecule is not random, but is a function of the particular biochemistry within the cow [4]. Of particular importance is the placement of the short-chain fatty acids preferentially on the *sn*-3 position [2,5]. Also important is the placement of myristic and palmitic somewhat preferentially on the center, or *sn*-2, position of the TAG. This has important consequences to both physical properties and health implications for milk fat.

Detailed chemical analysis of milk fat is far from complete. Gresti et al. [6] used a combination of analytical and predictive techniques to identify the major TAG present in milk fat. Over 250 individual TAG molecules were identified from the chromatograms and assigned fatty acid composition. The ten most prevalent TAG in milk fat, according to this analysis, are listed in Table 3. The concentration of the most prevalent TAG (C4:0-C16:0-C18:1) in milk fat, according to Gresti et al. [6], was only 4.2% on molar basis. Kemppinen and Kalo [7] found that the concentration of the most prevalent TAG (C16:0-C16:0-C4:0) was 5%. Despite the disparity in these two studies, clearly there are no single

Table 2 Positional Arrangement of Fatty Acids on Triglyceride

Fatty acid	*sn*-1	*sn*-2	*sn*-3
C4:0	—	—	35.4
C6:0	—	0.9	12.9
C8:0	1.4	0.7	3.6
C10:0	1.9	3.0	6.2
C12:0	4.9	6.2	0.6
C14:0	9.7	17.5	6.4
C16:0	34.0	32.3	5.4
C18:0	10.3	9.5	1.2
C18:1	30.0	18.9	23.1
C18:2	1.7	3.5	2.3

Source: Ref. 3.

TAG that dominate the chemical composition of milk fat. Characterizing the TAG in milk fat according to levels of unsaturation gives the breakdown shown in Table 4. According to this analysis, trisaturated TAG make up nearly one third of all TAG in milk fat. However, this does not distinguish between long- and short-chain fatty acids. If the TAG are analyzed according to degree of unsaturation and chain length, the breakdown shown in Table 5 is obtained. Although

Table 3 Concentration (mole%) of the 10 Most Prevalent Triacylglycerols (TAG) in Milk Fat

TAG[a]	Composition
C4:0; C16:0; C18:1	4.2
C4:0; C16:0; C16:0	3.2
C4:0; C14:0; C16:0	3.1
C14:0; C16:0; C18:1	2.8
C16:0; C18:1; C18:1	2.5
C4:0; C16:0; C18:0	2.5
C16:0; C16:0; C18:1	2.3
C16:0; C18:0; C18:1	2.2
C6:0; C16:0; C18:1	2.0
C4:0; C14:0; C18:1	1.8

[a] Not in positional order.
Source: Ref. 6.

Table 4 Distribution (mole%) of Triacylglycerol Classes in Milk Fat Based on Degree of Unsaturation

Class[a]	Composition[b]
S, S, S	32.4
S, S, U	32.6
S, U, U	10.6
S, S, D	2.5
Other	5.2

[a] S = saturated; U = monounsaturated; D = diunsaturated fatty acids.
[b] Based on analysis of 255 molecular species.
Source: Ref. 6.

this information is useful, the actual positional arrangement of the fatty acids on the TAG is critical to an understanding of crystallization behavior and fractionation. Since fractionation is based on molecular properties (melting point, solubility, etc.), a detailed understanding of the composition of milk fat is necessary to promote efficient separation. Despite numerous recent studies on characterization

Table 5 Distribution (mole%) of Triacylglycerol Classes in Milk Fat Based on Degree of Unsaturation and Fatty Acid Chain Length

Class[a]	Composition[b]
Short, S, S	21.5
S, S, U	15.9
Short, S, U	14.2
S, S, S	8.2
S, U, U	7.1
Short, U, U	2.7
U, U, U	1.1

[a] S = saturated; U = monounsaturated; short = C4:0, C6:0 C8:0.
[b] Based on analysis of 110 most abundant triacylglycerols.
Source: Ref. 6.

of milk fat fatty acid and TAG composition [6,8–24], detailed knowledge of the specific TAG composition in milk fat is insufficient for predicting crystallization.

However, the wide range of molecular constituents and the resulting wide range in melting points [25] is what makes milk fat an attractive target for fractionation, particularly by crystallization techniques. It has been suggested [26–28] that milk fat can be considered to contain three main component fractions, as seen in Table 6. A high-melting fraction (HMF), composed of TAG containing only long-chain saturated fatty acids, constitutes about 5–10% of milk fat. This component has a melting point at about 50°C. The TAG that contain two long-chain fatty acids and either one short-chain or a cis-unsaturated fatty acid make up the middle-melting fraction (MMF). This fraction constitutes about 25% of milk fat and has a melting point between 35 and 40°C. The low-melting fraction (LMF), which makes up the bulk of milk fat (65–70%), contains TAG with one long-chain saturated fatty acid and two short-chain or cis-unsaturated fatty acids. The melting point of the LMF is less than 15°C. Depending on the crystallization conditions and the initial composition of the milk fat, fractionation of these components from one another can be accomplished relatively efficiently. However, there are many factors that influence fractionation, and it is possible to produce numerous fractions with a wide range of physical properties by manipulating fractionation conditions. One potential problem for milk fat fractionation is the variability of milk fat composition across different regions and seasons [29–30]. Natural fluctuations in milk fat composition, due primarily to feeding practices, result in milk fats with different crystallization and fractionation behavior. These differences may be significant enough to cause problems in fractionation during certain times of the year.

Table 6 Three Primary Classes of Triacylglycerols Found in Milk Fat, Based on Melting Point Ranges

Component	Composition	Amount (%)	Melting point (°C)
High melting	long-chain saturated acids	5	>50
Middle melting	two long-chain saturated acids and one short-chain or unsaturated acid	25	35–40
Low melting	one long-chain saturated acid and two short-chain or unsaturated acids	70	<15

Source: Ref. 28.

Table 7 Techniques Used for Fractionation of Milk Fat

Technique	Mechanism
Melt crystallization	Melting point
thin layer separation	
solution-based separation	
Solvent crystallization	Solubility
organic solvent solution	
supercritical carbon dioxide	
Short-path distillation	Volatility

There have been many attempts to fractionate milk fat over the past three to four decades [1]. Over the years, many different techniques have been studied, as shown in Table 7. These range from supercritical carbon dioxide extraction to short-path molecular distillation. One of the first techniques for fractionation of milk fat was based on dissolution in an appropriate solvent (e.g., acetone) followed by crystallization at different temperatures to produce fractions with different chemical composition and physical properties. Currently, the primary technique used for fractionating milk fat, as with many fats, is based on crystallization from the melt. That is, a melted fat is cooled under appropriate conditions to cause a solid fraction to form in a slurry and this slurry is separated from the liquid phase to produce two fractions with different composition and melting profile. To control the fractionation process, it is necessary to have an understanding of the parameters that influence the mechanisms and kinetics of crystallization of milk fat.

I. MILK FAT CRYSTALLIZATION

When milk fat cools from the melt, it quickly solidifies into a semisolid mass with properties that depend on the conditions of solidification. Depending on the conditions, certain molecular components of milk fat crystallize when the melt is cooled. However, milk fat is such a complex mixture of TAG that it is not easy to predict which components crystallize under different conditions. Thus, our understanding of the mechanisms and kinetic rates of milk fat crystallization under different conditions is severely lacking.

In all crystallization processes, there are several steps that must take place. First, a thermodynamic driving force must be generated within the system to allow crystallization to proceed. Once the system reaches an appropriate state of nonequilibrium, nucleation of crystals from the liquid state can occur. After nucleation, the crystals grow until a thermodynamic equilibrium is attained, after which no further change in crystal phase volume occurs unless the temperature is changed. However, rearrangement of the crystalline microstructure (i.e., size, shape, polymorph, etc.), without a change in phase volume, can still occur over time as either internal energy of the crystals or their surface free energy continue to approach a thermodynamic minimum.

A. Crystallization Driving Force

Most natural fats are considered melt systems, where cooling the fat sufficiently below its melting point causes crystallization to occur. Freezing of water is a common example of crystallization of a melt. When water is cooled sufficiently below its freezing point, it eventually crystallizes (except under unusual circumstances). However, natural fats, and particularly milk fat, contain a range of TAG with a range of melting points. Thus, it is not clear what to define as the melting point. Often, the final melting point (or clear point) of the fat is used, and this may provide some understanding of the driving force for crystallization of a certain component of the fat. However, it does not provide an understanding of the driving force for crystallization of species with lower melting points nor does it help to understand compound crystal formation. This leaves us without a good indicator of the thermodynamic equilibrium point for a natural lipid system.

To further complicate things, some of the components of a fat (the low-melting components) act as a solvent for the higher-melting components. With a simple binary system of two pure TAG, such as stearic-oleic-stearic and oleic-stearic-oleic, for example [31], a phase diagram documents the solubility of one component in the other in both solid and liquid phases. Each polymorphic form of the crystalline material has its own equilibrium with the low-melting component [32]. However, for natural fat systems, the wide range of components present eliminates this simple option for describing the phase behavior. There are too many components, especially in milk fat, to deal with in a reasonable fashion.

Despite this problem, the phase mixing behavior of milk fat and its components is needed to help us better understand the driving force for milk fat crystallization. A simple approach is to treat the three main components of milk fat as separate entities and look at phase diagrams for mixtures of these components (HMF, MMF, and LMF). Although these are not true phase diagrams, they can be referred to as such for simplicity. This approach has been suggested for mixtures of natural fats and has been used to study milk fat [33].

Marangoni and Lencki [34] recently looked at the mixing behavior of mixtures of the three main components of milk fat (HMF, MMF, and LMF). These fractions were prepared by solvent fractionation, so the purity of these components was quite high. However, the specific composition of these fractions was dependent on the conditions used for fractionation. Solid fat content (SFC) of mixtures of the different fractions at varying levels was analyzed by pulsed nuclear magnetic resonance (NMR). As expected, the lower-melting components were found to decrease the amount of SFC in the mixture at any given temperature. Based on the shape of the isosolids diagrams, where lines of constant SFC are drawn as a function of temperature and mixture composition, some conclusions were drawn regarding the nature of interactions between the components. Mixtures of HMF and MMF showed monotectic mixing behavior with a solid solution of TAG in the crystalline phase. That is, SFC decreased continuously at all temperatures as the amount of MMF mixed with HMF increased until the melting profile for MMF was obtained. No evidence for eutectic behavior was observed. The data suggests a high degree of complementarity between HMF and MMF. That is, despite differences in melting points and molecular volume between the two fractions, the arrangement of fatty acids on the TAG is such that they are fully miscible in the solid state. This result may be taken as evidence for compound crystal formation in milk fat, which has been previously suggested [35–37]. Mixtures of LMF with HMF and MMF showed monotectic, partial solid solution phase mixing. Marangoni and Lencki [34] state that this behavior is ''characteristic of eutectic systems which shift to monotectic systems when differences in the melting points of the two components increase (e.g., 20°C and above) and the *high melting* component dissolves a substantial amount of the *low melting* component (20–30%).'' Thus, the LMF does not act simply as a solvent for MMF and HMF, but rather a large component of LMF may be incorporated into the MMF and HMF as crystallization occurs. Ternary phase diagrams were also constructed [34] for the three components of milk fat.

Isosolid diagrams for mixtures of HMF, MMF and LMF produced by melt (or dry) fractionation clearly demonstrated the excellent mixing properties of the components of milk fat. Lines of constant SFC decreased monotonically as the level of lower-melting component was increased, as seen in Fig. 1 for the mixture of HMF and LMF. Again, little evidence of eutectic behavior was found. Since these fractions were not as pure as those used by Marangoni and Lencki [34], there may have been substantial liquid entrainment during separation, so it is not surprising that they mixed well together.

The solubility of HMF in LMF has been estimated by using a turbidity technique. Crystalline HMF at various levels was added to the liquid LMF and the slurry maintained under gentle agitation at the study temperature (between 20 and 40°C) for seven days. At the effective solubility point, the HMF had all completely dissolved. A sharp demarcation in turbidity was observed near the point of HMF solubility in LMF. That is, at an HMF level just below some critical concentration, the solution was not turbid as all the HMF had dissolved. However, when this critical concentration was exceeded, and not all of the HMF had dissolved, the solution remained turbid. Thus, a sharp distinction in turbidity was

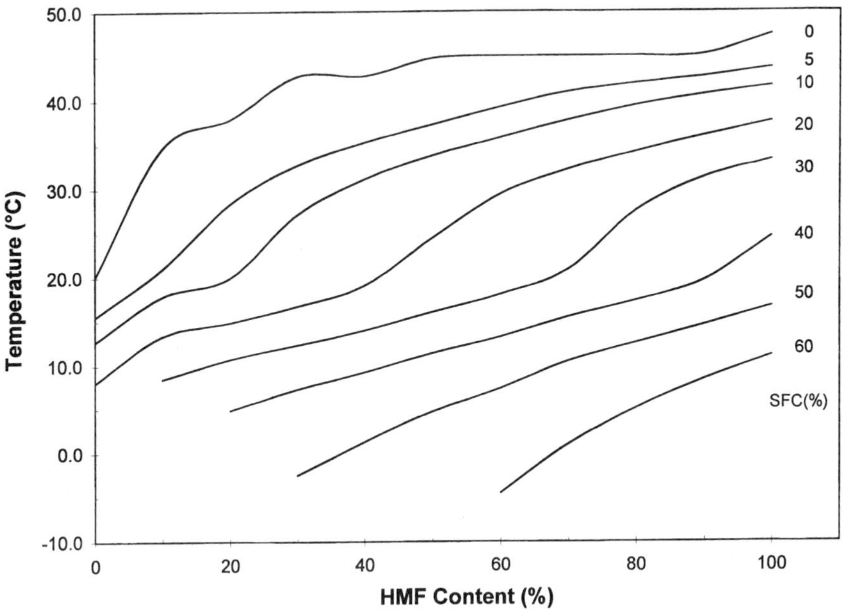

Fig. 1 Isosolid diagram for mixture of high-melting (HMF) and low-melting (LMF) milk fat fractions.

observed at this critical concentration. Note that this is not a true solubility concentration since nonequilibrium conditions may still have existed. Nevertheless, this measurement provides an estimate of how much HMF may remain soluble in LMF at any temperature. This solubility curve is shown in Fig. 2. As temperature increased, the amount of HMF dissolved in LMF increased from less than 2% at 25°C to nearly 20% at 40°C. When temperature is below 25°C, the solubility of HMF in LMF is very low and massive crystallization occurs. Note that the solubility of HMF in canola oil is less than that in LMF. This is not surprising since the TAG molecular structures between HMF and canola oil are less compatible than those of HMF and LMF.

Under certain conditions, the solubility curve in Fig. 2 can be used to estimate the supersaturation driving force for crystallization. This may be true for milk fat crystallized in the first step of fractionation, where a relatively small amount of milk fat (10–20%) is crystallized at relatively high temperatures (25–30°C). Note that the phase behavior is more complicated than described by the simple system in Fig. 2, which is based on HMF in LMF, due to the presence of the MMF in natural milk fat.

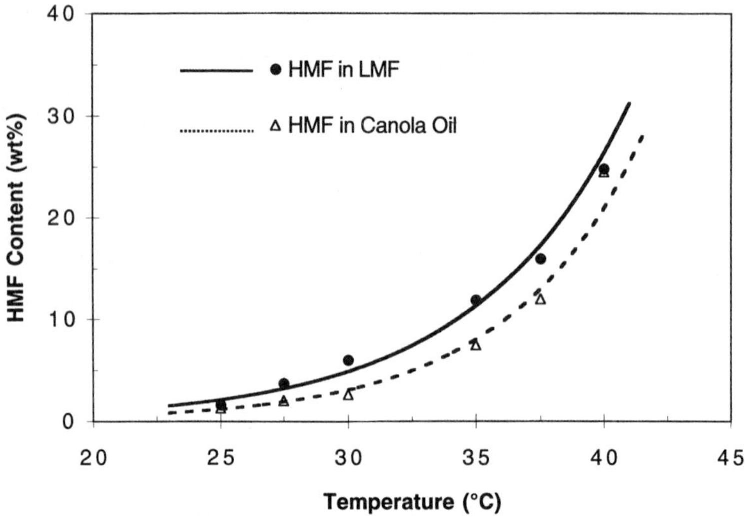

Fig. 2 Effective solubility (weight percent) of high-melting milk fat fraction into a low-melting milk fat fraction or canola oil. (From Hartel [133].)

This phase diagram approach may be further modified based on compositional analysis of the milk fat components. Thus, instead of using weight percent of HMF solubilized, the content of specific TAG present in the HMF was used. Specifically, the ratio of higher-molecular-weight TAG (typically with higher melting point) to lower-molecular-weight TAG (lower melting point) was used to characterize various mixtures of HMF in LMF. Figure 3 shows the solubility curve of HMF in LMF based on the ratio of high-acyl-carbon-number TAG (C46–52) to low-acyl-carbon-number TAG (<C40). Thus, milk fats with different ratios of HMF to LMF can be compared on this diagram. In Fig. 3, the difference in supersaturation for two milk fats with different initial compositions, which might be representative of summer and winter AMF, can be seen. The rationale for more rapid crystallization of a milk fat with higher content of high-acyl-carbon-number TAG is easily recognized in this approach. Although this approach also has its limitations, this is an interesting direction to pursue to further our understanding of the driving forces for crystallization of milk fat and other complex natural fats.

In summary, milk fat and other complex natural fats exhibit a complex phase behavior based on the chemical composition. These fats exhibit aspects of both melt behavior and solution behavior. Future efforts should focus on the relative aspects of solubility and melt behavior since, under certain circumstances, this simplified approach may be used to help control crystallization.

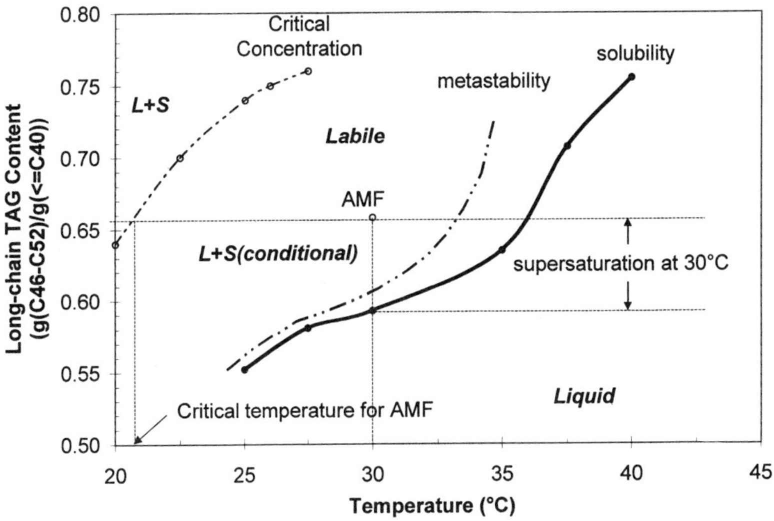

Fig. 3 Effective solubility, expressed as a ratio of high-acyl-carbon triacylglycerols (TAG) to low-acyl-carbon TAG, of high-melting milk fat fraction into a low-melting milk fat fraction. (From Hartel [133].)

B. Polymorphism

As for all fats, milk fat TAG can crystallize into different polymorphic forms based on the arrangement of the molecules within the crystal lattice. Three main polymorphs are generally characterized (α, β', and β) in fats, although sometimes an even more unstable form (γ) is found [25]. There also may be multiple forms within a category (i.e., β_1 and β_2) with slightly different melting points. Fats demonstrate monotropic polymorphism, meaning that the less-stable polymorphs form first but then transform into more stable forms. The series γ, α, β', and β represents increasing order of stability, as indicated by an increase in density, melting point, enthalpy of fusion, and degree of contraction. In the polymorphs with lower stability (α), the molecules have significantly more rotational freedom within the crystal lattice than found in the most stable polymorph (β).

Milk fat primarily forms into the α and β' polymorphs due to the complexity of molecular composition [26,38]. The less stable α polymorph transforms to the more stable β' state, especially when temperature is somewhat elevated (room temperature). The presence of a β polymorph has been seen under certain conditions [37–39], primarily associated with the high-melting fraction or milk fat that has been stored for long times.

A combination of differential scanning calorimetry (DSC) and X-ray powder diffraction (XRD) spectroscopy were used to characterize the different polymorphs found in milk fat crystallized under different conditions [40]. In contrast to some previous work [37–39], these researchers found no evidence for a β polymorph. However, the γ, α, and β′ forms were found to crystallize depending on the cooling rate and final temperature. Milk fat fractions were also found to remain in the β′ polymorph for long times [26,40].

C. Nucleation

The formation of a crystalline lattice from a molten state requires that an energy barrier for formation of the solid-liquid surface be overcome. According to classical homogeneous nucleation theory [41–42], this energy barrier is overcome when the number of molecules aggregating in the liquid reaches a certain critical size. Above this size, the molecular clusters survive as nuclei and grow into stable crystals depending on the environmental conditions. Nucleation is enhanced by the presence of certain impurities (dust, vessel walls, crystals of other materials, etc.) that provide a measure of surface energy and allow a cluster with fewer molecules to attain the critical size for stability. In most natural systems, the abundance of nucleating sites makes heterogeneous nucleation the primary mechanism for birth of new crystals. In fats, secondary nucleation is thought to be an important mechanism [37], where the presence of a crystalline fat phase induces additional nucleation in the melt. However, few studies have been done to quantify this type of secondary nucleation in lipids.

When the first milk fat crystals form upon cooling of a milk fat melt, it is likely that the highest-melting TAG play a major role. That is, the molecules with the highest melting point become supersaturated first and are the first to nucleate. It is also likely that these nuclei are catalyzed by foreign sites [37]. That is, milk fat nucleation is a heterogeneous nucleation event, where a foreign (not milk fat TAG) surface provides a measure of energy to help the TAG molecules overcome the energy barrier needed to form a stable nucleus. The exact nature of this nucleation site is a matter of debate, although there has been much recent discussion about the role of minor lipids with high melting points (e.g., phospholipids) as the active site for nucleation. The presence of MAG has been thought to promote nucleation [37].

Due to the wide range of molecular components in milk fat, there is no single species that dominates nucleation. Rather, it is groups of similar TAG that crystallize together. For example, the trisaturated, long-chain TAG have approximately the same melting point and will crystallize at about the same time. Depending on the temperature of crystallization, other groups of TAG with similar molecular orientation may fit loosely into the crystal lattice and form compound crystals with the trisaturated TAG. The speed of cooling and crystallization also play a role in which TAG fit together into the crystal lattice. This principle can be demonstrated by the difference in solid fat content obtained dependent on the path of crystallization [37]. Slow cooling allows similar TAG molecules to crystallize preferentially, whereas rapid cooling allows TAG molecules with diverse structures to cocrystallize in the lattice and form compound crystals. Thus, the final solid fat content at a given temperature is higher when the melt is cooled rapidly than when it is cooled slowly. The formation of compound crystals also helps to explain why milk fat remains in the β′ form rather than converting to more stable β polymorph.

The temperature profile and degree of supercooling during crystallization have significant effect on the nature of milk fat crystals formed [37,43]. When crystallization kinetics are slow, the differences in TAG composition between liquid and crystalline state are maximized [43]. Under conditions of significant supercooling, with faster crystallization kinetics, more compound crystals are formed [37,43].

Along these same lines, it has been suggested [44] that a semisolid state exists in milk fat crystals based on measurement made by nuclear magnetic resonance (NMR). The presence of a component of milk fat with an intermediate mobility (between crystal and liquid) suggests that not all of the milk fat is fully crystalline. These intermediate components may be molecular chains of TAG within the crystal lattice that have higher mobility (''liquid-like'' end groups). Based on these measurements [44], this semisolid state makes up less than 7% in milk fat. One can easily imagine that these liquid-like groups arise from the incorporation of mismatching TAG into the crystalline structure. Thus, these end groups are not bonded very well into the lattice and have greater mobility than expected. One would also expect that this phenomenon would be most prevalent in the least stable polymorphs (α) and become less important for higher-stability polymorphs.

To help distinguish the process of nucleation in milk fat, two different nucleation processes have been compared. In the first process, supercooled milk fat was cooled and continually agitated to promote nuclei formation and subsequent growth of crystals for separation. In the second process, the milk fat was quickly supercooled to a greater extent (low temperature), subjected to intense agitation for a short time, and then the quiescent melt was warmed up to an incubation temperature only slightly below the melting point of the milk fat. In the latter process, there was a prenucleation state, following the inducing action, where the melt remained clear and no distinct crystals were evident (either visually or microscopically). This is a period when TAG molecules are ordering into prenucleation embryos. This was followed by a period where the melt became slightly cloudy as embryos were growing into nuclei. These nuclei then grew by consuming the embryos in their vicinity until structure transformation was complete. The number and type (or composition) of nuclei formed were dependent on the specific energy level used in the inducing action as well as the TAG composition of the original melt. Interestingly, the melting point of the nuclei formed was dependent on the nature of the initial formation of nuclei. If nuclei were allowed to form spontaneously in the melt and grow under static conditions, the melting point of the crystals remained the same throughout the duration of crystallization. However, if nuclei were initiated by a brief inducing action and then allowed to grow under static conditions, the melting point of the nuclei increased with time. This transformation was accompanied by an increase in long-chain TAG and a decrease in short-chain TAG within the crystals. These results suggest that the shearing action causes a different group of TAG to combine into the initial nucleus than when the milk fat nucleates without shear.

The kinetics of nucleation of a milk fat model system, based on driving force as described in Fig. 3, were obtained at 25 and 27.5°C. By mixing HMF with LMF at different ratios, different supersaturations were generated. The induction time for onset of nucleation, as determined by a turbidity technique, decreased as the content of HMF in LMF increased. Figure 4 shows this trend in terms of the supersaturation driving force based on the effective solubility

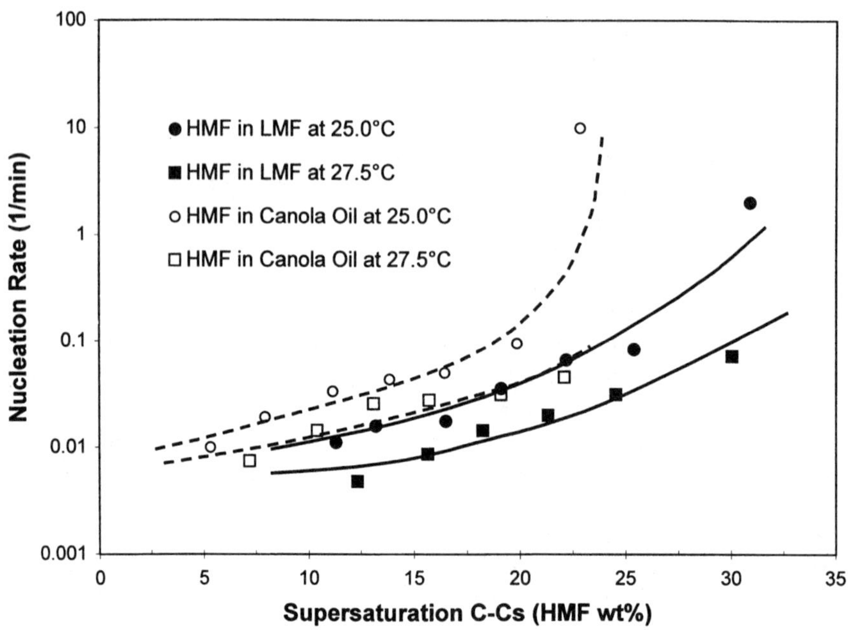

Fig. 4 Nucleation rate, expressed as inverse of induction time for onset of nucleation as measured by a turbidity technique, of a milk fat model system of high-melting in low-melting milk fat fractions. Effective solubility, C_s, as obtained from Fig. 3.

concentration from Fig. 3. Nucleation rate, expressed as the inverse of the induction time for onset of nucleation, increased with increasing supersaturation driving force, as expected.

The nucleation rate was also found to depend on the nucleation temperature and the Reynolds number characteristic of the agitation. As one might expect, the number of nuclei increased as nucleation temperature decreased (from 34.0 down to 26.5°C) and increased as the Reynolds number of agitation increased. The yield of crystalline material (solid fraction) also increased as nucleation temperature decreased and agitation intensity increased.

Nucleation kinetics of milk fat and a milk fat model system composed of mixtures of high- and low-melting fractions were also studied [45]. A spectrophotometric turbidity technique was used to characterize the onset of crystallization at isothermal conditions for temperatures from 25 to 40°C. The kinetics of nucleation of the intact milk fat (less than 20% high-melting fraction) were slower than any of the mixtures of 30–70%, 40–60%, and 50–50% high-melting in low-melting milk fat fractions. Induction time for onset of crystallization decreased as the percentage of high-melting component in the system increased. In all cases, it was the β′ polymorph observed by x-ray diffraction. The activation free energies for nucleation, found from the Fisher-Turnbull equation, were quite low (1.0 to 10 kJ/mol), although they were in the same range found for other fat systems. The free energy for nucleation increased as crystallization temperature increased. Also, the free energy for nucleation decreased, at any given temperature, as the percentage of high-melting component in the system increased.

The minor lipid component of milk fat has also received recent attention as a potential influence on nucleation. It has long been suggested that polar lipid components with high melting point (e.g., phospholipids) form structures in solution prior to onset of TAG crystallization. Somehow, these polar lipid structures influence formation of TAG crystalline nuclei. Removal of the minor lipids (MAG, DAG, free fatty acids, sterols, phospholipids, etc.) naturally present in AMF caused the onset time for nucleation to decrease (Fig. 5) and the activation free energy for nucleation to decrease [45]. That is, nucleation occurred more rapidly upon removal of the minor lipids in milk fat. Other studies [46–47] have also found that removal of minor lipids from milk fat results in faster nucleation and that as more minor lipids are added, generally the nucleation rate is inhibited. However, the effect appears to be system specific and no general rules have yet been found. This is an active area of research and the effects of specific minor lipids on crystallization of milk fat (and other fats) has great potential for modification of crystalline fat microstructure.

D. Crystal Growth

Once nuclei have formed, they continue to grow as stable crystals. The nature of their growth is dependent on the conditions in the surrounding environment.

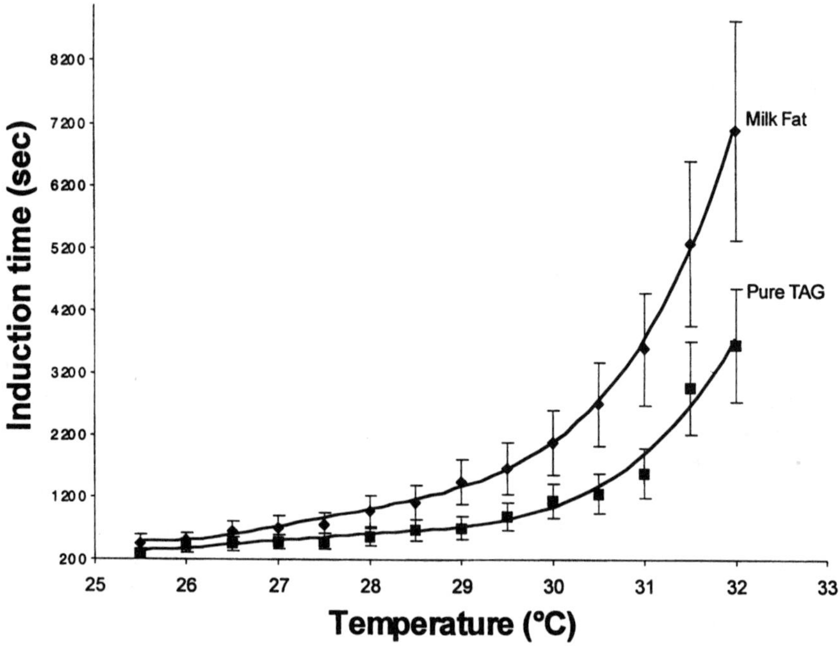

Fig. 5 Comparison of induction time for onset of nucleation of milk fat, as measured by a turbidity technique, at different temperatures with purified milk fat triacylglycerols. (From Herrera et al. [45].)

Due to the similarity between all TAG in milk fat, whether HMF or LMF, there is undoubtedly substantial mixed crystal formation, where different TAG fit more or less easily into the lattice structure. The nature of this mixed crystal formation depends greatly on the operating conditions in the crystallizer, as discussed previously.

The change in solid fat content (by pulsed NMR) during crystallization of milk fat or a milk fat model system under isothermal conditions was also studied [45]. The solid fat content data were analyzed with the Avrami-Erofeev equation [48] to obtain kinetic parameters for crystallization. In all cases, the Avrami exponent was found to be approximately 3, although the half-time for crystallization decreased as the percentage of high-melting component increased.

Comparison of crystallization rate (which also includes some contribution from nucleation) was found to correlate well with increasing supersaturation as defined in Fig. 3. For model systems with different ratios of high-acyl-carbon-number (C46–C52) to low-acyl-carbon-number TAG (<C40), crystallization rate increased monotonically with increased supersaturation (Fig. 6). As ex-

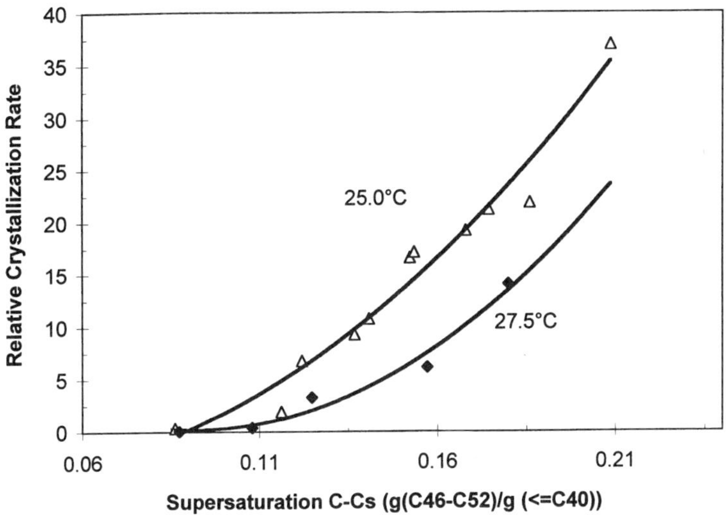

Fig. 6 Crystallization rate, expressed as relative slope of change in turbidity, of a milk fat model system of high-melting in low-melting milk fat fractions. Effective solubility, C_s, as obtained from Fig. 3. (From Hartel [133].)

pected, for equivalent supersaturation based on chemical composition, a decrease in crystallization temperature resulted in an increased crystallization rate. Such a relationship only provides a model for describing the rate of crystallization of milk fat for different compositions, and does not yield insight to the mechanisms of crystal growth. Further work is necessary to characterize milk fat crystallization kinetics based on a thermodynamic measure of driving force.

E. Crystal Network Structure

The arrangement of the milk fat crystals within the food product (like butter) contributes significantly to product texture. The crystalline microstructure, or the number, mean size, and distribution of sizes, shape, polymorphic form, and interacting networks between crystals, affects the physical and mechanical properties of the product, such as hardness, spreadability, etc. The nature of this crystalline microstructure is determined by the composition of the fat and the conditions under which it is crystallized [4,49]. For example, butter made from winter milk fat has traditionally been harder than that made from summer milk fat, probably due to the higher levels of saturated fatty acids in winter fat. However, butter can be softer or harder depending on the manufacturing process.

In lipid foods it has proven extremely difficult to measure the crystalline microstructure. Traditional microscopy techniques, including polarized light microscopy and electron microscopy, have proven of limited value in characterizing the distinct nature of individual crystals in lipid-based foods. Recently, confocal scanning laser microscopy (CSLM) has been used to identify the individual crystals, without loss of identity, within milk fat samples [49]. This technique is based on fluorescence of a dye that is either diffused into the product or cocrystallized with the solid fat to provide a distinct boundary between solid and liquid fat for observation. This distinct boundary is easily characterized by fluorescence measurement techniques, and the images obtained show distinct crystal microstructures within the intact sample. The use of laser light allows penetration within the sample to depths of 30–40 μm. Further advances in multiphoton excitation allow penetration up to 80 μm and provide greater potential for microstructure analysis. However, observation of lipid crystal network structure even by confocal microscopy is somewhat limited. The resolution of CSLM is only about 1 μm and the penetration depth within the sample is still quite limited. Thus, not all of the details of network structure can be obtained by this technique. Despite these limitations, CSLM can be used to provide details of the network structure that have not been previously obtained.

The lipid crystalline microstructures in a model milk fat system were related to the mechanical properties of the product [50–51]. High- and low-melting milk fat fractions were mixed at different ratios, crystallized under agitation at 25°C, and then cooled quiescently to 10°C for observation. The crystalline microstructure of these products was evaluated with CSLM, and the mechanical properties (storage modulus) were measured with dynamic mechanical analysis (DMA). Figure 7 shows a CSLM image of two samples crystallized at 25°C with different agitation rates. The images clearly show primary crystals (denser) formed at 25°C surrounded by a matrix of secondary crystals (more diffuse). Since the solid fat content (as measured by pulsed NMR) of these samples was 41%, there was substantial liquid fat surrounding these crystalline matrices. Significant differences in crystalline microstructure and storage modulus (related to hardness) were observed for crystallization at different processing conditions. For example, higher agitation rates in the initial stage of crystallization resulted in smaller and more numerous crystals (Fig. 7). These smaller crystals in this lipid matrix led to lower storage modulus [50–51]. Clearly, both the chemical composition and

---

Fig. 7 Confocal scanning laser microscopy images of a 50–50% blend of high-melting fraction in low-melting fraction of milk fat crystallized at 25°C, cooling rate of 5.5°C/min, and agitation rates of (a) 50 rpm, (b) 100 rpm, and (c) 200 rpm. Samples were stored 24 hr at 10°C. Depth: 15 μm from surface. (From Herrera and Hartel [51].)

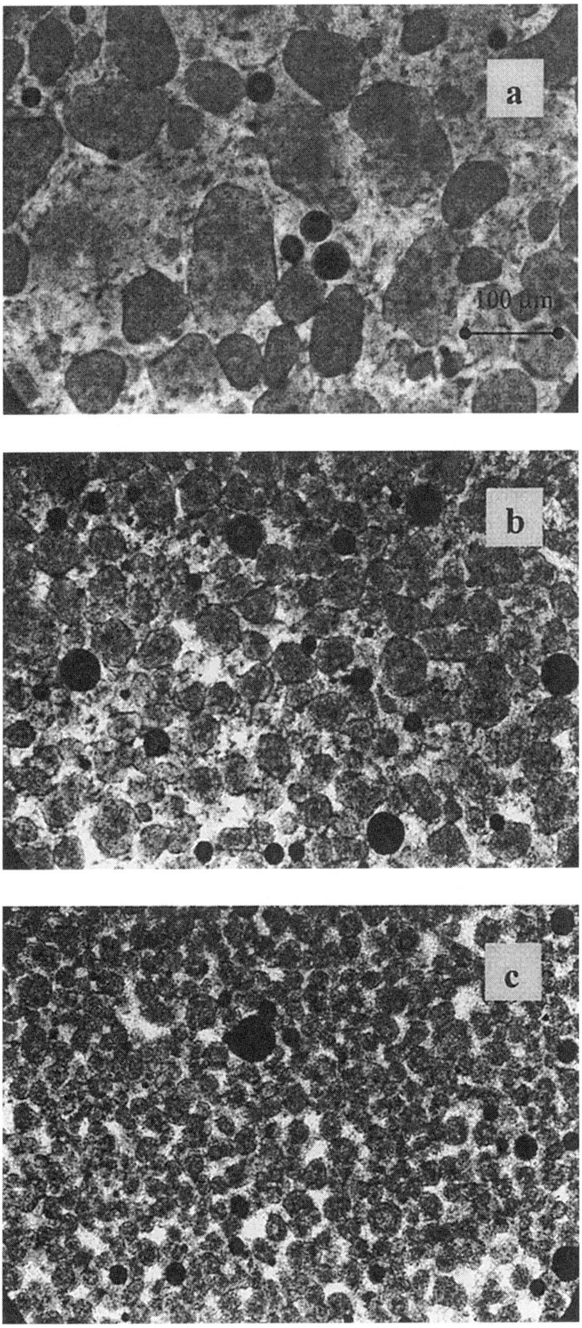
a
100 μm
b
c

crystallization conditions can significantly impact the textural properties of a food through influence on the crystalline microstructure. Further work in this area is necessary to fully document these interrelationships.

II. FRACTIONATION TECHNOLOGIES

In recent years, most of the work on milk fat fractionation has been done in the absence of solvent and based on crystallization by manipulation of the temperature. However, there are different techniques by which the milk fat can be crystallized and separated. These can be loosely classified according to how crystallization is accomplished. In suspension crystallization, the heat of crystallization is removed through the liquid surrounding the crystals, whereas in solid-layer crystallization, the heat of crystallization is removed through the crystallizing layer itself.

Controlling crystallization of milk fat to promote efficient separation is at the heart of fractionation technologies. In its simplest representation, the molten milk fat must be brought to some condition of thermodynamic nonequilibrium where formation of nuclei from the liquid mass is promoted. These crystals must then be grown to the desirable size and shape to give efficient separation in some way. To provide efficient separation, the crystallization process must be controlled to produce crystals that are easily separated from the remaining liquid.

A. Suspension Crystallization

Suspension crystallization of milk fat involves cooling a molten fat to an appropriate crystallization temperature, allowing crystallization to occur for an adequate period of time, and then filtering off the slurry that has formed, as shown schematically in Fig. 8. Commercially, this process has been enhanced by controlling the cooling profile to control nucleation and growth of milk fat crystals. In this way, the efficiency of crystallization and separation can be enhanced to produce milk fat fractions of high yield with the highest possible melting point. However, one of the primary problems with this technique is that a substantial amount of the liquid fat is entrained in the solid fraction during separation. High liquid entrainment means inefficient separation and fractions with less desirable properties. For example, the entrainment of LMF in HMF upon fractionation causes the yield to increase but also causes the melting point and solid fat content (SFC) at any temperature to be lowered. Thus, techniques for optimizing the separation process are necessary.

Typically, separation of the milk fat slurry is accomplished by filtration. In the early development of milk fat fractionation, low filtration pressures (5 bar) were used. It is now more and more common to use high-pressure filtration (30

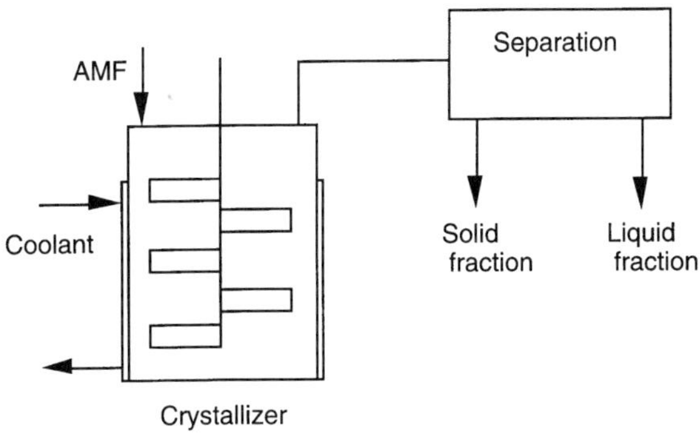

Fig. 8 Schematic diagram of anhydrous milk fat (AMF) fractionation by melt crystallization.

bar) to obtain the most efficient separation. The higher filtration pressure squeezes more of the entrained liquid out of the filter cake and gives higher efficiency at the price of slightly lower yield at a given temperature.

To promote efficient separation of HMF from the remaining liquid, for example, both composition and crystallization conditions play an important role. The composition of milk fat can vary due to differences in feeding source, lactation period, time of year (related to both feeding source and lactation period), breed of cow, and the process used to produce the anhydrous milk fat (AMF) from which milk fat fractions are produced. Thus, different milk fats may crystallize in different ways and give different products upon fractionation. Also, the conditions under which milk fat is crystallized can play a major role in the efficiency of fractionation. Cooling rate from the molten state to the crystallization temperature, the temperature to which the sample is cooled, and the agitation rate during crystallization are all important variables for controlling crystallization and obtaining the most efficient fractionation.

Recently, milk fat fractionation efficiency was evaluated based on filtration rates in a laboratory, low-pressure (5 bar) filtration device [52]. Molten milk fat was cooled from 60 to 28°C under agitation to promote crystallization with controlled agitation. The rate of filtration was measured by weighing the amount of liquid collected over time when the slurry was filtered. A plot of inverse of filtration rate against mass of filtrate, according to Darcy's law for filtration, resulted in a straight line over a portion of the filtration process, from which a measure of filtration resistance was determined. Note that this filtration resistance is correlated inversely with filtration time. The amount of liquid entrained in the filter cake was measured with a colorimetric technique, which assumes that all the yellow color in milk fat remains in the liquid phase [53]. Physical properties and chemical composition of the fractions were also measured. The results are summarized in Fig. 9, which shows the relationship between agitation intensity and filtration resistance, melting point of the solid fraction and percentage of liquid entrained in the filter cake. The agitation intensity is based on agitator tip speed, which is a function of both agitator diameter and agitation rate (RPM). Also shown in Fig. 9 is a representation of the state of the crystal dispersion in the slurry. At low agitator tip speeds (small diameter impellers and/or low agitation rates), the crystal size distribution contained many large aggregates along with smaller milk fat crystals. These did not filter very well and filtration resis-

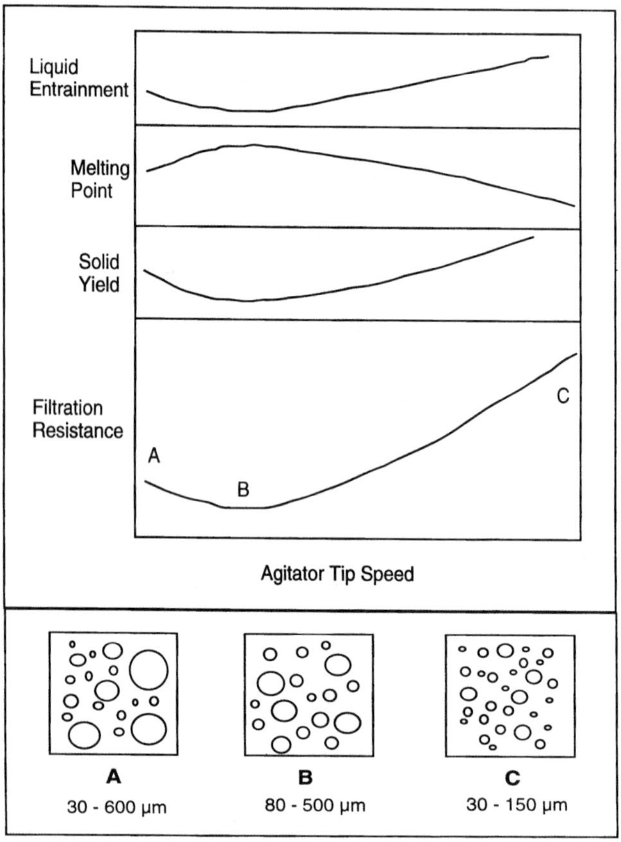

Fig. 9 Schematic diagram showing effect of agitator tip speed during crystallization on filtration resistance, yield, melting point of solid fraction, and extent of liquid entrainment in solid fraction. Approximate nature of crystal size distribution of milk fat crystals also shown. (Adapted from Patience et al. [52].)

tance was relatively high. At some optimal agitator tip speed, the crystal size distribution had many uniformly sized crystals that filtered efficiently and produced a solid fraction with the highest melting point (lowest level of liquid entrainment). At higher agitator tip speeds (either by large-diameter impellers or higher stirring rates), the crystals became too small to filter efficiently. Even though the crystals were fairly uniform in size at this condition, their small size caused difficulties in filtration. Thus, the melting point of the solid fraction was lower and the amount of liquid entrained was substantially higher.

Based on the principles of controlling the three steps of crystallization (generation of supersaturation, nucleation, and growth), a new technique for fractionating milk fat has been devised [54]. In this system, the molten milk fat is cooled rapidly to the proper temperature, where nucleation is induced by intense agitation for a brief period of time. Since conditions that give optimal nucleation are not those required for optimal growth of crystals, the nuclei must be brought into a different environment to allow them to grow at the most efficient rate. In this process, the nuclei are brought to a warmer temperature, and only moderate agitation is employed to promote rapid growth of the desired crystal size and shape. If temperatures are kept too low during crystal growth, mixed crystals are formed with lower melting points and less efficient separation. Using this technology, a continuous slurry of easily separated milk fat crystals is supplied to filtration device for separation. This technology also has been evaluated for fractionation of other fats (palm oil, tallow, etc.).

Recently, a continuous fractionation apparatus based on crystallization under shear conditions also has been proposed [55]. Here, crystallization of milk fat was enhanced through controlled shear conditions with separation of solid phase accomplished by cross-flow filtration.

B. Solid-Layer Crystallization

In many cases, it has been found that removing heat through a solid layer of crystallizing material provides better control of crystallization and separation. The chemical process industry has been using these techniques for years to purify various chemicals. The primary advantage of solid-layer separation techniques is that there is no handling or separation of solids. Separation of solid from liquid is accomplished simply by draining and no filtration is needed. These solid-layer fractionation techniques have been applied to milk fat in recent years with mixed results [56–57].

In the simplest case, a cold surface (a cold finger) is submerged into a vessel containing molten milk fat. By carefully lowering the temperature of the metal surface, with refrigerant circulating within a cylinder of metal submerged in the milk fat, a purified layer of crystalline milk fat can be separated from the liquid, as shown in Fig. 10. To maintain constant temperature at the crystallizing

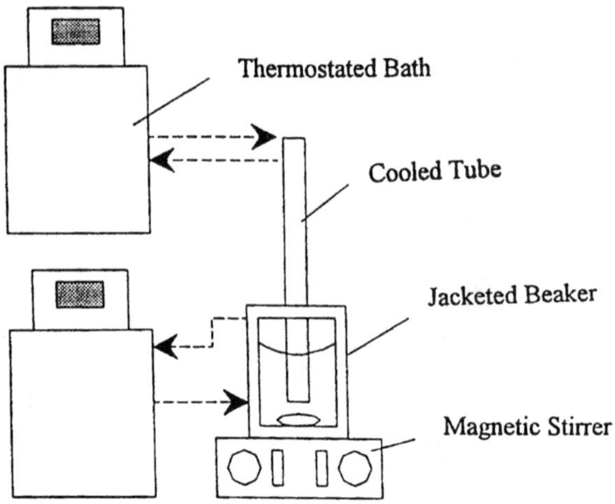

Fig. 10 Principle of solid-layer separation operated in static mode. (From Peters-Erjawetz et al. [57].)

surface despite the continual buildup of the layer, a temperature profile is employed. Typically, the temperature of the inside of the tube is reduced at a rate correlated to the rate of growth of the layer, since the heat transfer resistance is directly proportional to thickness of the layer. The layer is built up on the cold surface until the desired separation is complete. The actual physical separation step is quite simple—the cold surface is removed from the liquid fat and allowed to drain off. Once the liquid melt has drained away sufficiently, the metal surface is warmed and the solid fat layer melts off the metal surface. The high-melting fraction, now in liquid form, is collected and pumped away for further processing. The process as described is called static layer crystallization since the liquid fat is not moving during crystallization.

It is often more common to find a dynamic solid-layer process, where the liquid melt is continually pumped in a thin film along the cold surface. The remaining liquid is collected at the bottom of the tube and recirculated to the top for further crystallization. The dynamic process can be seen schematically in Fig. 11. The advantage to using this process is that the liquid flow promotes mass transfer of crystallizing species to the solid layer and can give higher crystallization rates.

In both static and dynamic solid-layer crystallization, separation efficiency can be enhanced by postcrystallization treatment of the solid layer prior to melting it off the tube. Postcrystallization treatment can involve sweating at intermediate temperature to promote removal of any liquid fat entrained in the solid layer or

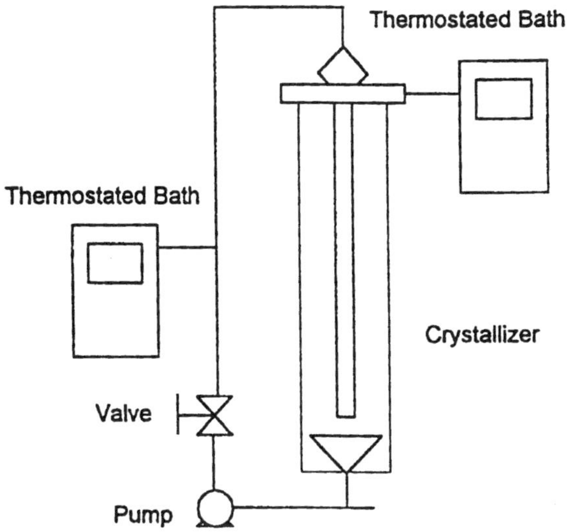

Fig. 11 Principle of solid-layer separation operated in dynamic mode. (From Peters-Erjawetz et al. [57].)

mechanical pressing of the surface to squeeze out the liquid. Sweating can be accomplished by heating the outside of the tube with warm gas or by warming the inside of the tube with heating medium. If the inside of the tube is warmed too much, adhesion of the milk fat layer to the metal tube is reduced and the solid-layer can slough off the tube and be contaminated with the liquid draining out. This reduces the effective separation efficiency of the technique.

In the past few years, solid-layer technologies have been evaluated for separation of milk fat [56–57]. The results clearly showed that this technology could be used to separate milk fat into similar fractions as produced by commercial suspension technologies. Fractions with similar yields and with similar chemical and physical properties were produced. However, the growth rates on the solid layers were on the same order of magnitude as those found in suspension crystal growth. Therefore, the advantages of solid-layer technology were not as clear as is often the case with other crystallizing species (in the chemical process industry, for example). When the costs of fractionation by the two technologies were compared, the high surface area required to obtain sufficient yields by solid-layer crystallization resulted in unreasonably high capital costs for the equipment. Washing and mechanical squeezing also were found to increase the efficiency of separation, but were not studied in detail.

Separation efficiency in solid-layer fractionation of milk fat can be enhanced by using sweating techniques to purify the solid layer after crystallization [57]. It was shown that sweating with either a warm gas on the outside or warm water on the inside of the tube caused an increase in melting point of the solid fraction by 3–4°C. The advantage of using warm gas on the outside was that the solid layer did not slough off the tube. Postcrystallization treatment has the potential to enhance the economic competitiveness of solid-layer fractionation, although further work is probably needed to optimize a system for production of certain milk fat fractions.

III. PROPERTIES OF MILK FAT FRACTIONS

The raw material used for milk fat fractionation is anhydrous milk fat (AMF). AMF is defined by the U.S. Department of Agriculture [58] as 99.8% milk fat, obtained from fresh cream or butter. Fresh cream is the preferred starting material in order to maximize the flavor and quality. Anhydrous milk fat is produced by concentrating cream to 70–80% fat, followed by a phase inversion to produce a water-in-oil emulsion. The product is then further concentrated, washed, and vacuum dried.

The chemical composition of milk fat is one of its most unique characteristics, because of its complexity, its high level of short-chain fatty acids, and its fairly ordered TAG structure. The native chemical composition is a fundamental property that results from, and varies due to, feed, environmental factors, breed, health, and stage of lactation of the cow. The chemical composition, in turn, influences the physical and functional properties, the flavor profile, nutritional aspects, and the storage stability of milk fat.

A. Effects of Fractionation of Milk Fat Composition

The fatty acid and TAG compositions of milk fat fractions depend on many aspects of the fractionation process, including the driving force or type of fractionation (i.e., thermal-driven crystallization processes or molecular weight/solubility-driven extraction processes), the conditions used to fractionate the milk fat, the type of filtration (pressure or vacuum), and the starting materials (intact milk fat or milk fat fraction).

The effects of fractionation on the chemical composition of milk fat are broadly generalized as follows. In fractions obtained by thermally driven processes, the high-melting fractions contain greater amounts of long-chain saturated fatty acids and higher-numbered TAG, and the lower-melting fractions contain more short- medium-, and long-chain unsaturated fatty acids and lower-numbered

TAG. In general, there is a gradual change in composition from high-melting to low-melting fractions. These changes are documented in Tables 8 and 9 [28]. In fractions obtained by molecular weight or solubility-driven processes, the high-melting fractions contain long-chain saturated and unsaturated fatty acids, and the low-melting fractions contain the short- and medium-chain fatty acids.

The chemical compositions of milk fat fractions obtained from melt crystallization (thermal process) have been compared [59] with those obtained from supercritical (SC) CO_2 extraction (separation based on molecular weight). The high-melting fraction obtained from SC CO_2 extraction had a high concentration of long-chain, unsaturated fatty acids compared to the high-melting fraction obtained from melt crystallization. This is not surprising since the molecular weight of oleic (C18:1) acid is the same as stearic (C18:0) acid, although the melting points are substantially different. Cholesterol generally was concentrated in the fractions with lower melting points regardless of the fractionation technique used. However, carotenoids showed different behavior between melt crystallization and SC CO_2 extraction of milk fat [59]. The carotenoids were found almost exclusively in the high-melting fraction obtained by SC CO_2 extraction, whereas carotenoids were not fractionated at all in the different fractions obtained by melt crystallization. In general, the carotenoids remain with the liquid phase during crystallization, so this latter result is a little surprising. However, in melt crystallization, a substantial amount of liquid (50–70%) remains entrained in the filter cake [52], and perhaps this is the reason why little segregation of carotenoids in fractions produced by melt crystallization was found [59].

Table 8 Fatty Acid Composition (wt%) of Typical Milk Fat Fractions

Fatty acid	Milk fat	HMF[a]	MMF[a]	LMF[a]
C4:0	4.0	—	2.1	4.9
C6:0	2.8	0.1	1.5	3.0
C8:0	1.3	0.1	0.8	1.5
C10:0	2.6	0.6	2.0	3.1
C12:0	3.0	1.6	2.5	3.4
C14:0	10.4	10.5	10.2	10.5
C16:0	26.1	40.2	35.1	22.7
C18:0	13.3	31.6	22.0	9.2
C18:1	24.4	8.9	14.9	28.0
C18:2	1.7	0.6	0.9	2.1

[a] HMF = high-melting fraction; MMF = middle-melting fraction; LMF = low-melting fraction.

Source: Ref. 28.

Table 9 Triacylglycerol Composition (wt%) of Typical Milk Fat Fractions

TAG[a]	Milk fat	HMF[b]	MMF[b]	LMF[b]
24	0.3	—	—	0.4
26	0.3	—	—	0.4
28	0.5	—	—	0.5
30	0.9	—	—	1.4
32	2.2	—	0.1	2.9
34	5.4	0.1	1.1	7.0
36	10.6	0.2	6.2	12.9
38	12.9	0.3	10.3	15.0
40	10.0	0.6	7.6	11.7
42	6.2	1.9	7.0	6.2
44	5.7	5.2	8.5	4.8
46	6.6	12.4	9.9	5.3
48	9.0	20.7	12.2	6.7
50	12.3	28.3	17.5	9.2
52	11.3	22.4	14.9	9.4
54	5.5	7.3	4.6	6.1
56	0.3	0.5	0.1	—

[a] Acyl carbon number.
[b] HMF = high-melting fraction; MMF = middle-melting fraction; LMF = low-melting fraction.
Source: Ref. 28.

In most fractionation processes, particularly melt crystallization, the chemical composition of middle-melting fractions is very similar to intact AMF. More variation in composition is seen between fractions obtained by multiple-step fractionation than by single-step processes. A detailed summary of milk fat fraction composition was reported by Kaylegian and Lindsay [1].

B. Effect of Milk Fat Fractionation on Flavor

Milk fat flavor is unusual because it involves many compounds, some of which are found as precursors, that form flavor compounds in the presence of heat, water, acid, or lipases. This is further complicated by the fact that milk fat has several characteristic but distinct flavor profiles described as fresh, cultured, cooked or drawn, browned, and burnt.

Milk fat flavor is a combination of its inherent chemical composition, its solvent property to transfer flavors from feed sources, the cow's metabolism, compounds generated from other reactions such as oxidation or thermal degradation, and microbial and enzymatic reactions [60]. Classes of compounds important to milk fat flavor include lactones, methyl ketones, low-molecular-weight and branched-chain fatty acids, aldehydes, and other minor compounds [1,60].

The flavor and flavor potential of milk fat fractions have been reported as lactone, methyl ketone, and aldehyde concentrations [1,60]. The production of AMF for use in milk fat fractionation results in an overall decrease in flavor potential of AMF compared to fresh milk fat. General trends observed were that the lactone and methyl ketone concentrations in very-high-melting and high-melting fractions were lower than in intact milk fat and higher than intact milk fat in middle-melting and low-melting fractions.

Kaylegian and Lindsay [1] noted that flavor compounds in milk fat fractions obtained by crystallization processes tend to partition into the liquid fractions that are produced later in the fractionation process, whereas in fractionation using supercritical fluid extraction the flavor compounds were greater in the first extracted fraction obtained. This was verified in the work of Bhaskar et al. [59].

C. Nutritional Aspects

In the past, the nutritional aspects of milk fat have focused on the saturated fat and cholesterol content of milk fat, largely from a negative point of view. However, as total fat and cholesterol are put into perspective in a healthy diet, milk fat is a nutrient with some very positive aspects. The metabolic and anticarcinogenic properties of milk fat are becoming more widely known, even to the extent of being listed specifically as a neutraceutical ingredient in a popular trade magazine article [61].

Milk fat is generally considered to be a saturated fat, but 10% of the saturated fatty acids are short-chain fatty acids. These fatty acids are metabolized differently than other saturated fatty acids and do not contribute to hypercholesterolemia [62–63]. A study on the metabolism of milk fat and milk fat fractions in rats showed that ingestion of low-melting fractions produced plasma lipid profiles similar to a corn oil diet, and ingestion of high-melting fractions produced plasma lipid profiles similar to a palm oil diet [64]. These results suggest that fractionation may improve the nutritional profile of milk fat.

Supercritical fluid extraction has been used specifically to remove cholesterol from milk fat [65–67]. Supercritical carbon dioxide extraction with a silica gel column has been used to remove cholesterol from milk fat simultaneously during fractionation [68].

Individual fatty acids in milk fat are recently recognized as having important biological functions. Conjugated linoleic acid (CLA) is an effective anticarcinogenic agent in animal models, and is found in high concentrations in milk fat [69–71]. Butyric acid is important in cellular functions, particularly cell death, and is considered to be an anticarcinogenic agent [71–72]. The molecular and genetic effects of butyric acid on cells and their role in bowel disease and cancer have also been described [72]. Milk fat fractions enriched in these fatty acids may prove to be nutritionally desirable food ingredients.

D. Oxidative Stability

The oxidation of milk fat leads to a deterioration of quality and is detected by the presence of off-flavors and the formation of polymers. Factors affecting the oxidative stability of milk fat include saturated fatty acid content, availability of oxygen, storage temperature, exposure to light, metallic contamination, and naturally occurring antioxidants. In general, the oxidation rate of fats, as measured by peroxide value or carbonyl content, increases as the unsaturated fatty acid content increases. Consequently, low-melting milk fat fractions tend to oxidize sooner than intact milk fat, and high-melting milk fat fractions tend to oxidize slower than intact milk fat [1].

Typically, polymer formation in fats and oils occurs when they are subjected to high temperatures, such as in frying, although not all fats undergo this polymerization at the same rate. Both solid and liquid milk fat fractions were more stable toward polymer formation than intact milk fat after 16 hr at 185°C [73]. In addition, the degradation properties of a low-melting milk fat fraction did not differ significantly from intact milk fat in a frying situation [74]. The addition of the low-melting milk fat fraction to sunflower oil improved the frying stability of the sunflower oil, and may show synergistic effects with other oils in this application [74].

E. Effect of Milk Fat Fractionation on Physical Properties

The physical properties of milk fat and milk fat fractions are often expressed in terms of melting and crystallization behaviors and textural or rheological properties. These properties are temperature dependent and result from complex interactions between the solid and liquid components and the transition between phases. The basic melting and crystallization properties of fats are determined by the chemical composition. The physical properties are also influenced by the conditions employed during processing and the temperature history.

The dropping point of AMF ranges from 32 to 36°C. Milk fat has a broad melting range; it is completely solid at −40°C, 50% solid at 5°C (refrigerator temperature), 17% solid at 20°C (room temperature), and fully melted at 38°C (body temperature). Typical solid fat melting profiles for milk fat and milk fat fractions are shown in Fig. 12. Typical differential scanning calorimetry (DSC) melting profiles of milk fat or milk fat fractions obtained from melt crystallization (Fig. 13) demonstrate how the peaks for the three main glyceride species in milk fat shift with fractionation temperature. Even in the highest melting fraction, there

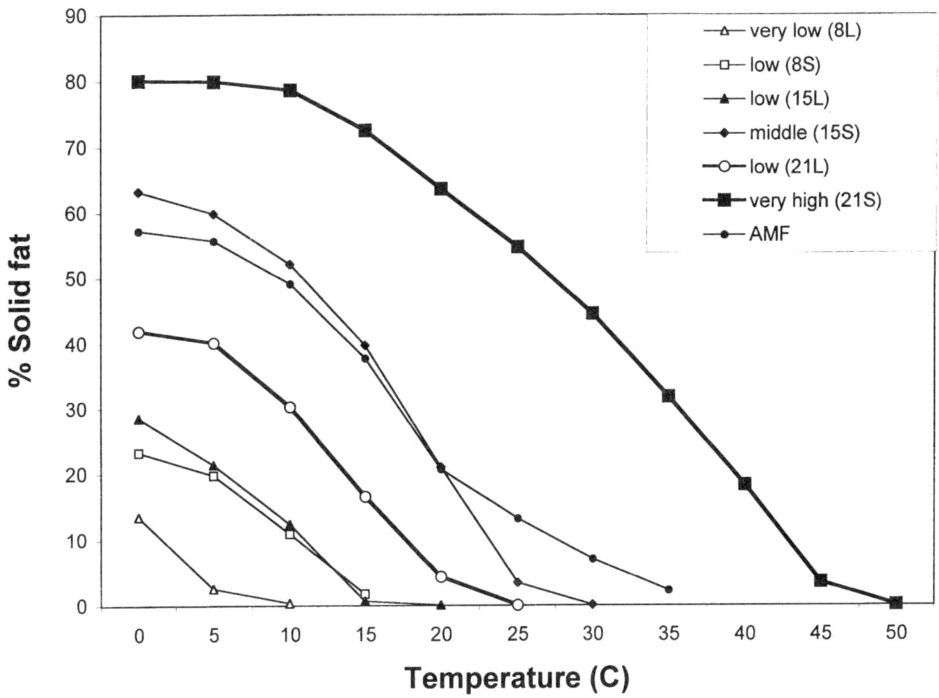

Fig. 12 Solid fat content curves for milk fat and milk fat fractions. The number in parentheses represents the temperature (°C) of fractionation. S indicates solid fraction and L indicates liquid fraction at that temperature.

is a peak denoting the presence of some of the middle or low-melting components. Also note that the peak temperature for the high-melting fraction increases dramatically (from about 35°C to about 45–50°C) when the lower-melting components are removed. This demonstrates the softening effect of the lower-melting glycerides on the higher-melting glycerides.

Milk fat fractions are often defined by their melting point: low-melting fractions melt below 25°C, middle-melting fractions melt between 25 and 35°C, high-melting fractions melt between 35 and 45°C, and very-high-melting fractions melt above 45°C [1]. The effects of fractionation on physical properties are related to the changes in chemical composition that occur during fractionation. In some cases, notable variations in physical properties can occur with only minor changes apparent in the chemical composition.

The melting behavior of milk fat fractions is dependent on the type of fractionation technology, the conditions used to fractionate and filter, the number

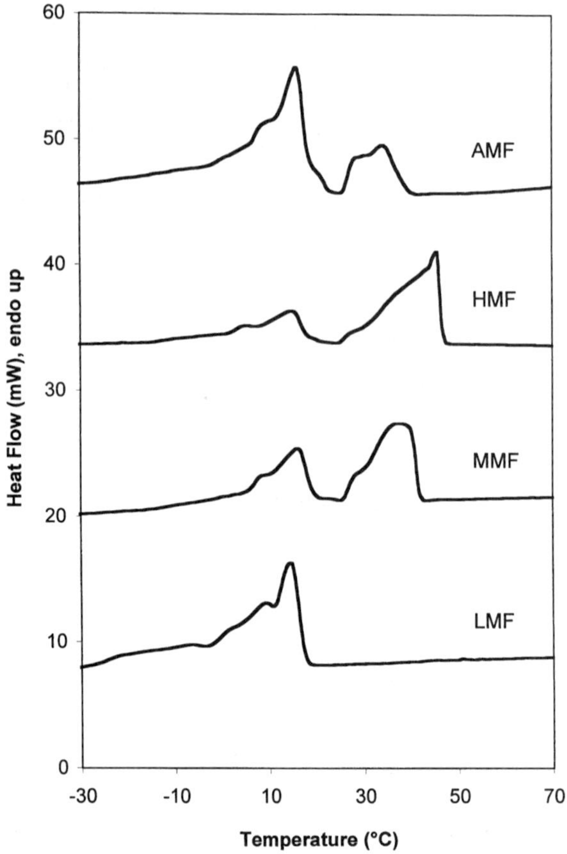

Fig. 13 Melting profiles, measured by differential scanning calorimetry, for milk fat and milk fat fractions. AMF = anhydrous milk fat; HMF = high-melting milk fat fraction; MMF = middle-melting fraction; LMF = low-melting fraction.

of fractionation steps used, and the nature of the starting materials. Milk fat fractions generally exhibit increased melting points and solid fat content profiles as fractionation temperature increases. Fractions obtained from single-step fractionation processes exhibit a gradual decrease in the melting profile as the fractionation temperature decreases. Fractions obtained from multiple-step process exhibit more varied profiles: the first fractions obtained have gradual melting curves similar to single-step fractions, 50–70% solid at 0°C, and melted by 40°C; the middle fractions have steeper melting profiles, 50–60% solid at 0°C, and melted by 30°C; and the last fractions obtained have shallow curves, less than 20% solid at 0°C, and melted by 10°C (Fig. 12). A detailed summary of the melting behavior of milk fat fractions has been reported by Kaylegian and Lindsay [1].

F. Textural Characteristics

Butter (i.e., intact milk fat) is generally considered too hard to spread from the refrigerator, but too soft for pastry manufacture, and is not considered to be plastic. These features of butter have led to much research on ways to improve its performance. The desired textural characteristics will depend on the application, and, consequently, properties should be tailored as needed to produce the best possible spread or ingredient.

Textural properties, such as hardness, spreadability, and plasticity, result from a combination of the melting properties and the processing conditions, and the presence of other ingredients (e.g., the aqueous phase of butter). These characteristics are quantified by hardness force, penetration depth, yield value, and other rheological measurements.

Plasticity in fats is described as an ability to be pliable, or kneadable, without getting greasy and oiling off. Plastic fats are characterized by a three-dimensional network consisting of strong primary bonds between crystals and weak secondary van der Waals bonds [75]. The primary bonds are broken irreversibly upon deformation, but the weak bonds reform easily after deformation resulting in viscoelastic behavior [76]. Rheological properties of plastic fats are related to the phase volume of crystals, their spatial arrangement, and the formation of primary bonds between crystals [76–78]. Key parameters in creating a plastic fat include having the correct solid-to-liquid fat ratio, generating small β' crystals (5 μm or less), the correct crystal network and microstructure, and a good water droplet distribution [1,75,79–85].

Improving the plasticity of milk fat is accomplished by selecting the correct blend of milk fat and milk fat fractions and fine-tuning processing parameters in churns and scraped-surface heat exchangers. Milk fat fractions are blended based on their melting profiles and chemical properties to achieve the desired melting profile for the application (e.g., a high-melting pastry butter or low-melting spreadable butter). Butter hardness was correlated with the content of myristic (C14), palmitic (C16:0), and oleic acids (C18:1), and with four major groups of TAG [86], whereas milk fat plasticity was correlated with the presence of long-chain trisaturated TAG [87].

Texturization and tempering parameters can greatly affect the crystallization properties to yield milk fat products that range from hard or brittle to pliable and plastic to greasy. Parameters such as the amount of cooling and crystallization, mechanical work, and resting time can be adjusted to create the appropriate texture for desired applications [76,83,88–90]. Recent studies on milk fat microstructure have increased our knowledge of the relationship between processing, microstructure, and performance of plastic fats [78–80,91].

IV. APPLICATIONS OF MILK FAT FRACTIONS

A. Bakery Products

Butter is used in premium baked goods because of its unique, desirable flavor and because it is a statement of quality. However, the functional performance of butter, primarily melting profile and degree of plasticity, is less than optimal for some applications. Traditionally, in the United States, the only two products available to the bakery industry are butter and AMF, essentially differing by the amount of water they contain. The availability of milk fat fractions, blending, and texturization technology allows the dairy industry to tailor the melting profiles, plasticity, and firmness of milk fat ingredients to meet the specific needs of individual bakery applications [1,82].

1. Laminated Pastries

Laminated pastries are flaky products characterized by discrete layers of fat and dough. A roll-in fat needs to be plastic in order to form thin continuous films (70 μm to 3 mm) during the rolling and folding processes. These films form barriers between dough layers to prevent gluten interaction and yield a flaky structure. The fat must have a sufficiently high melting point (36–42°C) so that it remains solid during production, but not be so hard and brittle that it ruptures the dough layers. The desired melting point is higher for fats used to make crispy products, such as puff pastries, than for more tender products, such as croissants or Danish pastries.

Traditional butter has a lower melting point and is less plastic in texture than is considered optimal for making pastries. The melting profile of intact milk fat is sufficiently high at cold temperatures, but it is too soft at most bakery temperatures. Doughs become soft and difficult to handle, and the liquid fat can be absorbed by the doughs and decrease the flakiness of the finished pastry. Blends of high-melting milk fat fractions with intact milk fat and other milk fat fractions, followed by appropriate texturization, create butters that are very suitable to pastry applications [1,82–83,90,92–93]. In countries where milk fat fractionation is common, the production of pastry butters is an important outlet for the high-melting fractions.

2. Short Pastries

Butter cookies, shortbread biscuits, and pie crusts are categorized as short pastries and are characterized by a tender, flaky crumb, or ''short'' texture. Fats function in short pastries to form barrier layers to prevent gluten interactions, and are dispersed throughout the dough in the form of small sheets and globules. A soft, plastic fat is desirable because it is more dispersable in the dough. Plasticity is particularly important in cookie applications where the butter is creamed with the sugar to incorporate air during the initial dough mixing stages. The liquid fat functions to lubricate the dough and affects cookie spread. The liquid fat also lubricates the finished product, enhancing its eating quality.

Lower-melting milk fat fractions and intact milk fat are blended to a target melting point of 21 to 32°C and a solids content of 18% to 30% at 20°C and 3% to 10% at 30°C for short pastry applications [1,94]. If the fat is too soft, the dough is overly fragile and greasy, leading to excessive cookie spread. Milk fat ingredients for short pastries can be either anhydrous shortenings or emulsified butters. An additional benefit of using low-melting fractions is the inhibition of fat bloom commonly associated with the storage of shortbread and butter cookies [1,82,93].

3. Cakes

Cakes are tender products characterized by a foam structure. Fat functions to stabilize small air cells that are required for good structure and volume in cakes, and provide lubricity and flavor to the finished product. The fat must have good aeration properties to thoroughly stabilize the air cells at the early stages of mixing so that a stable air-in-fat-in-water emulsion is formed when the liquid is incorporated into the batter. During the baking process, it is essential that the fat melt before the cake begins to rise, thus allowing the air bubbles to migrate into the aqueous phase and expand, due to temperature and chemical leavening effects, to form the proper structure and volume. A fat that is not plastic or has too high of a melting point will result in cakes with an uneven grain structure and low volume. Vegetable cake fats are generally anhydrous products (i.e., shortenings) to maximize the fat content and aeration properties, and they contain emulsifiers to improve dispersion of the fat and air bubbles and improve the batter liquid-holding capacity [84].

Traditional butter and anhydrous milk fat exhibit poor plasticity and have a slightly higher melting point than is desirable for cake applications. Texturization of intact milk fat greatly improves its performance in cakes [88]. Low- and middle-melting milk fat fractions are used to produce milk fats for cakes with target melting points of 28 to 32°C. The combination of softer milk fat and plasticization produces a milk fat shortening that exhibits improved performance, which can be further enhanced by the addition of emulsifiers to the milk fat [1,90,92–93].

A low-melting milk fat fraction was used [95] to improve the melting characteristics and functional properties of bakery shortenings based on palm oil (PO). Shortenings made with hydrogenated or interesterified PO and AMF or LMF were characterized for melting behavior, crystal characteristics, and textural properties. Shortenings made with PO and LMF formed β crystals, but shortenings made with modified POs and AMF formed β′. PO is naturally a β-forming fat and was not affected by the presence of 40% LMF. The shortenings were not performance tested in bakery products.

4. *Icings*

A blend of low-melting milk fat fraction and AMF was used to produce a buttercream icing with good aeration properties that performed better than regular butter during creaming and piping [92].

B. Confectionery Products

Milk and milk components are a major ingredient in many confections. The use of milk fat in confections is primarily for flavor reasons, although textural impacts may be important as well. Probably the most important applications of milk fat in confections are in chocolate, caramel, and toffee. In caramel and toffee, the milk fat imparts a high-quality dairy flavor. In chocolate, the presence of milk fat also imparts a dairy flavor, but other effects (texture, shelf stability, etc.) may be just as important.

1. *Chocolate*

Milk fat has several important impacts on chocolate, besides being added for the dairy flavor. Even in dark chocolates, small amounts (<2–3%) of AMF may be used to modify the texture of the chocolate. This low level of milk fat is sufficient to soften the chocolate substantially. The degree of softening can be controlled by the amount of AMF added. The softening effect is related to the phase compatibility of the two fats [33,96–98]. Addition of high-melting milk fat fractions causes less softening of chocolates than addition of low-melting fractions [96–98]. The phase behavior of mixtures of cocoa butter and milk fat fractions has also been studied by Md. Ali and Dimick [99] and Sabariah et al. [100]. The solubilization of cocoa butter TAG by the low-melting components of milk fat was clearly shown by Simoneau and German [101].

The addition of even low levels of milk fat requires that tempering conditions be modified to account for the slower crystallization of cocoa butter due to milk fat inhibition. In general, lower tempering temperatures are needed as milk fat level increases. Milk fat and middle- to high-melting milk fat fractions crystallize much more rapidly than cocoa butter at temperatures between 15 and 30°C [98]. However, the addition of milk fat or fractions to cocoa butter, at even levels of 10%, substantially delays the time for onset of cocoa butter crystallization [98,102–103]. Despite this substantial decrease in crystallization kinetics, acceptable chocolates with up to 40% replacement of cocoa butter by milk fat or fractions can be accomplished through control of crystallization conditions [104–105]. Interestingly, the minor lipids in milk fat and milk fat fractions were also found to impact crystallization kinetics of cocoa butter [106]. Removal of the minor lipids from milk fat inhibited the onset of cocoa butter crystal formation, as did the addition of twice the normal level of minor lipids. The normal levels of minor lipids in milk fat resulted in the most rapid onset of crystallization of cocoa butter when both intact milk fat and milk fat fractions were added at the 10% level.

Milk fat also has a significant effect on the shelf stability of chocolate, particularly to fat bloom that occurs during storage [97]. Over time, the cocoa butter in chocolate can recrystallize and form a grayish-white haze on the surface of the piece. The addition of AMF, and particularly HMF, to chocolate has long been known to inhibit the formation of fat bloom during storage in chocolate [97,107–108], although the exact mechanism(s) of this inhibition is not clearly understood. Bricknell and Hartel [109] followed the polymorphic transition of cocoa butter to the most stable state during storage of chocolate by making a chocolate with an amorphous sugar phase. Their results clearly show that HMF delays the polymorphic transition of cocoa butter, although other effects appear to be more important in controlling visual bloom formation. The minor lipids in milk fat also impact the rate of bloom formation in dark chocolate [106]. Both removal of the minor lipids from milk fat and addition of twice the level of minor lipids in the intact AMF resulted in more rapid bloom formation than when milk fat with the intact level of minor lipids was added to chocolate. These minor lipids influenced the crystalline microstructure of cocoa butter in model systems, and this may explain their impact on bloom in chocolate [110].

2. Compound Coatings

The potential for application of milk fat fractions in confectionery coatings based on palm kernel oil (as a replacement for cocoa butter in chocolate) has been evaluated in recent years [111–113]. The coating industry does not utilize whole fat dairy ingredients in compound coating formulations, despite the potential flavor benefits, due to the negative physical effects that result from incorporation of milk fat. Although milk fat or milk fat fractions (with the exception of the high-melting components) and palm kernel oil have been found to be nearly fully compatible fats [112–113], the use of milk fat ingredients to replace a portion of palm kernel oil in coatings results in softening (depending on the melting point of the fraction) and extensive blooming of the product. Both milk fat and palm kernel oil form relatively stable β′ polymorphs, although palm kernel oil does gradually transform to a more stable β form over long-term storage [113]. The addition of milk fat fractions to palm kernel oil also affects the onset of crystallization of the mixture and results in different crystalline microstructures. Even the presence of the minor lipids in milk fat affect crystalline microstructures of the palm kernel oil, as visualized by using confocal microscopy [113]. According to these recent studies, the potential application of milk fat ingredients in compound coatings does not look promising.

C. Dairy Products

Milk fat is a natural component in most dairy foods, and its chemical and physical properties can limit its functional performance in some applications. Milk fat fractions are added to traditional dairy foods or recombined to create tailored dairy foods.

1. Spreadable Butter

Butter is the original spread. It has a flavor that is unique and exhibits a cooling sensation as it melts in the mouth, making it an organoleptic standard to which all other spreads are compared. However, butter is notoriously firm at refrigerator temperatures, and, therefore, inconvenient for many consumers. Historically, many processing strategies have been tried to improve the spreadability of butter with limited success. Milk fat fractionation provides a way to make a butter that spreads at refrigerator temperature. In countries where milk fat fractionation is common, spreadable butter is often the main product produced from milk fat fractions.

Desirable functional properties of butter and spreads include flavor, cold spreadability, and the ability to maintain structure and consistency at room temperature without melting and oiling off. Target properties for spreadable butters include melting point of 28 to 30°C, with a solids level between 30% and 40% at 5°C and at least 10% at 20°C [1].

Spreadable butters are made by either the addition of low-melting fractions, at least 20% of the total fat, to cream and churning, or by combining 70–80% very-low-melting fractions with 20–30% high-melting fractions and processing on scraped-surface heat exchangers [1,114–115]. Ingredients added to recombined butter include buttermilk, skim milk, or water and skim milk powder, and salt. In addition to selecting the proper melting profile, the texturization parameters employed during manufacture also affect the spreadability. Spreadable butter exhibits a good buttery flavor because the characteristic flavor compounds and flavor precursors tend to concentrate in the low-melting fractions.

D. Dairy Spreads

The dairy spread category includes products made from all milk fat but with lower total fat than butter, butter-margarine blends, and specialty butters like high-melting butter. The desired functional attributes for most dairy spreads are the same as for spreadable butter: flavor, spreadability, and stand-up properties.

A high-melting butter was churned from cream made with high-melting milk fat fractions [116] and evaluated for structural and rheological characteristics [90,117–119]. Increased concentration of high-melting TAG led to a butter microstructure with increased crystallinity, larger crystals, more birefringence, and a finer aqueous phase [90]. The high-melting butter had higher consistency values, viscoelastic parameters, and lower stress growth coefficients than regular butter [90,117–119]. Changing the composition of the fat influences the molecular arrangement within the crystal lattice and the strength of interactions between crystals because of the amount of crystalline fat present.

Low-melting and middle-melting milk fat fractions were used to produce spreadable low-fat butters, and high-melting fractions were used as hard stocks and blended with vegetable oils to produce butter-margarine blends [1,115]. Functional ingredients such as emulsifiers and thickening agents were required for some of the blends, particularly lower-fat formulas. Low-fat butters and butter-margarine blends were processed using scraped-surface heat exchangers.

E. Recombined Dairy Products

Low- and middle-melting milk fat fractions are used commercially for recombined dairy products [1,82]. The use of milk fat fractions to either soften or firm recombined butter, depending on the end use and market, has been recommended [115]. For use in recombined ice cream, they recommend the lower-melting fractions which may be easier to reincorporate into the mixes. High-melting fractions are suggested for recombined cream to improve the stand-up properties.

1. Dairy Powders

Low-melting fractions used in milk powders improved the dispersability and reconstitution of milk powders, and high-melting fractions improved flowability of milk powders. High-melting fractions were used to make butter powders for use in baking. The powders showed good flow and compaction properties [120]. All dairy powders made with milk fat fractions lacked flavor [1].

2. Cheese

Fat in cheese contributes to flavor and flavor development, structure, and texture. Milk fat fractions have been evaluated in cheeses as an alternate application for fractions. Natural cheeses made with high-melting milk fat fractions compared with control cheeses, and processed cheeses made with very-high-melting fractions were firmer than the control [1]. Up to 40% of the total milk fat in processed cheeses was replaced with high-melting milk fat fractions without changing the organoleptic qualities.

Rosenberg and Fang [121] studied milk gels made with different milk fat fractions and determined that the rheological properties could be modified by changing the melting characteristics, particle size distribution, and proportion of the fat. This study was further expanded to low-fat cheese curd systems, and products made with low- or middle-melting milk fat fractions showed promising results for improving the texture of low-fat cheese [122].

3. *Ice Cream*

Milk fat serves several important functions in ice cream, including providing dairy flavor and creaminess. Milk fat also imparts desirable textural attributes to ice cream that result in a smooth texture and body [123]. Milk fat, in the form of emulsified and agglomerated droplets, coats the surface of the air cells and provides body (stand-up) as the ice cream melts. The use of modified vegetable fats with the same melting profile does not impart the same flavor or textural properties in ice cream as milk fat. Abd El-Rahman et al. [124–125] evaluated the physical and sensory attributes of ice creams made with low- and high-melting milk fat fractions. From a physical standpoint, the high-melting fraction imparted better stand-up properties upon melting, due to the higher solid fat content. The low-melting fraction exhibited the poorest flavor profile of the ice creams studied, and this was attributed to the higher peroxide value of this fraction.

Milk fat fractions were found to impact the processing kinetics of ice cream [126]. The rate of solidification during the aging step was dependent on the type of milk fat fraction used. Higher-melting fractions solidified faster and reached a higher level of solid fat. The level of solid fat in the ice cream mix affected the rate of overrun development during freezing, with the lower-melting milk fat fractions requiring longer batch freezing times to reach maximum overrun. However, the extent of fat destabilization during freezing was higher for the lower-melting fractions. Apparently, the extent of destabilization of the milk fat emulsion is a function of the amount of solid fat present in the droplets. Further details on the effects of the fat phase on quality and stability of ice cream may be found in Chapter 9.

F. Specialty Applications

Milk fat fractions may exhibit advantageous functional properties in applications that are not always suitable for intact milk fat. Functional benefits in these applications may come from the chemical characteristics, such as fatty acid profile, or from physical characteristics, such as melting point.

1. Flavors

The low-melting fractions are of interest as starting materials for the production of natural and enzyme-modified butter flavors because of their increased flavor and flavor potential [1,60]. A patented process exists where low-melting milk fat fractions are added during the manufacture of cultured dairy products to enhance flavor development [127].

2. Emulsifiers

Glycerolysis reactions have been used to make MAG and DAG emulsifiers from intact milk fat [128] and high-melting milk fat fractions [129]. The emulsifier made from milk fat fraction was a 60:40 blend of MAG to DAG, had a different composition compared with commercial emulsifiers, had a lower melting range, and imparted a lower surface tension at the oil/water and oil/skim milk interfaces in test systems. Preliminary results suggest that the butterfat emulsifiers are more effective at lower concentrations than commercial emulsifiers [129].

3. Salad Dressings/Sauces

Liquid milk fat fractions were used in mayonnaise formulas with some success [92]. Rajah [92] reported on a European patent where a low-melting milk fat fraction was used to stabilize sauces.

4. Frying Oils

High-melting and low-melting fractions were evaluated as frying oils. Low-melting fractions produced acceptable deep-fried products with a buttery flavor [1,92]. The milk fat fractions exhibited reasonable stability as a frying oil, but light protection was recommended to maintain good shelf life. Foods pan-fried in low-melting milk fat fractions were preferred by sensory panelists over foods fried in butter or sunflower oil [93].

Polymer formation in fats and oils occurs when they are subjected to high temperatures, such as in frying. Both solid and liquid milk fat fractions were more stable toward polymer formation than intact milk fat after 16 hr at 185°C [73]. The degradation properties of a low-melting milk fat fraction did not differ significantly from intact milk fat in a frying situation [74]. The addition of the low-melting milk fat fraction to sunflower oil improved the frying stability of the sunflower oil and may show synergistic effects with other oils in this application [74].

5. *Edible Films*

High-melting lipids are often added to edible films to improve their water vapor properties. The performance of several lipid sources in whey protein films was investigated [130], with better water vapor permeability found with films made from high-melting milk fat fractions and beeswax than for films made with carnauba and candelilla wax. Films made with very-high-melting milk fat fractions exhibited better barrier and rheological properties than films made with high-melting fractions [131].

6. *Nutritional Foods*

The unique metabolic properties of the short-chain fatty acids and the content of fatty acids such as CLA and butyric acid positions milk fat as a unique ingredient in nutritional foods. Intact milk fat was used as a C8 and C10 fatty acid source in a U.S. patent describing the production of structured lipids for nutritional foods, and it was recommended that milk fat fractions with higher levels of short- and medium-chain fatty acids be further evaluated for use in nutritional foods [132].

V. SUMMARY

Fractionation provides significant potential for production of value-added ingredients from milk fat. Over the past three to four decades of study, researchers have documented the potential for utilizing milk fat components as value-added ingredients in a wide variety of food and nonfood applications. However, there remains much we do not understand. Crystallization of milk fat still remains somewhat of a mystery due to the wide number of triacylglycerol and minor lipid species that interact. Control of milk fat crystallization is still uncertain due to this lack of understanding. Fractionation processes may be improved dramatically as our understanding of the mechanisms and kinetics of milk fat crystallization grow. In the same sense, application of milk fat fractions in food products requires an understanding of how the milk component interacts with the other components in the product. Improved understanding of these interactions will lead to better control of milk fat in food applications.

ACKNOWLEDGMENTS

The assistance of Baomin Liang and David Illingworth in reviewing this chapter is greatly appreciated.

REFERENCES

1. Kaylegian, K. E., and R. C. Lindsay, *Handbook of Milkfat Fractionation Technology and Applications*, American Oil Chemists Society Press, Campaign, IL (1995).
2. Jensen, R. G., A. M. Ferris, and C. J. Lammi-Keefe, *J. Dairy Sci.*, *74*: 3228–3243 (1991).
3. Christie, W. W., in *Advanced Dairy Chemistry 2*: *Lipids*, 2nd ed. (P. F. Fox, ed.), Chapman and Hall, London, pp. 1–36 (1995).
4. German, J. B. and C. J. Dillard, *Food Technol.*, *52(2)*: 33–38 (1998).
5. Itabashi, Y., J. J. Myher, and A. Kuksis, *J. Am. Oil Chem. Soc.*, *70(12)*: 1177–1181 (1993).
6. Gresti, J., M. Bugaut, C. Maniongui, and J. Bezard, *J. Dairy Sci.*, *76*: 1850–1869 (1993).
7. Kemppinen, A., and P. Kalo, *J. Am. Oil Chem. Soc.*, *75(2)*: 91–100 (1998).
8. Angers, P., E. Tousignant, A. Boudreau, and J. Arul, *Lipids*, *33(12)*: 1195–1201 (1998).
9. Robinson, N. P. and A. K. H. MacGibbon, *J. Am. Oil Chem. Soc.*, *75(8)*: 993–999 (1998).
10. Fraga, M. J., J. Fontecha, L. Lozada, and M. Juarez, *J. Agric. Food Chem.*, *46*: 1836–1843 (1998).
11. Ruiz-Sala, P., M. T. G. Hierro, I. Martinez-Castro, and G. Santa-Maria, *J. Am. Oil Chem. Soc.*, *73(3)*: 283–293 (1996).
12. Kalo, P., A. Kemppinen, and I. Kilpelinen, *Lipids*, *31(3)*: 331–336 (1996).
13. Hawke, J. C. And M. W. Taylor, in *Advanced Dairy Chemistry 2*: *Lipids*, 2nd ed. (P. F. Fox, ed.), Chapman and Hall, London, pp. 37–88 (1995).
14. Spanos, G. A., S. J. Schwartz, R. B. van Breeman, and C-H. Huang, *Lipids*, *30(1)*: 85–90 (1995).
15. Wolfe, R. L., C. C. Bayard, and R. J. Fabien, *J. Am. Oil Chem. Soc.*, *72(12)*: 1471–1483 (1995).
16. Lipp, M., *Food Chem.*, *54*: 213–221 (1995).
17. Laakso, P. and H. Kallio, *J. Am. Oil Chem. Soc.*, *70(12)*: 1161–1171 (1993).
18. Laakso, P. and H. Kallio, *J. Am. Oil Chem. Soc.*, *70(12)*: 1172–1176 (1993).
19. Kermasha, S., S. Kubow, M. Safari, and A. Reid, *J. Am. Oil Chem. Soc.*, *70(2)*: 169–173 (1993).
20. Kalo, P. and A. Kemppinen, *J. Am. Oil Chem. Soc.*, *70*: 1209 (1993).
21. Laakso, P. H., K. V. V. Nurmela, and D. R. Homer, *J. Agric. Food Chem.*, *40*: 2472–2482 (1992).
22. Maniongui, C., J. Gresti, M. Bugaut, S. Gauthier, and J. Bezard, *J. Chrom.*, *543*: 81–103 (1991).
23. Kuksis, A., L. Marai, and J. J. Myher, *J. Chrom.*, *588*: 73 (1991).
24. Barron, L. J. R., M. T. G. Hierro, and G. Santa-Maria, *J. Dairy Res.*, *57*: 517–526 (1990).
25. Hagemann, J. W., in *Crystallization and Polymorphism of Fats and Fatty Acids* (N. Garti and K. Sato, eds.), Marcel Dekker, New York, pp. 9–96 (1988).
26. Van Aken, G. A., E. ten Grotenhuis, A. J. van Langenvelde, and H. Schenk, *J. Am. Oil Chem. Soc.*, *76*: 1323–1331 (1999).

27. Deffense, E., *J. Am. Oil Chem. Soc.*, *70(12)*: 1193–1201 (1993).
28. Timms, R. E., *Austr. J. Dairy Technol.*, *35(2)*: 47–53 (1980).
29. Hinrichs, J., U. Heinemann, and H. G. Kessler, *Milchwissenschaft*, *47(8)*: 495–498 (1992).
30. Bornaz, S., G. Novak, and M. Parmentier, *J. Am. Oil Chem. Soc.*, *69*: 1131–1135 (1992).
31. Koyano, T., I. Hachiya, and K. Sato, *J. Phys. Chem.*, *96*: 10514–10520 (1992).
32. Wesdorp, L. H., Ph.D. dissertation, Technical University of Delft, The Netherlands (1990).
33. Timms, R. E., *Lebensm. Wiss. Technol.*, *13*: 61–65 (1980).
34. Marangoni, A. G. and R. W. Lencki, *J. Agric. Food Chem.*, *46(10)*: 3879–3884 (1998).
35. Mulder, H., *Neth. Milk Dairy J.*, *7*: 149–176 (1953).
36. Walstra, P. and E. C. H. van Beresteyn, *Neth. Milk Dairy J.*, *29*: 238–241.
37. Walstra, P., in *Food Structure and Behaviour* (J. M. V. Blanshard and P. Lillford, eds.), Academic Press, New York, pp. 67–86 (1987).
38. Woodrow, I. L. and J. M. de Man, *J. Dairy Sci.*, *51*: 996–1000 (1968).
39. Schaap, J. E., H. T. Badings, D. G. Schmidt, and E. Frede, *Neth. Milk Dairy J.*, *29*: 242–252 (1975).
40. ten Grotenhuis, E., G. A. van Aken, K. F. van Malssen, and H. Schenk, *J. Am. Oil Chem. Soc.*, *76(9)*: 1031–1039 (1999).
41. Boistelle, R., in *Crystallization and Polymorphism of Fats and Fatty Acids* (N. Garti and K. Sato, eds.), Marcel Dekker, New York, pp. 189–226 (1988).
42. Mullin, J., *Crystallization*, 2nd ed., Butterworths, London (1993).
43. Breitschuh, B. and E. J. Windhab, *J. Am. Oil Chem. Soc.*, *75*: 897–904 (1998).
44. Le Botlan, D., L. Ouguerram, L. Smart, and L. Pugh, *J. Am. Oil Chem. Soc.*, *76(2)*: 255–261 (1999).
45. Herrera, M. L., M. de Leon Gatti, and R. W. Hartel, *Food Res. Int.*, *32*: 289–298 (1999).
46. Illingworth, D. and R. W. Hartel, Paper presented at International Society of Fat Research, Brighton, UK (1999).
47. Wright, A., R. W. Hartel, S. Narine, and A. Marangoni, *J. Am. Oil Chem. Soc.*, *77(5)*: 463–476 (2000).
48. Ng, W. L., *J. Chem.*, *28*: 1169–1178 (1975).
49. Marangoni, A. G. and R. W. Hartel, *Food Technol.*, *52(9)*: 46–51 (1998).
50. Herrera, M. L. and R. W. Hartel, *J. Am. Oil Chem. Soc*, *77(11)*: 1189–1195 (2000).
51. Herrera, M. L. and R. W. Hartel, *J. Am. Oil Chem. Soc*, *77(11)*: 1197–1204 (2000).
52. Patience, D., R. W. Hartel, and D. Illingworth, *J. Am. Oil Chem. Soc.*, *76(5)*: 585–594 (1999).
53. Evans, A. A., *N. Z. J. Dairy Sci. Technol.*, *11*: 73–78 (1976).
54. Hartel, R. W., B. Liang, and Y. Shi, U.S. Patent No. 6,140,520, Oct. 31, 2000.
55. Breitschuh, B., M. Drost, and E. J. Windhab, *Chem. Eng. Technol.*, *21*: 425–428 (1998).
56. Tiedtke, M., J. Ulrich, and R. W. Hartel, *Crystal Growth of Organic Materials* (A. S. Myerson, D. A. Green, and P. Meenan, eds.), Amer. Chem. Soc., pp. 137–144 (1996).

57. Peters-Erjawetz, S., J. Ulrich, M. Tiedtke, and R. W. Hartel, *J. Am. Oil Chem. Soc.*, *76(5)*: 579–584 (1999).
58. U.S. Department of Agriculture, 21 CFR part 58: 325 (1997).
59. Bhaskar, A. R., S. S. H. Rizvi, C. Bertoli, L. B. Fay, and B. Hug, *J. Am. Oil Chem. Soc.*, *75(10)*: 1249–1264 (1998).
60. Urbach, G. and M. H. Gordon, in *Fats in Food Products* (D. P. J. Moran and K. K. Rajah, eds.), Blackie Academic & Professional, London, pp. 347–406 (1994).
61. Pszczola, D. E., *Food Technol.*, *52(3)*: 30–37 (1998).
62. Gurr, M., *Lipid Technol.*, *4*: 93 (1992).
63. Ney, D. M., *J. Dairy Sci.*, *74*: 2002 (1991).
64. Lai, H.-C., J. B. Lasekan, C. C. Monsma, and D. M. Ney, *J. Dairy Sci.*, *78*: 794–803 (1995).
65. Bradley, R. L., Jr., *J. Dairy Sci.*, *72*: 2834 (1989).
66. Huber, W., A. Molero, C. Pereyra, and E. Martinex de la Ossa, *Int. J. Food Sci. Technol.*, *31*: 143–151 (1996).
67. Lim, S. and S. S. H. Rizvi, *J. Food Sci.*, *61(4)*: 817–820 (1996).
68. Kankare V. and M. Alkio, *Agric. Sci. Finl.*, *2*: 387–393 (1993).
69. Gurr, M., *Lipid Technol.*, *7(6)*: 133–135 (1995).
70. Parodi, P. W., *Aust. J. Dairy Technol.*, *49*: 93–97 (1994).
71. Parodi, P. W., *Aust. J. Dairy Technol.*, *51*: 24–32 (1996).
72. Smith, J. G. and J. B. German, *Food Technol.*, *49(11)*: 87–90 (1995).
73. Kupranycz, D. B., M. A. Amer, and B. E. Baker, *J. Am. Oil Chem. Soc.*, *63*: 332 (1986).
74. Harmer, W. R. and C. Wijesundera, *Aust. J. Dairy Technol.*, *51(2)*: 108–111 (1996).
75. Joyner, N. T., *J. Am. Oil Chem. Soc.*, *30(11)*: 526–535 (1953).
76. Moran, D. P. J., in *Fats in Food Products* (D. P. J. Moran and K. K. Rajah, eds.), Blackie Academic & Professional, London, pp. 155–211 (1994).
77. Johansson, D. and B. Bergenstahl, *J. Am. Oil Chem. Soc.*, *72(8)*: 911–920 (1995).
78. Marangoni, A. G. and D. Rousseau, *J. Am. Oil Chem. Soc.*, *73(8)*: 991–994 (1996).
79. Heertje, I., J. van Eendenburg, J. M. Cornelissen, and A. C. Juriaanse, *Food Microstructure*, *7*: 189–193 (1988).
80. Juriaanse, A. C. and I. Heertje, *Food Microstructure*, *7*: 181–188 (1988).
81. Lane, R. in *The Technology of Dairy Products*, 2nd ed. (R. Early, ed.), Blackie Academic & Professional, London, pp. 158–197 (1998).
82. Munro, D. S., P. A. E. Cant, A. K. H. MacGibbon, D. Illingworth, and P. Nicholas, in *The Technology of Dairy Products*, 2nd ed. (R. Early, ed.), Blackie Academic & Professional, London, pp. 198–227 (1998).
83. Pedersen, A., in *Utilizations of Milkfat*, Bull. No. 260, Internat. Dairy Fed., p. 10 (1991).
84. Podmore, J., in *Fats in Food Products* (D. P. J. Moran and K. K. Rajah, eds.), Blackie Academic & Professional, London, pp. 213–254 (1994).
85. Walstra, P., T. van Vliet, and W. Kloek, in *Advanced Dairy Chemistry 2: Lipids*, 2nd ed. (P. F. Fox, ed.), Chapman and Hall, London, pp. 179–246 (1995).
86. Bornaz, S., J. Fanni, and M. Parmentier, *J. Am. Oil Chem. Soc.*, *70(11)*: 1075–1079 (1993).
87. Simoneau, C., Ph.D. thesis, Univ. of California, Davis (1994).

88. Black, R. G., *Aust. J. Dairy Technol., 30(2)*: 60–63 (1975).
89. Dungey, S. G., S. Gladman, G. Bardsley, M. Iyer, and C. Versteeg, *Aust. J. Dairy Technol., 51(2)*: 101–104 (1996).
90. Keogh M. K. and A. Morrissey, *Irish J. Food Sci. Technol., 14(2)*: 69–83 (1990).
91. Shukla, A. and S. S. H. Rizvi, *Milchwissenschaft, 51(3)*: 144–148 (1996).
92. Rajah, K. K., in *Fats in Food Products* (D. P. J. Moran and K. K. Rajah, eds.), Blackie Academic & Professional, London, pp. 277–318 (1994).
93. Versteeg, C., L. N. Thomas, Y. L. Yep, M. Papalois, and P. S. Dimick, *Aust. J. Dairy Technol., 49*: 57–61 (1994).
94. Tolboe, O., in *Milkfat and Its Modifications*, Contributions at a LIPIDFORUM Symposium (R. Marcuse, ed.), Scandinavian Forum for Lipid Research and Technology, Goteborg, Sweden, p. 43 (1984).
95. Noraini, I., M. S. Embong, A. Aminah, A. R. Md. Ali, and C. H. Che Maimon, *Fat Sci. Technol., 97(7/8)*: 253–260 (1995).
96. Bystrom, C. E. and R. W. Hartel, *Lebensm.-Wiss. Technol., 27(2)*: 142–150 (1994).
97. Hartel, R. W., *J. Am. Oil Chem. Soc., 73(8)*: 945–953 (1996).
98. Metin, S., Ph.D. thesis, University of Wisconsin, Madison, WI (1997).
99. Md. Ali, A. R. and P. S. Dimick, *J. Am. Oil Chem. Soc., 71(8)*: 803–806 (1994).
100. Sabariah, S., A. R. Md Ali, and C. L. Chong, *J. Am. Oil Chem. Soc., 75(8)*: 905–910 (1998).
101. Simoneau, C. and J. B. German, *J. Am. Oil Chem. Soc., 73(8)*: 955–961 (1996).
102. Metin, S. and R. W. Hartel, *J. Thermal Anal., 47(5)*: 1527–1544 (1996).
103. Metin, S. and R. W. Hartel, *J. Am. Oil Chem. Soc., 75(11)*: 1617–1624 (1998).
104. Dimick, P. S., G. R. Ziegler, N. A. Full, and S. Yella Reddy, *Aust. J. Dairy Technol., 51*: 123–126 (1996).
105. Yella Reddy, S., N. Full, P. S. Dimick, and G. R. Ziegler, *J. Am. Oil Chem. Soc., 73(6)*: 723–727 (1996).
106. Tietz, R. A. and R. W. Hartel, *J. Am. Oil Chem. Soc, 77(7)*: 763–772 (2000).
107. Campbell, L. B., D. A. Anderson, and P. G. Keeney, *J. Dairy Sci., 52*: 976–979 (1969).
108. Timms, R. E. and V. E. Parekh, *Lebensm-Wiss. Technol., 13*: 177–181 (1980).
109. Bricknell, J. and R. W. Hartel, *J. Am. Oil Chem. Soc., 75(11)*: 1609–1616 (1998).
110. Hartel, R. W., *Manuf. Confect., 79(5)*: 89–99 (1999).
111. Buning-Pfaue, H. and A. Bartsch, *J. Thermal Anal., 35*: 671–675 (1989).
112. Ransom-Painter, K. L., S. D. Williams, and R. W. Hartel, *J. Dairy Sci., 80(10)*: 2237–2248 (1997).
113. Schmelzer, J., M. S. thesis, University of Wisconsin (1998).
114. Deffense, E., *Fat Sci. Technol., 13*: 1 (1987).
115. Illingworth, D. and T. G. Bissell, In *Fats in Food Products* (D. P. J. Moran and K. K. Rajah, eds.), Blackie Academic & Professional, London, pp. 111–154 (1994).
116. Shukla, A., A. R. Bhaskar, S. S. H. Rizvi, and S. Mulvaney, *J. Dairy Sci., 77*: 45 (1994).
117. Shukla, A. and S. S. H. Rizvi, *J. Texture Studies, 26*: 299–311 (1995).
118. Shukla, A. and S. S. H. Rizvi, *J. Food Sci., 60(5)*: 902–905 (1995).
119. Shukla, A. and S. S. H. Rizvi, and J. A. Bartsch, *J. Texture Studies, 26*: 313–323 (1995).

120. Onwulata, C. I., P. W. Smith, and V. H. Holsinger, *J. Food Sci., 60(4)*: 836–840 (1995).
121. Rosenberg, M. and C. Fang, *J. Dairy Sci., Supp. 78*: 135(Abst. D120) (1995).
122. Rosenberg, M., in *National Milkfat Technology Forum Proceedings*, Dairy Management, Inc., Rosemont IL, p. 73 (1998).
123. Arbuckle, W. S., *Ice Cream*, 4th ed., Van Nostrand Reinhold, New York (1986).
124. Abd El-Rahman, A. M., S. A. Madkor, F. S. Ibrahim, and A. Kilara, *J. Dairy Sci., 80*: 1926–1035 (1997).
125. Abd El-Rahman, A. M., S. I. Shalabi, R. Hollender, and A. Kilara, *J. Dairy Sci., 80*: 1936–1940 (1997).
126. Adleman, R. L., M. S. thesis, University of Wisconsin, Madison, WI (1998).
127. Mehnert, D. W., U.S. Patent 5,462,755 (1995).
128. Yang, B. and K. L. Parkin, *J. Food Sci., 59(1)*: 47–52 (1994).
129. Campbell-Timperman, K., J. H. Choi, and R. Jiminez-Flores, *J. Food Sci., 61(1)*: 44–53 (1996).
130. Shellhammer, T. H., and J. M. Krochta, *J. Food Sci., 62(2)*: 390–394 (1997).
131. Shellhammer, T. H. and J. M. Krochta, *Trans. Am. Soc. Agric. Eng., 40(4)*: 1119–1127 (1997).
132. Babayan, V. K., G. L. Blackburn, and B. R. Bistrian, U.S. Patent 4,952,606 (1990).
133. Hartel, R. W., *Crystallization in Foods*, Aspen Publishing, Gaithersburg, MD (2001).

12
Crystallization Properties of Cocoa Butter

Kiyotaka Sato
Hiroshima University, Higashi-Hiroshima, Japan

Tetsuo Koyano
Meiji Seika Kaisha, Ltd., Saitama, Japan

I. INTRODUCTION

Cocoa butter (CB) is a vegetable fat [1] and most abundantly employed as a confectionery fat [2,3]. The vegetable oil involving high amounts of solid fat fractions, such as illipe butter, shea nut oil, sal fat, palm oil, and kokum fats, are added to modify the physical property of CB-based confections or employed as main bodies of chocolate fats to replace CB [2]. Hydrogenated vegetable oils, such as soybean and cottonseed oils, are also employed for confectionery coating fats. Processing of the confectionery fats involves three major physical controls: crystallization of fat crystals, morphology and network of the fat crystal particles, and rheological variations of the chocolate mass. CB reveals the optimal properties necessary for producing confections among the all-vegetable fats displayed above, and thereby CB has attracted the greatest attention in confectionery science and technology research.

Chocolate consists of a polycrystalline system of confectionery fats in which fine particles of coca mass, sugar, and milk powders are homogeneously dispersed, and small amounts of water and food emulsifiers are also included. The key factor relevant to all three processes is polymorphic crystallization of the fats. The quality of chocolate products, such as gloss, snap, heat resistance, and fat bloom stability, is predominantly influenced by the CB crystals [4]. Eventually, the CB crystals maintain optimal physical properties in a specific polymorphic form, namely Form V in accordance with the nomenclature of Wille and Lutton [5]. The conversion from Form V to Form VI induces a serious deterioration called fat bloom.

Closely connected to fat crystallization, changes in viscosity are also critical. Viscosity of the confectionery fats is mainly affected by recipe, fat content, moisture content, particle size distribution of the solid distributed in the mass, the state of conching, and the type and quantity of the emulsifiers employed. In particular, the ratio of the solid and liquid fat present in chocolate, [solid]/[liquid fat], directly affects the viscosity of the chocolate. Namely, if the ratio is low, the viscosity of the chocolate becomes low, and vice versa. For example, milk chocolate with a total fat content of 37% has the viscosity values shown in Fig. 1a, as measured by a Brookfield-type viscometer (T. Koyano, unpublished work). These values increased rapidly with time below about 30°C, after the crystal seeds were formed by tempering, although not shown here.

During manufacturing the tempering process is performed with a tempering machine [6,7], in which molten and conched chocolate is subjected to cooling/reheating/recooling processes (Fig. 1b). The liquid fat in the chocolate mass de-

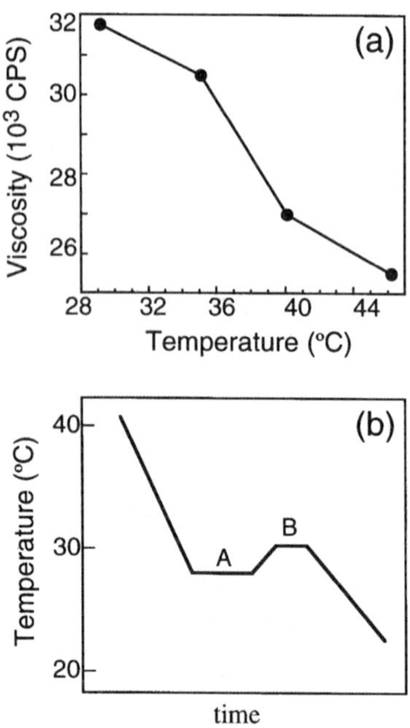

Fig. 1 (a) Viscosity variation of dark chocolate with temperature (°C) and (b) temperature variation (°C) during a tempering process.

creases during the tempering process and the seed crystals are formed, so [solid]/[liquid fat] is increased and the viscosity of the chocolate almost increases. The temperatures during tempering are precisely determined in accordance with the recipe of the chocolate. So, the temperature during the tempering process should be correctly controlled. In the actual tempering process, chocolate is first cooled to 26–28°C (A in Fig. 1b), then reheated to 30–31°C (B in Fig. 1b) using a heat exchanger under shear. However, if the viscosity of the chocolate becomes too high, it is difficult to uniformly control the temperature, because diffusion of the heat is minimized in highly viscous states. Furthermore, in the case of high-viscosity chocolate, shear force generates frictional heat, which disturbs the correct measurement of temperature. If the viscosity of the chocolate is too low, chocolate sometimes flows the shortest way in the heat exchanger, so the temperature is not well controlled. Thus, it is quite important to coordinate tempering conditions and viscous properties of chocolate.

This chapter discusses fundamental and application aspects of crystallization behavior of CB and the polymorphic nature of cocoa butter and its crystallization behavior under various external influences. The mechanisms of fat bloom are also highlighted.

II. POLYMORPHIC NATURE OF COCOA BUTTER

A. Polymorphic Crystallization of Cocoa Butter

Cocoa butter exhibits several polymorphic forms. Table 1 compares two representative nomenclatures of CB polymorphs [5,8], together with the polymorphic forms of SOS (1,3-distearoyl-2-oleoyl-*sn*-glycerol) [9], a major component triac-

Table 1 Cocoa Butter Polymorphs Based on the Definition of Wille and Lutton [5] and Vaeck [8] Compared with Five Polymorphs of SOS

Wille and Lutton	Melting points (°C)	Major XRD short spacing spectra (nm)	Vaeck	SOS
I	17.3	0.419, 0.37	γ	sub-α
II	23.3	0.424	α	α
III	25.5	0.425, 0.386	β'	β'
IV	27.5	0.435, 0.415	β'	β'
V	33.8	0.458, 0.367	β	β_2
VI	36.3	0.458, 0.365	β	β_1

Source: Ref. 9.

ylglycerol (TAG) of CB. Figure 2a shows X-ray diffraction (XRD) short spacing spectra of the five polymorphs of CB taken during successive conversion from Form I to Form V during heating after chilling the molten CB liquid at −5°C. In addition, precise XRD short spacing spectra of Form V and Form VI are shown in Fig. 2b. Although CB is composed of various TAGs, in particular 80 wt% of CB contains SOS, POP (1,3-dipalmitoyl-2-oleoyl-*sn*-glycerol), and POS (1,3-stearoyl-palmitoyl-2-oleoyl-*rac*-glycerol), each polymorph of CB exhibits a unique melting point and crystal structural properties. This may be ascribed to

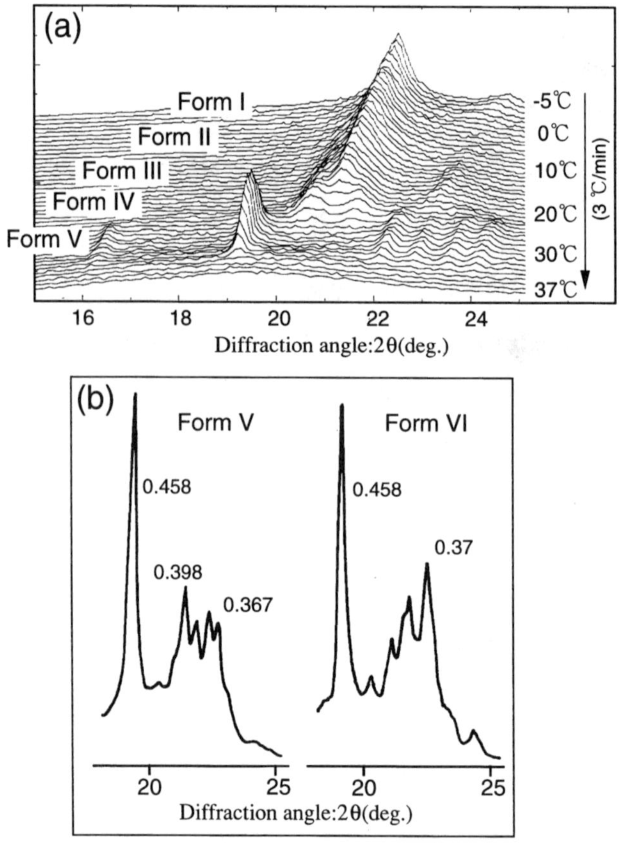

Fig. 2 X-ray diffraction short spacing spectra of (a) Form I through Form V of cocoa butter after crystallization from liquid at −5°C and subsequent heating, and (b) Form V and Form VI of cocoa butter.

cooperative interactions of three major TAGs of CB, SOS, POS, and POP (see below).

Thermal and chemical analyses of CB crystals formed by static crystallization at 26–33°C showed that the first visible crystals of CB contain POP, POS, and SOS [10]. The SOS fraction increased in concentration with increasing crystallization temperature. Further reports unveiled that, at the early stages of static crystallization around 26°C, cocoa butter starts with precrystallization of high-melting fractions composed of complex lipids and saturated acid TAGs [11,12]. An Amsterdam University group has recently performed isothermal crystallization experiments of CB under varying conditions, using a real-time XRD technique [13–16]. Figure 3 shows the crystallization behavior when CB was cooled from 60°C linearly in 120 sec to different solidification temperatures [16]. It is shown that the CB polymorphs, categorized in accordance with Vaeck's nomenclature [8], crystallized in less stable forms at lower solidification temperatures over a shorter period of induction time, and vice versa. A simultaneous in situ observation of XRD using synchrotron radiation and differential scanning calorimetry (DSC) has presented all polymorphs of CB, as reported by Ollivon et al. [17]. In this study, further clarification was made concerning the presence of a liquid crystalline phase in Form I, and high-melting fractions (melting point around 50°C) comprising saturated fatty acid moieties associated with SOS and POP.

As to the tempering behavior, it was reported that metastable forms other than Form V simultaneously crystallize during the first cooling (stage A in Fig. 1b); then the metastable forms convert to Form V during reheating (stage B in

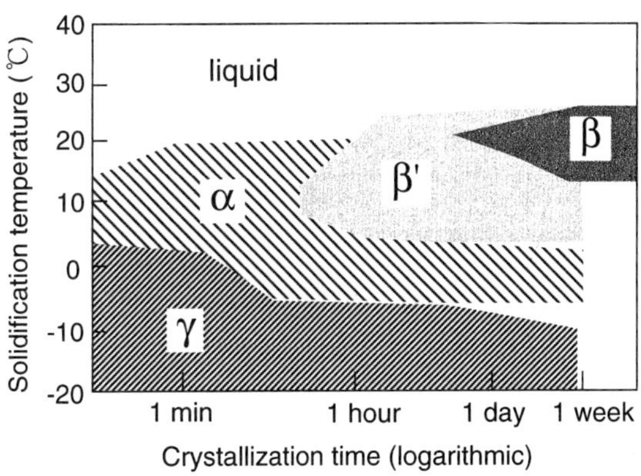

Fig. 3 Occurrence domains of cocoa butter polymorphs examined by dynamic crystallization [16].

Fig. 1b). The Form V crystals thus nucleated act as seed crystals during the recooling process [18,19]. In addition, experimental data on the amount of the seed crystals present at stage B ranged from 0.5% [20] to 5% [19]. Our studies [21,22] showed that (a) the crystals formed at stage A were a mixture of Forms V and IV, yet only Form V crystals are present at stage B, (b) fat blooming was induced in the solidified fat samples after the filtration of crystals using membrane filters having a pore size of around 220 nm for the tempered cocoa butter, and (c) the amount of crystallized fats at stage B was estimated far less than 1% by measuring viscosity values. Similar analyses, however, based on torque values and viscosity of dark chocolate during a temperature cycle of 50.0–26.1–30.5–33.3°C reported the formation of the seed crystal of CB about 1.15% [23].

To conclude, the crystallization of CB is quite polymorph-controlled, and the final stage of the fat crystallization in the chocolate production is aimed at the formation of Form V, which is the second stable polymorphic form.

B. Polymorphic Structures of Cocoa Butter

The arguments discussed in a previous section in relation to the crystallization and maintenance of the specific polymorph of cocoa butter have raised a long-lasting question; what types of individual polymorphs are present in CB crystals and how is each polymorphic form characterized in comparison to the others at a crystal structure level? Table 1 summarizes two major reports on cocoa butter polymorphs [5,8]. Even up to the 1970s, observation of CB polymorphs with XRD [24], electron microscopy [25], and DSC [26] supported the existence of the six forms claimed by Wille and Lutton [5]. However, the XRD short spacing spectra apparently did not differentiate the forms. In particular, two combinations of Forms III/VI and V/VI showed similar diffraction patterns. Pulsed nuclear magnetic resonance (NMR) singly detected four forms: γ, α, β', and β [27]. Thereby, Merken and Vaeck claimed the presence of four polymorphs [28].

Ambiguity in the polymorphic definition of CB may be ascribed to the fact that CB is a mixture of major TAGs of SOS, POS, and POP, and that fat segregation and solid-state transformation may occur simultaneously. Attention has been primarily focused on the differentiation of Forms V and VI [5], since they are closely related to tempering and fat bloom processes (see below). Two typical interpretations have so far been proposed; one is based on fractional changes in the main TAG components of SOS, POS, and POP, which together represent about 80% of CB [28], and the other involves the difference in polymorphic structures of CB crystals [29].

Here described are fundamental studies predicting the polymorphic structures of CB crystals examined by the authors. We have claimed that Forms V and VI are two different polymorphic forms, performing physical analyses of melting and XRD diffraction experiments of the ternary mixture phases of POP, POS, and SOS [30,31]. Further, fat segregation during fat blooming was explained by solubility differences of β forms of POP, POS, and SOS [9,32,33].

Figure 4 shows the XRD short spacing spectra of two β forms of SOS [9] (see Chapter 5 of this book). Subtle but definite differences are seen in small peaks denoted by arrows, particularly in the relative intensity of the 0.367-nm (β_2) and 0.365-nm (β_1) peaks. The difference is fairly informative, since it must reflect different crystal structures between the two forms, even if the differences were minimized. To compare with CB, the XRD patterns of β_2 and β_1 of SOS are quite similar to those of Forms V and VI, respectively, as shown in Fig. 2b. As described in Chapter 5, further clarification of the two β forms using FT-IR [34,35] and solid-state NMR techniques [36] have indicated that the differences between the two β forms are revealed in the molecular conformations of the oleic acid chains of SOS.

The presence of the two β forms was observed in the mixtures of POP, POS, and SOS, using the pure samples [30,31]. Table 2 summarizes the XRD long spacing values and DSC melting points of β_2 and β_1 forms of the three mixtures, in which mixture A was formed so that the relative concentration ratios of the three main TAGs mimic that of CB. The melting point differs by about 2°C for all the mixtures, and the XRD short spacing spectra of β_2 and β_1 forms of mixture A were quite similar to Forms V and VI of CB, respectively. For further clarification, ternary mixture systems consisting of POP, POS, and SOS were examined for 73 mixtures, with variation in the concentrations of each com-

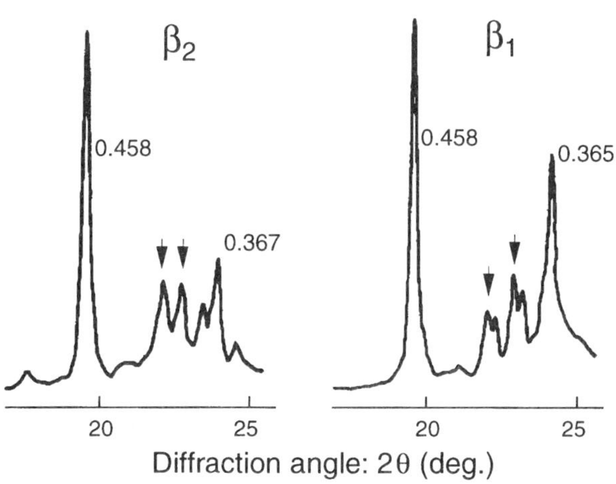

Fig. 4 X-ray diffraction short spacing spectra of two β forms of SOS [9].

Table 2 Long Spacings (LS, nm) and Melting Points (T_m, °C) of Two β Polymorphs Observed in Three Mixtures of POP, POS, and SOS

	β_2		β_2	
Mixture	LS	T_m	LS	T_m
A	6.30	33.0	6.30	35.3
B	6.30	32.6	6.30	34.6
C	6.10	33.2	6.10	35.5

A: POP (18.2%), POS (47.8%), SOS (34.0%).
B: POP (44.5%), POS (16.3%), SOS (39.2%).
C: POP (58.1%), SOS (41.9%).

ponent of 5 or 10 wt% [31]. The XRD and DSC measurements were performed for thermally stable forms, which were prepared by melting the mixed samples, crystallizing at 20°C, and incubating at 30°C over six months.

Figure 5a shows the trajectory of the melting points over the whole concentration range of the ternary mixture systems. The melting points become higher as the SOS-rich concentration ranges are approached, with the lowest temperatures occurring around the concentration ratio of POP:POS = 1:1. The mixing behavior in terms of eutectic or miscible properties was assessed by observing the profiles of the DSC melting peaks: e.g., the eutectic property is reflected on the splitting in the melting peaks, whereas the miscible nature is revealed in single melting peaks. The pattern formed is shown in Fig. 5b, where the single peaks were observed in blank areas, and the eutectic feature was observed in dashed areas. The latter was divided into two, clearly split peaks and broad peaks, both of which are of eutectic nature. The two results in Fig. 5 indicate that CB must be placed in a miscible region.

Based on the above data, the polymorphic correspondence between CB and SOS is shown in Table 1. The main points are that (a) differentiation of Forms III and IV is unclear, most probably due to the influences of β′ forms of POP, and (b) the occurrence of the two β forms is the origin of Forms V and VI of CB.

Note the thermodynamic stability of the β forms of POP, POS, and SOS, which was determined by solubility measurements using single crystals in tetradecane solvent (Fig. 6) [33]. Solubility of SOS β is always lowest in a range of temperatures examined, yet solubility values of the β forms of POS and POP are reversed around 13°C. POS β has a lower solubility below 13°C, whereas POP β is less soluble above that temperature. This results in fractional changes to concentration ratios of POP, POS, and SOS when CB crystals are left for long

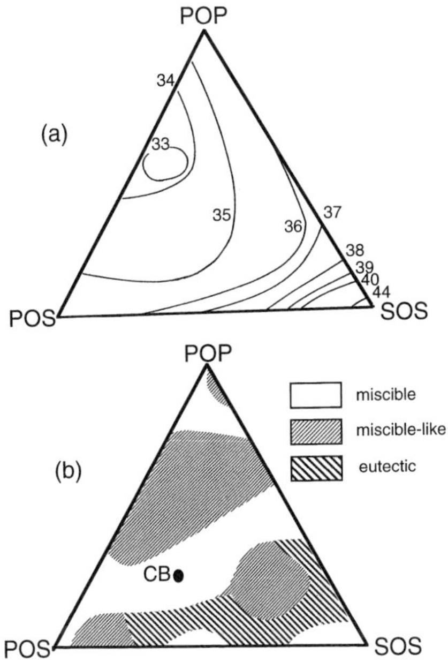

Fig. 5 (a) Melting temperature variation (°C) and (b) phase behavior of the mixtures of SOS, POS, and POP [31].

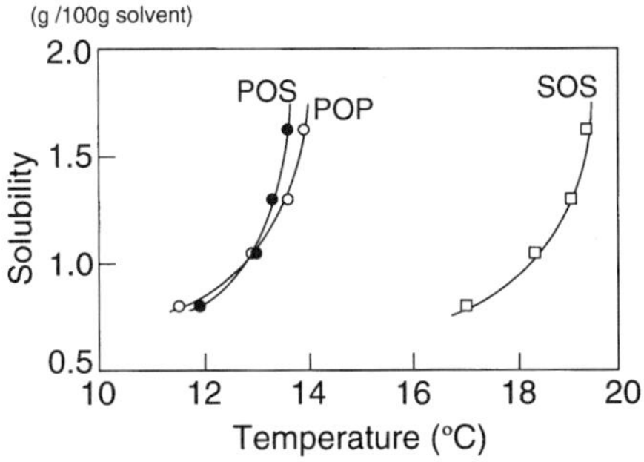

Fig. 6 Solubility variations of the stable forms of POS, POP, and SOS with temperature (°C) in tetradecane [33].

duration above 13°C. Thermal incubation changes the concentration ratios toward increasing concentration for SOS and POP compared to POS. The result of Fig. 6 indicates the changes in fractional concentrations of POP, POS, and SOS during fat bloom. After the formation of fat bloom, the concentrations of SOS and POP are increased compared to POS when the growth of Form VI occurs because of the lower solubility values of SOS and POP. This will be discussed later in relation to the fat bloom mechanism, which is essentially based on the polymorphic transformation of CB crystals.

Note that the phase behavior shown in Fig. 5 is useful for materials design of cocoa butter equivalent (CBE) [37].

III. CONTROLLING OF COCOA BUTTER CRYSTALLIZATION

Many attempts have been made to control the polymorphic crystallization of CB, having the aim of achieving (a) acceleration or retardation of the crystallization of CB, (b) better controlled polymorphic crystallization in Form V, (c) improvement of fat bloom stability, and (d) no usage of the tempering process, if possible. The use of seed materials put in liquid CB before crystallization has been a long-run attempt. In addition, novel ideas of the use of ultrasonic sonication and shear forces have also provided visible results. This section discusses the work performed recently.

A. Crystal Seeding Effects

1. Polar Lipids Present in Cocoa Butter

Davis and Dimick observed that some minor components present in cocoa butter, such as glycolipids, phospholipids, and saturated acid TAGs, promote the crystallization of CB [11,12]. More recent studies have indicated that phospholipid species may show the crystal seeding effects for promoting the crystallization of CB [38–40]. Six cocoa butters were selected and divided into two groups: (1) hard butters showing rapid nucleation and harvested in Malaysia, Ivory Coast, and Ghana, and (2) soft butters showing slow nucleation and obtained in Ecuador, Dominican Republic, and Brazil. The TAG concentrations of the six CBs are shown in Table 3. The induction times of isothermal crystallization at 26.5°C, which were measured by dynamic conditions with a viscometer, for the hard and soft butters are shown in Table 4. Among the six origin CBs, lysophosphatidylcholine (LPC), phosphatidylcholine (PC), and phosphatidylinositol (PI) were isolated, characterized, and quantified as major phospholipid species. The phospholipids were present at less than 1 wt%, regardless of the origin-country source,

Table 3 Triacylglycerol Composition of Cocoa Butters from Different Origins

	Triacylglycerols (wt%)[a]													
Cocoa butter	PLO	PLP	OOO	POO	PLS	POP	SOO	SLS	POS	PPS	SOS	PSS	SOA	SSS
Malaysia	0.4	1.1	0.1	1.1	2.6	12.6	1.8	1.6	46.9	0.7	29.8	0.4	0.9	0.2
Ivory Coast	0.7	1.7	0.4	1.8	3.7	15.0	2.3	1.7	46.3	0.7	24.0	0.5	0.8	0.4
Ghana	1.0	1.8	0.8	2.0	3.6	14.5	2.8	2.0	42.8	0.8	26.3	0.4	1.0	0.2
Ecuador	0.5	1.6	0.7	2.7	3.1	14.1	3.3	1.6	45.4	0.8	24.8	0.6	0.8	0.3
Dominican Republic	0.7	1.8	0.6	3.8	4.2	14.6	4.4	1.8	42.8	0.7	22.8	0.5	1.0	0.4
Brazil (Bahia)	0.9	1.7	0.7	5.8	3.9	13.9	6.7	2.1	40.2	0.6	21.7	0.5	0.9	0.6

[a] P = palmitate, O = oleate, S = stearate, L = linoleate, A = arachidonate.

Source: Ref. 38.

Table 4 Crystallization Induction Times of Cocoa Butters under Dynamic Conditions at 26.5°C

Cocoa butter source	Induction time (min)
Malaysia	78 ± 10
Ivory Coast	80 ± 13
Ghana	95 ± 14
Ecuador	117 ± 18
Dominican Republic	277 ± 44
Brazil (Bahia)	300 ± 51

Source: Ref. 39.

yet the concentration ratios were varied from one source to others. Namely the concentrations of LPC and PI were high in soft butters, whereas PC was predominant in hard butters. In addition, the hard butters had higher concentrations of POS and SOS than did the soft butters.

The crystallization properties revealed in the soft and hard CB were interpreted by the roles of mesophases (lyotropic liquid crystalline phases) formed by the phospholipid species as shown in Fig. 7 [40,41]. The interpretation conceives that the mesophases may provide the template for catalytic nucleation of CB crystals, namely seeding effect. LPC and PI may construct H_I phase (micelle), because of the large sizes of their polar head groups compared to those of acyl moieties. In this phase, the molecular interactions between the fat molecules of CB and the acyl moieties of the phospholipids are steric-hindered, requiring more

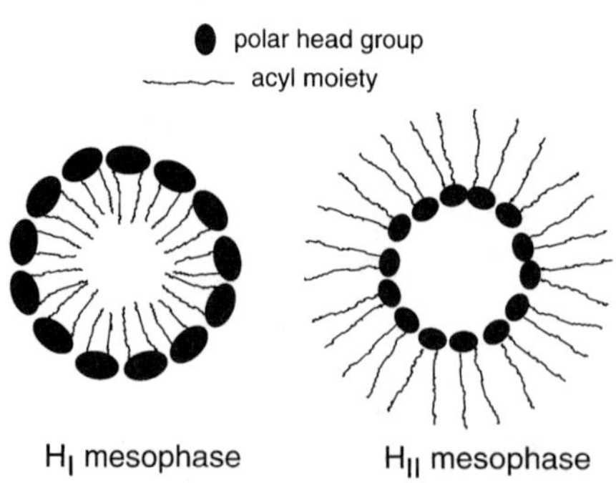

Fig. 7 Schematic models of mesophases of polar lipids present in cocoa butter [41].

time to orient into seed nuclei. Thus, the rate of nucleation assisted by the phospholipid precursors is minimized, as evidenced in the soft butters. In contrast, PC may form H_{II} mesophase (reversed micelle). In this micelle phase, the acyl moieties present in exterior regions are conformationally compatible with the fat molecules in CB, leading to short induction times for nucleation of hard butters. This means that increased PC concentrations may promote the nucleation and crystal growth of CB. In this regard, the use of commercial soy lecithin rich in PC, which is usually employed for viscosity control, seems also effective for a nucleation promoter [40].

2. *High-Melting Fats Added to Cocoa Butter*

The addition of fat crystals for seeding is an optimal technique to control the polymorphic crystallization of fats. It is suggested that specific molecular interactions between the crystal seed materials and the polymorphic forms of specific fats are a prerequisite for the seeding effect, in, for example, polymorphic correspondence, aliphatic chain matching, and thermal stability [42].

As for the polymorphic correspondence, it is highly required for the seed material to reveal the same polymorphic forms as those of the crystallizing materials, mostly β' or β. The crystal structures of the two polymorphs mostly differ in the subcell packing. Therefore, the evaluation of the polymorphic forms of the seed materials using XRD or other methods is prerequisite. Aliphatic chain matching means two things: chain length and the chemical nature of the fatty acid moiety. It is expected that differences in the chain length, which are expressed in the number of carbon atoms between the seed and crystallizing materials, may not exceed 4, in accordance with recent work on template effects of fatty acid thin films on the crystallization of long-chain *n*-alcohol crystals [43]. The chemical nature of the fatty acid moiety concerns saturation or unsaturation in such a manner that, when the saturated-unsaturated mixed-acid TAGs are the crystallizing material, the seed material without the unsaturated fatty acid moiety may be less effective. Finally, the thermal stability of the seed crystals requires the property that the seed material not melt or dissolve in the liquid phase of the crystallizing materials at the seeding temperature. Therefore, the melting point of the seed materials must be much higher than that of the crystallizing material.

A remarkable success was made with seeding BOB (1,3-dibehenoyl-2-oleoyl-*sn*-glycerol) β_2 polymorph [44–49]. The polymorphism of BOB and SOS was isomorphic, and the melting temperatures of the corresponding forms were higher in BOB than in SOS by about 10–15°C, because of the presence of n_c = 22 of behenic acid moiety in BOB and n_c = 18 of stearic acid moiety in SOS. The BOB β_2 seed crystal satisfied four conditions: (1) acceleration of the crystallization, (2) no usage of the tempering process, (3) polymorphic crystallization in Form V, and (4) improvement of fat bloom stability.

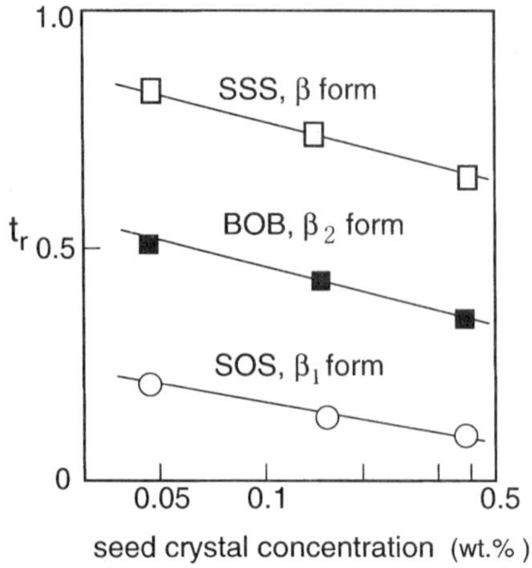

Fig. 8 Relative crystallization rate (t_r) of dark chocolate with three kinds of seed crystals at varying concentrations [46].

As for points (1)–(3), Fig. 8 shows the relative crystallization time (t_r) of seeded dark chocolate examined at 30°C. The crystallization time was measured with a rotational viscometer, and t_r was defined as a ratio of the crystallization times with and without seeding. Figure 8 shows that the stable form β_1 of SOS is most effective, stable form β_2 of BOB is intermediate, and β of SSS (tristearin) is least effective. Although β_1 of SOS is most effective, this seed crystal does melt at temperatures where the seed crystals are added in molten chocolate around 30–35°C. Regarding polymorphic control, Fig. 9 shows an in situ XRD experiment of the crystallization of CB with and without the addition of BOB β_2 seed crystals, when the liquid of CB was cooled from 35°C to 22°C. Without BOB β_2, cocoa butter crystallized at 22°C in Form III, but Form V quickly crystallized in the first with the addition of BOB β_2 at 22°C. Therefore, the tempering process is not necessary with use of BOB β_2 seed crystals.

Fat bloom stability was tested under two thermal treatments of the seeded chocolate at temperatures below and above 36°C, as shown in Table 5 [47]. In the 32/20 cycle test, the seeded chocolate was repeatedly kept at 32°C for 12 hr followed by 20°C for 12 hr. In addition, the 38/20 test underwent the thermocycle of 38°C (12 hr) and 20°C (12 hr). The seed crystals of SOS β_1 showed the same fat bloom stability effects as that using BOB β_2 in the 32/20 test, but no improvement was seen in the 38/20 test with SOS β_1. In contrast, in the 38/20 test, BOB_2 improved fat bloom stability at higher seed concentrations: for example, 5% of BOB β_2. This is ascribed to the low solubility of BOB in molten CB [50]. The

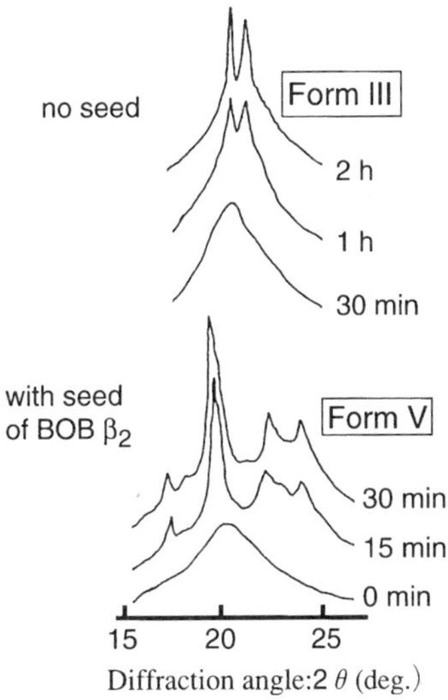

Fig. 9 In situ X-ray diffraction observation of crystallization of cocoa butter with and without the seed crystals of BOB β_2.

demolding property of chocolate after the crystallization process also prevailed in BOB β_2, similar to SOS β_1.

Using the properties in Table 5, various types of functional confections have been put on the factory production by using BOB β_2 seed crystals. Some examples are shown here. Crystallization is well controlled for the chocolates to which the conventional tempering process cannot apply. For example, softer chocolates containing low-melting fractions such as nut pastes and emulsion-type chocolates can be solidified by tempering with the presence of BOB β_2 seed

Table 5 Crystal Seeding Effects on Physical Properties of Dark Chocolate

Seed material	Optimal crystallization	Optimal demolding	Fat bloom stability	
			32/20[b]	38/20
SOS β_1	++++++[a]	++	++	−
BOB β_1	++++	++++++	+++++	+++++
SSS β	++	−	−	−

[a] + = effective, − = not effective.
[b] Temperature cycles (°C).

crystals. Additional processing of chocolate can be made at elevated temperatures around 33–37°C at reduced viscous conditions (Fig. 1). In particular, enrobing is easily carried out under this condition, because reduced viscosity is maintained over a long period without the crystallization or melting of CB. Further, chocolate with melting ranges around 20°C has been formulated, so it can be stored in a freezer (around −20°C) for the summer consumption. If once melted around normal temperature, crystallization in Form V occurs upon freezing with the help of BOB β_2 seed crystals.

Despite the excellent advantages of the BOB seed material, it is tremendously costly, because BOB is rarely present in natural fats and oils. Conversion from erucic acids to behenic acid is made by hydrogenation, which is followed by lipase-catalyzed esterification to produce BOB. Then crystallization of BOB and subsequent transformation from metastable γ or β' forms to β_2 form also needs further processing, which altogether make BOB β_2 quite costly. Some new techniques to produce BOB and its crystal powders with reduced cost is highly anticipated.

The use of other TAGs as seed materials has also been reported. Simultaneous usage of SSS seed crystals and sorbitan tristearate was found to effect fat bloom stability [51]. Ollivon et al. examined the effects of addition of SSS, stearic acid, and distearoyl-glycerol on the two-stage crystallization of dark chocolate [52]. In this work, dynamic crystallization was observed in a lab-scale scraped-surface heat exchanger by following the variations of torque with time during isothermal crystallization at temperatures between 24.9 and 31.0°C. As shown in Fig. 10, chocolate crystallization occurred in two steps at crystallization temperatures above 26.2°C, but in one step below that temperature. The first step

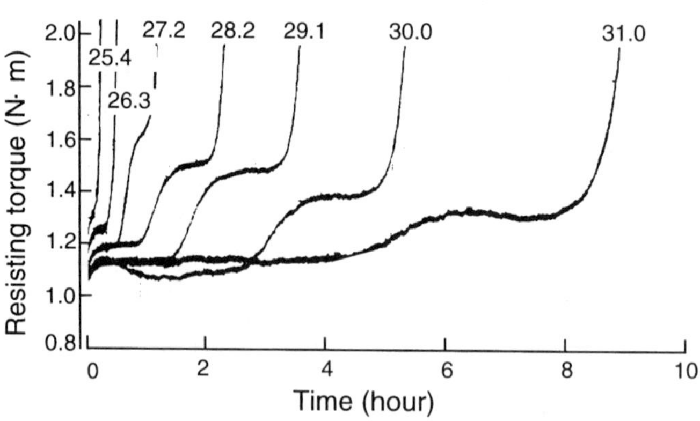

Fig. 10 Resisting torque values of dark chocolate during isothermal crystallization at different temperatures (°C) [52].

corresponded to crystallization of high-melting TAGs (about 1% containing SSS), and the second-step crystallization corresponded to the rest of TAGs in the β-type polymorph. The addition of SSS seed crystals preceded the first-step crystallization, but did not influence the main crystallization of the second step.

B. Effects of Ultrasound Stimulation

Ultrasound at high power and low frequencies can assist various processes in food technology involving crystallization [53]. Crystallization enhanced by high-power ultrasound (here called sonocrystallization) may be seen in both nucleation and crystal growth by creating fresh additional nucleation sites in the crystallization medium. Sonocrystallization effects on chocolate processing were recently reported, claiming that tempering is achieved by sonication [54].

For further clarification, sonocrystallization was examined with ultrasound (20 kHz, 300 W) during cooling of CB from liquid (T. Koyano, unpublished work). Ultrasound was applied to CB (from Ivory Coast, 250 g) at 32.3°C for up to 15 sec after cooling from 60°C. Ultrasound-treated samples were then aged at 20°C for 30 min and subjected to crystallization at 4°C. The polymorphic forms of crystallized CB were examined by X-ray diffraction.

Figure 11 shows that sonocrystallization induced the formation of Form V under limited conditions. Without sonication Form II was crystallized (Fig. 11a) as in the case of CB crystallization without tempering. Sonication for 3 sec pro-

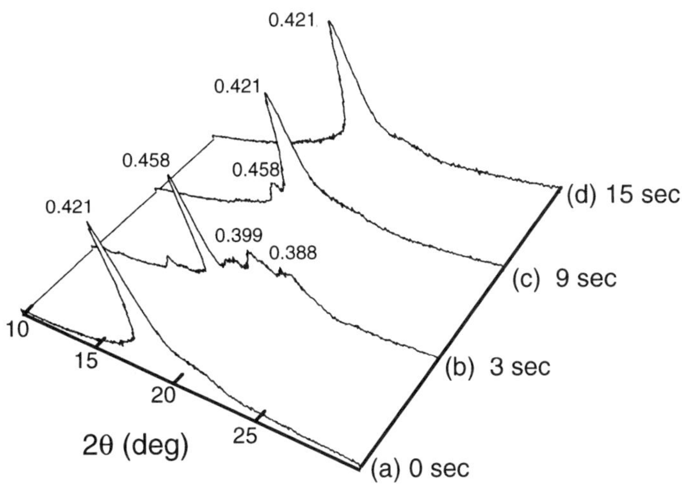

Fig. 11 X-ray diffraction short spacing spectra of cocoa butter under high-power ultrasound.

duced Form V as shown in Fig. 11b. In more detail, the temperature of CB sample rose by 0.6°C because of the transfer of mechanical energy into heat during sonication. Temperature variation of 60.0–32.3–32.9°C does not correspond to the tempering illustrated in Fig. 1b. Therefore, the result of Fig. 11b indicates that sonication directly induced the formation of crystal nuclei of Form V in CB liquid. Further treatment with ultrasound for 9 sec resulted in the formation of a Form II–Form V mixture (Fig. 11c), and singly Form II was again crystallized after sonication for 15 sec (Fig. 11d). The temperature of CB sample rose to 34.3°C (9 sec) and 36.2°C (15 sec) by sonication. Since the melting point of Form V of CB is around 33–34°C, the rise in temperature by sonication may induce the melting of Form V crystal nuclei when sonication time exceeds 9 sec. Long sonication time (15 sec) must completely dissolve the Form V crystal nuclei, and thereby the sonication effect diminished due to the thermal energy converted from ultrasound mechanical energy.

In the traditional tempering sequence (Fig. 1b), unstable Forms III or IV of CB are crystallized during the first cooling, and then the unstable forms transform to stable polymorphs during the reheating process. This polymorphic change occurs through melt-mediated transformation in which the stable Form V having triple-chain-length structure is formed soon after the melting of the unstable forms having double-chain-length structure. However, in the case of sonocrystallization, it is quite curious to see that the stable form is directly crystallized without the formation and subsequent melt mediation of the unstable forms. The mechanisms underlying the polymorph-controlling sonocrystallization process are not clear. There would be some combined effects inherent to sonocrystallization, most probably due to pressure-induced thermodynamic effects and chain-chain interactions leading to the formation of stable molecular packing of triple-chain-length and $T_{\parallel}$ subcell of Form V.

C. Effects of Shear Forces

It was indicated quite early [55] that mechanical working consisting of extrusion under pressure promotes a solid-to-solid transformation of fats from unstable to more stable polymorphic forms. The acceleration effect of shearing on the cocoa butter crystallization in Form V was then claimed by the induction time measurement of the occurrence of Form IV and Form V using a viscometer and DSC under varying shear rates (D), as shown in Fig. 12 [56]. It was observed that, with increasing rate of shearing, the polymorphic forms of crystallized cocoa butter converted from Form IV to Form V, as observed in the increase in the DSC melting points (Fig. 12a). Correspondingly, the induction time of the crystallization measured by a viscometric method decreased and the first-occurring polymorph converted from Form IV to Form V (Fig. 12b).

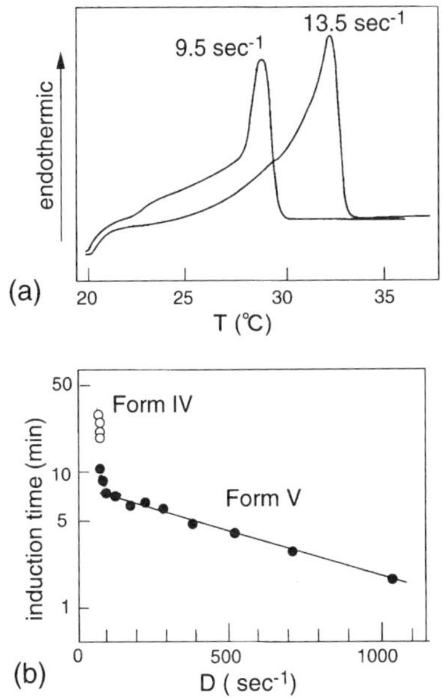

Fig. 12 (a) DSC endothermic peaks of cocoa butter under shear, and (b) induction time for crystallization of cocoa butter under varying shear (D) [56].

The shear effects on CB crystallization were visualized by the time-resolved XRD measurement using synchrotron radiation beam (SR-XRD) [57,58]. The occurrence of the polymorphic forms was observed by long spacing SR-XRD during isothermal crystallization after simple cooling of molten CB from 50°C to 22 and 20°C without shear (Fig. 13a) and with a shear rate of 12 sec^{-1} (Fig. 13b), respectively [58]. It seems that no direct crystallization in Form V was detectable in the two cases; instead Form III first crystallized, probably as a precursor form. Without shear, Form III converted to Form IV during isothermal crystallization. However, Form III turned to Form V under the shear effect, the rapid diminishing in the intensity of the XRD spectra being reversibly proportional to the increase in the XRD spectra of Form V.

The result of Fig. 13b indicates that the precursor crystals of Form III transform directly to Form V under shear, possibly through hydrodynamic interactions between small crystal particles of Form III, or through polymorphic transformation in Form III crystals under the shear force.

The application of shear effects on chocolate processing has recently been examined by several workers [59–61]. For example, the polymorphic forms were

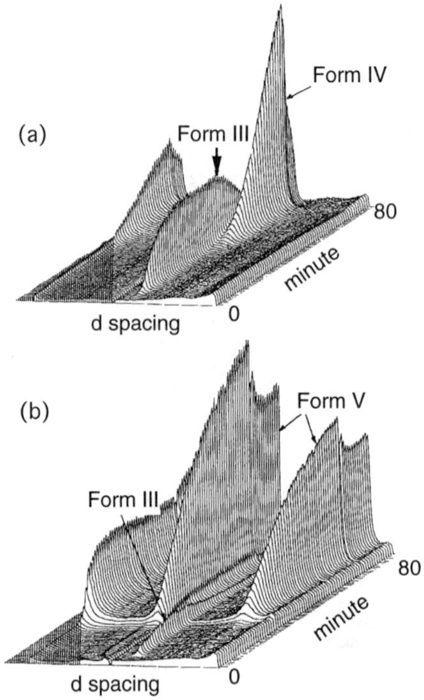

Fig. 13 Synchrotron X-ray diffraction spectra (small-angle region) of cocoa butter crystallization (a) at 22°C without shear and (b) at 20°C with shear (12 sec^{-1}) [58].

monitored after crystallization, showing that two polymorphic forms occurred in such a manner that the higher-melting form occurred at a higher shearing rate, low rewarming temperature, and slow DSC scan rates [59]. Windhab et al. showed that the shear stress effects were revealed under temperature control with a concentric cycling geometry which enabled uniformity of shear rate during tempering [60,61]. A new shear crystallizer integrated into a circular loop was tested on the crystallization behavior of cocoa butter and chocolate. The physical properties of shear-treated fats were examined with DSC, rheometry, and near-infrared (IR) spectroscopy. It was shown that viscosity is linearly dependent on crystal content and the mechanical energy input had a significant influence on viscosity, melting enthalpy, and slope at the second point of inflection of the temper curve of the fats as shown in Fig. 14. Also shown was improved cost performance with a high-shear crystallizer in terms of reduced residence, higher outlet temperature, and fat bloom stability.

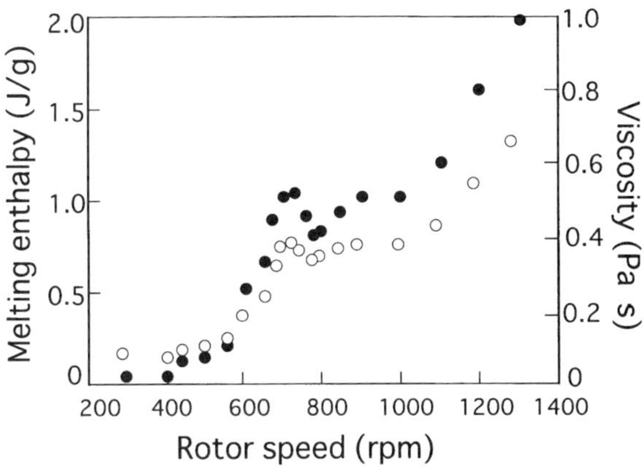

Fig. 14 Melting enthalpy values and viscosity values of cocoa butter under shear (data calibrated with respect to no-shear condition) [61].

The shear effects were also verified in confectionery coating fat made of hydrogenated and fractionated mixture of soybean and cottonseed oils, whose onset temperature of crystallization increased significantly with shear rate [62]. Like this, the effects of shear forces on the crystallization of CB and other confectionery fats have convincingly been visible, and further research on mechanistic interpretation of the crystallization processes and on commercial base application under shear are needed.

IV. FAT BLOOM PHENOMENA

A. Fat Bloom Formation

Fat bloom, which exhibits white-patterned surface and microfractured interior of end products, is a hazardous deterioration of confections. There are various features of fat bloom, as discussed recently by Hartel [63]. The main causes of fat bloom are improper tempering, uncontrolled recrystallization after melting of fats, long-term deterioration during storage, fat migration between interfaces of coating chocolate with fillings containing fats and oils, etc. Major concerns from industry and academia are the basic mechanisms of fat bloom formation as reviewed by Garti et al. in 1988 [64] and reported thereafter [65–75]. Having understanding of fat bloom mechanisms depending on various chocolates, such as plain or coating, researchers have attempted retardation of fat bloom by examining fat blending of CB with milk fats [76–81] and tallow [82], and by using emulsifiers, as recently reviewed by Wilson [83], or some other techniques.

As phenomenological aspects, fat bloom involves the following processes: (a) during long-term storage, Form V–VI transformation occurs in CB crystals [65]; (b) temperature control after crystallization and before packaging is quite important, and precooling before packaging is useful for retarding fat bloom [66]; (c) microstructures of fat crystals also influence fat bloom, as observed by porosity examination [68]; (d) as for the fat bloom in coating/filling chocolates, migration of fats and oils is critically important [72–75], and migration occurs not only from the filling to the coating but also in the opposite direction [68]; (e) ingredients such as sugar particles also effect fat bloom; for example, despite the Form V–Form VI transformation in CB, the control chocolate with amorphous sugar did not bloom, while the crystalline sugar did [69].

In this section, we discuss the basic mechanisms of fat bloom formation, solely focusing on the polymorphic transformation and crystallization of the stable forms of Forms V and VI of CB, with which various aspects discussed above are interpreted [70].

B. Basic Mechanisms in Fat Bloom

Figure 15 depicts the model of fat bloom phenomena, based on the concept that the transformation in the stable β-type polymorphs from Form V to Form VI is the main origin of the fat bloom. Two types of polymorphic transformation may occur, depending on temperature: a solid-state transformation in a crystalline par-

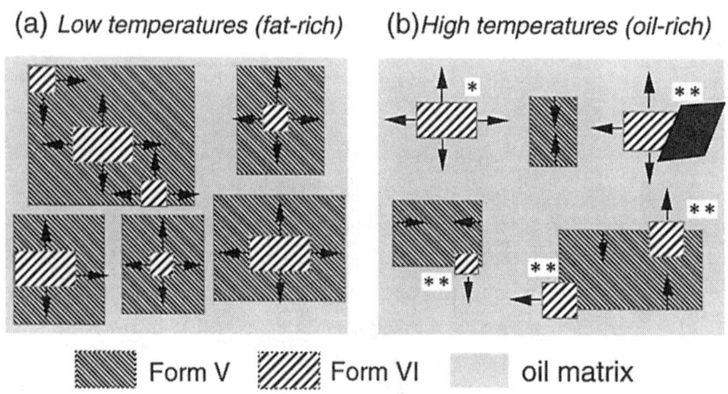

Fig. 15 Basic models of fat bloom formation. (a) Solid-state transformation and (b) oil-mediated transformation (spontaneous nucleation shown by *, and heterogenous nucleation shown by **).

ticle of Form V, and an oil-mediated transformation which is actually nucleation and crystal growth of Form VI from the surfaces of Form V crystals in a liquidus medium. The liquid oil matrix is formed by melting the low-melting fractions present in CB, such as SLS, PLS, SOO, POO, etc., which altogether comprise 9% (from Malaysia) through 21% (Brazil) (Table 3). In the latter case, mass transfer of TAG molecules of CB through the liquidus medium plays important roles.

At low temperatures, e.g., below ca. 22°C, a high value of solid fat content (SFC) is maintained (fat-rich) as depicted in Fig. 15a. In this phase, the solid-state transformation dominates, since the nucleation and growth of Form VI crystals proceed in the interior of Form V crystal particles. Since the oil-mediated transformation in the fat-rich phase is minimized because of low concentrations of liquidus fractions, the formation of long and needle crystals characteristic of Form VI is limited. Thereby, the fat bloom is not manifest in the sense that individual large needle crystals of Form VI are not observed, although the V–VI transformation proceeds over a long period.

In contrast, the oil-mediated transformation of CB occurs at high temperatures with lower SFC values (oil-rich) above ca. 25°C, as depicted in Fig. 15b. The crystal growth of Form VI can be initiated either by spontaneous nucleation in liquid or by heterogeneous nucleation at the surfaces of Form V crystals and catalytic ingredients such as sugar particles. During the former event, TAG molecules are detached from the dissolving crystals of Form V and form nuclei of Form VI crystals through volume diffusion in the oil matrix. The latter event of heterogeneous nucleation of Form VI may also occur at the low temperatures, but its rate is raised at the elevated temperatures. In the oil-rich phase, the growing crystal surfaces of the Form VI crystals are exposed to liquid, and the morphology of Form VI crystals develops. The equilibrium crystal shape of β form of TAGs was calculated and experimentally evidenced as a flat needle-like habit [84]. Their growth forms may further extend the needle axis, providing the slender crystals of Form VI that cause the fat bloom, as observed with electron microscopy [4,85]. Therefore, the mechanism depicted in Fig. 15b is the primary process of the fat bloom.

The above mechanisms are not limited in the transformation from Form V to Form VI. Even the transformation from Forms III and IV to Form V may occur when the polymorphic solidification is not controlled. In such a case, Form V crystals grow at the expense of Forms III and IV in the oil matrix, and the formation of larger Form V crystals causes the fat bloom. In any case, the concentrations of POP and SOS must increase compared with POS during the fat bloom, because of the differences in the solubility values in the range of temperatures at which the fat bloom phenomena occur, as measured in Fig. 6.

The present model raises the following indications for retardation of fat bloom.

(a) Reduction of the nucleation rate of Form VI through the solid-state transformation and oil-mediated transformation. In order to minimize the former process, lowering the storage temperature is, of course, most important, since the solid-state transformation is minimized. In addition, high crystallinity of Form V crystals should be maintained, since the solid-state transformation is promoted when lattice defects are introduced in the crystal [86]. Blending of special fats, which are incorporated in Form V crystals and effectively block the solid-state transformation into Form VI mechanistically, must also be critically useful.

The rates of crystal nucleation in the oil matrix, both primary and heterogeneous, are determined by temperature and supersaturation with respect to the nucleating materials (see Chapter 1 of this book). Then the lowering of temperature results in conflicting influences: increase in supersaturation for nucleation and growth with respect to Form VI which promotes the fat bloom, and decrease in thermal activation energy and increase in viscosity of the oil matrix, both of which retard the fat bloom. Heterogeneous nucleation may occur at the ingredient particles such as sugar particles, as Bricknell and Hartel observed [69].

(b) Reduction of the crystal growth of Form VI in the oil matrix is critical. For this purpose, two processes may be valuable: (1) to inhibit the crystal growth at the growing interfaces using, for example, impurities, and (2) to decrease the diffusion rates of TAG molecules in the oil matrix. Most effective impurities are food emulsifiers, which interact with the growing crystal surfaces of CB [64; Chapter 7 of this book]. The increase of the oil-matrix viscosity is also effective so that the volume diffusion of the TAG molecules onto the growing interfaces is retarded.

V. CONCLUSION

The crystallization phenomena of CB still exhibit unresolved characteristics, some of which are related to molecular-level understanding of the phase behavior of mixture states of principal and minor TAGs comprising CB. Previous work [87–90] indicates that it is highly possible to form molecular compound forming mixing phases, depending on specific combinations of TAGs, such as SOS/SSO, SOS/OSO, and other similar systems, and on the polymorphic forms. This property, together with eutectic and miscible properties, is closely related to fractionation of CB and fat blending with other fat resources. More precise determination of the polymorphic structures of Forms V and VI may lead to novel ideas of fat bloom inhibition with other fats or some ingredients with the idea of tailor-made structure modification. Although not elaborated here, interrelations between fat crystal network and rheological properties in CB-based confections are of extreme importance.

REFERENCES

1. Fatty Acids and Glycerides, ed. A. Kuksis, in *Handbook of Lipid Research*, vol. 1, ed. D. J. Hanahan, Plenum, New York, 1978, pp. 197–232.
2. R. E. Wainwright, in *Bailey's Industrial Oil and Fat Products*, 5th ed., Vol. 3, ed. Y. H. Hui, Wiley, New York, 1996, pp. 353–407.
3. *Lipid Technologies and Applications*, eds. F. D. Gunstone and F. B. Padley, Marcel Dekker, New York, 1997, pp. 391–432.
4. D. R. Manning and P. S. Dimick, *Food Microstructure*, *4*: 249 (1985).
5. R. L. Wille and E. S. Lutton, *J. Am. Oil Chem. Soc.*, *27*: 491 (1996).
6. S. T. Beckett, *Industrial Chocolate Manufacture and Use*, Blackie & Son, Glasgow, 1988, pp. 172–202.
7. C. O. Chichester, E. M. Mark, and B. S. Schweigert, *Advances in Food Research*, Vol. 31, Academic Press, New York, 1987, pp. 308–313.
8. S. V. Vaeck, *Manuf. Confect.*, *40*: 35, 71 (1960).
9. K. Sato, T. Arishima, Z. H. Wang, K. Ojima, N. Sagi, and H. Mori, *J. Am. Oil Chem. Soc.*, *6*: 664 (1989).
10. P. S. Dimick and M. D. Manning, *J. Am. Oil Chem. Soc.*, *64*: 1663 (1987).
11. T. R. Davis and P. S. Dimick, *J. Am. Oil Chem. Soc.*, *66*: 1488 (1989).
12. T. R. Davis and P. S. Dimick, *J. Am. Oil Chem. Soc.*, *66*: 1494 (1989).
13. K. van Malssen, R. Pescher, and H. Schenk, *J. Am. Oil Chem. Soc.*, *73*: 1209 (1996).
14. K. van Malssen, R. Pescher, and H. Schenk, *J. Am. Oil Chem. Soc.*, *73*: 1217 (1996).
15. K. van Malssen, R. Pescher, C. Brito, and H. Schenk, *J. Am. Oil Chem. Soc.*, *73*: 1225 (1996).
16. K. van Malssen, A. van Langevelde, R. Pescher, C. Brito, and H. Schenk, *J. Am. Oil Chem. Soc.*, *76*: 669 (1999).
17. C. Loisel, G. Keller, G. Lecq, C. Bourgaux, and M. Ollivon, *J. Am. Oil Chem. Soc.*, *75*: 425 (1998).
18. L. R. Cook and E. H. Meursing, in *Chocolate Production and Use*, Harcourt Brace Jovanovich, New York, 1982, pp. 401–423.
19. G. G. Jewell, 35th P.M.C.A. Production Conference, 63 (1981).
20. L. B. Campbell and P. G. Keeney, *Food Technol.*, *22*: 1150 (1968).
21. I. Hachiya, T. Koyano, and K. Sato, *J. Jpn. Oil Chem. Soc.* (*Yukagaku*), *37*: 431 (1988).
22. I. Hachiya, T. Koyano, and K. Sato, *J. Jpn. Oil Chem Soc.* (*Yukagaku*), *37*: 613 (1988).
23. C. Loisel, G. Keller, G. Lecq, B. Launay, and M. Ollivon, *J. Food Sci.*, *62*: 773 (1997).
24. G. M. Chapman, E. E. Akehurst, and W. R. Wright, *J. Am. Oil Chem. Soc.*, *48*: 824 (1971).
25. K. G. Berger, G. G. Jewell, and R. J. M. Pollitt, in *Food Microscopy*, ed. J. G. Vaughan, Academic Press, London, 1979, pp. 445–497.
26. A. Huyghebaert and H. Hendrickx, *Lebensm.-Wiss. Technol.*, *4*: 59 (1971).
27. E. Brosio, F. Conti, A. Di Nola, and S. Sykora, *J. Am. Oil Chem. Soc.*, *57*: 78 (1980).

28. G. V. Merken and S. V. Vaeck, *Lebensm.-Wiss. Technol.*, *13*: 314 (1980).
29. K. Sato, in *Advances in Applied Lipid Research*, vol. 2, ed. F. Padley, JAI Press, 1996, pp. 213–268.
30. N. Sagi, T. Arishima, H. Mori, and K. Sato, *J. Jpn. Oil Chem. Soc.* (*Yukagaku*), *38*: 306 (1989).
31. T. Koyano, Y. Kato, I. Hachiya, R. Umemura, K. Tamura, and N. Taguchi, *J. Jpn. Oil Chem. Soc.* (*Yukagaku*), *42*: 453 (1993).
32. T. Arishima and K. Sato, *J. Am. Oil Chem. Soc.*, *66*: 1614 (1989).
33. T. Arishima, N. Sagi, H. Mori, and K. Sato, *J. Am. Oil Chem. Soc.*, *68*: 710 (1991).
34. J. Yano, S. Ueno, K. Sato, T. Arishima, N. Sagi, F. Kaneko, and M. Kobayashi, *J. Phys. Chem.*, *97*: 17967 (1993).
35. J. Yano, K. Sato, F. Kaneko, D. M. Small, and D. R. Kodali, *J. Lipid Res.*, *40*: 140 (1999).
36. T. Arishima, K. Sugimoto, R. Kiwata, H. Mori, and K. Sato, *J. Am. Oil Chem. Soc.*, *73*: 1231 (1996).
37. D. J. Cebula and K. W. Smith, *J. Am. Oil Chem. Soc.*, *69*: 992 (1992).
38. S. Chaiseri and P. S. Dimick, *J. Am. Oil Chem. Soc.*, *72*: 1491 (1995).
39. S. Chaiseri and P. S. Dimick, *J. Am. Oil Chem. Soc.*, *72*: 1497 (1995).
40. C. M. Savage and P. S. Dimick, in The 49th P.M.C.A. Production Conference, 54 (1995).
41. P. S. Dimick, Compositional effect on crystallization of cocoa butter, in *Physical Properties of Fats, Oils and Emulsifiers*, ed. N. Widlak, AOCS Press, Champaign, 1999, pp. 140–163.
42. K. Sato, *Chem. Eng. Sci.*, in press.
43. H. Takiguchi, K. Iida, S. Ueno, J. Yano, and K. Sato, *J. Cryst. Growth*, *193*: 641 (1998).
44. I. Hachiya, T. Koyano, and K. Sato, *J. Jpn. Oil Chem. Soc.* (*Yukagaku*), *66*: 1757 (1989).
45. I. Hachiya, T. Koyano, and K. Sato, *J. Jpn. Oil Chem. Soc.* (*Yukagaku*), *66*: 1763 (1989).
46. I. Hachiya, T. Koyano, and K. Sato, *J. Am. Oil Chem.*, *66*: 1757 (1989).
47. I. Hachiya, T. Koyano, and K. Sato, *J. Am. Oil Chem.*, *66*: 1763 (1989).
48. I. Hachiya, T. Koyano, and K. Sato, *Food Microstructure*, *8*: 257 (1989).
49. T. Koyano, I. Hachiya, and K. Sato, *Food Structure*, *9*: 231 (1990).
50. Z. H. Wang, K. Sato, N. Sagi, T. Izumi, and H. Mori, *J. Jpn. Oil Chem. Soc.* (*Yukagaku*), *36*: 671 (1987).
51. O. Jovanovic, D. Karlovic, J. Jakovljievic, and B. Pajin, Prof. World Conf. Oilseed Edible Oils Process, AOCS, Champaign, 1998, pp. 135–140.
52. C. Loisel, G. Lecq, G. Keller, and M. Ollivon, *J. Food Sci.*, *63*: 73 (1998).
53. T. J. Mason in *Ultrasound in Food Processing*, eds. M. J. W. Povey and T. J. Mason, Blackie Academic & Professional, London, 1998, pp. 105–126.
54. EU Patent Application EP0765605A1.
55. R. O. Feuge, W. Ladmann, D. Mitcham, and N. V. Lovegren, *J. Am. Oil Chem. Soc.*, *39*: 310 (1962).
56. G. Ziegleder, *Int. Z. Lebensm. Technol. Verffarenstechn.*, *36*: 412 (1985).
57. P. N. M. R. van Gelder, N. Hodgson, K. J. Roberts, A. Rossi, M. Wells, M. Polgree,

and I. Smith, in *Crystal Growth of Organic Materials*, Conference Proceedings Series, eds. S. Myerson, D. A. Green, and P. Meenanet, Am. Chem. Soc., Washington, 1996, pp. 209–215.
58. S. D. MacMillan, K. J. Roberts, A. Rossi, M. Wells, M. Polgree, and I. Smith, in Proceedings of World Congress on Particle Technology, Brighton 1998, pp. 96–103.
59. A. G. F. Stapley, H. Tewkesbury, and P. J. Fryer, *J. Am. Oil Chem. Soc.*, *76*: 677 (1999).
60. S. Bolliger, B. Breitschuh, M. Stranzenger, T. Wagner, and E. J. Windhab, *J. Food Eng.*, *35*: 281 (1998).
61. S. Bolliger, Y. Zeng, and E. J. Windhab, *J. Am. Oil Chem. Soc.*, *76*: 659 (1999).
62. C. Garbolino, G. R. Ziegler, and J. N. Coupland, *J. Am. Oil Chem. Soc.*, *77*: 157 (2000).
63. R. W. Hartel, *Manuf. Confect.*, *79*: 89 (1999).
64. J. Schlichter-Aronhine, and N. Garti, in *Crystallization and Polymorphism of Fats and Fatty Acids*, eds. N. Garti and K. Sato, Marcel Dekker, New York, 1988, pp. 363–393, and references therein.
65. D. J. Cebula and G. Ziegleder, *Fett Wiss. Technol.*, *95*: 340 (1993).
66. H. Adenier, H. Chaveron, and M. Ollivon, Mechanism of fat bloom development on chocolate, in *Shelf Life Studies of Foods and Beverages. Chemical, Biological, Physical and Nutritional Aspects*, ed. G. Charalambous, Elsevier, London, 1993, pp. 353–389.
67. G. Ziegleder and H. Mikle, *Suesswaren*, *39*: 26 (1995).
68. C. Loisel, G. Lecq, G. Ponchel, G. Keller, and M. Ollivon, *J. Food Sci.*, *62*: 781 (1997).
69. J. Bricknell and R. W. Hartel, *J. Am. Oil Chem. Soc.*, *75*: 1609 (1998).
70. K. Sato, in *Confectionery Science*, Proc. Int. Symp, ed. G. Ziegler, Pennsylvania State University Press, 1997, pp. 155–176.
71. R. W. Hartel, IFT Basic Symp. Ser. 13, Phase/State Transitions in Foods, 1998, pp. 217–251.
72. G. Ziegleder, J. Geier-Greguska, and J. Grapin, *Fett Wiss. Technol.*, *96*: 390 (1994).
73. G. Ziegleder and H. Mikle, *Suesswaren*, *39*: 23 (1995).
74. G. Ziegleder, *Manuf. Confect.*, *77*: 43 (1997).
75. G. Ziegleder, *Fett/Lipid*, *100*: 411 (1998).
76. C. E. Bystrom and R. W. Hartel, *Lebensm.-Wiss. Technol.*, *27*: 142 (1994).
77. M. H. Lohman and R. W. Hartel, *J. Am. Oil Chem. Soc.*, *71*: 267 (1994).
78. C. Simoneau and J. B. German, *J. Am. Oil Chem. Soc.*, *73*: 955 (1996).
79. R. W. Hartel, *J. Am. Oil Chem. Soc.*, *73*: 945 (1996).
80. S. Y. Reddy, N. Full, P. S. Dimick, and G. R. Ziegler, *J. Am. Oil Chem. Soc.*, *73*: 723 (1996).
81. S. D. Williams, K. L. Ransom-Painter, and R. W. Hartel, *J. Am. Oil Chem. Soc.*, *74*: 357 (1997).
82. A. R. M. Ali, L. M. Moi, A. Fisal, R. Nazaruddin, and S. Sabaria, *J. Sci. Food Agric.*, *76*: 285 (1998).
83. E. J. Wilson, *Manuf. Confect.*, *79*: 83 (1999).
84. P. Bennema, L. J. Fogels, and S. de Jong, *J. Cryst. Growth*, *123*: 41 (1992).

85. J. D. Hicklin, G. G. Jewell, and J. F. Heathcock, *Food Microstructure*, *4*: 241 (1985).
86. K. Sato, *J. Phys. D*: *Appl. Phys. B*, *26*: 77 (1993).
87. D. P. Moran, *J. Appl. Chem.*, *13*: 91 (1963).
88. J. B. Rossel, *Adv. Lipid Res.*, *5*: 353 (1967).
89. T. Koyano, I. Hachiya, and K. Sato, *J. Phys. Chem.*, *96*: 10514 (1992).
90. L. J. Engstrom, *Fat Sci. Technol.*, *94*: 173 (1992).

13

Influences of Colloidal State on Physical Properties of Solid Fats

Heike Bunjes and Kirsten Westesen
Friedrich Schiller University, Jena, Germany

I. INTRODUCTION

Colloidally dispersed lipids are interesting carrier systems for poorly water soluble, bioactive substances. Beside their importance in fields such as cosmetics or food science, they are under intensive investigation as drug carriers, e.g., for the administration of lipophilic drugs into the bloodstream. Some of these systems may also serve as drug-targeting devices, i.e., for the delivery of drugs to specific target organs or tissues. Since colloidal lipid carrier systems often mimic physiological structures and are of similar composition as their physiological counterparts, a good physiological acceptance is expected for these types of carriers. Important examples for colloidal lipid drug carriers are the phospholipid-based liposomes, mixed micelles of lecithin and bile salts and colloidal fat emulsions containing phospholipid-stabilized droplets of liquid oils [1–6]. The latter also play an important role as a calorie source in parenteral nutrition [7,8]. These ''classic'' types of colloidal lipid carrier systems have been under investigation for several decades.

Since the beginning of the nineties, different novel types of nanoparticulate lipid drug carrier systems, based on solid lipids, have received much attention [9–21]. Compared with other lipidic carrier systems, particles with a solid matrix are more promising candidates for drug targeting since a solid matrix of the dispersed lipid could help to overcome the problem of rapid drug release often observed with the more fluid types of lipid drug carrier systems, e.g., with emulsions [22,23]. A slow release from the colloidal carrier is a prerequisite for drug targeting, since the drug has to remain bound to the carrier until it has reached the target. Nanoparticles based on solid lipids are, however, not only under investigation with respect to parenteral administration but are also promising for, e.g., ocular, pulmonary, peroral, and dermal applications [19,21,24–26]. Moreover, they receive much attention in the field of cosmetics [27,28].

Solid lipid nanoparticles can be prepared in different ways: A common method is to disperse a melted lipid, e.g., a triacylglycerol, into an aqueous dispersion of colloidal emulsion droplets in the heat and to subsequently crystallize the lipid droplets on cooling [13,18]. Other ways of preparation are the precipitation of lipidic nanoparticles from organic solvent-in-water emulsions by solvent evaporation [15,29], the precipitation from microemulsions [10], or cold homogenization of a solid lipid [18]. An important advantage of the melt-homogenization process is that it is easy, rapid, and reproducible to perform with well-known, commercial homogenization equipment. Narrow particle size distributions with mean particle sizes far below 200–300 nm can be obtained at comparatively high lipid concentrations (e.g., 10%). Moreover, there is no risk of residual organic solvents in the final product. The melt-homogenization process was originally described for the colloidal dispersion of acylglycerols like monoacid triacylglycerols or complex acylglycerol mixtures as present in hard fat suppository masses [13,18,30,31]. The preparation procedure can, however, easily be transferred to other low-melting, poorly-water-soluble organic substances such as waxes or solid paraffins [28,32,33], thereby opening other areas of use. Depending on the intended type of application, different pharmaceutical and cosmetic surfactants and surfactant blends, e.g., based on phospholipids, bile salts, poloxamers, tyloxapol, and other nonionic surfactants, can be used as stabilizers [13,18,34–38].

For intravenous administration, *colloidal* dimensions of the dispersed particles are an absolute prerequisite. Since large particles may block small blood vessels, thereby causing severe or even possibly lethal side effects such as thrombosis or embolism, the particles must not exceed a critical size. The disintegration of solid, crystalline lipids into particles of colloidal size by melt-homogenization causes a variety of often unexpected, interesting phenomena in the resulting dispersions. Some of these phenomena related to the colloidal state of the matrix lipids will be outlined below. In some cases, they form the basis for the development of modified, novel systems such as colloidal dispersions of supercooled melts [39,40].

II. SUPERCOOLING AND FORMATION OF DISPERSIONS OF SUPERCOOLED MELTS

In principle, the production of colloidal lipid suspensions by melt-emulsification is a simple process: The solid lipid, e.g., a (monoacid) triacylglycerol or an acylglycerol mixture, is heated above its melting temperature and lipophilic stabilizers, such as phospholipid mixtures, are dispersed in the lipid melt. The aqueous phase, containing water-soluble co-emulsifiers, is heated to approximately the same temperature as the lipid melt and a crude premix of the two phases is prepared. The predispersion is passed through a high-pressure homogenizer in the heat and the resulting colloidal melt-in-water emulsion is allowed to cool down. Cooling melt-emulsified dispersions based on solid lipids to room or refrigerator temperature does not, however, necessarily lead to the crystallization of the dispersed matrix acylglycerols and the formation of nanosuspensions. Melt-emulsified nanoparticle dispersions prepared from shorter-chain triacylglycerols like trimyristin ($T_{\text{melt}} \sim 56°\text{C}$) or from certain complex acylglycerol mixtures such as the suppository mass Witepsol H42 ($T_{\text{melt}} \sim 41°\text{C}$) do not display melting transitions upon heating in the DSC (Fig. 1)* or X-ray reflections due to crystalline material (Fig. 2) after cooling to and storage at room temperature [40–42]. TEM studies of freeze-fractured trimyristin dispersions reveal that such lipid particles are spherical and do not have a crystalline internal order (Fig. 3a). According to quantitative high-resolution ^{1}H NMR spectroscopy, the acylglycerol molecules exhibit a high molecular mobility even after cooling of the emulsified melt far below the crystallization temperature of its bulk material, indicating that the amorphous acylglycerols exist in the liquid state and do not form a solid amorphous phase (Fig. 4). Comprehensive DSC and X-ray diffraction studies reveal that, compared to the bulk material, the crystallization temperature of triacylglycerols is lowered significantly in the colloidal state and may be more than 20 to 30°C below that of the corresponding bulk material upon slow cooling (Fig. 5). That means that melt-homogenized, colloidal particles from, e.g., trimyristin or Witepsol H42 consist of a supercooled melt at room temperature. Trilaurin ($T_{\text{melt}} \sim 47°\text{C}$) nanodispersions can retain the emulsified state even at refrigerator

---

* The systems under discussion in the figures are based on different commercial, technical-grade triacylglycerols and acylglycerol mixtures (all from Condea, D-Witten): The monoacid triacylglycerols Dynasan 112 (D112, trilaurin), Dynasan 114 (D114, trimyristin), Dynasan 116 (D116, tripalmitin), Dynasan 118 (D118, tristearin), the triacylglycerol-rich mixed acylglycerols Softisan 154 (S154), Witepsol H35 (H35), Witepsol H42 (H42), Witepsol E85 (E85), and the acylglycerol mixture Witepsol W35 (W35) were used. The fatty acid fraction of the Witepsols (hard fat suppository bases) is composed of even saturated fatty acids with chain lengths mainly between 12 and 18 carbons. Softsan 154 mainly contains C16 and C18 chains. The dispersions (of usually 10% acylglycerol) were stabilized with soy bean lecithin–bile salt blends [2.4% Lipoid S100 and 0.6% sodium glycocholate (S100/SGC) or 1.6% Lipoid S100 and 0.4% sodium glycocholate (S100/SGCa)], a combination of soy bean lecithin with the nonionic surfactant tyloxapol [2% Lipoid S100 and 2% Tyl (S100/Tyl)], different concentrations of tyloxapol (Tyl) or other stabilizers specified in the text. All dispersions were prepared by melt-homogenization above the melting temperature of the lipid. For most dispersions under discussion, the z-average mean diameter obtained by photon correlation spectroscopy is in the range of about 100–200 nm.

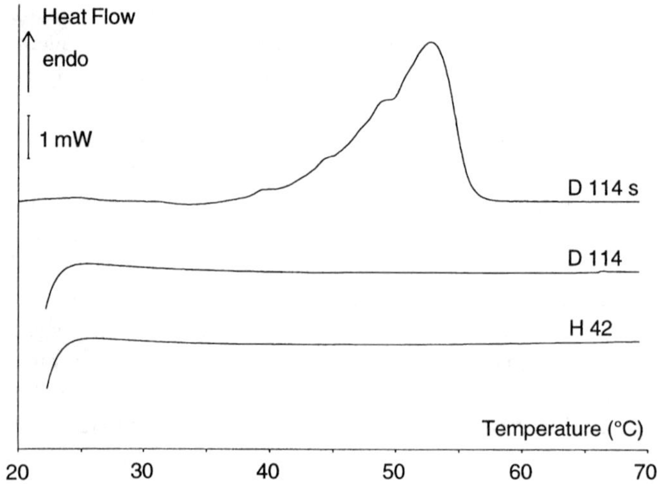

Fig. 1 DSC heating curves of colloidal dispersions (S100/SGC) of supercooled triacylglycerols after long-term storage at room temperature [Witepsol H42 (H42): >3.5 years; trimyristin (D114): >5 years]. The heating curve of a dispersion of crystalline trimyristin nanoparticles (D114 s) is also given for comparison.

temperature for several years without forming significant amounts of crystalline material (e.g., only about 2% crystallized material was found after more than two years of storage [40]).

Crystallization of a material from the melt usually requires that the melt be supercooled to a certain extent to induce the formation of crystal nuclei and subsequent crystal growth. Nucleation may occur spontaneously in the volume of a sample (homogeneous nucleation) or be facilitated by the presence of foreign material such as impurities within the material or container walls (heterogeneous nucleation). In bulk materials, where crystallization is usually a result of heterogeneous nucleation, the formation of a single crystal nucleus can, in principle, lead to the crystallization of the whole material. In contrast, the effect of a given number of nucleating impurities is limited to only a few droplets in an emulsion. For the crystallization of the remaining droplets which are free of impurities, homogeneous nucleation is required. Since homogeneous nucleation takes place at lower temperatures than heterogeneous nucleation, dispersed substances usually display much larger supercooling compared to bulk materials. This makes emulsions favorable model systems for the study of supercooling and nucleation phenomena [43].

For dispersed triacylglycerols, the occurrence of pronounced supercooling is a phenomenon known for many decades [44–46]. In colloidal acylglycerol

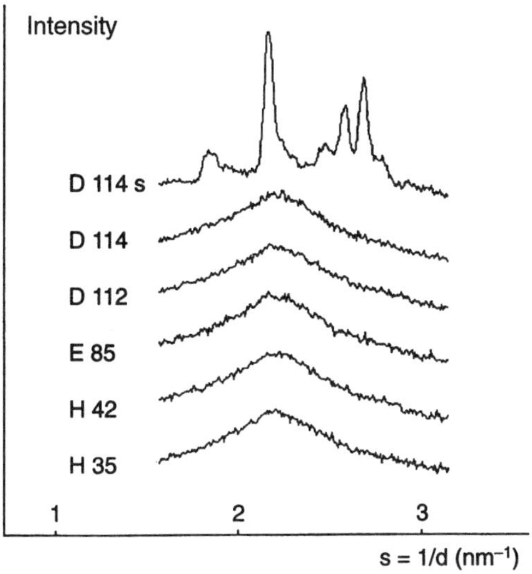

Fig. 2 Wide-angle synchrotron radiation X-ray diffractograms of colloidally dispersed supercooled triacylglycerols [trimyristin (D114), trilaurin (D112), and the hard fats Witepsol E85, H42, and H35]. The diffractogram of crystalline trimyristin nanoparticles (D114 s) is also given for comparison [$s = 1/d = 2 \sin \Theta/\lambda$, where 2Θ is the scattering angle and λ is wavelength (0.15 nm)]. [Reprinted from Ref. 40. Copyright OPA (Overseas Publishers Association N.V.) with permission from Gordon and Breach Publishers.]

dispersions, a comparatively large difference between the melting and crystallization temperatures is found for dispersed monoacid triacylglycerols (Fig. 5). In nanoparticles from complex triacylglycerol mixtures there is a lower supercooling tendency. Admixture of higher-melting triacylglycerols can also reduce the supercooling tendency of a low-crystallizing triacylglycerol (Fig. 6). This leads to the conclusion that the longer-chain fraction acts as nucleation center and thus induces crystallization of the supercooled triacylglycerol already at higher temperatures [41,42]. The addition of other compounds, such as drugs, can also influence the supercooling tendency [47].

In many studies on supercooling in triacylglycerol dispersions (which were often performed on comparatively crude dispersions with particle dimensions in the micrometer size range), the long-term stability of the supercooled melts was not a point of investigation. For colloidally dispersed supercooled triacylglycerol melts, a remarkably high stability on storage has been observed [40–42]: The emulsified triacylglycerols can retain the supercooled state for extremely long periods (months to years) provided that they are not allowed to cool below a critical temperature. For example, in dispersions of supercooled trimyristin and

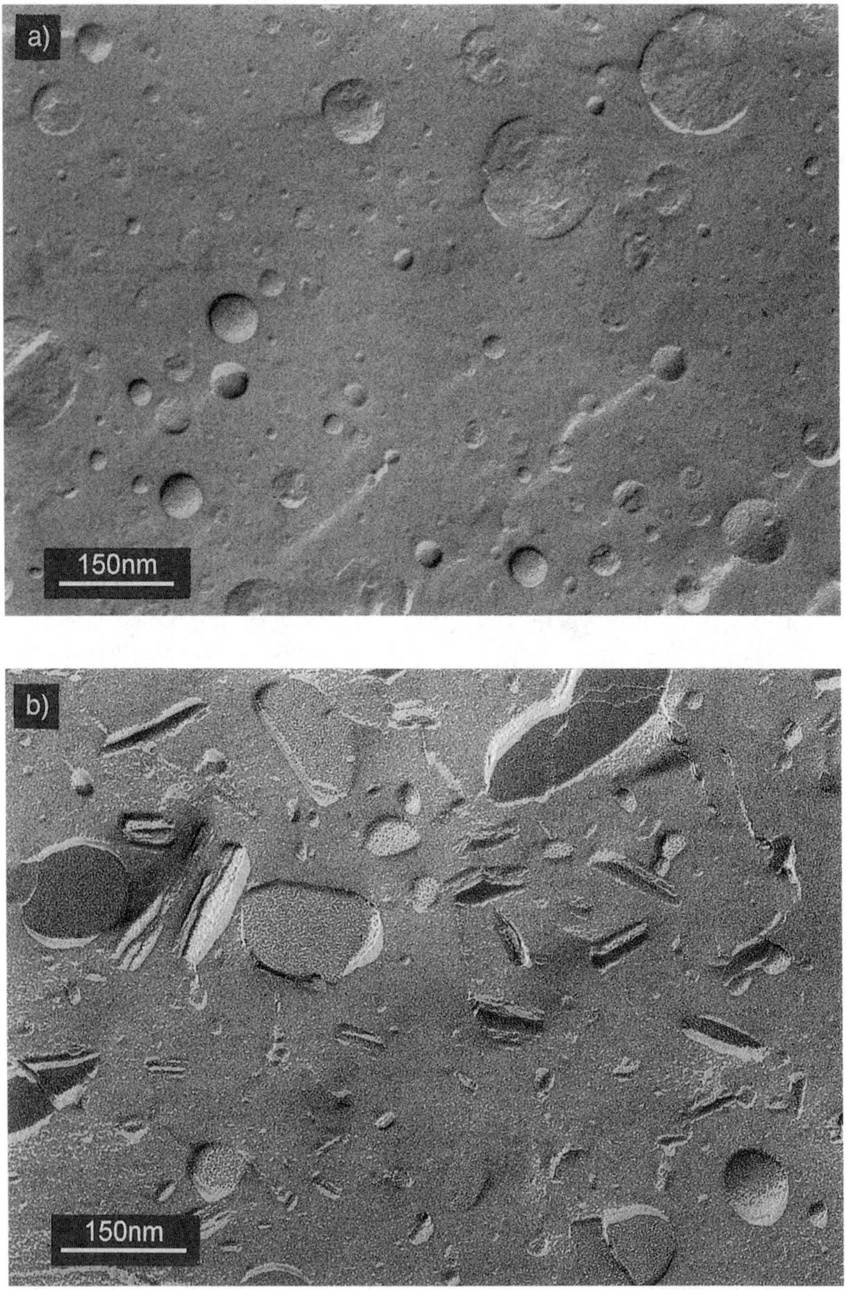

Fig. 3 Transmission electron micrographs of a freeze-fractured trimyristin (D114) dispersion (S100/SGC) stored at room (a) and at refrigerator temperature (b).

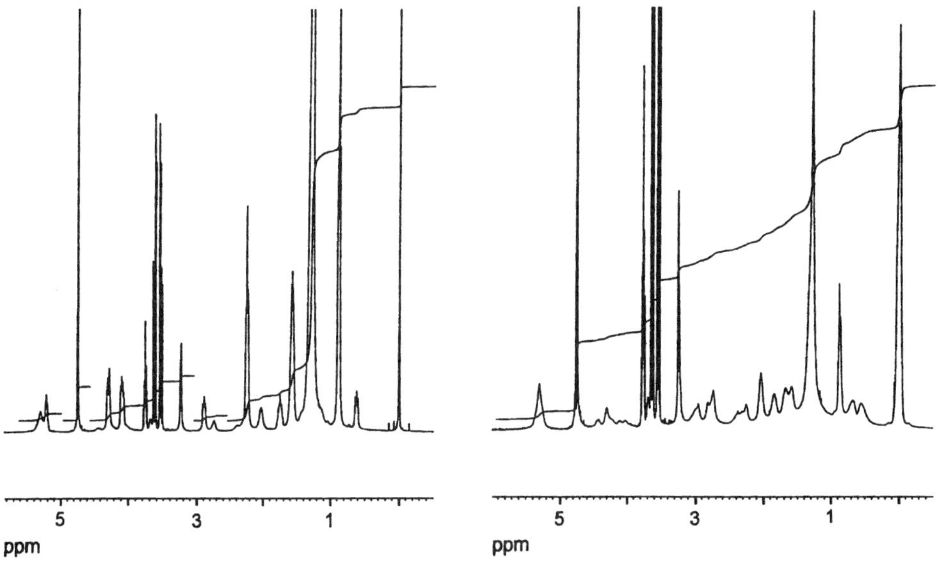

Fig. 4 ^{1}H NMR spectra of a trimyristin dispersion [D114 in D_2O (S100/SGC)] stored for one day at room (left) and at refrigerator temperature (right). Compared to that of the suspension (right), the spectrum of the emulsion (left) displays several additional or distinctly larger signals (around 0.9, 1.3, 1.6, 2.25, 4.1, 4.25, and 5.2 ppm) due to the presence of liquid trimyristin (the signal at 0 ppm is caused by an internal standard). (Reprinted from Ref. 42, Copyright 1997, with permission from Elsevier Science.)

the complex triacylglycerol mixture Witepsol H42 virtually no crystalline triacylglycerol could be detected by DSC after several years of storage at room temperature (Fig. 1). This behavior is not only in contrast to that of bulk triacylglycerols but also to the situation in dispersions with larger droplets: Coarser emulsion droplets of trimyristin, for instance, crystallize within less than 1 hr at 20°C [48]. Obviously, the crystallization temperature of microdisperse triacylglycerol droplets is higher than that of colloidal particles. For triacylglycerol particles in the lower micrometer size range a pronounced influence of the particle size on the crystallization tendency has been observed, with the mean crystallization temperature of the droplets decreasing linearly with the logarithm of the droplet diameter [45]. The mean crystallization temperature of, e.g., trimyristin droplets in the size range between 1 and 10 μm was found to be between about 15 and 25°C upon slow cooling. In the colloidal size range, however, the onset of crystallization in trimyristin dispersions is observed at lower temperatures (Fig. 5) and there are only minor influences of the particle size on the crystallization temperature of triacylglycerols, indicating homogeneous nucleation in these systems in contrast

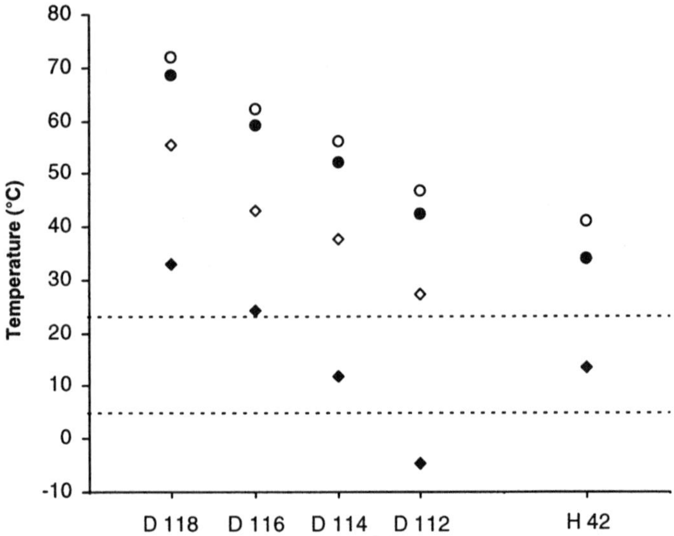

Fig. 5 Melting (circles) and crystallization temperatures (diamonds) of triacylglycerols [tristearin (D118), tripalmitin (D116), trimyristin (D114), trilaurin (D112), and the hard fat Witepsol H42) in the bulk (open symbols) and in colloidal dispersion (S100/SGC, full symbols] [heating with 2.5°C/min (T_{Peak}), cooling with 0.5°C/min (T_{Onset})]. For the H42 dispersion, the symbols indicating crystallization of the raw material and melting of the dispersion overlap. The dotted lines indicate typical storage conditions, i.e., room and refrigerator temperatures.

to mainly heterogeneous nucleation in coarser systems [41,45,49]. These results indicate that the preparation of stable dispersions of supercooled triacylglycerols is much facilitated in colloidal dispersions containing no micrometer-size particles or that the colloidal state of the droplets is even a prerequisite for the long-term stability of the supercooled melts.

In practice, the possibility of obtaining stable colloidal dispersions of supercooled melts by melt-homogenization of triacylglycerols is of major importance. The properties of emulsions of supercooled melts can be expected to be in many aspects similar to those of conventional fat emulsions, e.g., with respect to stabilizer requirements, as well as loading capacity for and release of incorporated bioactive substances. A major difference is in the fact that the matrix is, in principle, able to crystallize. The development of dispersions of supercooled melts thus requires very careful evaluation of potential instabilities due to undesired crystallization of the matrix material [e.g., gel formation (see below) or drug expulsion] which may occur highly retarded. For the determination of the physi-

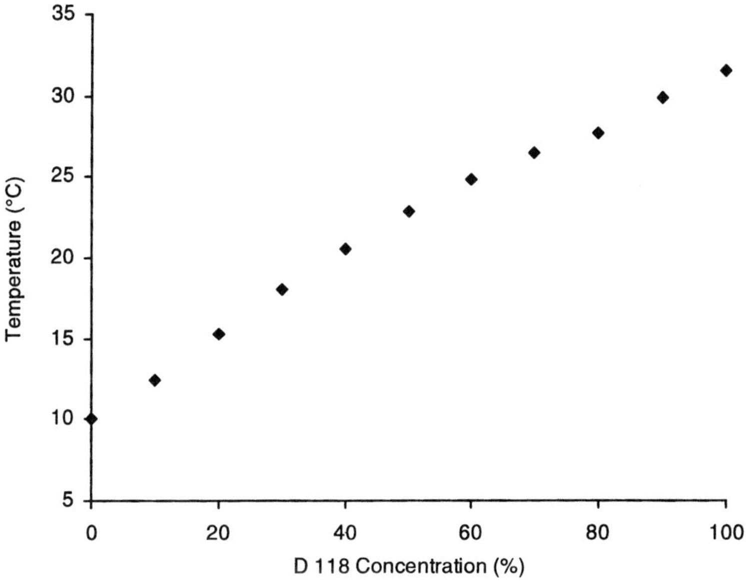

Fig. 6 Crystallization temperature of nanoparticles (S100/Tyl) from trimyristin/tristearin (D114/D118) mixtures in dependence on mixing ratio [cooling with 5°C/min. (T_{Onset})].

cal state of the nanoparticles obtained after melt-emulsification experimental techniques, which allow the investigation of native dispersions, have to be used since investigation procedures, which induce the loss of the colloidal particle state, may lead to artificial results. Dispersions of certain triacylglycerols, which can be obtained and stored as suspensions or emulsions of supercooled melts under convenient laboratory conditions (e.g., trimyristin dispersions), are interesting model systems to study the effect of the physical state of the nanoparticles on relevant dispersion properties, for instance with respect to drug incorporation or stabilizer requirements and distribution [41,42].

III. PHASE SEPARATION WITHIN COLLOIDAL TRIACYLGLYCEROL PARTICLES

As is obvious from Fig. 6, the addition of longer-chain triacylglycerols such as tristearin can be used to increase the crystallization tendency of shorter-chain triacylglycerol nanoparticles. This possibility of controlling the crystallization temperature in nanoparticles from shorter-chain triacylglycerols may be of high importance in technological processes. Concerning the internal structure of nanoparticles from binary mixtures of monoacid triacylglycerols, it has been observed that, in agreement with the limited mutual solubility of triacylglycerols in the bulk [50,51], phase separation of triacylglycerols can also take place in the colloidal state [41]: For dispersions of trilaurin/tristearin mixtures stabilized with phospholipid/bile salt blends, two small-angle X-ray reflections, which can be attributed to a trilaurin- and a tristearin-rich phase are observed (Fig. 7). These two reflections appear simultaneously upon crystallization of the nanoparticles from the melt. The course of the melting process of these particles (with two main transitions) points to eutectic behavior of the two triacylglycerols within the nanoparticles. These results indicate that, in spite of their small size, such particles from trilaurin/tristearin mixtures contain different crystalline domains. The situation in trimyristin/tristearin nanoparticles was found to be less obvious: Although such nanoparticles melt in a structured manner they display only a single broad small-angle reflection [41].

As indicated by the differences in shape of the small-angle reflection of trilaurin/tristearin nanoparticles, the internal structure of mixed nanoparticles seems to be influenced by the type of stabilizer used (Fig. 7). This effect requires further investigation. Tentatively, the loss of structuring of the small-angle reflection observed for dispersions stabilized with a phospholipid/tyloxapol blend might point to an increased intersolubility of the triacylglycerols, but it could

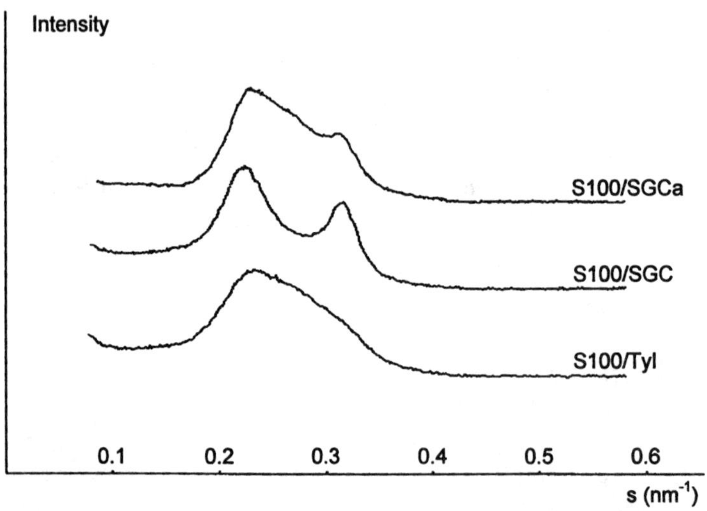

Fig. 7 Small-angle synchrotron radiation X-ray diffractograms of colloidal dispersions of trilaurin/tristearin (D112/D118) 1:1 mixtures stabilized with different surfactant blends (S100/SGCa, S100/SGC, S100/Tyl).

also be due to line broadening of the reflections as a result of a decrease in particle size or in the size of the crystalline domains within the nanoparticles.

IV. POLYMORPHIC TRANSITIONS AND OTHER AGING PHENOMENA

A. Polymorphic Transformations in Nanoparticulate Systems

For the preparation of triacylglycerol nanoparticles with a solid matrix, the dispersions have to be cooled below the critical crystallization temperature of the droplets after melt-homogenization (Fig. 5). Crystallization is, however, not the last step in the formation process of lipid nanosuspensions, since the colloidally dispersed triacylglycerols usually undergo polymorphic transitions after crystallization from the melt. For example, tripalmitin nanoparticles crystallize in the α form and subsequently transform to the stable β form as does the bulk material (Fig. 8). The time course of polymorphic transitions is, however, significantly

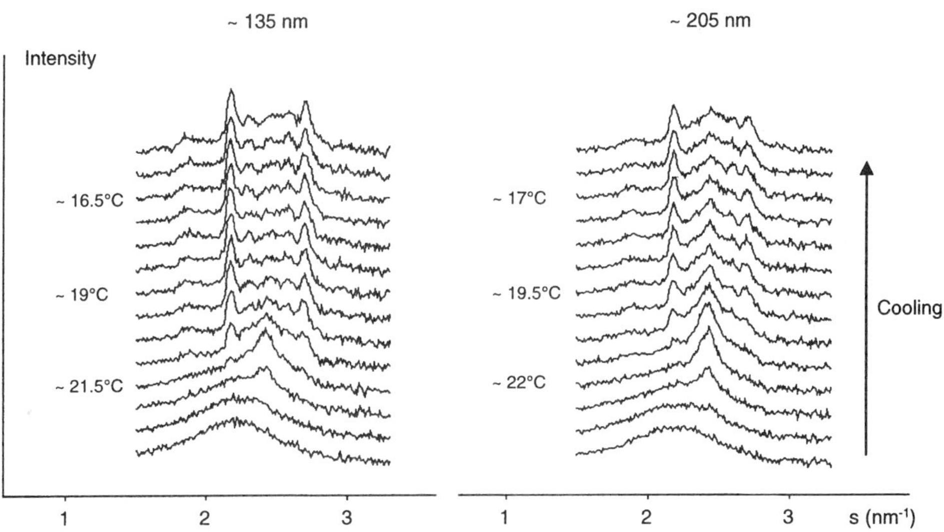

Fig. 8 Evolution of the wide-angle synchrotron radiation X-ray diffractograms of tripalmitin (D116) dispersions with the indicated approximate PCS *z*-average mean particle sizes [10% D116 stabilized with 3.2% Lipoid S100 and 0.8% sodium glycocholate (135 nm) or with 2.4% S100 and 0.6% sodium glycocholate (205 nm)] during crystallization of the particles from the melt (cooling rate ~0.31°C/min, diffractograms are shown in 2-min steps). The α form is more stable in the coarser dispersion (right panel).

altered in nanosuspensions. Colloidal monoacid triacylglycerols transform much more rapidly to the stable β form than the bulk triacylglycerols (Fig. 9). This behavior may result from the fact that the relaxation of lattice strain developing during the transition is facilitated in smaller crystalline domains due to the high surface-to-volume ratio, as suggested for bulk material by Dafler [52]. Moreover, the generally higher energy level of colloidally dispersed material may more readily provide the activation energy required for the transformation [53]. This may also explain the observation that the α form is less stable in smaller than in larger colloidal tripalmitin particles (Fig. 8) [49]. Comparable to observations on bulk triacylglycerols, polymorphic transitions proceed more rapidly in colloidally dispersed shorter-chain than in longer-chain triacylglycerols [41].

The polymorphic behavior of nanoparticles prepared from triacylglycerol-rich, complex acylglycerol mixtures such as pharmaceutically relevant hard fat suppository masses is different from that of colloidally dispersed monoacid triacylglycerols [42]. For the hard fats Witepsol H42 or E85, for example, the β′ form is very stable in the bulk and can be observed in the raw materials even after several years of storage at room temperature. In colloidal dispersions, however,

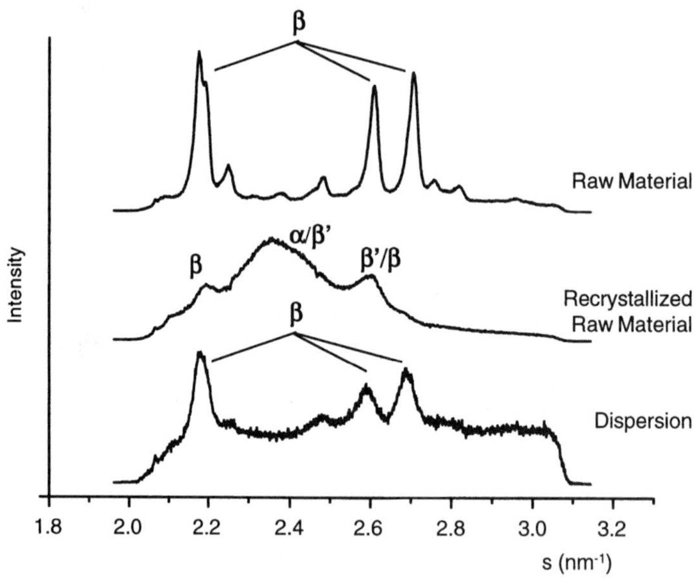

Fig. 9 Wide-angle X-ray diffractograms of trimyristin (D114) systems. The diffractograms of the recrystallized samples were recorded six days after crystallization from the melt and storage at refrigerator temperature. The diffractogram of the thermally untreated raw material is also given for comparison.

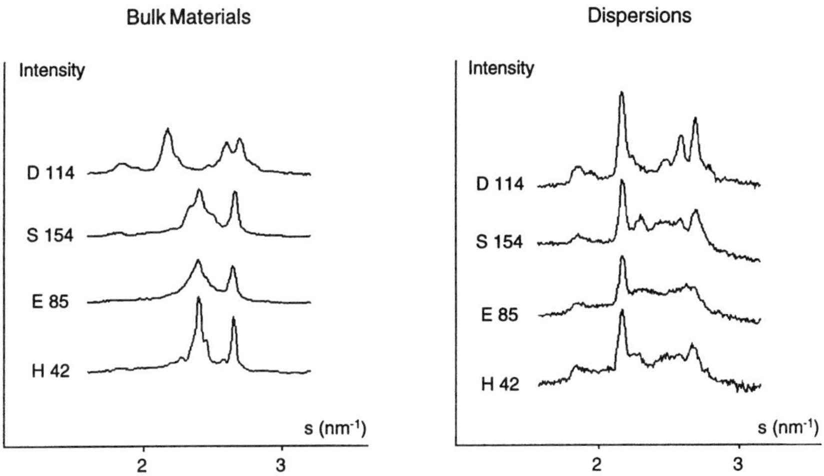

Fig. 10 Wide-angle synchrotron radiation X-ray diffractograms of crystalline triacylglycerols [trimyristin (D114), Softisan 154 (S154), and Witepsol E85 and H42] in bulk and dispersion (S100/Tyl). The dispersions had been stored at refrigerator temperature for two (D114) or six to seven weeks.

these hard fats transform to a crystal form which displays an X-ray diffraction pattern similar to that of the stable β form of monoacid triacylglycerols already after a short period (days to weeks) of cold storage* (Fig. 10). To our knowledge, this crystal form has hitherto not been described for the corresponding bulk lipids. For hard fat suppository masses in the bulk, the intermediate β (β_i) or β_2 form [54,55] is usually considered to be the stable form. Transformation of these acylglycerols from the β′ to the β_i form usually requires months to years of storage at room temperature [56]. Triacylglycerol-rich hard fat nanoparticles thus seem to transform into a more stable polymorph than their bulk material, indicating higher dynamics in the colloidally dispersed state in agreement with the behavior of colloidal monoacid triacylglycerols. Upon slow crystallization from the melt, the metastable α form in hard fat nanoparticles is, however, retained longer than in the bulk, where it transforms to the β′ form almost immediately after crystallization (Figs. 11, 12). The higher stability of the metastable α form in hard fat nanoparticles compared to the bulk materials probably results from the much lower crystallization temperature in the dispersed state. When the fat crystallizes

* For the hard fat Witepsol W35, however, which is rich in partial acylglycerols, a deviating behavior is observed: Solid nanoparticles from this complex acylglycerol mixture display an X-ray diffraction pattern which can be attributed to the intermediate β (β_i) form, indicating that the polymorphic behavior strongly depends on the type of matrix material [30,60].

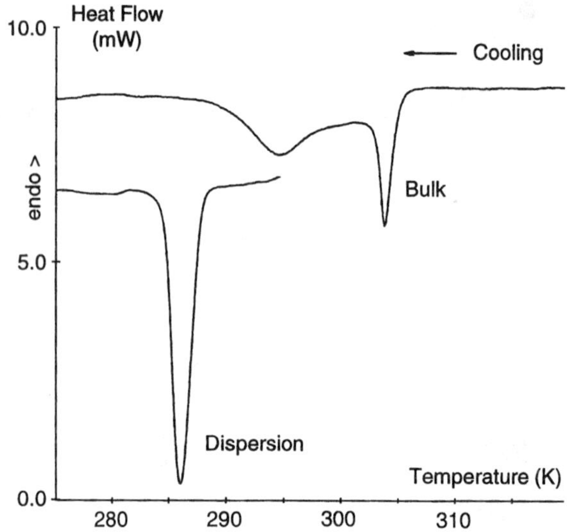

Fig. 11 DSC cooling curves (5°C/min) of the melted hard fat Witepsol E85 in bulk and dispersion (S100/Tyl). (Reprinted from Ref. 42, Copyright 1997, with permission from Elsevier Science.)

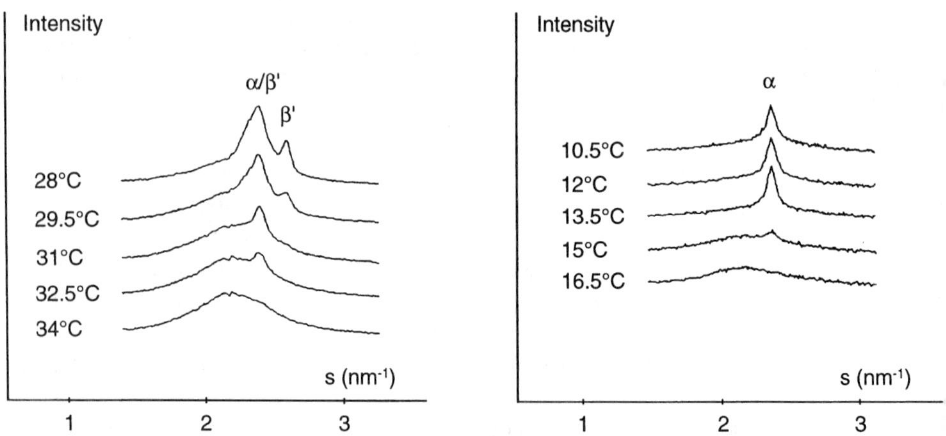

Fig. 12 Wide-angle synchrotron radiation X-ray diffractograms of the hard fat Witepsol E85 in the bulk (left) and in colloidal dispersion (S100/Tyl, right) during crystallization from the melt (cooling rate ~ 0.31°C/min). (Reprinted from Ref. 42, Copyright 1997, with permission from Elsevier Science.)

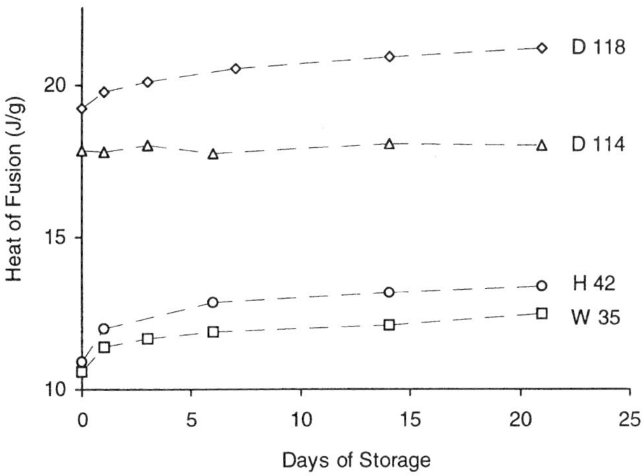

Fig. 13 Evolution of the heat of fusion in dispersions of solid lipid nanoparticles [from tristearin (D118), trimyristin (D114), and Witepsol H42 and W35 (S100/SGC)] upon storage (W35, H42, D114 dispersions stored at refrigerator temperature; D118 dispersion stored at room temperature).

at temperatures distinctly below the melting temperature of the α form, the α phase can be expected to be more stable than at temperatures close to its melting point.

The dynamic processes in solid lipid nanoparticles may continue for days or weeks after crystallization, as indicated by an increase in the heat of fusion, in particular for longer-chain and complex acylglycerols (Fig. 13) [30]. In addition to polymorphism, annealing processes leading to a higher crystalline order within the acylglycerol matrix may also contribute to this effect.

The results given illustrate that colloidally dispersed triacylglycerols are highly dynamic systems, not only during crystallization from the melt but also after solidification as a result of polymorphic transitions and aging processes. The type of crystal polymorph formed after a given period of storage cannot be deduced from the situation in the bulk material, since the kinetics of polymorphic transitions may be altered dramatically in the colloidal state. Moreover, the crystal modification finally obtained in the dispersion may even be completely different from that described in the literature or observed for the raw materials [42].

B. Effect of Surfactants

For triacylglycerols in the bulk, it is well known that the addition of surfactant may have pronounced effects on their polymorphic transitions [57–59]. Also for colloidally dispersed monoacid triacylglycerols, the time course of polymorphic transitions can be distinctly altered by the use of different types of surfactants as stabilizers [60,61]. For instance, hardly any reflections of the α form are observed upon crystallization of tripalmitin nanoparticles stabilized with polyoxyl 35 castor oil (Cremophor EL) (Fig. 14). On the other hand, nanoparticles stabilized with the bile salt sodium glycocholate display a comparatively high stability of the α form. In contrast to the situation in water-free bulk systems, the surfactant may partition between different locations (lipid phase, water phase, interface) in the colloidal lipid-in-water dispersions. The observation that hydrophilic surfactants, which are supposed to be distributed almost exclusively toward the aqueous phase and the particle interface, may significantly alter the transition kinetics suggests that polymorphic transitions in triacylglycerol nanoparticles start at the interface rather than in the core of the nanoparticles. These results indicate that the stabilizers do not only influence the colloidal state of the dispersion, e.g., with respect to particle size and stability, but may also have pronounced effects on the internal structure of the particles.

The comparatively high stability of the α form in sodium glycocholate-stabilized dispersions can be utilized to study the changes in matrix structure

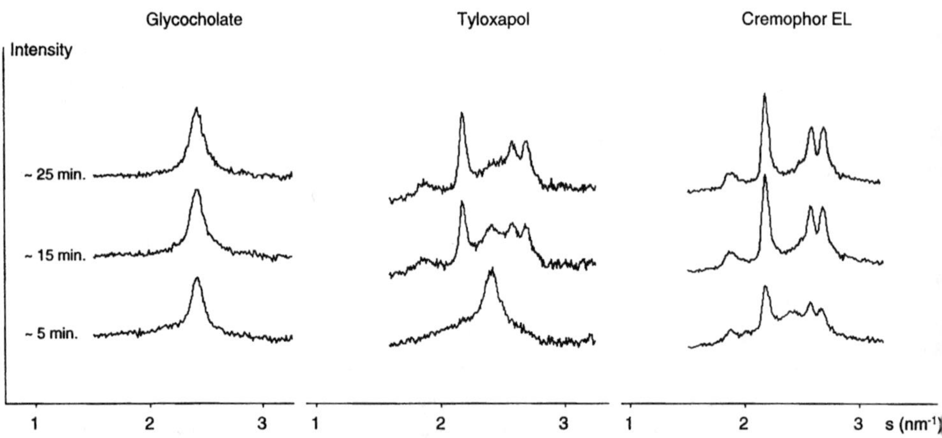

Fig. 14 Wide-angle synchrotron radiation X-ray diffractograms recorded during the crystallization of tripalmitin (D116) nanoparticles stabilized with different surfactants (glycocholate: 5% D116 with 1.2% sodium glycocholate; tyloxapol: 10% D116 with 3% tyloxapol; Cremophor EL: 10% D116 with 4% Cremophor EL). The particles were crystallized from the melt by cooling with ~0.31°C/min (the glycocholate-stabilized dispersion was monitored under isothermal conditions after a temperature of 20°C was reached). The curve labels indicate the approximate time after the onset of crystallization.

and overall shape that the particles may undergo upon polymorphic transitions. For tripalmitin dispersions stabilized with sodium glycocholate different types of the α form are obtained when crystallization is carried out under different conditions [60,61]. Slow cooling to room temperature leads to the formation of the common α form displaying strong small- and wide-angle X-ray reflections (Fig. 15). After rapid cooling to refrigerator temperature, however, the resulting α form displays a strong wide-angle but only an extremely weak small-angle X-ray reflection. This unusual α form, which seems not to contain a highly ordered layered structure, transforms into the regular α form upon slow heating. Electron microscopic investigations reveal a more isometric particle shape and a less ordered internal structure for nanoparticles in the α form than in those that had transformed into the stable β form (Fig. 16) [60–62]. Particles in the unusual α form are spherical and display a virtually unstructured core. The molecular reorganization upon polymorphic transitions in these particles is thus combined with pronounced alterations of the matrix structure and the particle shape. Such processes may be of high importance for the physical stability of the colloidal

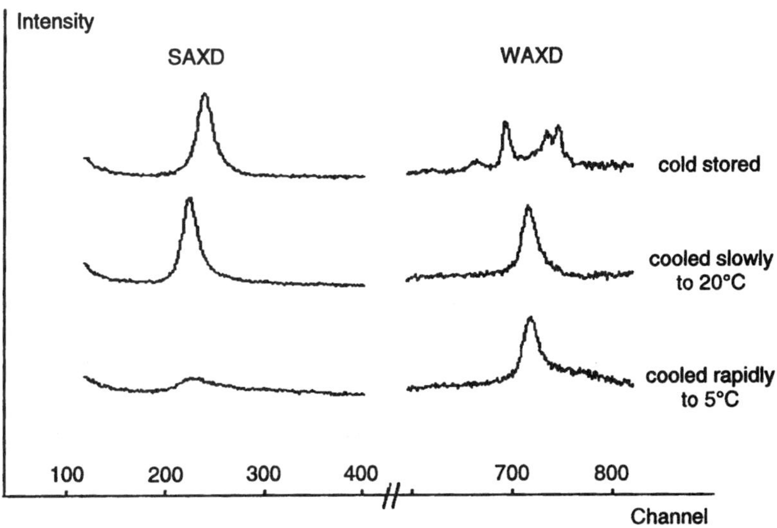

Fig. 15 Small- (SAXD) and wide-angle (WAXD) synchrotron radiation X-ray diffractograms of a sodium glycocholate-stabilized tripalmitin dispersion (5% D116, 1.2% sodium glycocholate). The native dispersion was heated to 85°C for 10 min and subsequently cooled slowly (~0.31°C/min) from 30 to 20°C or rapidly (>5°C/min) to 5°C. The diffractograms of the freshly crystallized dispersions were recorded approximately 30 min after the onset of crystallization.

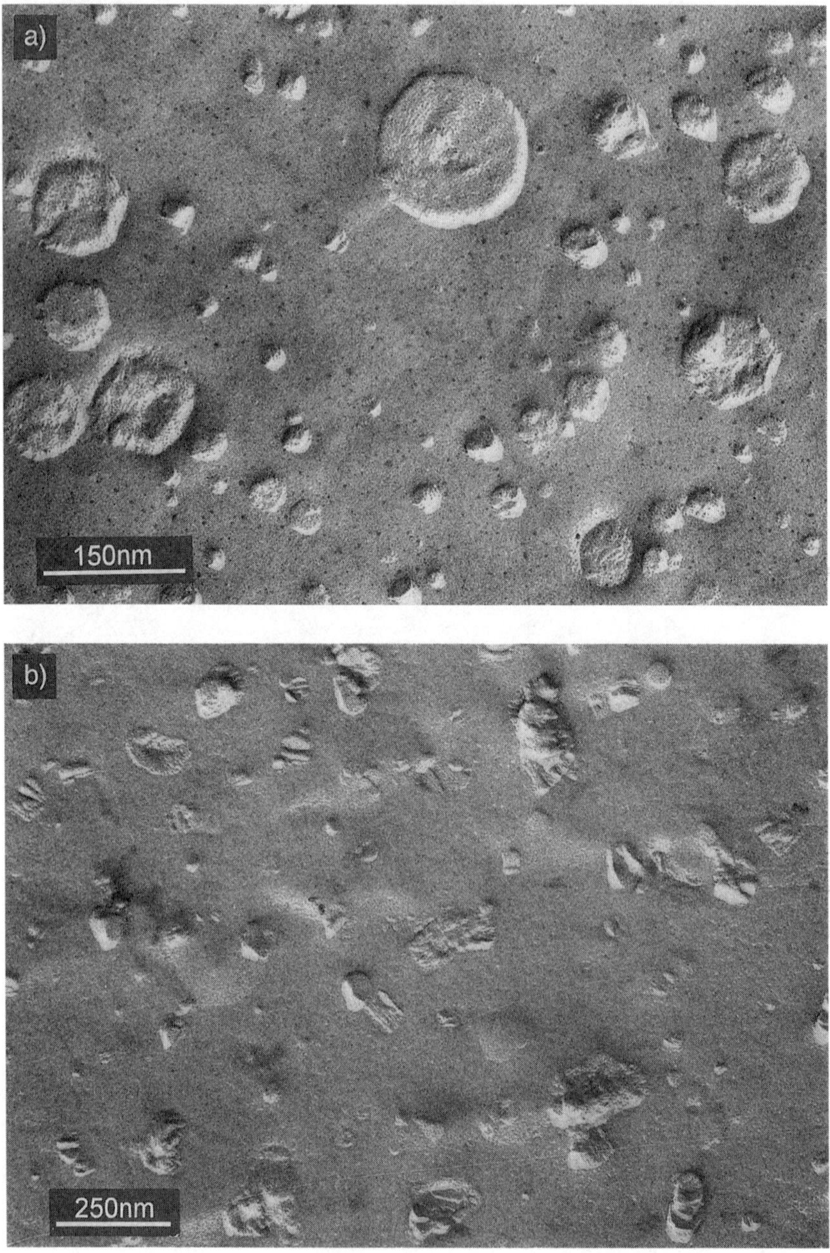

Fig. 16 Freeze-fracture electron micrographs of a sodium glycocholate-stabilized tripalmitin dispersion (5% D116, 1.2% sodium glycocholate) 30 min after rapid cooling from the melt to 5°C (a) or slow cooling (~0.31°C/min) to 20°C (b) (crystallization conditions are comparable to those of the particles described in Fig. 15) and after cold storage (c).

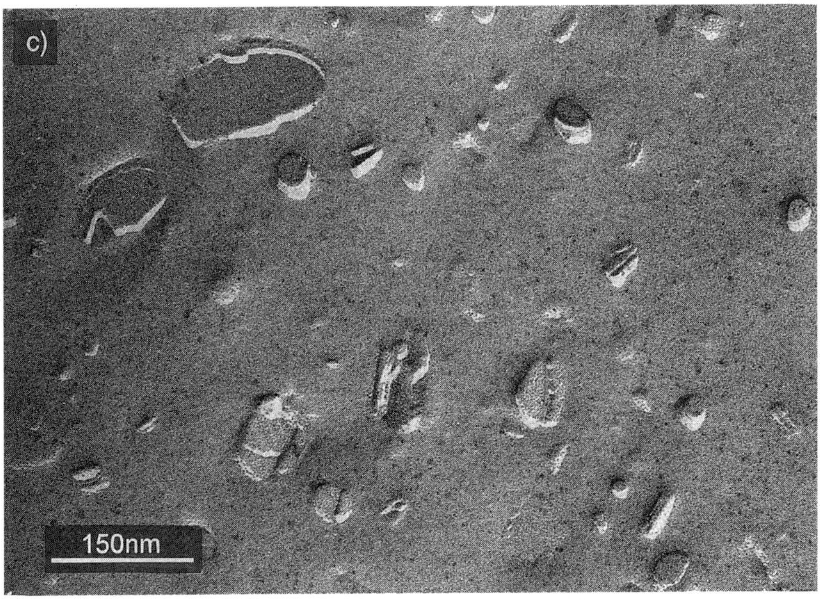

state of triacylglycerol nanoparticles but, possibly, also for their loading capacity for foreign compounds.

V. EFFECT OF PARTICLE SIZE

From the pharmaceutical point of view, particularly with respect to the biodistribution after intravenous administration, the preparation of very small (e.g., smaller than 100–150 nm) triacylglycerol nanoparticles may be advantageous. Recalling the pronounced differences of triacylglycerol properties in the bulk and in colloidal dispersion, the question arises whether the disintegration of triacylglycerols into very small colloidal particles results in further alterations of their physicochemical properties. Comparison of the thermal properties of small triacylglycerol nanoparticles to those in coarser colloidal dispersions in DSC heating runs (Fig. 17) reveals that the melting transition broadens and shifts to lower temperatures with decreasing mean particle size of the dispersions [49]. For small-size dispersions of monoacid triacylglycerols, several additional sharp transitions at lower temperatures appear in the DSC heating curves. The melting transition in these dispersions is thus not a continuous but a stepwise process. With increasing chemical complexity of the matrix triacylglycerol, the individual transitions in the melting endotherm become less sharp or even undetectable

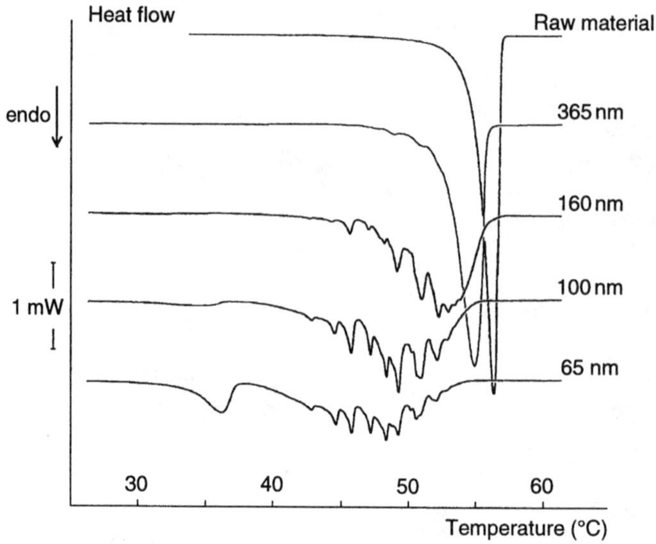

Fig. 17 DSC heating runs (0.04°C/min) of tyloxapol-stabilized trimyristin (D114) dispersions differing in approximate mean PCS particle size as indicated (the raw material was dispersed in a tyloxapol containing aqueous phase). (Reprinted with permission from Ref. 49, Copyright 2000, American Chemical Society.)

(Fig. 18). A decrease of the melting temperature with decreasing particle size is expected for colloidally dispersed materials according to the Gibbs-Thomson equation

$$-\frac{T_0 - T}{T_0} \approx \ln\frac{T}{T_0} = -\frac{2\gamma_{sl} V_s}{r\Delta H_{fus}}$$

where T is the melting temperature of a particle with radius r, T_0 is the melting temperature of the bulk material at the same external pressure, γ_{sl} is the interfacial tension at the solid-liquid interface, V_s is the specific volume of the solid, and ΔH_{fus} is the specific heat of fusion. The explanation for the occurrence of a considerable number of discrete melting transitions is, however, less obvious. A sequence of up to more than 10 sharp melting events [e.g., in small-size trimyristin dispersions (Fig. 17)] cannot be explained by the polymorphism of triacylglycerols. Moreover, the investigations were carried out with nanoparticles in the stable β polymorph as confirmed by X-ray diffraction (Fig. 19), and there were no indications for the contribution of submodifications to the stepwise melting event [49]. Instead, the unusual melting behavior can be attributed to the ultrastructure of the triacylglycerol nanoparticles. Solid triacylglycerol nanoparticles are of

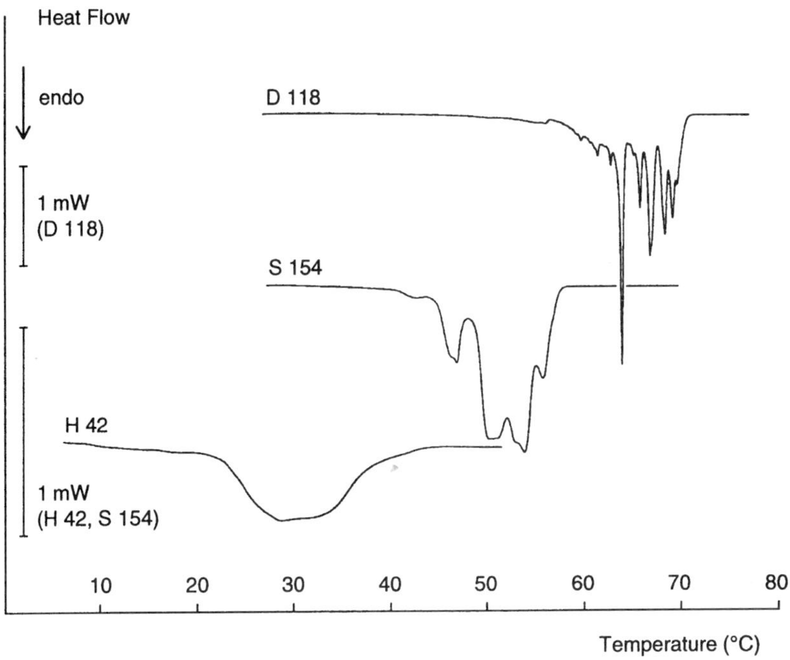

Fig. 18 DSC heating runs (0.04°C/min) of different colloidally dispersed triacylglycerol matrix materials (Tyl; mean PCS particle size about 100–110 nm). The complexity of the fatty acid chain composition increases from D118 (tristearin, almost pure C18 chains) over S154 (Softisan 154, mainly C16 and C18 chains) to H42 (Witepsol H42, mainly C12, C14, C16, and C18 chains).

platelet-like shape with a layered internal structure (Fig. 3b) similar to larger single crystals of triacylglycerols [63]. The triacylglycerol molecules are arranged in layers parallel to the large (001) face of the platelets. Due to the length of the triacylglycerol chains forming the molecular layers, the thickness of the nanoplatelets can only change in steps corresponding to the thickness of one molecular triacylglycerol layer. If the thickness of the nanoparticles is used as the basis for the size parameter r in the Gibbs-Thomson equation, a stepwise decrease of the melting temperature is obtained with decreasing particle size. Each discrete melting transition would thus be caused by the melting of a fraction of particles having identical thicknesses. Since the particle thicknesses can be expected to be less defined in nanoparticles from triacylglycerols with a more complex composition, the observation of a less structured melting event seems plausible for such systems. While simple model considerations based on the Gibbs-Thomson equation

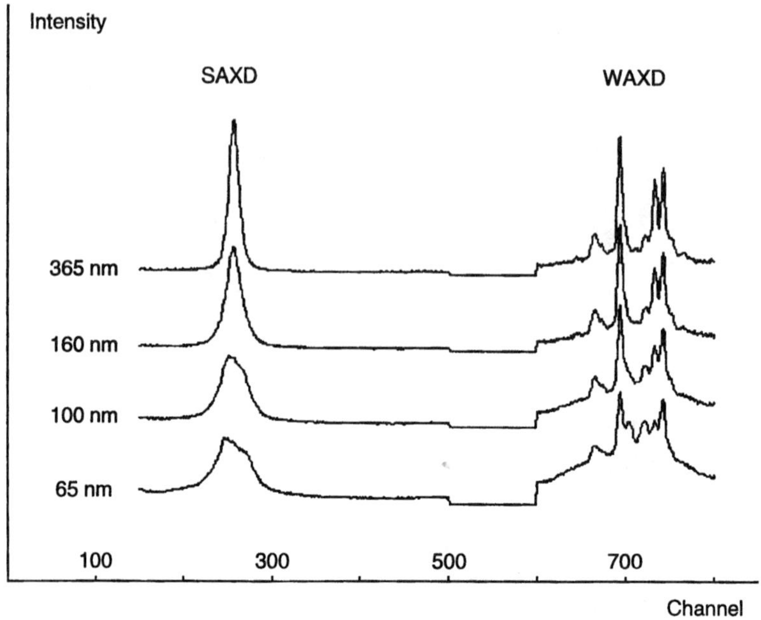

Fig. 19 Small- and wide-angle synchrotron radiation X-ray diffractograms of trimyristin (D114) dispersions (Tyl) differing in approximate mean PCS particle size as indicated. (Reprinted with permission from Ref. 49, Copyright 2000, American Chemical Society.)

can serve only as a rough approximation, the correlation of the single melting transitions in dispersions of monoacid triacylglycerols with discrete particle thicknesses could be confirmed experimentally with the aid of an advanced X-ray line shape analysis [64].

The X-ray diffraction pattern of the dispersions changes with decreasing particle size (Fig. 19). The small-angle reflection broadens and there is a slight shift of the peak position to smaller angles. In the wide-angle diffractogram, reflections around 0.44 and 0.40–0.41 nm appear or become more prominent with decreasing particle size. Dispersions with a mean particle size around 30 nm, which can be isolated by ultracentrifugation of small-size dispersions [49], display only broad reflections around 0.44, 0.41, and 0.37 nm besides, in most cases, a small reflection around 0.46 nm (Fig. 20). These reflections do not fit any of the patterns usually applied for the assignment of triacylglycerol polymorphs [65]. The structure underlying these reflections, which melts at comparatively low temperatures, has not yet been studied in detail. Tentatively, the extreme conditions in very small nanoparticles may influence the chain packing of

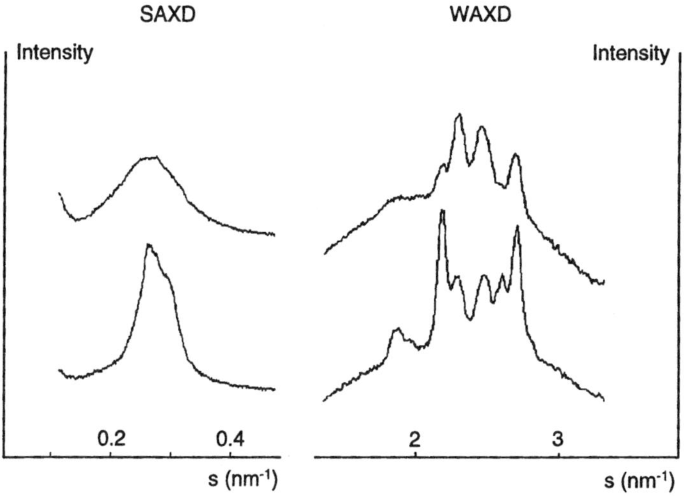

Fig. 20 Small- and wide-angle synchrotron radiation X-ray diffractograms of a trimyristin dispersion (10% D114 stabilized with 10% tyloxapol, PCS *z*-average diameter 65 nm, lower curves) and its small-size fraction isolated by ultracentrifugation of the corresponding emulsion for 18 hr (upper curves) (for details see Ref. 49). (Reprinted with permission from Ref. 49, Copyright 2000, American Chemical Society.)

the triacylglycerol molecules and thus lead to the formation of this unusual crystal form.

VI. GEL FORMATION

The specific structure of crystalline triacylglycerol nanoparticles is probably also the cause for an interesting instability phenomenon in certain triacylglycerol dispersions. In contrast to lipid oil-in-water emulsions, melt-emulsified colloidal lipid suspensions of comparable triacylglycerol concentration (e.g., 10%) cannot be effectively stabilized solely by phospholipid mixtures (Fig. 21). Attempts to stabilize colloidal suspensions of tripalmitin or trimyristin with phosphatidylcholine-rich phospholipid mixtures leads to the formation of semisolid systems on cooling of the dispersions below the crystallization temperature of the triacylglycerol particles. Alternatively, dispersions with a high tendency toward gel formation during storage or upon application of shear stress are obtained [66,67]. In dependence on their composition, the gels may be stable for years or reliquefy upon storage.

Fig. 21 Macroscopic appearance of tripalmitin systems. Both aqueous systems contain 10% D116 and 2.4% Lipoid S100 and, for the suspension, 0.6% sodium glycocholate.

When phospholipid-stabilized triacylglycerol nanodispersions are kept above the crystallization temperature of the dispersed lipid after melt-homogenization, gelation does not occur in the dispersions of supercooled melts. Gel formation is also prevented by the additional or exclusive use of mobile ionic or nonionic surfactants such as sodium glycocholate, tyloxapol, or poloxamers in sufficiently high concentrations [13,34,66]. The resulting liquid lipid nanosuspensions are physically stable on storage for years when optimized stabilizer compositions are employed. Obviously, the formation of crystalline particles dramatically alters the demands of the dispersion with respect to the stabilizers. DSC, X-ray diffraction, and rheological data confirm that the spontaneous gel formation on cooling of the dispersions of exclusively phospholipid-stabilized melts is correlated with the crystallization of the triacylglycerols [66,67]. Electron microscopic studies suggest that the resulting gels consist of a three-dimensional network of crystalline triacylglycerol. The crystalline network has been described as a ''house of cards'' structure with ''cards'' of an average thickness of several triacylglycerol layers only [66]. The maximized surfaces of the house of cards structure are oriented parallel to the large (001) faces of the solid nanoparticles, indicating that the network is formed by lateral aggregation or association of nanocrystals. From the transmission electron micrographs of stable dispersions

(Fig. 3), it is obvious that the specific surface area of triacylglycerol particles increases tremendously when the spherical droplets transform into platelet-like suspension particles. Colloidal triacylglycerol suspensions should thus not be regarded as ''lipid emulsions with solidified droplets.'' Since gel formation can be avoided when a second, more mobile stabilizer like sodium glycocholate is introduced into the compositions, it has been proposed that gelation results from a relative lack of highly mobile stabilizers in the aqueous phase upon the demand for additional stabilizers due to the sudden increase in specific surface area during crystallization [66]. Phospholipids in colloidal triacylglycerol emulsions are localized as monolayers in the interface of emulsion droplets and as vesicular structures in the aqueous phase [68]. They are comparatively immobile stabilizers which exhibit no significant solubility in aqueous media. Therefore, the addition of a mobile stabilizer, e.g., a micelle-forming surfactant with high critical micelle concentration, is required to allow the immediate stabilization of the surfaces freshly created during crystallization, thereby avoiding spontaneous gel formation.

VII. CONCLUSIONS

In acylglycerol nanoparticles, the properties of the crystalline lipids are superimposed by specific effects which are due to the colloidal state of the particles. This introduces a variety of interesting phenomena to the behavior of colloidally dispersed solid acylglycerols, which are expressed, for instance, in terms of stability, melting, and crystallization behavior. The examples outlined illustrate that it is very risky and often impossible to conclude from the behavior of bulk materials to that in the colloidal state. Even within the colloidal size range, the properties of triacylglycerol nanoparticles may vary distinctly in dependence on particle size. Wherever the development of colloidal formulations of solid lipids is intended, e.g., in the pharmaceutical, food, or cosmetic fields, potential effects such as pronounced supercooling, formation of stable dispersions of supercooled melts, significant depression of the melting temperature, as well as (long) lasting structural changes resulting, e.g., in sustained expulsion of incorporated foreign compounds, should be taken into consideration.

ACKNOWLEDGMENTS

Synchrotron radiation X-ray measurements at the European Molecular Biology Laboratory, Hamburg Outstation and the cooperation of M. H. J. Koch are gratefully acknowledged. The authors also thank S. Richter, A. Mohn, M. Drechsler, L. Kröhne, J. Mehlem, and B. Siekmann for their helpful assistance.

REFERENCES

1. D. D. Lasic, *Liposomes: From Physics to Applications*. Elsevier, Amsterdam (1993).
2. A. Sharma and U. S. Sharma, *Int. J. Pharm., 154*: 123 (1997).
3. D. D. Lasic, in *Microencapsulation* (S. Benita, ed.), Marcel Dekker, New York, 1996, p. 297.
4. M. A. Hammad and B. W. Müller, *Eur. J. Pharm. Biopharm., 46*: 361 (1998).
5. L. C. Collins-Gold, R. T. Lyons, and L. C. Bartholow, *Adv. Drug Deliv. Rev., 5*: 189 (1990).
6. R. J. Prankerd and V. J. Stella, *J. Parent. Sci. Technol., 44*: 139 (1990).
7. O. Schuberth and A. Wretlind, *Acta Chir. Scand. Suppl., 278*: 1 (1961).
8. A. Wretlind, *J. Parent. Enteral Nutr., 5*: 230 (1981).
9. P. Speiser, European Patent EP 0 167 825 (1990).
10. R. M. Gasco and S. Morel, *Il Farmaco, 45*: 1127 (1990).
11. R. M. Gasco, U.S. Patent 5,250,236 (1993).
12. A. Domb, U.S. Patent 5,188,837 (1993).
13. B. Siekmann and K. Westesen, *Pharm. Pharmacol. Lett., 1*: 123 (1992).
14. K. Westesen and B. Siekmann, U.S. Patent 5,785,976 (1998).
15. B. Sjöström and B. Bergenståhl, *Int. J. Pharm., 84*: 107 (1992).
16. J. S. Lucks, R. H. Müller, and S. König, *Eur. J. Pharm. Biopharm., 38*: 33S (1992).
17. R. H. Müller and J. S. Lucks, European Patent 0 605 497 (1996).
18. R. H. Müller, W. Mehnert, J. S. Lucks, C. Schwarz, A. zur Mühlen, H. Weyhers, C. Freitas, and D. Rühl, *Eur. J. Pharm. Biopharm., 41*: 62 (1995).
19. K. Westesen and B. Siekmann, in *Microencapsulation* (S. Benita, ed.), Marcel Dekker, New York, 1996, p. 213.
20. B. Siekmann and K. Westesen, in *Submicron Emulsions in Drug Targeting and Delivery* (S. Benita, ed.), Harwood Academic, 1998, p. 205.
21. R. H. Müller and S. A. Runge, in *Submicron Emulsions in Drug Targeting and Delivery* (S. Benita, ed.), Harwood Academic, 1998, p. 219.
22. B. Magenheim, M. Y. Levy, and S. Benita, *Int. J. Pharm., 94*: 115 (1993).
23. C. Washington and K. Evans, *J. Control. Rel., 33*: 383 (1995).
24. J. F. Pinto and R. H. Müller, *Pharmazie, 54*: 506 (1999).
25. V. Jenning, M. Schäfer-Korting, and S. Gohla, *J. Control. Rel., 66*: 115 (2000).
26. C. Santos Maia, W. Mehnert, and M. Schäfer-Korting, *Int. J. Pharm., 196*: 165 (2000).
27. R. H. Müller and A. Dingler, *Euro Cosmetics, 6*: 19 (1998).
28. A. Dingler, R. P. Blum, H. Niehus, R. H. Müller, and S. Gohla, *J. Microencapsulation, 16*: 751 (1999).
29. B. Siekmann and K. Westesen, *Eur. J. Pharm. Biopharm., 43*: 104 (1996).
30. K. Westesen, B. Siekmann, and M. H. J. Koch, *Int. J. Pharm., 93*: 189 (1993).
31. B. Siekmann and K. Westesen, *Colloids Surf. B: Biointerfaces, 3*: 159 (1994).
32. V. Jenning and S. Gohla, *Int. J. Pharm., 196*: 219 (2000).
33. T. de Vringer and H. A. G. de Ronde, *J. Pharm. Sci., 84*: 466 (1995).
34. B. Siekmann and K. Westesen, *Pharm. Pharmacol. Lett., 3*: 194 (1994).
35. B. Siekmann and K. Westesen, *Pharm. Pharmacol. Lett., 3*: 225 (1994).
36. A. Dingler, G. Lukowski, P. Pflegel, R. H. Müller, and S. Gohla, *Proceed. Int. Symp. Control. Rel. Bioact. Mater., 24*: 935 (1997).

37. R. H. Müller, S. Maaßen, H. Weyhers, and W. Mehnert, *J. Drug Targeting*, *4*: 161 (1996).
38. R. H. Müller, D. Rühl, S. Runge, K. Schulze-Forster, and W. Mehnert, *Pharm. Res.*, *14*: 458 (1997).
39. K. Westesen and B. Siekmann, European Patent 00/711151B1 (2000), U.S. Patent 6,197,349 (2001).
40. H. Bunjes, B. Siekmann, and K. Westesen, in *Submicron Emulsions in Drug Targeting and Delivery* (S. Benita, ed.), Harwood Academic, 1998, p. 175.
41. H. Bunjes, K. Westesen, and M. H. J. Koch, *Int. J. Pharm.*, *129*: 159 (1996).
42. K. Westesen, H. Bunjes, and M. H. J. Koch, *J. Control. Rel.*, *48*: 223 (1997).
43. D. Clausse, in *Encyclopedia of Emulsion Technology* (P. Becher, ed.), Vol. III, Marcel Dekker, New York, 1985, p. 77.
44. H. Mulder, *Neth. Milk Dairy J.*, *7*: 149 (1953).
45. L. W. Phipps, *Trans. Faraday Soc.*, *60*: 1873 (1964).
46. P. Walstra and E. C. H. van Beresteyn, *Neth. Milk Dairy J.*, *29*: 2935 (1975).
47. H. Bunjes, M. Drechsler, M. H. J. Koch, and K. Westesen, *Pharm. Res.*, in press.
48. C. Simoneau, M. J. McCarthy, R. J. Krauten, and J. B. German, *J. Am. Oil Chem. Soc.*, *568*: 481 (1991).
49. H. Bunjes, M. H. J. Koch, and K. Westesen, *Langmuir*, *16*: 5234 (2000).
50. B. Rossell, in *Advances in Lipid Research* (R. Paoletti and D. Kritchevsky, eds.), Vol. V, Academic Press, New York, 1967, p. 353.
51. E. Timms, *Prog. Lip. Res.*, *23*: 1 (1984).
52. R. Dafler, *J. Am. Oil Chem. Soc.*, *54*: 249 (1977).
53. H. Whittam and H. L. Rosano, *J. Am. Oil Chem. Soc.*, *52*: 128 (1975).
54. W. Hoerr, *J. Am. Oil Chem. Soc.*, *37*: 539 (1960).
55. K. Larsson, *Act. Chem. Scand.*, *20*: 2255 (1966).
56. K. Thoma, P. Serno, and D. Precht, *Pharm. Ind.*, *45*: 420 (1983).
57. N. Garti, in *Crystallization and Polymorphism of Fats and Fatty Acids* (N. Garti and K. Sato, eds.), Marcel Dekker, New York, 1988, p. 267.
58. J. Schlichter Aronhime, S. Sarig, and N. Garti, *J. Am. Oil Chem. Soc.*, *65*: 1144 (1988).
59. H. M. A. Mohamed and K. Larsson, *Fat Sci. Technol.*, *94*: 338 (1992).
60. H. Bunjes, Thesis, University of Jena, 1998.
61. H. Bunjes, M. H. J. Koch, and K. Westesen, in preparation.
62. K. Westesen, M. Drechsler, and H. Bunjes, in *Food Colloids: Fundamentals of Formulation* (E. Dickinson and R. Miller, eds.), *Royal Society of Chemistry*, 2001, p. 103.
63. W. Skoda, L. L. Hoekstra, T. C. van Soest, P. Bennema, and M. van den Tempel, *Kolloid Z. Z. Polym.*, *219*: 149 (1967).
64. T. Unruh, H. Bunjes, K. Westesen, and M. H. J. Koch, *J. Phys. Chem. B*, *103*: 10373 (1999).
65. W. Hoerr and F. R. Paulicka, *J. Am. Oil Chem. Soc.*, *45*: 793 (1968).
66. K. Westesen and B. Siekmann, *Int. J. Pharm.*, *151*: 35 (1997).
67. J. Mehlem, T. Unruh, and K. Westesen, *Proc. Int. Symp. Control. Rel. Bioact. Mater.*, *27*: 8128 (2000).
68. K. Westesen and T. Wehler, *Colloids Surf. A*, *78*: 115 (1993).

14

Separation and Crystallization of Oleaginous Constituents in Cosmetics

Sweating and Blooming

Hajime Matsuda and Michihiro Yamaguchi
Shiseido Laboratories, Yokohama, Japan

Hidetoshi Arima
Kumamoto University, Kumamoto, Japan

I. INTRODUCTION

Cosmetics are composed of oleaginous raw materials (oils, waxes, hydrocarbons, higher fatty acids, higher alcohols, esters, silicones, etc.), surfactants, humectants, polymers, ultraviolet absorbents, antioxidants, sequestering agents, coloring materials (pigments, lakes, and dyes), fragrance materials, and biologically active substances (vitamins, hormones, amino acids, etc.). Oleaginous raw materials have been widely employed to improve various characteristics of cosmetics (e.g., gloss, fluidity, molding, and hardness), they are important for the preparation of oleaginous solid cosmetics, especially point-makeup cosmetic pencils such as lipstick, rouge, eye shadow, and eyeliner.

In general, for preparations including oil, waxes, and higher fatty acids care must be taken to ensure stability and to prevent quality deterioration, including the color change and deterioration on the surface of oleaginous cosmetic pencils due to sweating, blooming, softening, etc. [1]. Sweating and blooming are defined as phenomena in which droplets and powder are generated on the surface of cosmetic pencils, respectively. These phenomena are caused by solid-liquid separation. However, the mechanisms responsible for sweating and blooming are still unclear.

Sweating and blooming mechanisms had been studied in cosmetic pencils; consequently our results indicated two different mechanisms: the former was due to the migration of oil expanded in the wax matrix to the surface of cosmetic pencils due to changes in temperature and humidity of the atmosphere [2,3], whereas the latter involved the solubility of higher fatty acid in oil. The process of blooming involves (1) dissolution of the higher fatty acid in oil, (2) separation (crystallization) of the higher fatty acid from oil, and (3) transformation of higher fatty acid crystals on the surface of cosmetic pencils [4–6]. This chapter concerns the mechanisms of sweating and blooming with respect to the separation of oils from wax and the crystallization of fatty acids.

II. SWEATING OF LIPSTICKS

Figure 1 shows the sweating phenomenon on lipstick. A number of droplets which decrease the quality of the lipstick can be seen on the surface. The mechanism by which sweating is generated should be determined to solve this problem.

A. A Binary Oil-Wax System

Since the predominant constituents in lipsticks are wax, oil, and pigments, and since sweating is associated with changes in these constituents during storage, we first attempted to elucidate the mechanism of sweating of lipsticks in a simplified binary oil-wax system without pigments.

In general, the weight ratios of wax, oil, and pigments in lipsticks are 15–25%, 65–80%, and 5–10%, respectively [7]. Under our experimental conditions, wax and oil were mixed in a ratio of 15:85 (w/w), melted at 80–85°C, poured into a metal mold (i.d. = 12 mm, length = 35 mm), and then cooled to 25°C to prepare the lipsticks.

1. Temperature Dependence of Sweating

Candelilla wax and glyceryl tri-2-ethylhexanoate (GTEH) were employed in this study. When the lipsticks were placed at 25°C and 60–70% relative humidity (R.H.), sweating was observed. A similar sweating phenomenon was observed when a solid paraffin was employed as a wax and glycerol di-2-heptylundecanoate (GDHU), trimethylopropane triisostearate (TPTS), isostearylpalmitate (ISP), or olive oil were used as the oil. Next, the temperature dependency of sweating was evaluated. Sweating was more marked at 28°C, by increasing the temperature, number, and diameter of generated droplets (Fig. 2). However, de-

Fig. 1 Sweating on a lipstick.

creasing the temperature after the generation of sweating caused disappearance of the droplets, which were generated again by increasing the temperature (Fig. 2). On the other hand, polarizing and phase-contrast microscopic analysis revealed that lipsticks have a number of wax matrices [8–10] in which oil is dispersed homogeneously (Figs. 3, 4). Taken together, these observations indicate that sweating is a phenomenon in which droplets migrate to the surface and/or penetrate inside lipsticks with fluctuations in temperature similarly to the breathing of skin. These studies also indicated that the threshold temperature for sweating is about 20°C. Thus, it was concluded that sweating is very sensitive to temperature.

2. Analysis of Constituents of Sweat Droplets

We analyzed the constituents of the droplets migrating to the surface of lipsticks composed of candelilla wax and GDHU by using gas chromatography (GC).

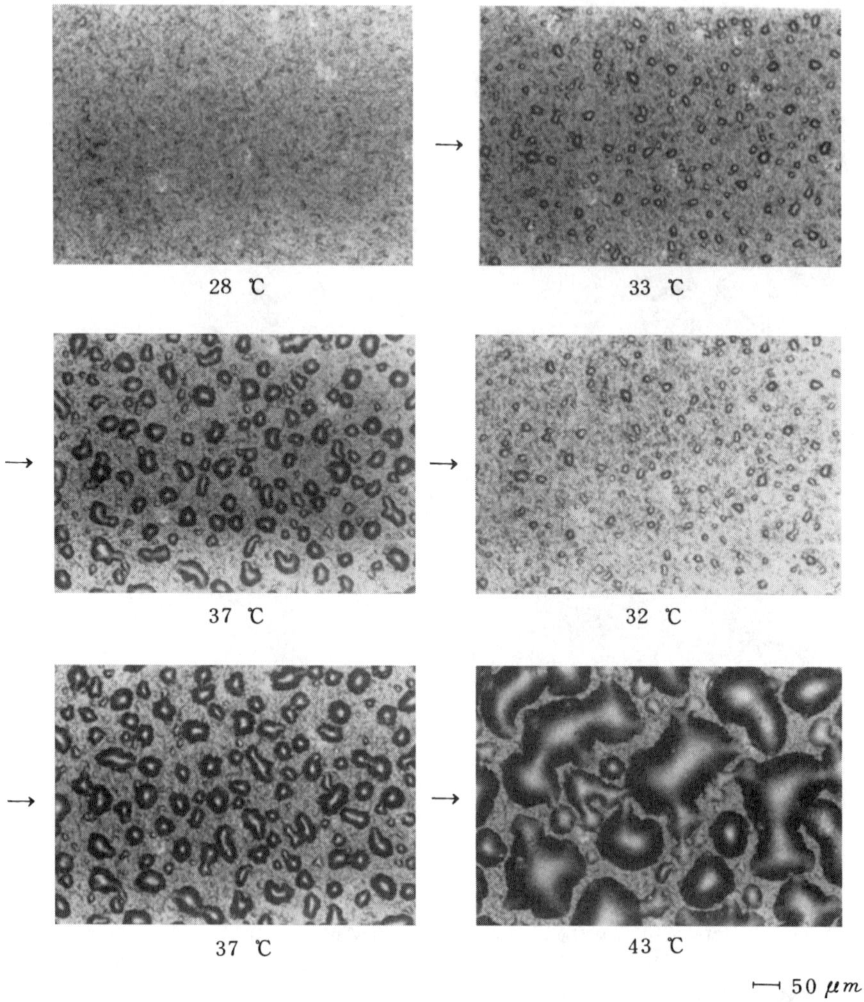

Fig. 2 Growth of droplets.

The predominant component was GDHU, and the others were C_{24}-C_{42} esters in candelilla wax. These findings indicated that sweating is a solid-liquid separation phenomenon in lipsticks composed of oil-wax mixtures.

3. Relationship between Sweating and Expansion of Oil

It was assumed that sweating involves the expansion of oil due to increase in temperature. To test this hypothesis, we studied the relationship between sweating

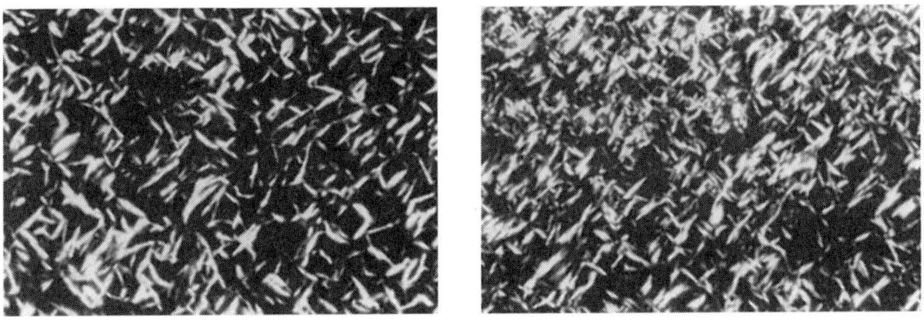

Wax : Oil = 15 : 85 Wax : Oil = 30 : 70

10 μm

Fig. 3 Wax in a lipstick observed under a polarizing microscope.

and the expansion coefficients of waxes and oils using a deratometer in from 31 to 40°C. The expansion coefficients of all studied types of oils were about double those of candelilla wax and solid paraffin; i.e., the expansion coefficients of candelilla wax and GTEH were about 300 and 560 ml/g/°C, respectively. This difference in expansion coefficients was important, since a significant fraction of the constituents in lipsticks comprises oil. In addition, the observed value of the oil extent that overflowed from the wax matrix in the wax-oil system was identified with the values calculated from the expansion coefficients of oils under these

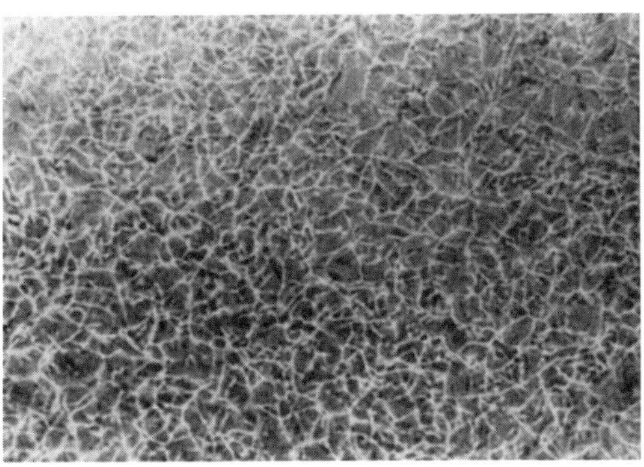

10 μm

Fig. 4 Wax on the surface of a lipstick observed under a phase-contrast microscope.

experimental conditions. Therefore, it is likely that sweating is generated due to the expansion of oil by increasing the temperature.

4. Effect of Wax Amount on Sweating

We studied the effect of wax type in the lipsticks on sweating. When solid paraffin was employed, sweating was observed but the number of droplets generated was less than in the candelilla wax-oil system. Less sweating was generated when the wax matrix was dense, while the converse is true in systems in which the matrix is sparse. Thus, the difference in sweating between solid paraffin and candelilla wax may be explained by the density of wax matrix on the basis of the disparity of their chemical compositions. Indeed, the former has a denser matrix compared with the latter.

Next, the effect of wax amount on sweating was examined. By comparison of lipsticks in which the wax fraction was 85% and 70%, it was confirmed that more wax in the lipstick decreased the number of droplets. This suggested that lipsticks with less wax have a decreased matrix density and, consequently, generate more sweat. Therefore, there seems to be a close relationship between sweating and oil-wax matrix density.

B. A Ternary Oil-Wax-Pigment System

The sweating phenomenon of lipsticks prepared without pigments was discussed in the previous section. However, in commercial preparations pigments are usually used in lipsticks. Pigments are classified as organic, inorganic, perlescent, and functional (boron nitrite, hybrid fine powder, photochromic pigments). All types of pigments have their own associated oil absorbency and are used in order to be dispersed in oils of lipsticks. On the other hand, Dweck reported that when the combination of pigments added to lipstick exceeded 20% (w/w), the likelihood of sweating was greatly increased [1]. However, there is little detailed information regarding these phenomena. Thus, we examined the sweating of lipsticks in a ternary oil-wax-pigment system.

1. Effect of Pigments on Sweating

The effect of pigments on the incidence of sweating was studied by employing lithol rubin BCA, a representative organic pigment, candelilla wax, and GDHU. The weight ratio of the wax, oil, and pigment was 10:15:85. Sweating was accelerated by inclusion of the organic pigment in the lipsticks. The influence of the other types of pigments was evaluated under the same conditions. When organic pigments such as lithol rubine BNA, lake red CBA, and lithol rubine BBA were used, sweating was generated at 89% R.H., and not at 22% or 50% R.H. In contrast, when inorganic pigments such as titanium dioxide and black oxide of iron were added, no sweating was observed under any of the examined conditions. The effect of perlescent or functional pigments on sweating could not be discussed due to the lack of experimental data.

2. Relationship between Moisture Sorption and Sweating

To gain insight into the different extents of sweating mechanisms between organic and inorganic pigments, the moisture sorption of pigments were measured during storage at 25°C and 22%, 50%, and 89% R.H. after pretreatment of pigments at 80°C overnight. The inorganic pigments failed to increase their own weight at any of the studied R.H. In contrast, organic pigments were more hygroscopic than inorganic pigments in the order lithol rubine BCA > lithol rubine BNA > lake Red CBA > lithol rubine BBA. In addition, a similar tendency was observed in lipsticks containing these pigments. Thus, the order of the extent of pigment moisture sorption was identical to that of the incidence of lipstick sweating. This suggests that greater lipstick sweating is attributable to increased water content on the basis of the hygroscopic properties of the used pigments. In addition, these studies indicated that sweating is apt to occur when the water content of lipsticks exceeds 0.35% (w/w). Similar phenomena were observed in the study employing general hygroscopic reagents such as calcium chloride and magnesium chloride. On the other hand, the use of more hygroscopic waxes than candelilla wax also led to more sweating. In addition, it was confirmed that the total weight of droplets generated on the surface of the lipstick was identical to the increased water content of lipsticks containing pigments. These findings support the positive correlation between sweating and the hygroscopic properties of lipsticks.

Sweating became marked when the lipsticks were preserved in an ethyl alcohol atmosphere, which is more soluble in an oil-wax system, than in water in systems with or without pigments. These observations indicated that sweating is a result of the excretion of oil on the lipstick surface due to the absorption of moisture from the atmosphere.

3. Contribution of Polymorphism of Waxes to Sweating

It is well known that waxes such as solid paraffin are polymorphic [11]. Thus, the effects of pigments on the crystal forms of solid paraffin were examined. Our studies with a scanning electromicroscope (SEM) and powder X-ray diffraction apparatus, however, did not indicate that pigments give rise to structural changes in the wax matrix of lipsticks. Therefore, it is unlikely that polymorphism contributes to sweating at least under present experimental conditions. The results described supported the suggestion that sweating is caused by moisture sorption of the lipsticks containing pigments, in addition to the expansion of oils under temperature increase.

C. Proposed Mechanism of Sweating

According to the previously mentioned results, a mechanism of sweating can be proposed as shown in Fig. 5. Thus, with increasing temperature and relative humidity in atmosphere, droplets migrate to the surface of lipsticks through channels in the wax matrix due to the expansion of oil or the compression effect of the extrinsic water on the oil in the wax matrices by moisture sorption. It is likely that the lyophilicity between wax and oil and the composition ratio of wax to oil are involved in sweating (data not shown). These findings will lead to improvements in lipstick preparation; i.e., methods to maintain adequate water in the lipstick will prevent sweating.

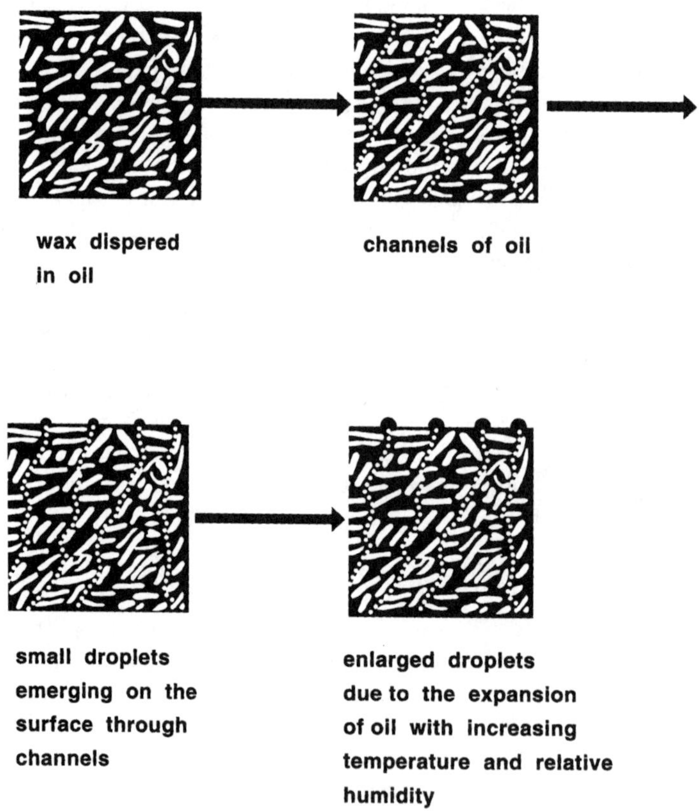

Fig. 5 Proposed mechanism of lipstick sweating.

III. BLOOMING OF COSMETIC PENCILS

Blooming occurs on the surface of cosmetic pencils such as eyeliner, eyebrow pencils, and lipsticks. Representative time courses of blooming, i.e., the formation of crystalline white powder on the surface of cosmetic pencils, are shown in Fig. 6. In general, cosmetic pencils are composed of waxes, oils, higher fatty acids, higher alcohols, esters, and hydrocarbons as the base in addition to inorganic

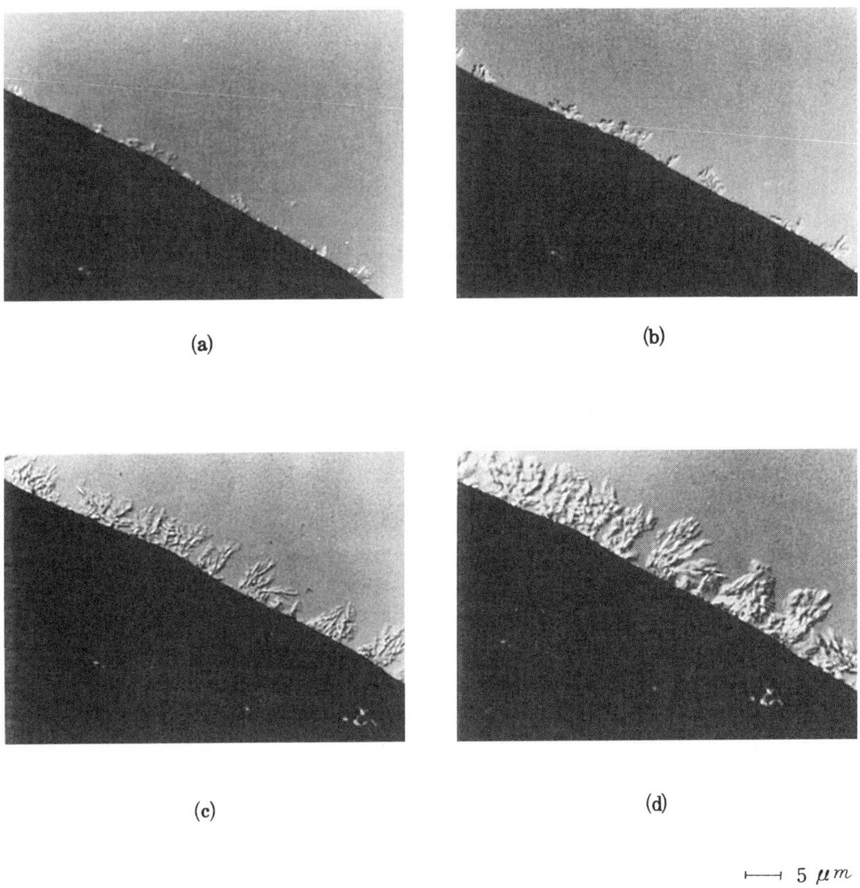

Fig. 6 Process of blooming on cosmetic pencils consisting of glyceryl tri-2-ethylhexanoate, stearic acid, solid paraffin, and black oxide of iron: (a) one day, (b) three days, (c) one week, (d) two weeks.

pigments. Furthermore, such pencils lacking higher fatty acids do not show blooming. Therefore, higher fatty acids are involved in blooming. Studies of blooming have mainly been performed in chocolate [12,13] and butter industries [14]. However, chocolate and butter comprise only solid oleaginous components such as waxes, not oils. In addition, the details of the mechanism by which blooming occurs in cosmetic pencils are not well known. We showed that stearic acid, a representative higher fatty acid, is associated with blooming in cosmetic pencils. In addition, the effects of oil type on blooming and then the relationship between cooling rate and annealing during the manufacturing process and blooming will be discussed.

A. Contribution of Higher Fatty Acid to Blooming

1. Blooming

In this study, solid paraffin, GTEH, black oxide of iron, and stearic acid were employed as wax, oil, pigment, and higher fatty acid at a weight ratio of 15:20: 50:15. These components were heated at 80–85°C and then molded at 25°C.

The presence of stearic acid in the mixture resulted in blooming within seven days at 25°C and 50% R.H. Blooming was also observed in preparations containing no stearic acid. In addition, the incidence of blooming increased with increasing amounts of stearic acid in the system. Thus, it was confirmed that blooming was dependent on the level of stearic acid in the system. Furthermore, GC analysis of the white powder generated on the surface of cosmetic pencils revealed that the predominant constituent was stearic acid. Hence, we conclude that stearic acid is directly involved in the mechanism of blooming.

2. Effect of Oil on Blooming

The effect of the oil levels on blooming was studied. Interestingly, the incidence of blooming decreased by reducing the essential oil content. This suggested that oil is required for blooming. In addition, these findings suggest that oil acts as a solvent for stearic acid. Second, we studied the effect of oil type on blooming, using 11 types of oil. When GDHU, diisostearyl malate (DISM), GTEH, 2-heptylundecyl palmitate (HUP), and glyceryl triisostearate (GTIS) were employed, blooming was observed. In contrast, when liquid lanolin, castor oil, polyoxyethylene polyoxypropylene pentaerythritol ether (PPPE), or polyoxypropylene butyl ether (PPBE) was employed, no blooming was observed. Those oils which failed to generate blooming have inorganic-organic balance values similar to that of stearic acid (0.42), high viscosity (450 cps at 30°C), or relatively high molecular weight. These results suggested that the solubility of stearic acid in oils is significantly involved in blooming.

3. Effect of Waxes on Blooming

As described, blooming occurs in mixtures of wax-oil-pigment-stearic acid. On the other hand, our preliminary study indicated that no blooming occurred in stearic acid alone or in oil-stearic acid mixtures. In addition, blooming was little affected by the addition of pigments. Thus, wax appears to play an important role in blooming. Here, the effect of wax type on blooming was studied in mixtures of stearic acid, oil, wax, and black oxide of iron in a weight ratio of 15:15:20:50. Cosmetic pencils were prepared as described above. Blooming was accelerated by using a hydrogenated castor oil (HCO) of which the iodine value was 5 (HCO-IV5), carnauba wax and candinia wax, compared with solid paraffin and observed within one day at 25°C and 50% R.H. On the other hand, when HCO-IV46 and candelilla wax were employed, no blooming was observed. Therefore, the difference in blooming between HCO-IV5 and HCO-IV46 may be explained by the solubility of stearic acid in the oil.

4. Relationship between the Solubility of Stearic Acid in Oil and Blooming

To confirm the above hypothesis, the solubilities of stearic acid in various oils were compared. Our experimental data indicated that the oils which generated blooming dissolved stearic acid to a lesser extent; the solubility of stearic acid in GDHU, DISM, GTEH, HUP, and GTIS was 0.47%, 0.41%, 0.30%, 0.26%, 0.20% (w/w), respectively. In contrast, stearic acid was highly soluble in those oils which did not generate blooming; the solubilities in liquid lanolin, PPPE, PPBE, and castor oil were 0.80–0.96% (w/w). In addition, this study indicated that the threshold of solubility of stearic acid in oils was about 0.5% (w/w). The effect of wax type on the solubility of stearic acid in oil was also examined. Those waxes that generated blooming decreased the solubility of stearic acid in oil in a similar manner as described above. The addition of HCO-IV5, carnauba wax, or solid paraffin significantly decreased the solubility. In contrast, the solubility of stearic acid in GTEH was 0.30% (w/w) in the absence of wax, and the value did not change by the addition of a wax such as HCO-IV46, candelilla wax, or solid paraffin. It is clear that some waxes are involved in blooming. It is noteworthy that the decreasing effect of stearic acid solubility by some waxes enhanced the incidence of blooming. This is consistent with the influence of oil type on blooming in cosmetic pencils. Thus, blooming was considered to be related to the decrease in the stearic acid solubility in oil with some added waxes. On the other hand, the use of liquid lanolin, in which the solubility of stearic acid is about 10-fold higher than that in GTEH, did not lead to blooming despite the presence of HCO-IV5. Thus, blooming can be estimated by measuring the solubility of stearic acid in the wax-oil system.

5. *Sugar Esters as Inhibitors of Blooming*

It was confirmed that lipophilic fatty acid sugar esters inhibit blooming on the surface of cocoa butter [15]. Thus, the effects of lipophilic fatty acid sugar esters in cosmetic pencils on blooming were examined. When Sugarwax A-10E (sugar ester), a commercial-grade lipophilic fatty acid sugar ester, was added to the ternary oil-wax-pigment system at concentrations over 0.5% (w/w), no blooming was observed. These results explain the direct proportion between the addition of sugar esters and the solubility of stearic acid in the oil. As expected, sugar esters increased the solubility of stearic acid with increasing temperature, i.e., by more than twofold at 25°C (Fig. 7). Thus, there is a negative correlation between the solubility of stearic acid and the incidence of blooming in multicomponent systems.

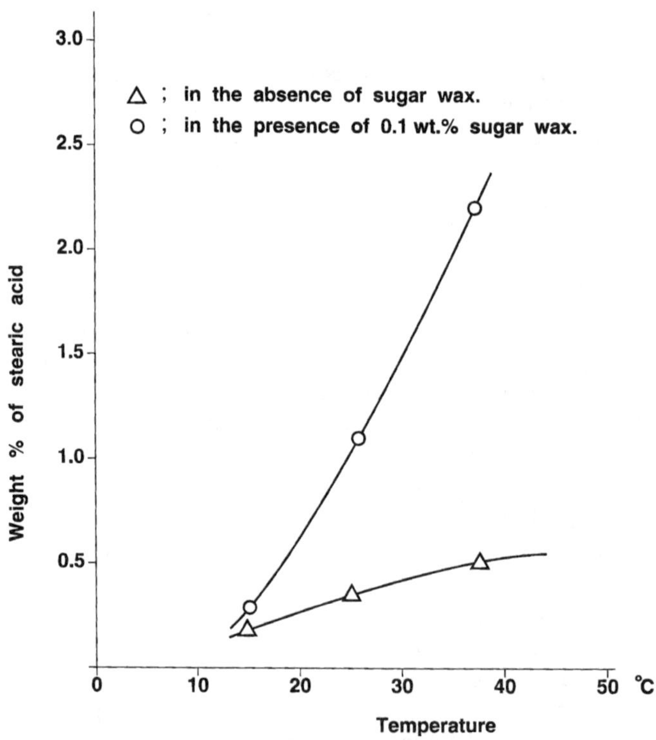

Fig. 7 Effects of sugar wax on the solubility of stearic acid in glyceryl triethylhexanoate.

6. Crystal Formation in Blooming

As described above, the predominant constituent of the white powder generated on the surface of cosmetic pencils is stearic acid. In addition, SEM study suggested that the white powder observed on the surface of cosmetic pencils is dendrites (Fig. 8). On the other hand, the crystalline stearic acid has three crystal forms, A, B, and C, defined as metastable, stable, and metastable stearic acid, respectively [16–18]. Taken together, it was suggested that the crystal form of stearic acid transforms during blooming. Thus, it was necessary to study the blooming phenomena crystallographically.

Figure 9 shows the powder X-ray diffractograms of stearic acid prepared by various methods: In sample 1, stearic acid crystals were heated to 85°C, and then melt was immediately cooled to 0°C. In sample 2, stearic acid powders were generated on the surface of cosmetic pencils. As shown in Fig. 9, the diffraction patterns of samples 1 and 2 indicate that they were B form and A form, respectively. Therefore, the B form was observed immediately after sample preparation, and it was gradually transformed to A form on the surface of the cosmetic pencils. This was supported by the observation that A-form growth dominates B-form growth [19]. In addition, these findings may also be applicable to cosmetic pencils that contain other higher fatty acids [20].

Next, the effects of the addition of sugar esters, which inhibit blooming, to a mixture of stearic acid and GTEH on the crystallization of stearic acid were examined. The ratio of the intensity of long spacing ($2\theta = 6.4°$) to short spacing ($2\theta = 21.4°$) was dependent on the concentration of sugar ester (Fig. 10). In the absence of sugar ester, the ratio was about 0.9, indicating that the crystals of stearic acid grew in the direction of long spacing, whereas the addition of sugar ester decreased the ratio in a concentration-dependent manner. The results indicated that sugar wax inhibited the growth of stearic acid crystals in the direction of long spacing. A similar inhibitory effect of sugar wax on crystal growth was observed in GTEH solution containing 0.1% (w/w) paraffin wax and 2.5% (w/w) stearic acid. It is evident that blooming is associated with crystal growth in the direction of long spacing. In addition, it is considered that the polymorphism of waxes is one of the determining factors in the mechanism of bloom formation in mixtures of oils, waxes, and pigments.

Next, we studied whether the difference in wax type affects the crystallization of stearic acid. Infrared absorption spectroscopic and GC analyses revealed that the predominant constituent of the bloom was stearic acid in the GTEH–HCO–black oxide of iron system. Powder X-ray diffraction analysis revealed that the crystal form of stearic acid in the bloom was A form and its shape was dendric. Thus, wax type has little influence on the crystal form of stearic acid in the bloom. These findings were supported by the observation that the addition

(a)

(b)

Fig. 8 Scanning electron micrographs of blooming on the surface of cosmetic pencils: (a) scale, 50 μm, (b) scale, 5 μm.

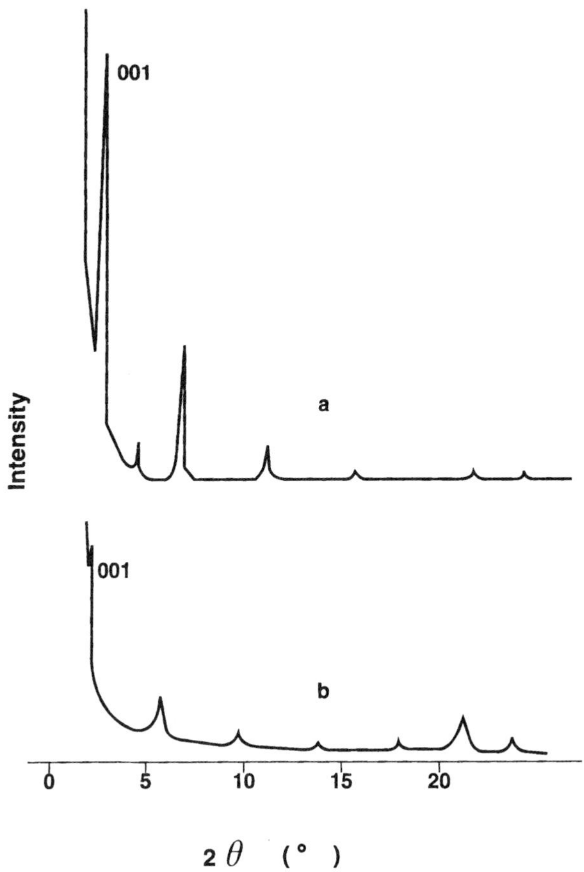

Fig. 9 Powder X-ray diffraction patterns of stearic acid (a, B form) and blooming of stearic acid (b, A form).

of pigments failed to affect the extent of blooming. Interestingly, blooming was not influenced by relative humidity, indicating that the mechanism of the bloom generation differs from that of sweating.

In conclusion, these findings indicated that bloom generation was related to the solubility of stearic acid in oils, shape, and the layer growth of stearic acid crystals.

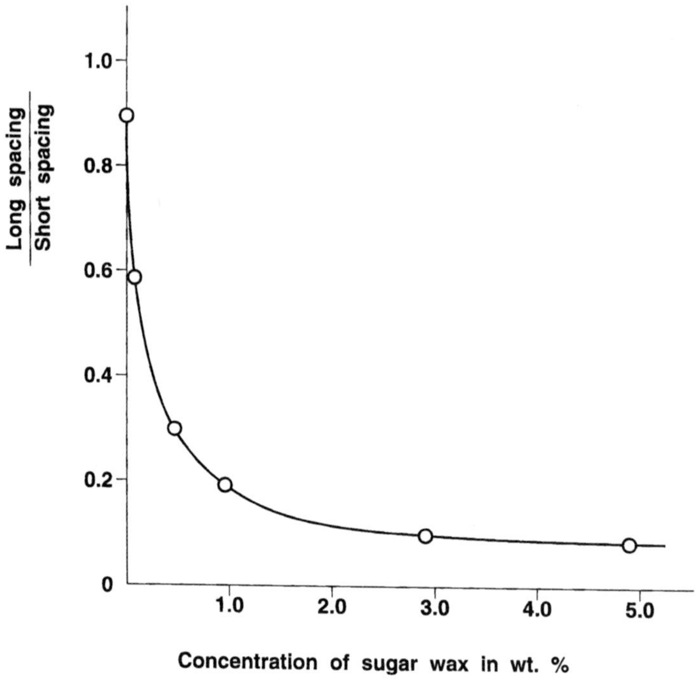

Fig. 10 Dependence of the ratio of the intensities of long spacing to short spacing on the concentration of sugar wax.

B. Effect of Cooling Rate on Blooming

The cooling rate during the preparation of cosmetic pencils also has an effect on bloom generation. We studied the effect of cooling rate on blooming in cosmetic pencils composed of HCO, stearic acid, GTEH, and black oxide of iron in a weight ratio of 20:15:15:50. When the mixture was melted at 80–85°C and rapidly cooled to 25°C, blooming was observed within seven days, whereas no blooming occurred when the mixture was cooled at a slower rate. These findings indicated that cooling rate is involved in bloom generation. Next, the shapes of the generated crystals were compared between the rapidly and slowly cooled samples, in which stearic acid was dispersed in GTEH. Almost immediately after rapid cooling, rhombic and thin leaflet crystals with a diameter of about 5–8 μm were observed. On the other hand, coarse crystals of 10–30 μm diameter were observed in the slowly cooled sample. In addition, after one month, crystal layer growth was observed in the rapidly cooled sample, as reported by Gilmer and Jackson [21]. These results suggested that the crystals generated in the rapidly cooled samples were of the metastable form of stearic acid (A form), whereas those generated in slowly cooled samples were of the stable form (B form). Thus, it was likely that more blooming would occur when the crystals generated in the cosmetic pencils were more unstable. These findings indicate that cosmetic pencils should be slowly solidified.

C. Effect of Annealing on Blooming

In systems in which stearic acid was added to mixtures of oil, wax, and pigments, we examined the relationship between annealing in the extrusion molding manufacturing process and blooming. Extrusion molding was performed as follows: mixtures of wax-oil-pigment that melted at 85°C were solidified at 25°C, the solid mixtures were kneaded with a roll mill, the flake mixtures were then molded into pencils by pressing, and the pencils were annealed at the indicated temperatures.

1. *Effect of Annealing Temperature on Blooming*

Cosmetic pencils composed of solid paraffin, GTEH, black oxide of iron, and stearic acid in the weight ratio of 15:20:50:15 were annealed at temperatures from 25 to 80°C. Blooming over 47°C, especially at 65°C, was observed on the entire surface of the cosmetic pencils. In addition, it was confirmed that the white powder crystals increased in size with increasing annealing temperature.

Powder X-ray diffraction analysis demonstrated that the crystal habit of stearic acid generated in samples annealed at temperatures from 25 to 65°C in the stearic acid–GTEH (1:1 w/w) system was monoclinic, indicating C-form crystals. In contrast, white powder crystals were triclinic when the same cosmetic pencils were annealed at over 55°C, indicating the A form. Taken together, these observations indicate that blooming involves transformation of the crystal habit of stearic acid from C form to A form after annealing. However, the mechanism by which C form was transformed to A form is unknown.

2. *Solubility of Stearic Acid in Oils at Higher Temperatures*

To gain insight into the mechanism by which blooming is dependent on annealing temperature, the effect of high temperature on the solubility of stearic acid in GTEH was studied. The solubility of stearic acid in GTEH increased with increasing temperature. This result may indicate that the solubility of stearic acid is related to the incidence of blooming even at higher temperature and that a positive correlation between the extent of recrystallization of stearic acid and blooming exists. Thus, blooming seems to be generated readily in systems in which crystals grow to about 200 μm.

These observations led us to conclude that annealing temperature markedly affected blooming and affected the transformation of stearic acid crystals from C form to A form. Blooming is likely a solid-liquid separation phenomenon, sensitive to many physicochemical factors such as temperature, solubility, and crystal form. Further studies are required to elucidate the details of the blooming mechanism.

IV. CONCLUSION

Sweating and blooming are solid-liquid separation phenomena, the generation of which must be avoided during the storage of oleaginous solid cosmetic pencils such as lipsticks, eyeliner, and eyebrow pencils. We studied the mechanisms by which sweating and blooming are generated on the surface of cosmetics. Unexpectedly, our results revealed that the mechanisms of these two phenomena were different. Sweating seems to be due to expansion of oils in the wax matrix and/or shriveling of the wax matrix due to increased temperature and humidity. In contrast, blooming involves the solubility of higher fatty acids, a modulator of the consistency and hardness of oleaginous cosmetics, in oils, and is associated with a series of processes such as dissolution–separation (deposition)–transformation of crystals of higher fatty acids.

These findings provide a rational basis for further studies of interfacial phenomena in oleaginous cosmetic pencils.

REFERENCES

1. A. C. Dweck, *Cosmetics and Toiletries*, *96*: 1 (1981).
2. H. Matsuda, K. Uehara, K. Tokubo, and M. Tanaka, *SIKIZAI*, *57*: 130 (1984).
3. H. Matsuda, H. Nakajima, K. Tokubo, and M. Tanaka, *SIKIZAI*, *57*: 492 (1984).
4. H. Matsuda, M. Yamaguchi, and M. Tanaka, *SIKIZAI*, *57*: 64 (1985).
5. H. Matsuda, M. Yamaguchi, and M. Tanaka, *SIKIZAI*, *58*: 515 (1985).
6. H. Matsuda, M. Yamaguchi, and M. Tanaka, *SIKIZAI*, *59*: 651 (1986).
7. M. S. Balsam and E. Sagarin, in *Cosmetics Science and Technology* (M. S. Balsam and E. Sagarin, eds.), Wiley-Interscience, New York, 1974, p. 365.
8. M. Katoh, *J. Electron Microsc.*, *28*: 198 (1979).
9. E. S. Lutton, *J. Am. Oil Chem. Soc.*, *27*: 276 (1950).
10. C. W. Hoerr, *J. Am. Oil Chem. Soc.*, *45*: 793 (1968).
11. S. V. Vaeck, *Manufacturing Confectioner*, *6*: 35 (1960).
12. K. Becker, *Fette Seifen Anstrichmittel*, *59*: 636 (1957).
13. T. Kleinert, *Int. Chocolate Rev.*, *16*: 201 (1961).
14. L. B. Campbell, D. A. Andersen, and P. G. Keeney, *J. Dairy Sci.*, *59*: 976 (1969).
15. M. Teshirogi, *Proc. Res. Jpn. Sugar*, *14*: 95 (1964).

16. N. Garti, E. Wellner, and S. Sarig, *J. Cryst. Growth*, *57*: 577 (1992).
17. E. Wellner, N. Garti, and S. Sarig, *Cryst. Res. Technol.*, *16*: 1283 (1981).
18. N. Garti, E. Wellner, and S. Sarig, *Kristjech*, *15*: 1303 (1980).
19. Y. Maeyashiki, M. Okada, and K. Sato, *Jpn. J. Appl. Phys.*, *21*: 781 (1982).
20. E. Vonsydow, *Acta Chem. Scand.*, *9*: 1685 (1955).
21. G. H. Gilmer and K. A. Jackson, in *Crystal Growth and Materials* (E. Kaldis and H. J. Scheel, eds.), Elsevier, Amsterdam, 1976, p. 79.

15

Crystallization Properties and Lyotropic Phase Behavior of Food Emulsifiers

Relation to Technical Applications

Niels Krog
Danisco Cultor, Brabrand, Denmark

I. INTRODUCTION

Natural as well as synthetic polar lipids are used as emulsifiers in foods, cosmetics, plastics, and pharmaceuticals. The molecular organization in the solid state of polar lipids is governed by their amphiphilic nature. The polar head groups are oriented in sheets, allowing interactions involving hydrogen bonds, dipol-dipol or ionic interactions, and the layers of polar groups are separated by the hydrocarbon chains forming a bilayer structure, similar to the structure of biomembranes. The crystal structure of pure, racemic 1-monoglycerides was first described by Larsson in 1966 [1] and was later reviewed by Small (1986) [2] and Larsson (1994) [3].

Food-grade emulsifiers are esters of polyvalent alcohols (glycerol, propylene glycol, sorbitol/sorbitan, or sucrose) and edible fatty acids. Monoacylglycerides and their organic acid derivatives are by volume the most commonly used polar lipids in food products, and their crystallization properties and phase behavior in aqueous systems are important to their functional effects in many food products.

When polar lipids are mixed with water, the water has a strong affinity to the polar groups and will penetrate the polar sheets, forming lyotropic, liquid crystalline mesophases, which have been known to man in the form of soap preparations for more than 2000 years. However, the molecular structure of liquid crystalline phases was first known about 40 years ago. The first descriptions of the mesomorphic behavior of monoglycerides were given by Lutton (1965) [4] and Larsson (1967) [5]. Comprehensive reviews of liquid crystalline structures of polar lipids have been given by Tiddy (1980) [6], Small (1986) [2], and Larsson (1994) [3].

The phase behavior of commercial food emulsifiers, primarily monoglycerides and their derivatives, has been described by Krog (1997) [7] and is reviewed here. The formation of liquid crystalline structures in aqueous systems is important for interactions with carbohydrates or proteins, and it facilitates foam formation in aerated foods, etc. Understanding the physical properties of polar lipids is a key to selecting the optimal emulsifier, or combinations thereof, for a given application in the food industry as well as other technical applications.

II. CRYSTALLIZATION PROPERTIES OF MIXED FATTY ACID MONOGLYCERIDES AND THEIR ORGANIC ACID ESTERS

A. Monoglycerides

Industrially produced monoglycerides vary in composition with respect to the distribution of mono-, di-, and triacylglycerides and with respect to their fatty acid profile. The content of monoglycerides may vary from 30–45% in products made as equilibrium compositions by interesterification of glycerol and fats or oils to 90–95% in concentrated monoglycerides made by a high-vacuum, short-path distillation process [7]. Such products are normally referred to as distilled monoglycerides and contain 90–95% 1-monoglycerides, 3–5% 2-monoglycerides, 2–4% diglycerides, and less than 1% free glycerol and free fatty acids.

The fatty acid profile of monoglycerides may vary considerably depending on the source of the raw materials used. The most common manufacturing process is interesterification of edible animal or vegetable fats and oils, which may be fully or partly hydrogenated to achieve a specific fatty acid profile. An alternative method is esterification of fatty acids with glycerol. By selecting fatty acids with a specific chain length, monoglycerides with a narrow fatty acid profile can be produced. The typical fatty acid composition of the fats and fatty acids employed in the production of industrial monoglycerides is shown in Table 1. The most commonly used fatty acids are palmitic (C16) and stearic (C18) acid present in products made from hydrogenated fats, which are the dominant raw materials used worldwide. In addition to the raw materials shown in Table 1, many other

Table 1 Main Fatty Acid Composition of Commercial Fats and Fatty Acids

Fatty acid chain length	Hydrog.[a] lard (%)	Hydrog.[a] soybean oil (%)	Hydrog.[a] palm oil (%)	Liquid palm oil (%)	Sunflower oil (%)	Commercial single fatty acids (90%)				
						Lauric C12	Myristic C14	Palmitic C16	Oleic C18:1	Behenic C22
C12, lauric	—	—	—	—	—	98	1	—	—	—
C14, myristic	2	—	1	1	—	2	97	1	—	—
C16, palmitic	30	11	45	45	6	—	2	96	1	—
C18, stearic	67	89	53	6	5	—	—	3	3	3
C18:1, oleic	—	—	—	37	20	—	—	—	90	—
C18:2, linoleic	—	—	—	10	69	—	—	—	6	—
C20, arachidonic	1	—	1	1	—	—	—	—	—	3
C22, behenic	—	—	—	—	—	—	—	—	—	94

[a] Hydrogenated

fats and oils may be used, such as tallow, cottonseed oil, and low-erucic rapeseed oil.

The mixed fatty acid profile and multicomponent composition of industrial monoglycerides affect their melting and crystallization properties as well as their polymorphic behavior in comparison with pure, synthetic monoglycerides. In general, the melting and crystallization temperatures of mixed fatty acid monoglycerides are lower and the polymorphic behavior is less complicated than that of pure components.

Table 2 shows X-ray diffraction data and the melting points of distilled monoglycerides with fatty acid profiles as presented in Table 1. The melting points are measured as peak temperatures by differential scanning calorimetry (DSC).

All monoglycerides crystallize after melting in a metastable α-crystal form, characterized by a single X-ray short spacing between 4.1 and 4.15 Å, which shows hexagonal chain packing. This molecular packing is similar to that of pure monoglycerides with parallel fatty acid chains, forming bimolecular layers separated by layers of the polar groups. The α-crystal form will transform during storage at ambient temperature to the stable β-crystal form, which has a higher melting point and is characterized by several X-ray lines in the short-spacing region, with the strongest line at 4.5–4.6 Å. The X-ray long spacings vary according to the chain length of the monoglycerides, in compliance with a double-chain-length (DCL) packing mode. The difference between the long spacings of the α and β forms is due to more vertical chain packing in the α form than in the β form. Unlike pure 1-monoglycerides, industrial distilled monoglycerides do not form any β′-crystal form.

If the α-crystal form is cooled to 35–50°C below its crystallization point, it transforms into a sub-α crystal form, which has orthorhombic chain packing, characterized by strong X-ray short spacings at 4.13 Å and several spacings from 3.9–3.7 Å. The transition temperature varies according to the fatty acid profile from about 16°C for hydrogenated palm oil monoglycerides and 33°C for hydrogenated soybean oil to about 50°C for monobehenin made from 90% behenic acid. The more complex the fatty acid profile, the lower is the temperature for the formation of the sub-α-crystal form. This is illustrated by the difference in the sub-α melting point of monoglycerides from hydrogenated palm oil (16°C) and that of 1-monopalmitin (35°C) as shown in Table 2. The transition from α- to sub-α-crystal form is a reversible, solid-state transformation, and the enthalpy change can be measured by DSC.

When cooled rapidly after melting, the α form crystallizes approximately 5°C below the melting point shown in Table 2 due to a supercooling effect, and a similar reduction in temperature is observed during the transition from the α- to the sub-α-crystal form on cooling.

Table 2 X-Ray Diffraction Data of Mixed Fatty Acid Distilled Monoglycerides

Monoglyceride	Long spacings (Å)		Main short spacings (Å)		Melting points[b] (°C)		
Crystal form	α	β	α	β	Sub-α	α	β
Monolaurin, 90% C12	—	37.3	4.15	4.57–4.32–4.00–3.84–3.71–2.44	16	45	61
Monomyristin, 90% C14	41.0	40.6	4.15	4.55–4.33–3.91–3.81–3.71–2.44	25	56	67
Monopalmitin, 90% C16	47.0	44.7	4.15	4.51–3.91–3.84–3.67–2.43	35	66	73
Monoolein, 90% C18:1	—	48.5	—	4.60–4.38–4.31–4.04	—	30	34
Monobehenin, 90% C22	57.3	57.5	4.15	4.50–3.94–3.84–3.74–2.43	56	82	85
Saturated monoglycerides[a] (1) (hydrogenated soybean oil)	54.0	51.4	4.13	4.55–3.94–3.86–3.78–2.43	37	71	75
Saturated monoglycerides[a] (2) (from hydrogenated lard)	53.2	49.8	4.15	4.52–4.35–3.93–3.84–2.43	20	66	72
Saturated monoglycerides[a] (3) (hydrogenated palm oil)	51.6	47.0	4.15	4.55–4.33–3.89–2.43	16	68	72
Unsaturated monoglycerides[a] (4) (palm oil)	—	46.5	—	4.55–4.31–4.03–3.86	8	(48)	60

[a] Danisco Cultor products: (1) DIMODAN PS, (2) DIMODAN P, (3) DIMODAN PVP, (4) DIMODAN BP.
[b] Differential scanning calorimetry (DSC) peak temperatures.

B. Organic Acid Esters

The esterification of the OH groups in monoglycerides with organic acids in the form of acetic, lactic, diacetyl tartaric, or citric acid changes the lipophilic/hydrophilic properties of monoglycerides considerably [7]. Acetylated monoglycerides become more lipophilic than the monoglyceride itself, while esterification of monoglycerides with diacetyl tartaric acid or citric acid makes the final product more hydrophilic and anionic active. The physical properties of monoglycerides are, thus, changed considerably by esterification with organic acids, resulting in a lower melting point, changed polarity, and different phase behavior in water.

X-ray diffraction data and the melting points of typical organic acid esters of monoglycerides are shown in Table 3. All the organic acid esters selected have fatty acid profiles which are very similar to those given for hydrogenated lard in Table 1.

Long spacings for acetic, lactic, and diacetyl tartaric acid esters indicate a single-chain-length (SCL) packing mode with penetrating fatty acid chains, and the increase in long spacings from 32.9 Å for acetic acid esters to 41 Å for diacetyl tartaric acid esters may be due to the increasing size of the polar group. A diacetyl tartaric acid ester of 1,2-dipalmitin (1,2-dipalmitin-DATE) has a long spacing of 54.6 Å, which corresponds to a DCL packing mode. The short spacings indicate a very complex crystal structure, which is not known in detail. In general, diacyl esters with large polar groups have a tendency to crystallize in a DCL form, as shown for phospholipids [2]. Long spacings for citric acid esters indicate a DCL packing mode such as that of nonesterified monoglycerides. An explanation of this may be that citric acid esters contain a higher proportion of diacylglycerides than the other organic acid esters, which may contribute to the formation of the DCL structure.

The short spacings (4.10–4.13 Å) show that all organic acid esters are α crystalline. The melting points are considerably lower than those of corresponding α-crystal forms of nonesterified monoglycerides, showing a higher degree of molecular disorder (entropy) in the crystals. The exact crystal structure of organic acid esters is not known since no pure components have been analyzed.

It should be mentioned that the composition of each type of organic acid ester of monoglycerides may vary considerably in different commercial samples due to variations in the raw materials and production methods used. The products described here are commercial samples made according to the specifications given by the European regulation for food-grade emulsifiers.

C. Mixed Fatty Acid Esters of Other Polyols or Lactic Acid

The products in this category are usually made using mixtures of saturated fatty acids with C16–C18 chain length in the ratio 1:1. Such polyol esters or lactic

Table 3 X-Ray Diffraction Data of Organic Acid Esters of Monoglycerides (C16/C18 ratio, 35:65)

Organic acid esters of monoglycerides	Long spacings (Å)	Chain packing mode	Short spacings (Å)	Crystal form	Melting point (°C)
Acetic acid esters (monoacetylated)	32.9	SCL	4.10	α	39
Lactic acid esters	39.5	SCL	4.13	α	42
Diacetyl tartaric acid esters (DATEM)	41.0	SCL	4.11	α	43
1,2-Dipalmitin-DATE[a]	54.6	DCL	4.21–3.76	—	—
Citric acid esters	60.3	DCL	4.11	α	60

[a] Diacetyl tartaric acid ester of pure 1,2-dipalmitin.

acid esters are solid at an ambient temperature, and their X-ray powder diffraction data and melting points are shown in Table 4. Polyglycerol monostearate is a very complex product, containing many different components with variations in the polyol, which is predominantly tetraglycerol, but minor amounts of di-, tri-, penta-, and hexaglycerol are present. This mixture of polyols is esterified with fatty acids in a ratio selected to yield a maximum content of monoesters, typically 40–50%, combined with a mixture of components with a higher degree of esterification. The polyglycerol esters are stable in the α-crystal form, and the long spacings suggest a DCL structure.

Propylene glycol esters can, like monoglycerides, be concentrated by molecular distillation and contain 90% or more monoesters. They are stable in the α-crystal form and have the lowest melting point (39°C) among the polyol esters shown in Table 4. The long spacings of 50.7 Å are similar to those of monoglycerides, and propylene glycol monostearate can co-crystallize with saturated monoglycerides, stabilizing the α-crystal form, which is important in many food applications.

Sorbitan esters of fatty acids are made from dehydrated sorbitol, which is esterified with fatty acid to form products in which the main component is either fatty acid monoesters or triesters. None of these products have a high degree of purity with regard to the main components. All sorbitan esters are stable in the α-crystal form, but the long spacings indicate a different molecular packing. A concentrated sorbitan monostearate containing 80% monoesters has a long spacing of 33.8 Å, corresponding to an SCL packing mode which is in contrast to sorbitan monodistearate and sorbitan tristearate. The sorbitan monodistearate crystallizes as monoglycerides in bimolecular layers with the hydrocarbon chains separated from the polar groups. In the case of sorbitan tristearate, the long spacing are very similar to those of triglycerides with the same fatty acid profile, and this indicates that sorbitan tristearate crystallizes in a DCL form similar to that of triglycerides.

Fatty acid esters of lactic acid are neutralized with sodium hydroxide to form sodium stearoyl lactylate. Like the polyol esters, sodium stearoyl lactylate crystallizes in an α-crystal form with an SCL packing mode. This may, however, not be a stable state due to the content of different components, such as free fatty acids, monolactylic esters, and dilactylic esters. This mixture of components may segregate into two crystal forms, both α-like, on storage at a temperature close to the melting point. The segregation of components results in an increase in melting point to about 65°C, caused by a high content of sodium salts of monolactoyl fatty acid esters together with free fatty acids [7]. The high-melting fraction crystallizes in a DCL form with long spacings of 50–54 Å and several short spacings from 4.7–3.7 Å. In some cases, depending on the composition of the high-melting fraction, it gives one short spacing at only 4.1 Å, showing that this may also exist in an α-crystal form. The optimal crystal form for sodium stearoyl

Table 4 X-Ray Diffraction Data of Fatty Acid Esters of Polyols and Lactic Acid (C16/C18 ratio, 50:50)

Product	Long spacings (Å)	Chain packing mode	Short spacings (Å)	Crystal form	Melting point (°C)
Polyglycerol monostearate	64.2	DCL	4.13	α	56
Propylene glycol monostearate	50.7	DCL	4.15	α	39
Sorbitan monodistearate	54.5	DCL	4.11	α	54
Sorbitan monostearate	33.8	SCL	4.11	α	—
Sorbitan tristearate	49.8	DCL	4.13	α	56
Sodium stearoyl lactylate	37.6 (49.7)	SCL (DCL)	4.10	α	38

lactylate is the low-melting α-crystal form that follows crystallization from a molten state. Due to the complex composition of sodium stearoyl lactylate, this crystal form is an amorphous state with α-crystalline properties and optimal for dispersibility in aqueous systems.

D. Structural Parameters of Polar Lipids

X-ray diffraction analysis gives direct information about crystal form of polar lipids. However, more details about the molecular packing conditions can be obtained from X-ray data, as shown by Small [2]. When a homolog series of monoglycerides based on single fatty acids with varying chain lengths is analyzed, several structural parameters can be calculated from the data of long spacings as shown in Fig. 1.

An estimated dimension of the polar head group in the longitudinal direction of the molecules in the lipid bilayer is given by the intercept of the regression

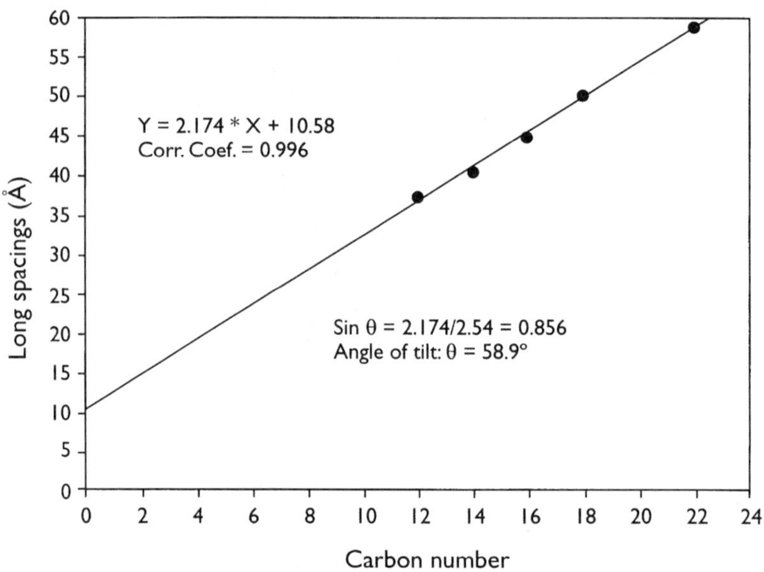

Fig. 1 X-ray diffraction long spacings (Å) for a homolog series of saturated single fatty acid monoglycerides in β-crystal form plotted against the fatty acid chain length (carbon number) of the monoglycerides. The angle of tilt of the molecules in the crystalline state can be calculated from the slope of the line (slope/2.54), and the intercept of the regression line with the *X*-axis corresponds to the dimension of two head groups in bimolecular lipid layers, as in the case of monoglycerides.

line in Fig. 1 with the *X*-axis (carbon number = zero). For lipids crystallizing in a DCL structure the intercept value (Å) corresponds to the dimension of two polar groups.

From the slope of the regression line, which is the increase in the long spacing value per $-CH_2-$ group, the angle of tilt of the hydrocarbon chains toward the plane of the methyl end groups can be calculated. If the hydrocarbon chains were oriented perpendicular to the plane of the end group, the angle of tilt would be 90°, because the increase in the long-spacing value per $-CH_2-$ groups is then identical to the linear distance (2.54 Å). Then slope/2.54 = sin θ, where θ is the angle of tilt.

Structural parameters of various emulsifiers calculated by the method demonstrated in Fig. 1 is shown in Table 5. The results show that the industrial distilled monoglycerides based on fully hydrogenated soybean oil (90% stearic acid, β-crystal form) have an angle of tilt slightly bigger than that of pure 1-monoglycerides with the same chain length [2]. The more vertical orientation of the fatty acid chains in the crystalline state of the mixed fatty acid monoglycerides is probably due to a higher degree of disorder (entropy) in the hydrocarbon region. The angle of tilt and long-spacing value is higher for mixed fatty acid monoglycerides in the α-crystal form than in the β-crystal form.

The calculated size of the polar head group of distilled monoglycerides is very similar to that of pure 1-monoglycerides. The diacetyl tartaric acid esters show a polar head group size of 9.3 Å, which is between the size of the head group of phosphatidylcholin (10.4 Å) and that of phosphatidylethanolamine (7.9 Å), according to Small [2].

Table 5 Structural Data of Polar Lipids

Lipid type	Long spacings d (Å)	Crystal form	Size of polar group d_p (Å)	Angle of tilt (°)	Chain packing mode
Pure 1-monopalmitin[a]	45.0	β	5.5	56	DCL
Monostearate (90% C18)	51.4	β	5.3	59	DCL
Monostearate (90% C18)	54.0	α	5.3	(72)	DCL
DATEM (C18)	41.0	α	9.3	90	SCL
1,2-Dipalmitin-DATE	54.6	—	9.3	(62)	DCL
Sorbitan monostearate	33.8	α	—	90	SCL
Sorbitan monodistearate	54.5	α	5.4	(68)	DCL
Sorbitan tristearate	49.8	α	5.4	79	DCL
1,2-DMPC[a]	54.9	—	10.4	78	DCL
1,2-DLPE[a]	47.7	—	7.9	90	DCL

[a] From Ref. 2.

The crystalline packing mode of sorbitan esters depends on the degree of esterification. The concentrated sorbitan monostearate crystallizes in an SCL mode with interdigitated fatty acid chains, while the mixed sorbitan monodistearate as well as the sorbitan tristearate crystallize in a DCL packing mode.

The long spacing of sorbitan tristearate is around 5 Å shorter than that of the sorbitan monodistearate, and a regression analysis of long spacings of sorbitan triesters with fatty acid chain lengths from C12 to C22 gives an estimated polar head group size of 5.4 Å. This concurs with the hypothesis that there is only one polar group in the bilayer of the sorbitan tristearate crystals and reveals some similarity between the crystal structure of sorbitan tristearate and that of long-chain saturated triglycerides.

A proposed structure is that the sorbitan group is oriented in the middle of the molecule, with one of the fatty acids extended to one side and the two other fatty acids to the opposite side, similar to the structure of triglycerides. The fact that sorbitan tristearate can co-crystallize with triglycerides and prevent crystal transformations from the metastable β' form to the higher-melting β form may support this concept.

III. PHASE BEHAVIOR OF FOOD EMULSIFIERS IN AQUEOUS SYSTEMS

A. Mixed Fatty Acids Monoglycerides

Binary phase diagrams of distilled saturated monoglycerides based on hydrogenated lard were first published by Krog and Larsson in 1968 [8]. In such mixed fatty acid monoglycerides, the fatty acid profile is dominated by stearic acid (67%) and palmitic acid (30%), and it was found that their phase behavior in water was very similar to that of pure 1-monopalmitin. The mesophases formed were quite identical for the two types of monoglycerides. It was also observed that the swelling capacity of the dominating lamellar phase of the industrial monoglycerides was dependent on the pH of the aqueous phase used. At a pH below 6, limited swelling occurred, while at pH 7 optimal swelling took place, forming a translucent dispersion. The effect of changing the pH was linked to the minor content of free fatty acids present in industrial monoglycerides, which, when neutralized into sodium salts, introduced ionic repulsion between the lipid bilayers in the lamellar phase. This was later confirmed with regard to lamellar phases [9] and α-crystalline gel phases of industrial monoglycerides [10]. Figure 2 shows phase diagrams of distilled monoglycerides from (a) hydrogenated lard, (b) sunflower oil, (c) elaidic acid (90%), and (d) oleic acid (90%). The phase behavior of saturated monoglycerides, as shown in Fig. 2a, is typical for most mixed C16/C18 fatty acid monoglycerides.

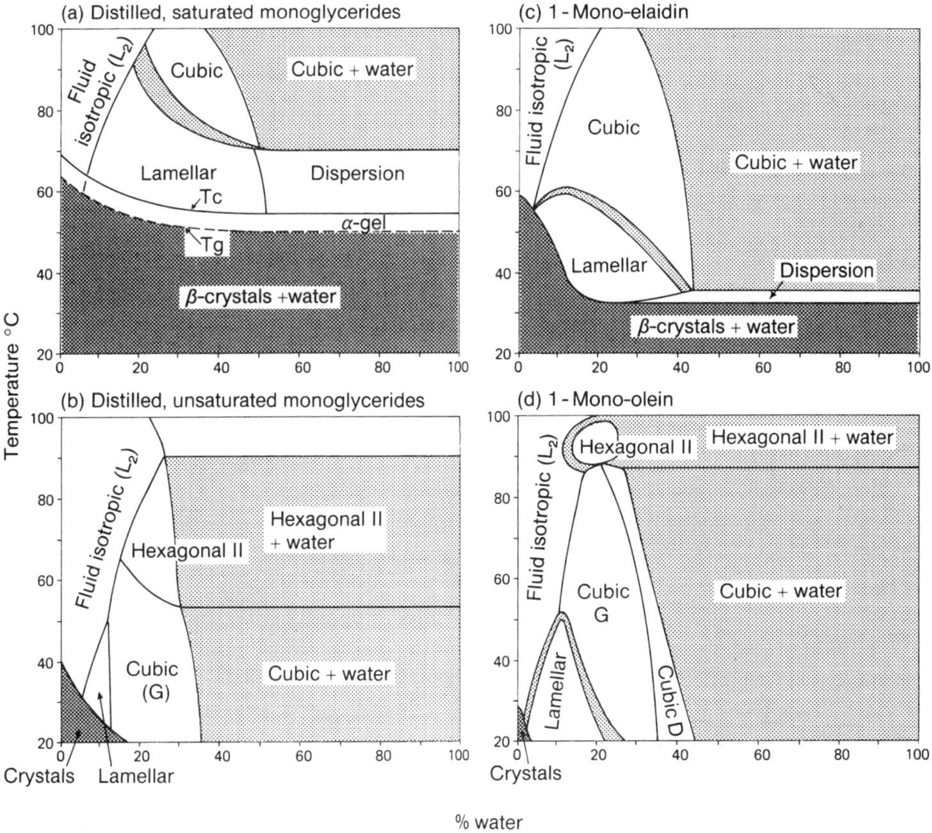

Fig. 2 Phase diagrams of (a) distilled monoglycerides from hydrogenated lard (DIMODAN P, Danisco Cultor); (b) distilled, unsaturated monoglycerides from sunflower oil (DIMODAN LS, Danisco Cultor); (c) distilled monoelaidin; and (d) distilled monoolein (RYLO MG 19, Danisco Cultor). Abbreviations: T_c = Krafft temperature; T_g = transition temperature for β crystals + water to α gel; cubic G = gyroidal structure; cubic D = diamond structure. (From Ref. 7.)

The transition from crystals + water to a lamellar phase when heating the mixture of monoglycerides and water to above the Krafft point is not the same with mixed fatty acid monoglycerides as with pure 1-monoglycerides. The 1-monoglycerides transform directly from β-crystals in water to a lamellar phase, while low-angle X-ray diffraction patterns versus temperature (DPT camera) have demonstrated [7] that mixed, saturated fatty acid monoglycerides transform via an α-crystalline gel phase to a lamellar mesophase, as expressed by the T_g line in Fig. 2a. The penetration of water molecules into the layers of polar head groups actually takes place below the Krafft point. This behavior is not found with unsaturated monoglycerides, which do not form α-crystalline gels.

Figure 2a shows that saturated monoglycerides are dispersible in water between 55 and 70°C in the form of lamellar aggregates or vesicles, a state referred to as a dispersion. The swelling properties of the saturated monoglycerides in the dispersed state are highly dependent on the presence of some ionized molecules in the lipid bilayer and on a low ion concentration (salts) in the water phase [9]. The purity of the monoglycerides is another important parameter; the content of di- and triacylglycerides must be very low, preferably not above 3%.

The swelling of polar lipids in water is due to a hydration repulsion force created by the strong affinity between water and the polar head groups of the lipids [11]. When the hydration repulsion force is in balance with the van der Waals attraction forces between the lipid bilayers, the swelling stops. This means that the water layer thickness, which can be measured by X-ray diffraction methods, becomes constant, and if the water content of the system is further increased the lamellar phase will disperse as aggregates or vesicles.

Without the neutralization of the free fatty acids present in the monoglycerides, the swelling behavior of saturated distilled monoglycerides in water at 60°C is limited. The water layer thickness does not exceed 16–20 Å, as found with pure 1-monoglycerides [5]. On the contrary, with the neutralization of the free fatty acids by adjusting the pH to 7, the swelling of the lamellar dispersion phase is proportional to the water content. With 75% water the water layer thickness is increased to about 120 Å. In highly diluted lamellar dispersions of saturated monoglycerides (e.g., 95% water) containing neutralized fatty acids, the water layer thickness can be about 1000 Å or more [10]. In the phase diagram of saturated, distilled monoglycerides (Fig. 2a), the borderline between the lamellar region and the dispersion phase is only indicative of a change in fluidity to a low-viscous state. Dispersions containing 90% water or more are fluid and pumpable.

The needed concentration in the lipid bilayer of ionized, amphipathic molecules to increase the swelling of lamellar phases by an electric repulsion effect between bilayers is very low (0.5–1 mol%), but the repulsion effect will be shielded if the ion concentration of the water phase (e.g., Na^+, Cl^-) is above 0.1–0.3% [9,10]. The addition of salts to a lamellar phase will, thus, reduce swelling to a minimum, corresponding to a water layer thickness around 20 Å, as found with pure synthetic 1-monoglycerides [5].

The unsaturated monoglycerides (Fig. 2b) show very different behavior, forming viscous, cubic phases when mixed with water, even at low temperatures. The cubic phase has a limited swelling capacity and is not dispersible in excess water except under specific conditions in presence of high-polar co-emulsifiers.

Industrial monoglycerides are also produced from partly hydrogenated vegetable oils which may contain a certain amount of monounsaturated fatty acids in *cis* or *trans* configuration (oleic versus elaidic acid). The phase diagrams of monoelaidin and monoolein are shown in Figs. 2c and 2d. The purity of the elaidic and oleic acids used for these products was 90% (Table 1). The monoelaidin shows a reduced lamellar phase and a lower Krafft point compared to saturated monoglycerides, and the monoolein shows higher temperature stability in the cubic phase than unsaturated, mixed fatty acid monoglycerides.

B. Organic Acid Esters

The esterification of monoglycerides with organic acids, such as acetic or lactic acid, changes their polarity and hydration properties considerably. Neither acetic acid esters nor lactic acid esters of monoglycerides form any mesomorphic phases in water. Acetic acid esters are less polar than corresponding monoglycerides and do not swell in water. Lactic acid esters of monoglycerides contain too many lipophilic components to swell into mesomorphic phases, but, in the α-crystalline state, water can penetrate between the polar groups, forming an α-crystalline gel phase with water layers of 16–20 Å [7,12]. When this gel phase is heated above the melting point of the lactic acid ester, an emulsion is formed. Similar behavior is also found in propylene glycol esters of fatty acids and monoacetylated monoglycerides.

Esters of dicarboxylic acids, such as diacetyl tartaric acid esters of monoglycerides (DATEM), contain a free carboxyl group, and their swelling properties in water depend strongly on the ionization of this carboxyl group. At a low pH, no swelling occurs, but when pH is above 4.5 a lamellar phase is formed above the Krafft point, which is about 45°C [7] for DATEM based on saturated monoglycerides and 20°C for DATEM based on unsaturated monoglycerides. The lamellar phase of DATEM is very stable, and no other mesophases are formed at temperatures up to 100°C. Once the DATEM has swollen in water, it remains in the lamellar liquid crystalline form even when stored for a long time at low temperatures.

Commercial citric acid esters of monoglycerides contain too many lipophilic components (diacylglycerol esters) to form mesomorphic phases in water.

C. Polyglycerol Esters

Medium-chain-length fatty acid (e.g., C12) esters of polyglycerol with an average degree of tetraglycerol polymerization form lamellar mesophases in water, and spherical aggregates with a lamellar structure (vesicles) are formed in diluted dispersions. Long-chain-length fatty acid (C16–C18) esters of polyglycerol form reversed hexagonal phases above 60°C [7]. The Krafft point of these esters is identical to the melting point, due to their α-crystalline structure. In combination with ionic emulsifiers (sodium stearate, stearoyl lactylates), polyglycerol monostearate forms dispersions of lamellar aggregates, which, on cooling, transform into stable α-crystalline gels. Fatty acid esters of polyols with a higher degree of polymerization are more hydrophilic than tetraglycerol esters, but, due to their complex composition, they form predominantly reversed hexagonal liquid crystalline structures in water above their Krafft point [13]. The formation of reversed hexagonal structures in water is typical of polar lipids with a mixed composition of hydrophilic and lipophilic components.

Polyglycerol esters with higher purity with respect to their degree of polymerization as well as fatty acid composition behave differently in aqueous systems compared to mixed polyol fatty acid esters. Concentrated diglycerol monoesters of fatty acids (e.g., 85% purity) with chain lengths C12–C18 have been examined, and their phase behavior in water shows the existence of lamellar phases only, combined with the formation of multilamellar vesicles in diluted dispersions [14].

Figure 3 shows the T_c line (Krafft temperature) of diglycerol monoesters of saturated fatty acids with chain lengths from C12–C18, and oleic acids (*cis*-C18:1).

The Krafft temperature varies from 1°C for diglycerol monooleate to 48°C for diglycerol monostearate. Above the T_c line lamellar phases are formed by all diglycerol monoesters at water above 10%. The swelling behavior of the lamellar phases increases with a water content of up to 40–50% for diglycerol monostearate and 35% for diglycerol monooleate. At higher concentrations of water, stable dispersions of multilamellar vesicles are formed. The lamellar phases of diglycerol monoesters are very stable at higher temperatures (<80°C), and it is interesting to observe that no cubic or reversed hexagonal phases are formed with diglycerol monoesters. X-ray diffraction of lamellar phases shows *d* spacings from 58 Å for diglycerol monooleate to 66 Å for diglycerol monopalmitate and stearate. This corresponds to a water layer thickness of between 22 and 30 Å, respectively, which is the maximum swelling of diglycerol esters in water in the absence of anionic-active emulsifiers, which may increase the swelling properties of diglycerol esters. When cooled below the Krafft point, the saturated diglycerol monoesters form α-crystalline gels with limited swelling, unless anionic-active emulsifiers (sodium stearate, stearoyl lactylate, or DATEM) are present. Addition of anionic-active co-emulsifiers increases the swelling properties of diglycerol monoesters in a similar fashion to that of monoglycerides with neutralized fatty acids.

When the degree of esterification is increased to produce mixtures of mono-, di-, and triacyl esters, the lipophilic properties gradually increase, and, as a result, mesomorphic phases with cubic or reversed hexagonal structures are

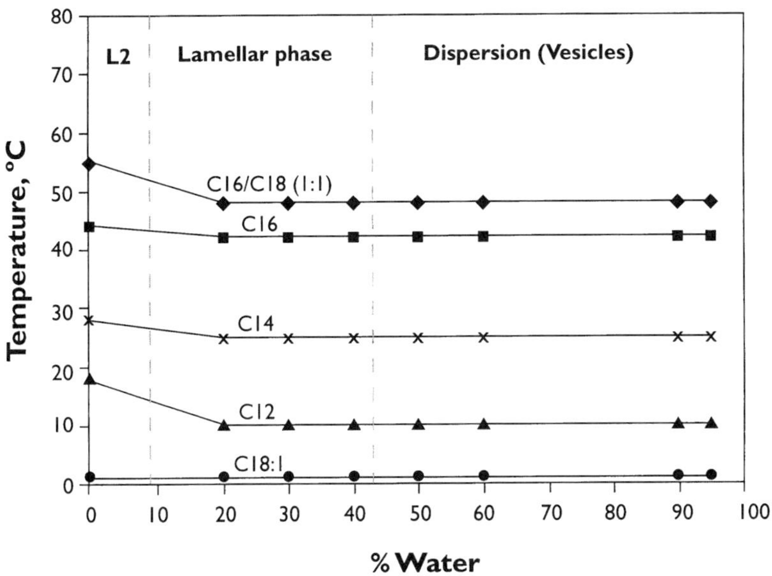

Fig. 3 Schematic phase diagram of diglycerol ester/water mixtures showing Krafft temperatures of C12, diglycerol monolaurate; C14, diglycerol monomyristate; C16, diglycerol monopalmitate; C16/C18, diglycerol monopalmitate/diglycerol monostearate (1:1); and C18:1, diglycerol monooleate (RYLO PG 29, Danisco Cultor). The composition of the fatty acids used as shown in Table 1. All diglycerol esters form lamellar phases above the Krafft temperature at water concentrations between 10% and 43%. At higher water concentrations, lamellar dispersions or liposomes are formed.

formed in water. Special compositions can, thus, be made to form a specific mesomorphic phase or phase mixtures in water. The ternary phase diagram of diglycerol monooleate (DGMO):glycerol monooleate (GMO):water have been studied [15], and it was found that the limited swelling capacity of the lamellar and cubic phases formed by GMO:water was increased by addition of diglycerol monooleate, as shown in Fig. 4.

In the ternary DGMO:GMO:water system (20:30:50) a cubic phase exists, and the rheological properties of the cubic phase are changed significantly compared to those of a cubic phase of GMO:water (35:65) system. The storage modulus, G', is decreased with increasing content of DGMO in the ternary mixture. The cubic phases of lipid:water systems are of interest as a vehicle for encapsulation of ingredients intended as slow-release preparations for foods (flavors, enzymes, etc.) [16] or pharmaceutical drugs [17,18].

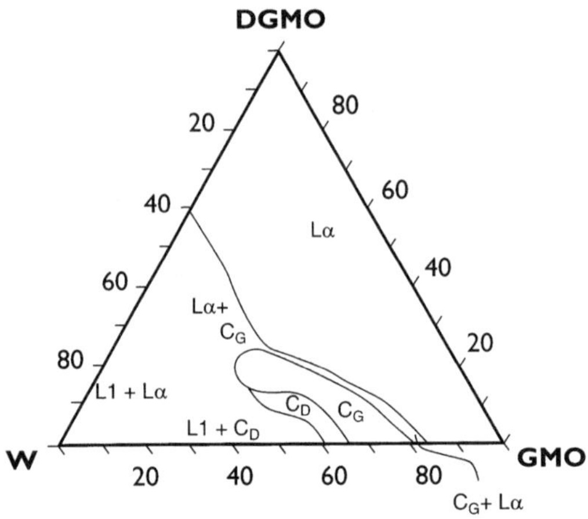

Fig. 4 Ternary phase diagram of glycerol monooleate (GMO)/diglycerol monooleate (DGMO)/water (W) at 25°C. Lα = lamellar; Cg = cubic gyroid, Cd = cubic diamond; L1 = isotropic fluid. (Reprinted with permission from Ref. 15, copyright 2000, American Chemical Society.)

D. Sodium Stearoyl Lactylate (SSL)

The phase behavior of SSL in water varies according to the system pH. This is due to the fact that SSL contains a considerable amount of free fatty acids (C16–C18) which are not esterified to lactic acid or neutralized completely to form sodium salts. Thus at pH below 5, SSL forms reversed hexagonal phases in water at temperatures above the melting point (45°C). At higher pH, near neutral or higher, SSL forms lamellar phases in a temperature range of 45–85°C [7]. SSL is often used in combination with monoglycerides or polyglycerol esters in systems having neutral pH. Lamellar phases of such combinations are applied in aerated foods.

IV. APPLICATIONS OF LIQUID CRYSTALLINE PHASES IN FOOD EMULSIONS, FOAMS, AND STARCH-BASED FOODS

A. Liquid Crystalline Phases in Emulsions

The history of liquid crystalline phases as a stabilizing factor in emulsions dates back to 1954 [19]. Ten years later the formulation of cosmetic emulsions was related to the phase equilibria of emulsifier-water systems [20]. The stabilizing effect of interfacial liquid crystalline layers around oil droplets was later studied systematically by Friberg and Mandell [21]. The first applications of liquid crystalline phases in emulsions were found in cosmetic and technical emulsions, and the concept was later applied to food emulsions containing stearoyl lactylates [7], lecithin [22], and polysorbates [23].

Egg yolk lecithin or pure phosphatidylcholin in combination with monoglycerides can stabilize oil-in-water emulsions, and birefringent layers around the oil droplets show the formation of lamellar multilayers on the surface of oil droplets. Similar structures may be found in some emulsions (e.g., salad dressings) containing lecithin [24]. In general, liquid crystalline interfacial films are rarely found in food emulsions because the concentration of emulsifiers in foods is not high enough to form liquid crystals at interfaces. Furthermore, most food emulsions contain proteins (e.g., milk proteins) which are the primary stabilizing factor in such emulsions. In bakery products, however, liquid crystalline phases may be formed by polar flour lipids (phospholipids, galactolipids), which play an important role in the baking quality of wheat flour [25].

B. α-Crystalline Gels and β-Crystalline Hydrates in Foods

Knowledge of the preparation of liquid crystalline phases of distilled monoglycerides was first used commercially in the bakery industry and starch-based foods. Lamellar dispersions of saturated distilled monoglycerides and α-crystalline gel phases was used in fat-free sponge cakes and in starch-based foods.

Interactions of starch components, especially amylose, with monoglycerides are important factors in reducing the firmness of bread [26] or improving the texture of starch-based foods such as potato flakes, puddings, etc. Since the starch components are dissolved in the aqueous phase in food systems, it is very important that the complexing lipids be water dispersible. In the case of monoglycerides, which are effective amylose-complexing agents, the lamellar dispersions are the optimal physical state for interactions in the water phase. In the production of dried potato flakes, the application of saturated, distilled monoglycerides in the form of lamellar dispersions was among the first industrial applications of such preparations about 40 years ago.

The amylose-complexing effect of monoglycerides decreases in the order monopalmitin > monoelaidin > monoolein in aqueous solutions [27]. The low complexing efficiency of monoolein in aqueous systems was due to the formation of cubic mesophases, which reduce the monomer concentration of monoolein in water and, thus, the ability to react with amylose. There was no difference in the complexing effect when the monoglycerides and starch were dissolved in a solvent (dimethyl sulfoxide).

The commercial production of so-called hydrates, an aqueous suspension of monoglycerides in β-crystalline form, involves mixing monoglycerides and water into a lamellar phase, followed by cooling under shear while reducing pH to 3.0. This results in a fast transition from the α-gel phase formed on cooling to a mixture of β crystals in water. Such hydrates are semisolid or plastic, depending on the concentration of monoglycerides, and are produced commercially on a large scale.

The specific surface area of the β crystals in monoglyceride hydrates is extremely large compared to that of nonhydrated, powdered monoglyceride products, ensuring an optimal distribution of the monoglyceride crystals in the bread dough during mixing. This is important to the reaction with dissolved amylose, which takes place when the monoglycerides transform into a lamellar dispersion during the baking process at a temperature above 45–50°C.

C. Formation of α-Crystalline Gel Phases in Whippable Emulsions and Low-Fat Products

Emulsifiers such as lactylated monoglycerides or propylene glycol monostearate can form α-gel phases with water, even though they do not form mesomorphic phases. This is utilized in whippable emulsions to improve the aeration properties and foam texture. The formation of α gels is related to the α-crystal form of emulsifiers. When α-stable emulsifiers are mixed with water, they absorb water due to hydration of the polar groups, and the crystals will swell, forming an α-gel phase [7,12,24]. In liquid nondairy creams this may take place on the surface of the fat globules and promote fat particle aggregation, which is essential for good whipping properties. The absorption of water may also result in the disintegration of emulsified fat particles in reconstituted, powdered toppings, forming a crystalline matrix, increasing viscosity, and enhancing the incorporation of air during aeration [28]. The fat crystals formed during reconstitution of the topping powder in water or milk adsorb to the surface of the air cells when the mixture is aerated, and the adsorbed fat crystals stabilize the air cells, preventing coalescence.

Emulsifiers in the form of lactic acid esters of monoglycerides or propylene glycol monostearate do not co-crystallize with high-lauric fats (hydrogenated palm kernel or coconut oil). The high-lauric fats crystallize in a β′-crystal form, while the emulsifiers crystallize in bimolecular sheets with α-crystalline hydrocarbon chains. When the topping powder is mixed with water or milk, water penetrates the polar regions of the emulsifier crystals, forming an α-gel structure which disintegrates the fat phase, providing the concentration of the emulsifier phase is sufficiently high.

The application of α-crystalline gel or β-crystalline hydrates (coagel) as structuring agents in foods is common in low-calorie foods with reduced fat content [29]. Vesicles of α-gel phases improve the aeration capacity of vegetable fat–based imitation creams. Such preparations are made by mixing a low-fat nondairy cream with a dispersion of vesicles of saturated monoglycerides and stearoyl lactylates.

The tridimensional network of β crystals formed by saturated monoglycerides in water has a structure similar to that of fat crystals in margarine, spreads, or shortenings and provides a similar fatty mouthfeel when replacing fats in a number of low-calorie foods [30].

REFERENCES

1. K. Larsson, *Acta Cryst., 21*:267–272 (1966).
2. D. M. Small, In *The Physical Chemistry of Lipids.* Marcel Dekker, New York, 1986, pp. 386–392, 475–492.
3. K. Larsson, In *Lipids—Molecular Organisation, Physical Functions and Technical Applications.* Oily Press, Dundee, Scotland, 1994.
4. E. S. Lutton, J. *Am. Oil Chem. Soc., 42*:1068–1073 (1965).
5. K. Larsson, *Z. Phys. Chem.* (*Neue Folge*), *56*:173–198 (1967).
6. G. J. Tiddy, *Phys. Rep., 58*:1–46 (1980).
7. N. Krog, In S. E. Friberg and K. Larsson (eds.), *Food Emulsions,* 3rd ed., revised and expanded. Marcel Dekker, New York, 1997, pp. 141–188.
8. N. Krog and K. Larsson, *Chem. Phys. Lipids, 2*:129–143 (1968).
9. N. Krog and A. P. Borup, *J. Sci. Fd. Agric., 24*:691–701 (1973).
10. K. Larsson and N. Krog, *Chem. Phys. Lipids, 10*:177–180 (1973).
11. D. M. Le Neveu, R. P. Rand, V. A. Parsegian, and D. Gingell, *Biophys. J., 18*:209–230 (1977).
12. J. M. M. Westerbeek and A. Prins, In E. Dickinson (ed.), *Food Polymers, Gels and Colloids.* Cambridge, The Royal Society of Chemistry, 1991, pp. 147–158.
13. W. Hemker, *J. Am. Oil Chem. Soc., 58*:114–119 (1981).
14. J. Holstborg, B. V. Pedersen, N. Krog, and S. K. Olesen, *Colloids Surf. B: Biointerfaces, 12*:383–390 (1999).
15. P. Pitzalis, M. Monduzzi, N. Krog, H. Larsson, H. Ljusberg-Wahren, and T. Nylander, *Langmuir, 16*:6358–6365 (2000).
16. M. Leser and S. Vauthey, Food Composition Containing a Monoglycerides Mesomorphic Phase, Patent Application WO 99/47004 assigned to S. A. Nestlé, Vevey, CH, 1999.
17. A. Ganem-Quintanar, D. Quintanar-Guerrero, and P. Buri, *Drug Dev. Ind. Pharm., 26*:809–820 (2000).
18. E. Boyle and J. B. German, *Crit. Rev. Food Sci. Nutrition, 36*:785–805 (1996).
19. R. Salisbury, E. E. Leuallan, and L. T. Chavkin, *J. Am. Pharm. Assoc. Sci., Ed., 43*:117 (1954).
20. B. W. Burt, *J. Soc. Cosmetic Chem., 16*:465–477 (1965).
21. S. Friberg and L. Mandell, *J. Am. Oil Chem. Soc., 47*:149–152 (1970).
22. L. Rydhag and I. Wilton, *J. Am. Oil Chem. Soc., 58*:830–837 (1981).
23. N. Pilpel and M. E. Rabbani, *J. Colloid Interface Sci., 122*:266–273 (1988).

24. N. Krog, In A. G. Gaonkar (ed.), *Ingredient Interactions, Effect on Food Quality.* Marcel Dekker, New York, 1995, pp. 377–409.
25. A.-C. Eliasson and K. Larsson, In *Cereals in Bread-making, A Molecular Approach.* Marcel Dekker, New York, 1993, pp. 31, 293.
26. N. Krog, S. K. Olesen, H. Toernæs, and T. Joensson, *Cereal Foods World, 34*:281–285 (1989).
27. T. Riisom, N. Krog, and J. Eriksen, *J. Cereal Sci., 2*:105–118 (1984).
28. N. Krog, N. M. Barfod, and W. Buchheim, In E. Dickinson (ed.), *Food Emulsions and Foams.* The Royal Society of Chemistry, Cambridge, 1987, pp. 144–157.
29. European Patent Specification No. 0 558 523 B1 to Unilever N.V., P.O. Box 137, NL 3000 DK Rotterdam, The Netherlands, 1994.
30. I. Heertje, E. C. Roijers, and H. A. C. M. Hendriks, *Lebensm.-Wiss. Technol., 31*: 387–396 (1998).

Index